Library of Congress Cataloging-in-Publication Data

Luminescence of solids / edited by D.R. Vij.
p. cm.
Includes bibliographical references and index.
ISBN 978-0-306-45643-5 ISBN 978-1-4615-5361-8 (eBook)
DOI 10.1007/978-1-4615-5361-8
1. Luminescence. 2. Solids--Optical properties. 3. Inorganic compounds. I. Vij, D. R.
QC476.5.L86 1998
535'.35--dc21 98-18247
CIP

ISBN 978-0-306-45643-5

Originally published by Plenum Press,New York in 1998

10 9 8 7 6 5 4 3 2 1

Contributors

L. Yu. Brovko, Department of Chemistry, Lomonosov Moscow State University, Moscow, Russia.

B. P. Chandra, Department of Postgraduate Studies and Research in Physics, Rani Durgavati University, Jabalpur 482001, India.

Sean M. Cordry, Department of Physics and Astronomy, University of Mississippi, Mississippi 38677, United States.

Lawrence A. Crum, Applied Physics Laboratory, University of Washington, Seattle, Washington 98105, United States.

E. I. Dementieva, Department of Chemistry, Lomonosov Moscow State University, Moscow, Russia.

Debabrata Ghose, Saha Institute of Nuclear Physics, Bidhan Nagar, Calcutta 700064, India.

V. P. Gribkovskii, B. I. Stepanov Institute of Physics, Academy of Sciences of Belarus, Minsk, 220072, Byelorussia.

H.-E. Gumlich, Institut für Festkörperphysik, Technische Universität, D-10623, Berlin, Germany.

Rainer Hippler, Fakultät für Physik, Universität Bielefeld, D-33501, Bielefeld, Germany. Present address: Institut für Physik, Universität Greifswald, Domstrasse 10a, D-Germany.

R. Mauch, Heinrich Hertz, Institut, 10117 Berlin, Germany.

A. Meijerink, Department of Condensed Matter, Universität Utrecht, 3508 TA, Utrecht, The Netherlands.

Stefan Myhajlenko, Center for Solid State Electronics Research, College of Engineering and Applied Sciences, Arizona State University, Tempe, Arizona 85287, United States.

Luminescence of Solids

1

Theory of Luminescence

V. P. Gribkovskii

1. INTRODUCTION

Luminescence as a weak cold glow from rotting wood, certain insects, mushrooms, fish, microorganisms, sea algae, minerals, and, finally, polar lights has been known since ancient times. This interesting and initially mysterious phenomenon attracted the attention of many scientists during the last four centuries (Harvey, 1957). However, it was not subjected to systematic study until the middle of the 19th century. Stokes (1852) formulated the first law in the history of luminescence (Stokes' rule), which states that the wavelength of the luminescence is greater than the wavelength of the exciting radiation.

The term "luminescence" (the literal translation from Latin is "weak glow") was introduced into the literature by Wiedemann (1888). He also offered the first, although not entirely accurate, definition of luminescence as the excess emission over and above the thermal emission background. This definition reflected an important property, but it did not distinguish luminescence from other types of glow that are also excess emission over and above the thermal emission background, among which are reflected and refracted light, Rayleigh and Raman scattering, and Cherenkov and laser radiation. Vavilov (1951, 1952) suggested complementing Wiedemann's definition by adding the criterion of duration and using the term luminescence for the excess emission over and above the thermal emission of a body if this emission has a duration considerably exceeding the period of light oscillations.

On the basis of a detailed consideration of the processes that excite photon absorption and of the appearance of secondary photons within the framework of quantum electrodynamics, it has been shown that, as Vavilov thought, luminescence differs from Rayleigh scattering, from combination (Raman) scattering, and from resonance emission by the fact that there are intermediate processes between the absorption and the emission of the light (Apanasevich, 1977; Stepanov, 1991). The intermediate processes disturb the correlation between the characteristics of

---

V. P. Gribkovskii • B. I. Stepanov Institute of Physics, Academy of Sciences of Belarus, Minsk 220072, Byelorussia.

Luminescence of Solids, edited by D. R. Vij. Plenum Press, New York, 1998.

absorbed and emitted photons, so that the luminescence spectrum and indicatrix are determined by the properties of the substance. This holds true within the framework of linear optics and in the absence of interaction between the luminescence centers. Thus, luminescence may be defined as nonequilibrium radiation that is an excess over and above the thermal radiation background and arises in the presence of intermediate processes of energy transformation between absorption and emission. The duration of luminescence and the possibility of quenching it are direct consequences of these intermediate processes.

It may turn out that under certain conditions in some spectral region the nonequilibrium radiation of the body is less than the thermal emission background. This deficiency of radiation was termed negative luminescence (Stepanov, 1961). It was shown that the basic laws of luminescence are also valid for negative luminescence, but with the opposite sign. In particular, one needs sources of negative light fluxes, to excite it, i.e., bodies with temperatures lower than the ambient temperature.

Presently luminescence as a science is closely related to spectroscopy, which is the study of the general laws of the absorption and emission of radiation by matter. Both disciplines are based on the achievements of quantum mechanics, the quantum theory of radiation, and quantum electrodynamics. Many works, especially theoretical ones, can be equally related to both luminescence and spectroscopy. However, there are also essential differences in the range of the problems that are relevant to each. The focus of attention of spectroscopists is the study of absorption and emission of light at thermodynamic equilibrium when only thermal emission is observed and the luminescence is equal to zero. In this case, the processes are most often considered to be stationary. In luminescence, the subject of inquiry is only radiation that exceeds the thermal emission that is regarded as the background. Much consideration is given to the study of the time dependence of radiation intensity, the laws of luminescence rise and decay. A special class of problems is related to the study of the mechanisms of luminescence excitation and the elucidation of the dependence of luminescence characteristics on the method of excitation. Such problems do not arise in the case of thermal emission. In studying luminescence, the spectroscopic characteristics of matter are often regarded as initial data.

The development in 1960 of the first laser began a new stage in the development of optical science as a whole, including luminescence, and new possibilities for luminescence excitation were among the first to open up. New phenomena could be observed in known luminescent substances, namely, saturation and absorption of luminescence, induced dichroism, depolarization of luminescence with increasing excitation power, two-photon absorption, luminescence at forbidden transitions, hot luminescence, radiation of biexcitons (exciton molecules), electron–hole drops, induced self-transparence, self-focusing of light, and other nonlinear phenomena. It became possible to study ultrafast processes, in particular, to measure the time in which an equilibrium distribution of electrons within the energy band is established (10^{-11}–10^{-12} s).

On the other hand, the range of luminescent substances has been widened owing to the production of complex materials and, frequently, materials that do not exist in a natural state: hyperfine films of crystals, quantum-dimensional structures, stressed layers, and superlattices.

The spectrum, the power, and the quantum yield of luminescence are major characteristics of the active medium of any laser. Therefore, the investigation of luminescence and superluminescence has become an inherent part of laser physics and nonlinear optics. In these fields, the studies of excitation and emission processes have been conducted, from the very beginning, at a higher theoretical level than in traditional luminescence. The simplified models of substances were supplemented or replaced by more complex models that reflected the real properties of substances more fully and in greater detail. One began explicitly to take into account the density function of the state of the main and impurity bands, the electron wavevector selection rules, the absorption by free carriers, and electron–hole, electron–phonon, exciton–exciton and other kinds of collective interactions and nonlinear effects. The energy spectrum as well as the absorption and luminescence spectra of quantum-dimensional layers, quantum wires, and dots have been calculated, and the influence of the size of quantum dots on the properties of exciton luminescence are being studied.

In this chapter, an attempt is made to generalize and present sequentially the achievements that led to the establishment of the fundamental laws of luminescence. First, definitions are given, the physical meaning is elucidated, and phenomenological equations are given relating to the most important characteristics of luminescence: power, spectrum, polarization, energy and quantum efficiency, rise and decay time, excitation spectrum, and light sum. There is also a discussion of the theory of solids and quantum theory of luminescence.

The concluding section of the chapter is devoted to the calculation of luminescence from the most important theoretical models of matter, including the harmonic oscillator, the rotator, systems of particles with a discrete energy spectrum, models of parabolic bands, Gaussian bands, donor–acceptor pairs, excitons, biexcitons, and other models. Some of the important points of the theory of luminescence are not considered, but are noted with references to literature. In the chapters that follow the reader will find a significant supplement to this chapter, where theoretical calculations are used to elucidate the structure of radiating substances and the mechanism of luminescence.

2. CHARACTERISTICS OF LUMINESCENCE

2.1. Luminescence Power

By luminescence power is meant the amount of radiant energy of luminescence released by matter per unit time. If the substance's unit volume is taken, one speaks of the specific power of luminescence. Alongside the total, over the spectrum, specific power of luminescence W_l one can introduce the concept of the spectral density $W_l(\nu)$ of the quantity W_l. These are related to each other by the equation

$$W_l = \int_0^\infty W_l(\nu)d\nu \tag{1}$$

where the integration is carried out over the entire spectrum of luminescence. If the radiation spectrum consists of one band or line, then instead of Eq. (1) one can write $W_l = W_l(\nu)\Delta\nu$, where $\Delta\nu$ is the effective bandwidth. By definition, only the part of spontaneous radiation specific power W_{sp} that exceeds the thermal emission background pertains to luminescence. The energy distribution over the thermal emission spectrum was described by the Planck:

$$u^0(\nu) = \frac{8\pi h\nu^3}{c^3} \frac{1}{\exp(h\nu/kT) - 1} \tag{2}$$

where $h \approx 6.626 \times 10^{-34}$ Js $\approx 4.135 \times 10^{-15}$ eV s is the Planck constant, $k \approx 1.381 \times 10^{23}$ J K^{-1} is the Boltzmann constant, c is the velocity of light in vacuum, T is the absolute temperature on the Kelvin scale. The method for calculating the luminescence power was developed initially for the system of particles with two energy levels (Alentsev *et al.*, 1955). As applied to any pair of levels with numbers i (upper level) and j (lower level) of a multilevel system, the equation for W_{ij}^l can be given as

$$W_{ij}^l = n_i A_{ij} h\nu_{ij} - (n_j B_{ji} - n_i B_{ij}) u^0(\nu_{ij}) h\nu_{ij} \tag{3}$$

where n_i and n_j are the particle concentrations at the ith and jth levels, respectively, and A_{ij}, B_{ij}, and B_{ji} are Einstein coefficients for spontaneous and stimulated transitions from the ith to the jth level and back. The methods for calculating Einstein coefficients and their physical meaning will be discussed in Section 3.

The first term in Eq. (3) gives the total spontaneous emission power in the stationary regime and the second one is equal to the absorption power of the thermal emission background. Stimulated emission $n_i B_{ij} u^0(\nu_{ij}) h\nu_{ij}$ is considered as negative absorption of the background. Equation (3) proceeds from the assumption that precisely that portion of the spontaneous radiation that should belong to the thermal background is equal to the absorption power of this background, as the luminescing substance, although it is disturbed from thermodynamic equilibrium, is in equilibrium with the thermal radiation. For the thermal background to remain unchanged, the substance should give back to the environment as much radiation as it absorbs from it.

Equation (3) holds for systems with discrete energy levels and with monomolecular optical transitions. In the general form, the spectral power of specific luminescence is given by (Gribkovskii, 1975):

$$W_l(\nu) = W_{sp}(\nu) - v_g k(\nu) u^0(\nu) \tag{4}$$

where v_g is the group velocity of light, $k(\nu)$ is the absorption coefficient at frequency ν measured in cm^{-1}. This equation is valid for all methods of excitation and for all mechanisms of luminescence. The equations for luminescence powers in solids will be given in expanded form in Section 4.

2.2. Luminescence Spectrum

Graphs showing the dependence of luminescence power on the wavelength, frequency, or energy of radiation quanta are called luminescence spectra. In going from the frequency scale to the wavelength scale $\lambda = c/\nu$, the equality

$$W_l(\nu)d\nu = -W_l(\lambda)d\lambda \tag{5}$$

should hold. Taking into account the fact that $d\nu/d\lambda = -c\lambda^2$, one finds from Eq. (5) that

$$W_l(\nu) = \frac{W_l(\lambda)\lambda^2}{c} = \frac{W_l(c/\nu)c}{\nu^2} \tag{6}$$

If the luminescence band profiles in coordinates ν and λ are given in one and the same graph, their maxima will not coincide. According to Eq. (6), $\nu_m < c/\lambda_m$, where ν_m and λ_m are the coordinates of the maxima of the curves $W_l(\nu)$ and $W_l(\lambda)$. For this reason, e.g., the maximum of the solar radiation spectrum corresponds to the sensitivity of the human eye better on the frequency scale than on the wavelength scale.

2.3. Excitation Spectrum

The dependence of photoluminescence power on the frequency or wavelength of the exciting light $W_l(\nu_l)$ with its constant intensity at all frequencies is referred to as the excitation spectrum. Usually, it is an experimentally observable quantity and can be found for both the total, over the frequencies, luminescence power and for its individual parts. Simultaneous studies of excitation and absorption spectra and photoconductivity in solid bodies yield valuable information about the energy spectrum of matter and the optical and nonradiative transitions taking place in it.

2.4. Luminescence Polarization

As with other kinds of radiation, the polarization of luminescence consists of anisotropy of the electromagnetic oscillation distribution in the plane perpendicular to the beam. According to Newton, this "primary property of light" was described in sufficient detail in terms of classical electrodynamics, where matter is modeled by a set of electric and magnetic dipoles, quadrupoles, and rotators. The radiation of one elementary radiator has been found to be always polarized. In order for the radiation of a large number of particles to be polarized, either their selective excitation has to be realized or they have to have an anisotropic distribution in space. When viewed along the y-axis, the degree of luminescence polarization P is given by

$$P = \frac{I_z - I_x}{I_z + I_x} \tag{7}$$

where I_z and I_x are intensities of luminescence polarized along the z- and x-axis, respectively. As can easily be seen, the value of P varies from 0 to 1 at $I_z > I_x$.

The dependence of the degree of polarization on the orientation of the exciting light electric vector and on the observation angle is usually given in the form of polarization diagrams. Theoretically, a large number of equations have been obtained for the degree of polarization at various combinations of absorbing and emitting oscillators (Vavilov, 1952; Feofilov, 1959). As an example, consider the equation for the dipole–dipole combination:

$$P = \frac{\cos^2\eta - \sin^2\eta\cos^2\chi}{2 - \sin^2\eta\sin^2\chi} \tag{8}$$

where η is the angle between the exciting light electric vector and the x-axis, χ is the angle between the direction of observation in the $x0y$ plane and the x-axis. Equation (8) proceeds from the assumption that the dipoles' axes are oriented along the z-axis and the exciting light propagates along the x-axis. Comparison of experimental and theoretical dependences of polarization on angles η and χ permits determining the nature of elementary absorbing and radiating centers. Here classical electrodynamics still retains its significance. The limits of its applicability to the study of the phenomena of luminescence will be considered in Section 4.

2.5. Optical Indicatrix

The graph of the dependence of the electromagnetic radiation energy flux on the direction of its propagation is called the radiation indicatrix, and it is used mainly for describing elementary radiators. In the case of a harmonic oscillator, the energy flux passing through a unit area located at a distance R from the radiation source (Umov–Poynting vector) is equal to

$$\mathbf{S} = \frac{c}{4\pi}\,[\mathscr{E}\mathscr{H}] = \frac{e}{4\pi c^3 R_0^2}\left(\frac{\partial^2\mathscr{D}}{\partial t^2}\right)^2 \sin^2\vartheta \tag{9}$$

where

$$\mathscr{E} = -\frac{\omega^2}{c^2 R_0}\,[(\mathscr{D}\mathbf{e})\mathbf{e}], \qquad \mathscr{H} = -\frac{\omega^2}{c^2 R_0}\,(\mathscr{D}\mathbf{e}) \tag{10}$$

are the vectors of the electromagnetic and magnetic fields, $\mathbf{e}$ is the unit vector in the direction of light propagation, $\mathscr{D}$ is the oscillator dipole moment, ϑ is the angle between the vectors $\mathbf{e}$ and $\mathscr{D}$, and $\omega = 2\pi\nu$ is the oscillator's angular frequency.

The section of the oscillator indicatrix in the plane of the vector $\mathscr{D}$ is shown on Fig. 1. The spatial indicatrix can be obtained by turning a plane figure around the dipole axis. From Eq. (9) it is seen that the radiation intensity is maximal in the directions perpendicular to the vector $\mathscr{D}$ and is completely absent along the dipole axis. According to Eq. (10), the harmonic oscillator radiation in all the directions is linearly polarized, and the electric vector is perpendicular to $\mathbf{e}$ and in the plane of the vectors $\mathscr{D}$ and $\mathbf{e}$.

The radiation indicatrix of a plane electric rotator consisting of two unlike charges executing uniform rotary motion around their common center of gravity is given by the equation

$$S = \frac{\omega^4 e}{8\pi c^3 R_0}\,\mathscr{D}_0^2(1 + \cos^2\vartheta) \tag{11}$$

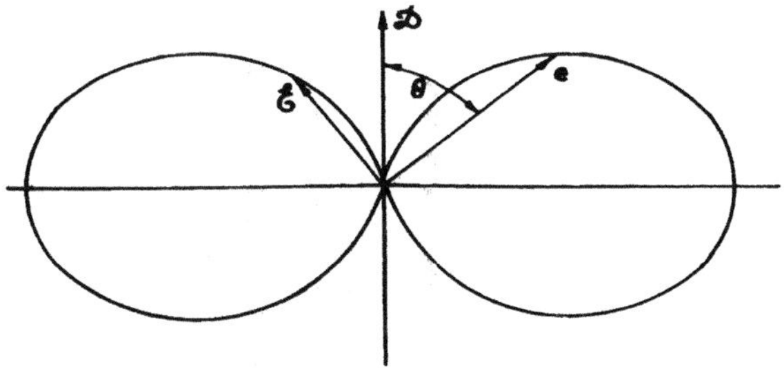

Figure 1. Polar diagram of dipole radiation from a harmonic oscillator (Feofilov, 1959).

where ϑ is the angle between the axis of rotation and the direction of observation. In deriving Eq. (11), the rotary motion of the charges was assumed to be the sum of two oscillatory motions in mutually perpendicular directions, and the rotator radiation was considered to be the radiation of two harmonic oscillators whose frequencies and amplitudes coincide with one another, with the phases differing by $\pi/2$. Unlike the harmonic oscillator, the radiation power of the rotator is maximal in the direction of its axis (Fig. 2). The radiation propagating along the rotator axis is circularly polarized. When viewed at a given angle with the axis $0 < \vartheta < \pi/2$, the light has elliptical polarization, which becomes linear when the angle ϑ reaches $\pi/2$.

The magnetic dipole radiation indicatrix coincides with the radiation indicatrix of the electronic dipole and the polarization plane is at an angle $\pi/2$ with the dipole axis. The radiation indicatrixes of the magnetic and electric rotators also coincide with one another. Both classical and quantum electrodynamics sometimes discuss radiators of higher orders as well: quadrupoles and octupoles (Feofilov, 1959; Gribkovskii, 1975).

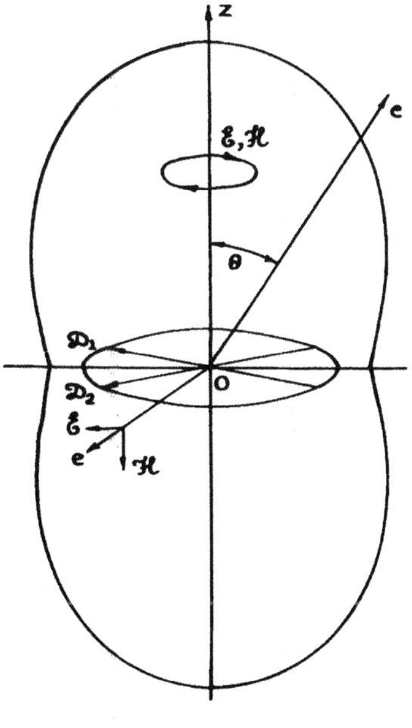

Figure 2. Polarization of radiation emitted by an electric rotator (Feofilov, 1959).

2.6. Energy and Quantum Yields

The energy yield or efficiency of luminescence, which is equal to the ratio of the luminescence energy to the absorbed excitation energy, is one of major characteristics of nonequilibrium radiation. At photoexcitation, one often uses the concept of the quantum yield, which is equal to the ratio of the number of luminescence quanta to the number of absorbed quanta. The concept of the luminescence yield was introduced into the literature by Vavilov (1924). He showed that there are substances that luminesce well with an energy yield close to unity. The theory of injection luminescence distinguishes between internal and external quantum yields of luminescence. The internal yield is equal to the ratio of the number of emitted radiation quanta to the number of recombinations of electrons and holes. The external yield takes into account only those radiation quanta that are emitted by the light-emitting diode and can be recorded by a detector. As a rule, the external quantum yield of injection luminescence is much smaller (sometimes by two to three orders of magnitude) than the internal quantum yield. This is due to the small angle of total internal reflection at the crystal–air boundary and the higher absorption coefficient of edge luminescence. Since recombination radiation propagates in all directions, a major portion of it is reflected back from the crystal surface and is absorbed in the device.

If the quantum mechanical system is excited optically and the exciting radiation energy density at frequency ν_{ij} is equal to u_{ij}, the energy yield of luminescence in the steady-state regime will be given by

$$\eta_l = \frac{W_l}{W_a} = \frac{\sum_{ij}[n_i A_{ij} - B_{ji} u_{ij}^0 (n_j - n_i g_j/g_i) h\nu_{ij}]}{\sum_{ij} B_{ji} u_{ij} (n_j - n_i g_j/g_i) h\nu_{ij}} \tag{12}$$

where g_j and g_i are the degrees of degeneracy of the jth and ith energy levels and W_a is the total absorption power. The summation is carried out over all combinations of the levels. Upon excitation by monochromatic light, the values of u_{ij} will be equal for all frequencies except the excitation frequency. In the numerator, the quantities n_i and A_{ij} can also be equal to zero for a number of values of i and j.

In the simplest case of optical transitions between levels 1 and 2, the energy and quantum yields η_{lq} of luminescence are equal and are given by

$$\eta_l = \eta_{lq} = \frac{A_{21}}{A_{21} + d_{21}[1 - \exp(-h\nu_{21}/kT)]} \approx \frac{A_{21}}{A_{21} + d_{21}} \tag{13}$$

Here d_{21} is the probability of nonoptical transitions $2 \to 1$. The product $d_{21}\exp(-h\nu/kT) = d_{12}$ is equal to the probability of nonoptical excitation of the system from level 1 to level 2. At room temperature $kT \approx 0.026\,\text{eV}$ and $h\nu$ for visible radiation varies from 1.7 to 2.9 eV. Therefore, the exponent value can be ignored and the last equality in Eq. (13) will hold, i.e., η_l and η_{lq} are equal to the ratio of the probability of spontaneous transitions to the sum of probabilities of spontaneous and nonradiative transitions.

Proceeding from the definition of the energy yield of luminescence, a relation analogous to Eq. (12) can also be written for solid bodies. In this case, it is necessary to introduce the concepts of the density function of states in the bands and the distribution function of electrons and holes over the bands, and to replace the summation by an integration over the frequencies (Section 4).

It follows from Eq. (13) that in a two-level system the quantum yield can vary from zero ($d_{21} \to \infty$) to unity ($d_{21} \to 0$). In systems with three or more energy levels, η_{lq} can be greater than unity, since, if, for example, $d_{31} = d_{12} = 0$, then by exciting by one light quantum $h\nu_{31}$, two quanta $h\nu_{32}$ and $h\nu_{21}$ can be obtained.

As to the quantum yield of luminescence of complicated molecules, Vavilov (1945) formulated a law according to which the quantum yield of photoluminescence will hold constant if the exciting wave is transformed, on the average, to a longer wave; and vice versa, when long waves are transformed to short ones the yield decreases sharply.

Vavilov's law has been verified by many authors for gaseous, liquid, and solid substances (Levshin, 1951; Borisevich, 1967).

The value $\eta_{lq} > 1$ seems quite natural and even obvious. At the same time, the possibility of obtaining a yield of luminescence greater than unity was for many years the subject of sharp debate (Chukova, 1980) involving many well-known scientists. It may be taken for granted that the value of $\eta_l > 1$ does not contradict the second principle of thermodynamics and is realized, in particular, in anti-Stokes excitation of luminescence.

In light-emitting diodes, under certain conditions, the energy of quanta of luminescence can be greater than the contact potential difference: $h\nu_1 > \Delta V$. Part of the energy they receive by the recombination of electrons and holes comes from the electric field applied to the p–n transition, and the rest is due to the energy of thermal warming, which leads to cooling in the region of recombination. The static electric field entropy is zero and that of electromagnetic radiation is greater than zero. Therefore, the decrease in entropy caused by the heat-to-light conversion is fully compensated for by its increase from the conversion of the electric field energy into radiation. The requirement that entropy be increased in all thermodynamic processes is satisfied. The maximum efficiency of electric-to-light energy conversion is (Weinstein, 1960)

$$\eta_l^m = \frac{T_{ef}}{T_{ef} - T} \tag{14}$$

where T is the temperature of the body, and T_{ef} is the effective temperature of a blackbody in which the thermal radiation entropy is equal to the luminescence entropy. Theoretically, at room temperature η_l^m can reach 160%. In experiments, however, the value of η_l has been found to be much lower than its thermodynamic limit.

2.7. Anti-Stokes Luminescence

Luminescent radiation whose wavelength λ_l is shorter than the wavelength of the exciting light, λ_{ex}, is referred to as anti-Stokes luminescence, for it does not satisfy the Stokes rule. In atomic systems, lines of Stokes and anti-Stokes photoluminescence can be observed. In complicated molecules and semiconductors, the absorption and luminescence lines partially overlap. In this case, a major part of the

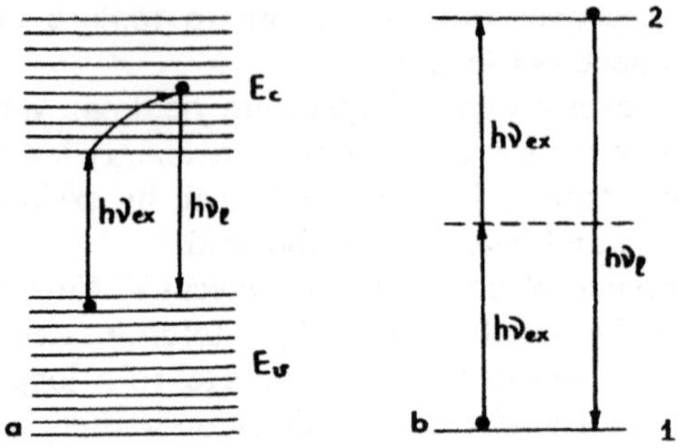

Figure 3. Two mechanisms responsible for anti-Stokes luminescence: (a) After being excited from the valence band E_v to the conduction band E_c an electron receives additional energy as a result of its interaction with the crystal lattice. (b) An electron is excited to the high-energy level 2 by two-photon (or by two-step) absorption.

radiation band pertains to Stokes luminescence and the remaining small part to anti-Stokes luminescence.

There are several reasons for the appearance of anti-Stokes luminescence:

1. The electron that has been promoted to an excited state under the action of the light quantum $h\nu_{ex}$ can receive additional energy as a result of energy exchange with other particles or with the crystal lattice and move to a higher level (Fig. 3). The transition of the electron from this level back to its original state leads to the light quantum $h\nu_l > h\nu_{ex}$. Thus, part of the heat is converted into nonequilibrium radiation. Simultaneously, reverse processes in which $h\nu_l < h\nu_{ex}$ are proceeding.
2. External radiation causes electrons to move upward not only from the ground, but also from excited states. A cascade of transitions to levels whose energies are considerably greater than $h\nu_{ex}$ arises (Auzel *et al.*, 1994).
3. At large excitation densities a nonlinear optical process of two-photon absorption is possible. The substance is excited to levels whose energy is equal to $2h\nu_{ex}$ (in principle, $3h\nu_{ex}$ and more).
4. As a result of the interaction between the excited centers, there may be summation of absorbed energy in one center.
5. Both of the above mechanisms in pure form and in combination can be observed experimentally. For example, when two centers interact, the energy of one center goes to the other with an increment that is due to the thermal energy (phonon) absorption.

The phenomenon of anti-Stokes luminescence or upward frequency conversion is used in infrared ray counters and light-emitting diodes, in particular, in those based on GaAs:Si, for obtaining radiation in the visible region (Galginaitis, 1971; Bergh and Dean, 1972).

2.8. Luminescence Rise and Decay

Any change in the conditions of excitation leads to a decrease or an increase in luminescence power. Changes in the luminescence intensity with time $I(t)$ are not instantaneous; rather the intensity increases or decreases in accordance with certain laws, the study of which yields additional information on the structure of matter and on the mechanism of radiation. In the simplest case of a two-level system, $I(t)$ is given by the equations (Stepanov and Gribkovskii, 1963)

$$I(t) = I_l^{st}[1 - \exp(-t/\tau)], \qquad 0 \leqslant t \leqslant t_l \tag{15}$$

$$I(t) = I_l(t_1)\exp[-(t - t_1)/\tau_0] = I_l^0 \exp(-t/\tau_0), \quad t \geqslant t_1 \tag{16}$$

where

$$\tau = [A_{21} + B_{21}(u + u^0) + B_{21}(u + u^0) + d_{21} + d_{12}]^{-1} \tag{17}$$

$$\tau_0 = [A_{21} + B_{21}u^0 + B_{12}u^0 + d_{21} + d_{12}]^{-1} \tag{18}$$

I_l^{st} is the luminescence intensity in a stationary regime, $t = 0$ and $t = t_1$ are the moments of switching on and off of the time-invariant intensity of excitation. The parameters τ and τ_0 determine the mean time of luminescence rise and decay, and τ decreases with increasing excitation energy density u. In the visible region, or for low temperatures, Eqs. (17) and (18) are simplified because one may put $u^0 = 0$, $d_{12} = 0$. If a quantum-mechanical system has three or more levels the rise and decay of luminescence will be given by the sum of exponents with different values of τ_i and τ_{j0}. In solids, the excitation and luminescence are described by nonlinear equations (bimolecular recombination). Therefore, in the general case, the dependence $I(t)$ is complicated and is expressed by the analytical functions of the exponent or second-order hyperbola type only in partial cases (Fok, 1964; Antonov-Romanovskii, 1966):

$$I(t) = I_l(0)/(1 + at)^2 \tag{19}$$

where a is the parameter measured in s^{-1}. At $t = 1/a$ the luminescence intensity decreases by a factor of four.

In the general case, the luminescence duration is calculated by the known equation

$$\bar{t} = \frac{\int_0^\infty t I_l(t)dt}{\int_0^\infty I_l(t)dt} \tag{20}$$

If luminescence decays exponentially, it follows from Eq. (20) that $\bar{t} = \tau$.

2.9. Light Sum

Luminescence of crystal phosphors reaches its steady state some time after the onset of excitation. At weak pumping τ_l can be as long as several minutes. During this period there is an accumulation of excited states in the crystal, which can also glow for extended periods after the excitation ceases. The number of light quanta $N(t)$ emitted by the crystal phosphor after the illumination has ceased is called the light sum Σ and is given by the integral

$$\Sigma = \int_0^\infty N_l(t)dt \tag{21}$$

The afterglow of crystal phosphors partially compensates for the difficiency of luminescence quanta Σ_d in the luminescence rise portion as compared with the steady-state regime. The ratio Σ/Σ_d depends on the excitation intensity and can be greater or less than unity because of the change in the quantum yield of luminescence that takes place during the excitation process. Sometimes the light sum is expressed in terms of the luminescence energy rather than in terms of the number of quanta:

$$W_l(t) = \int_v N_l(v, t)hvdv \tag{22}$$

2.10. Thermoluminescence

The presence of casual impurities and activator atoms in the crystal leads to the appearance of localized energy levels in the bandgap. Some of them are deep, i.e., they are located at a considerable distance from the top of the valence band or from the bottom of the conduction band. Such levels are often metastable and play the role of traps for charge carriers. For the electron at the metastable level to be able to recombine with a hole it should first be transferred to the conduction band. This process can be stimulated by an electric field, by infrared light, or by simply warming the crystal. The luminescence of a preexcited crystal phosphor stimulated by warming is termed thermoluminescence. The study of the temperature dependence of the afterglow intensity and spectrum permits one to draw conclusions about the depths at which impurity levels lie in solids (Levshin, 1951; Henisch, 1962).

2.11. Negative Luminescence

Luminescence arises as a result of the departure of matter from thermodynamic equilibrium. The following two cases are possible: (a) the spontaneous radiation power in a given spectral range is greater than the absorption power of thermal emission; (b) the spontaneous radiation power is less than the absorption power $v_g k(v)u^0(v)$. In the first case, the value of $W_l(v)$ in relation (4) is positive; in the second case it is negative. The concept of negative luminescence was introduced on this basis (Antonov-Romanovskii *et al.*; 1955; Stepanov, 1961). For a system of particles with discrete energy levels, the condition for the appearance of negative luminescence can be given as

$$n_i/g_i < (n_j/g_j)\exp(-hv_{ij}/kT), \qquad i > j \tag{23}$$

At thermodynamic equilibrium ($n_i = n_i^0, n_j = n_j^0$) the right and left sides of Eq. (23) are equal to each other. Consequently, to satisfy inequality, it is necessary to either reduce the population of the upper level or increase that of the lower level as compared with the equilibrium values of n_i^0 and n_j^0. Then, a gap appears in the thermal emission background at frequency v_{ij}, whereas at positive luminescence a spike would be observed in this region.

In semiconductors, negative luminescence arises under the condition (Gribkovskii, 1975)

$$\Delta F = F_e - F_h < 0 \tag{24}$$

where F_e and F_h are Fermi quasi-levels for electrons in the conduction band and for holes in the valence band, respectively. This relation is also valid for complex molecules for which F_e and F_h have the meaning of chemical potentials of excited and unexcited particles. It can also be obtained for a system of particles with discrete energy levels if the populations of the levels are given in the form

$$n_i = ng_i \exp(F_j - E_i)/kT, \qquad n_j = ng_j \exp(F_j - E_j)/kT \tag{25}$$

where n is the total number of particles, and F_i and F_j are parameters depending on the excitation intensity. At thermodynamic equilibrium, if $\Delta F_i = 0$, then

$$F_i = F_j = F_0 = kT \ln(n_l/ng_l) \tag{26}$$

and Eqs. (25) give the Boltzmann distribution of particles over the levels.

According to Eq. (4), the negative luminescence power cannot be greater than the thermal emission background in a given spectral range. It is, therefore, practically impossible to observe it in the visible region of the spectrum at room temperature while the power of positive luminescence can reach a large value. With increasing temperature or a shift to the infrared region the thermal emission background increases sharply and the conditions for the appearance of negative luminescence are created.

To excite negative luminescence, one usually uses sources of negative fluxes of light, i.e., bodies whose temperature is lower than the sample being investigated (Stepanov, 1961).

When the substance is excited by the radiation from hot sources or lasers, along with intense illumination in one spectral range, relatively weak negative luminescence in another section of the spectrum may arise. In semiconductors, negative luminescence is effectively excited by crossed electric and magnetic fields (Malyutenko *et al.*, 1990).

The use of negative light fluxes permits measuring all the usual spectroscopic characteristics of the substance, e.g., the absorption and scattering coefficients, the position and profile of absorption and radiation bands, the quantum yield, and polarization of luminescence and dichroism.

2.12. Hot Luminescence

In crystals and complex molecules, as a rule, during the lifetime of excited-state molecules within each energy band or electron-vibration band, a quasi-equilibrium distribution of charge carriers over the energy levels sets in. In the first case, it is described by the Fermi–Dirac relation, whereas in the second case it is given by the Boltzmann function. With the advent of lasers generating light pulses of picosecond duration and high-sensitivity radiation detectors capable of detecting even individual quanta, it became possible to study the processes associated with hot electrons, including hot luminescence. This term designates luminescence whose bands lie in that part of the spectrum where usual luminescence is observed at higher temperatures (Saari and Rebane, 1969; Rebane *et al.*, 1973).

The mechanism of hot luminescence excitation in semiconductors at band-to-band transitions consists of the following (Fig. 4). If the sample is excited by an

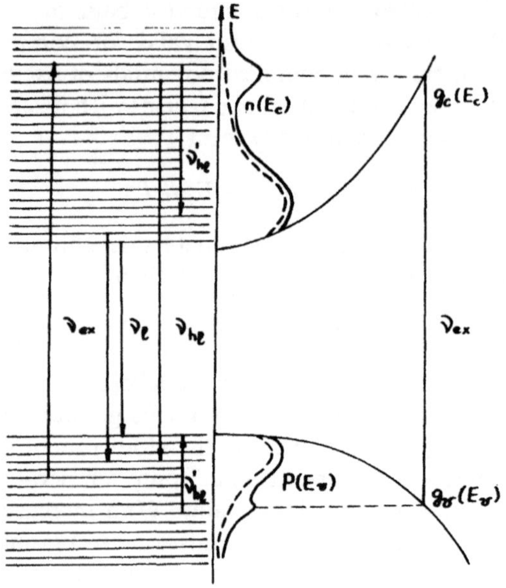

Figure 4. Ordinary (ν_l) and hot luminescence at band-to-band (ν_{hl}) and interband (ν'_{hl}) transitions in solids. At the right side the graphs of density of state functions for conduction $[g_c(E_c)]$ and valence $[g_v(E_v)]$ bands are shown; $n(E_c)$ and $p(E_v)$ are graphs of electron and hole energy distributions. Broken lines show carrier distribution under conditions of equilibrium (Gribkovskii, 1975).

exciting light whose quantum energy is much greater than the band gap $h\nu_{ex} > E_g$, the electrons will be taken to high energy levels in the conduction band, and holes will be formed in the depth of the valence band. Hot electrons and holes form some additions to those carriers that have already lost their excess energy and are distributed over the bands in accordance with the Fermi–Dirac relation. Recombining, these additional carriers emit hot luminescence photons. Hot luminescence also occurs when electrons are in a quasi-equilibrium state but their effective temperature is higher than that of the lattice (Zimin *et al.* 1990). The hot luminescence intensity depends on the ratio of the relaxation time to the excited-state lifetime. The energy exchange among gas-phase molecules is usually slight, so that in such systems there is almost always a considerable departure from the equilibrium distribution of particles over the energy levels (Borisevich, 1967). The intensities of hot and usual luminescence are often comparable.

In solids the time during which equilibrium distribution of carriers within one band is achieved is 10^{-12}–10^{-11} s and the excited-state lifetime is greater by two to three orders of magnitude. Therefore, the intensity of hot luminescence at band-to-band transitions will be quite weak. At the same time, when excitons are annihilated its intensity is sometimes comparable to the intensity of usual luminescence. The study of hot luminescence yields valuable information about the thermalization rate of excited states in solids, atoms, and molecules.

2.13. Superluminescence or Amplified Luminescence

The luminescence spectrum reaching the radiation detector differs, as a rule, from the luminescence spectrum of an infinitely small volume or from primary (spontaneous) luminescence determined solely by the energy spectrum of matter and by the probabilities of spontaneous transitions. On its way to the detector the luminescence flux S_l is absorbed, scattered, and reflected at the matter–air boundaries. Absorbed luminescence excites secondary luminescence, which in turn excites ternary luminescence, and so on. If between some energy levels an inverse population arises owing to the luminescence excitation, the substance will amplify the luminescence instead of absorbing it. In the absence of a cavity, or in a cavity at a pump below the threshold, the radiation stimulated by a luminescence flux is not related to lasing radiation and it is not luminescence, for luminescence, by definition, is a part of spontaneous radiation ([Eq. (4)]). This stimulated radiation, or stimulated radiation together with spontaneous luminescence, is designated as superluminescence or amplified (enhanced) luminescence.

Superluminescence always accompanies the laser action and most often produces a negative effect on the threshold and power of generation. In studying the mechanism of luminescence, it is necessary to find the spectrum of primary luminescence using the experimentally measured radiation spectrum. The relation between these two spectra depends on many factors: in particular, on optical characteristics, the size and shape of the sample being investigated; the excitation energy distribution over the volume; the degree of overlap of the absorption and luminescence bands; and the quantum yield of luminescence. For the spectral density of the luminescence flux $S_{21}(\nu)$ leaving the second face of a rod with length l and sectional area s the following relation has been obtained (Gribkovskii, 1975):

$$S_{21}(\nu) = S_{21}^0(\nu)\,\frac{(1 - R_1R_2)[\exp(k - \rho)l - 1][R\exp(k - \rho)l + 1]}{l(1 + R_1)(k - \rho)[1 - R_1R_2\exp 2(k - \rho)l]} \qquad (27)$$

where

$$S^0(\nu) = \xi W_1(\nu)sl(1 - R_2)(1 + R_1)/(1 - R_1R_2) \qquad (28)$$

is the luminescence flux leaving the second face provided that $k - \rho = 0$, R_1 and R_2 are the reflection coefficients of the first and second faces of the rod, k is the absorption-amplification coefficient of the substance, ρ is the coefficient of internal loss (scattering) of radiation at its passage between the rod faces, ξ is the coefficient of frequency-independent proportionality, and $W_l(\nu)$ is, as before, the spectral density of power luminescence. The contour $W_l(\nu)$ coincides with the luminescence spectrum not distorted by luminescence absorption and amplification. It can be observed experimentally provided that $R_1R_2\exp 2(k - \rho)l \ll 1$ (i.e., in a sufficiently thin layer with optically bleached surfaces). From comparison of $S_{21}(\nu)$ and $S_{21}^0(\nu)$ it is easy to find the value of $k - \rho$ for all the frequencies within the luminescence band.

2.14. Near- and Far-Field Patterns

Among the characteristics of luminescence may be the near- and far-field patterns that are widely used in the physics of injection lasers. The near-field pattern characterizes the radiation luminance distribution over the laser mirror, while the far-field pattern gives the luminance distribution on a rather distant screen. When the injection current is much less than the generation threshold, both patterns are created solely by the luminescence flux. Near the threshold, superluminescence begins to play an appreciable role. After the threshold is overcome, bright dots appear against the luminescence background in the near-field pattern and a mode structure arises in the far-field pattern, which is sometimes inherent in superluminescence as well. Correlation has been established between the near-field pattern and the internal quantum yield of lasing emission (Voronin *et al.*, 1971; Gribkovskii, 1988).

3. FUNDAMENTALS OF THE QUANTUM THEORY OF RADIATION

3.1. Classical, Quantum-Mechanical (Semiclassical), and Quantum-Electrodynamic Methods

It is well known that luminescence obeys laws that lie outside the framework of classical theory and to some extent contradict it. The classical theory of radiation, however, has been successful in considering the reflection, refraction, and scattering of light as well as other optical phenomena, correctly describing many of the important characteristic features of luminescence. For this reason, classical calculations in the theory of luminescence coupled with quantum ideas about the character of absorption and emission are still being used.

The classical theory of luminescence uses elementary sources of radiation: electric and magnetic dipoles, quadrupoles, rotators, and so on (Feofilov, 1959, Curie, 1960). They generate mainly one type of radiation. As an example, consider the solution of the problem for a harmonic oscillator, i.e., an electron that is in a potential well $V(x)$:

$$V(x) = kx^2/2 \tag{29}$$

where x is the deflection of the charge from the equilibrium position, and k is a quasi-elastic constant. In this case, when the electron is displaced from the equilibrium position it will be subjected to the action of the force

$$F = -dV(x)/dx = -kx \tag{30}$$

If one lets the dipole interact with linearly polarized light whose electric vector makes an angle ϑ with the dipole axis, then the equation of dipole motion can be given by

$$\frac{d^2x}{dt^2} + 2\gamma\frac{dx}{dt} + \omega_0^2 x = \left(\frac{e}{m}\right)\mathscr{E}_0 \exp(i\omega t)\cos\vartheta \tag{31}$$

where $\mathscr{E}_0$ is the amplitude of the electric field, $\omega = 2\pi\nu$ is the angular frequency of light $\omega_0 = (k/m)^{1/2}$ is the oscillator's natural oscillation frequency, e is an elementary charge, and m is the electron mass. The term $2\gamma dx/dt$ takes into account the back action of the dipole-emitted radiation on the dipole motion and medium friction ($2\gamma m dx/dt = 2\gamma_r m dx/dt + 2\gamma_t m dx/dt$ is the sum of the radiant and viscous frictional force). The general solution of Eq. (31) is given by

$$x(t) = C_1 e^{i\omega_t t - \gamma t} + C_2 e^{-i\omega_t t - \gamma t} + \frac{(e/m)\mathscr{E}_0 \cos}{\omega_0^2 - \omega^2 + 2i\gamma\omega} e^{i\omega t} \tag{32}$$

where $\omega_l^2 = \omega_0 - \gamma^2$, and C_1 and C_2 are integration constants. If at the initial instant of time the oscillator was at equilibrium, then, determining C_1 and C_2 from the equations $x(0) = 0$ and $dx(0)/dt = 0$, and substituting them into Eq. (32), one obtains the expressions for oscillations with increasing amplitude. Rejecting the first two damped terms, one arrives at steady-state behavior of forced oscillations with the exciting light frequency ω.

If one takes the values of $x(t)$ and $dx(t)/dt$ at any given time during steady-state oscillations as the initial conditions, and sets $\mathscr{E}_0 = 0$, one can get the oscillation damping process.

The dipole absorption power is equal to the work done by the lightwave per unit time. With the aid of complex and complex-conjugate (marked with asterisks) values it is given by

$$W_a(\omega, \vartheta) = \frac{(dx/dt + d^2x^*/dt)\, e(\mathscr{E} + \mathscr{E}^*) \cos\vartheta}{4} \tag{33}$$

from which it follows, for the steady-state regime, that

$$W_a(\omega, \vartheta) = \frac{e^2 \mathscr{E}_0^2 \gamma \omega^2/m}{(\omega_0^2 - \omega^2) + 4\gamma^2\omega} \cos^2\vartheta \tag{34}$$

The latter expression defines the absorption line profile. The smaller the constant γ, the sharper its maximum. To obtain complete emission of the oscillator, it is necessary to integrate Eq. (9) over a unit radius sphere. In the steady state this gives

$$W_r(\omega) = \frac{e^4 \omega^4 \mathscr{E}_0^2}{3c^3 m^2} \frac{\cos^2\vartheta}{(\omega_0^2 - \omega^2) + 4\gamma^2\omega} \tag{35}$$

According to Eq. (35), the stimulated emission power is proportional to the energy density of incident radiation ($\mathscr{E}_0^2 = 8\pi u$) and the fourth power of the frequency. The energy yield is equal to

$$\eta_e = \frac{W_r(\omega)}{W_a(\omega)} = \frac{e^2\omega^2}{3m\gamma c^2} = \frac{\gamma_r}{\gamma_r + \gamma_t} \tag{36}$$

where it is taken into account that $\gamma_r = e^2\omega^2/3mc^3$. As one can see, η_e is independent of the frequency of the exciting light. The properties of elementary sources of radiation are considered in detail in monographs on electrodynamics and on the theory of luminescence (Stepanov and Gribkovskii, 1963; Feofilov, 1959).

In quantum mechanics matter is regarded as a quantum system, and radiation is described by classical equations. In order to find the power, spectrum, polarization, and direction of light propagation, one should first solve the stationary Schrödinger equation

$$H\Psi = (T + U)\Psi = E\Psi \tag{37}$$

where H is the energy operator representing all the forms of motion inside the system under consideration and its interaction with the surrounding medium, E is the energy, and Ψ is the wavefunction. The operator H is equal to the sum of the operators of kinetic T and potential U energy. Equation (37) for the function Ψ is solvable only for a certain set of energy values E_1, E_2, E_3. Just as in a string of length L with fixed ends, standing waves with a discrete set of wavelengths $[\lambda_1 = L/2, \lambda_2 = L/3, \lambda_3 = L/(1+i)]$ can set in. In solids, some groups of energy values merge together and form allowable energy bands.

On finding the energy spectrum and corresponding wavefunctions one can solve the nonstationary Schrödinger equation:

$$i\hbar\partial\Psi(x, t)/\partial t = [H_0(x) + V(x, t)]\Psi(x, t) \tag{38}$$

where $H_0(x)$ is the energy operator of the system without an external field, $V(x, t)$ is the matter–field interaction operator, and $\hbar = h/2\pi$. Equation (38) is generally solved by perturbation theory methods under the assumption that the interaction operator is relatively small. The wavefunction is sought as a superposition of wavefunctions $\Psi(x)$ in the absence of interaction when $V(x, t) = 0$:

$$\Psi(x, t) = \sum_i C_i(t)\psi_i(x) \exp(-iE_i t/\hbar) \tag{39}$$

where $C_i(t)$ are denoted by time-dependent expansion coefficients. If one assumes that at the initial instant of time $t = 0$ the system was in the state with the number k, i.e.,

$$C_i(0) = 1 \qquad (i = k), \qquad\qquad C_i(0) = 0 \qquad (i \neq k) \tag{40}$$

then, in due course, the probability of finding it in other states as well will appear. This enables one to find the probabilities of induced transitions among all the energy levels. There are no spontaneous transitions in quantum mechanics. On solving Eq. (38) one can only see what is happening to the matter. To draw conclusions about the radiation, it is necessary to use the conservation of energy law. For instance, if the system has undergone a transition from level j to a higher level i, then, there has been absorption of a light quantum $h\nu_{ij} = E_i - E_j$. When a reverse transition takes place, a similar light quantum is emitted. In 1916 Einstein postulated that in order for a gas of identical molecules with discrete energy levels to be at thermodynamic equilibrium with thermal emission, the transitions from upper to lower levels should be of two types, induced and spontaneous, and the probabilities of induced B_{ij} and spontaneous A_{ij} transitions are related by the relation

$$A_{ij}/B_{ij} = 8\pi h\nu_{ij}^3/c^3 \tag{41}$$

Between the years 1927 and 1930, P. A. M. Dirac developed a rigorous quantum-electrodynamic theory of radiation, derived ratio (41), and disclosed the deep physical meaning of the probabilities of induced and spontaneous transitions (see Section 3.5). In quantum mechanics, ratio (41) is used to find the probabilities of spontaneous transitions from calculated values of B_{ij}.

In quantum electrodynamics, the interaction of matter with a field is described by the Schrödinger time equation

$$i\hbar\partial\Psi/\partial t = (H_m + H_r + V)\Psi \tag{42}$$

where H_m is the energy operator of the matter, H_r is the field-energy operator, and V is the interaction operator. The solution of Eq. (42) permits following not only the motion of the mechanical system, but also the appearance or disappearance of each light quantum separately (see Section 3.5).

By finding the eigenvalues of energy, eigenwavefunctions, and probabilities of optical transitions, one solves spectroscopic problems to a considerable extent. For luminescence, however, this is obviously insufficient. The luminescing matter functions as a complicated mechanism converting excitation energy to other kinds of energies. In the process, many states turn out to be occupied and various optical and nonoptical transitions take place simultaneously. Under the conditions of nonstationary excitation the picture is still more complicated. Its description cannot be obtained on the basis of rigorous wave equations, and approximated solutions permit following only elementary acts of one- two- or three-photon absorption and emission. In general, it is likely that the more rigorous the theory, the simpler the problem has to be for it to be solvable. The phenomena of luminescence are found to be extremely complicated and difficult to resolve in the framework of an exact theory. For this reason in the theory of luminescence the method of rate or balance equations is widely used, which is sometimes referred to as the probabilistic method of calculation. Examples of employing this method to calculate luminescence of various models of matter are given in Section 4.

3.2. Electronic States in an Ideal Crystal

The state of electron motion in solids would be known exactly if one could manage to solve Schrödinger's equation [Eq. (37)] and find the eigenwavefunctions and eigenvalues of energy (Bassani and Parravicini, 1975).

Historically, the first and simplest model of a solid body was proposed by Sommerfeld. He believed that the potential energy of an electron inside a crystal was constant throughout and that there was a high enough potential barrier on the surface. The electrons, moving freely in the crystal, cannot leave it, since they are held by the repulsive forces from the barrier. In this case one obtains the equation for plane waves. The presence of an impermeable barrier means that at the crystal's boundaries the wavefunction is equal to zero. Consequently, the energy of an electron closed in a potential box has a discrete set of values. For the nth level it is equal to

$$E_n = \frac{\hbar^2(k_x^2 + k_y^2 + k_z^2)}{2m} = \frac{\hbar^2\mathbf{k}^2}{2m} = \frac{(2\pi/L)^3\hbar^2 n^2}{2m} \tag{43}$$

where $\mathbf{k} = k_x, k_y, k_z$ is the wavevector of the electron and of its components on the coordinate axes, L is the linear size of the crystal chosen in the form of a cube, and $n^2 = n_x^2 + n_y^2 + n_z^2$ are quantum numbers determining the value of the electron's wavevector:

$$k_{xn} = 2\pi n_x/L \tag{44}$$

Counting the number of states per unit energy interval $g(E)$ or angular frequency $g(\omega)$ leads to

$$\begin{aligned} g(E) &= (1/2\pi^2)V(2m/\hbar^2)^{3/2}\sqrt{E} \\ g(\omega) &= V\omega^2/\pi^2 v^2 v_g \end{aligned} \tag{45}$$

where $v = \omega\lambda/2\pi$ and v_g are the phase and group velocities of the wave. The last equation also gives the density of electromagnetic field states. If one takes into account that $g(\omega)\Delta\omega = g(\nu)d\nu$, and in a vacuum $v^2 v_g = c^3$, then for a unit volume

$$g(\nu) = 8\pi h\nu^3/c^3 \tag{46}$$

This factor is contained in the Planck function, expressed by Eq. (2).

The expression for the density of states [Eq. (45)] obtained with the aid of an elementary model of matter is widely used in the solid state theory. It turns out that if one replaces m in Eq. (45) by the effective mass of the electron m_e, one can determine the density of states in a real crystal.

Qualitatively new properties of the motion of electrons appear in Kronig's and Peni's one-dimensional model, which is an infinite chain of potential rectangular wells. For this model the Schrödinger equation can be written in the form

$$[(\delta^2 - \varkappa^2)/^6 2\varkappa\delta] \sin\varkappa a' sh\delta b + \cos\varkappa a' ch\delta b = \cos\varphi \tag{47}$$

where $\delta^2 = 2m(V_0 - E)/\hbar^2$, $\varkappa^2 = 2mE/\hbar^2$, $a = a' + b$, a' is the width of the potential well, V_0 and b are the height and width of the barrier, $\cos\varphi$ is the phase factor by which the wavefunctions differ for identical regions of the crystal. Solutions of Eq. (47) are only possible under the condition that the absolute value of the left-hand side does not exceed unity. A set of possible values of energy forms the bands alternating with bands for which there are no solutions of the Schrödinger equation. The width of allowed bands increases with increasing a, and bandgaps get narrowed and vanish. As the energy is increased, there is a transition to free motion of electrons (Anselm, 1962).

In a three-dimensional strictly periodic crystal, the motion of electrons is described by the Bloch function

$$\Psi(r) = U(r)\exp(ikr) \tag{48}$$

and $U(r)$ has translational symmetry

$$U(r) = U(r + \mathbf{a}_n) \tag{49}$$

where $\mathbf{a}_n = n_1\mathbf{a}_1 + n_2\mathbf{a}_2 + n_3\mathbf{a}_3$ is the translational vector of the direct lattice, and $\mathbf{a}_1, \mathbf{a}_2, \mathbf{a}_3$ are the basis vectors. The most important property of Bloch functions is their periodicity in **k**-space or in the space of a reciprocal lattice whose translational vector components $\mathbf{b}_i$ are equal to

$$\mathbf{b}_1 = \frac{2\pi[\mathbf{a}_2\mathbf{a}_3]}{\Omega_0}, \qquad \mathbf{b}_2 = \frac{2\pi[\mathbf{a}_3\mathbf{a}_1]}{\Omega_0}, \qquad \mathbf{b}_3 = \frac{2\pi[\mathbf{a}_1\mathbf{a}_2]}{\Omega_0} \tag{50}$$

where $\Omega_0 = (a_1[a_2, a_3])$ is the volume of the direct lattice's elementary cell. This property makes it unnecessary to study the wavefunction in the whole of **k**-space; it is enough to confine attention to the first Brillouin band, which is a set of all nonequivalent vectors **k**, none of which can be shortened by subtracting any translational vector of the reciprocal lattice.

In the general case, the dependence of electron energy on the wavevector has a complicated form, and classification of electronic states in solids is rather unwieldy and is carried out by group theory methods (Jones, 1960; Boukaert *et al.*, 1936). Formally, the quantum-mechanical motion of an electron can be reduced to the Newton equation $\mathbf{v}_g = \mathbf{P}/\boldsymbol{m}_e$, where $\mathbf{P}$ is the electron momentum, and $\boldsymbol{m}_e$ is the effective mass of the electron, which is a second-rank tensor with components

$$m_{ij} = \partial^2 E/\partial k_i k_j \tag{51}$$

where by i and j are implied x, y, z. In the vicinity of extreme points, i.e., of the valence band top and conduction band bottom, the dependence of energy on the wavevector (in a one-dimensional case by parabola) can be expressed approximately by

$$E = E_0 + \hbar^2 k_x^2/2m_x + \hbar^2 k_y^2/2m_y + \hbar^2 k_z^2/2m_z \tag{52}$$

where $m_i = \partial^2 E/\partial k_i^2$, $(i = x, y, z)$. If m_e is a tensor, then in Eq. (45) for the density of states, $m^{3/2}$ should be replaced by $(m_x m_y m_z)^{1/2}$, or it is necessary to preserve the form of the equation, assuming that $m_e = (m_x m_y m_z)^{1/3}$.

3.3. Lattice Vibrations: Phonons

Vibrations of a crystal lattice have a certain impact on all the physical processes proceeding in a solid body. The vibrational motion of atoms is the main contribution to the heat capacity of a solid and serves as the primary mechanism for heat conduction. The mobility of electrons in a crystal and, consequently, its heat conductivity is determined to a considerable extent by lattice vibrations. This is also true for band-to-band, impurity, and exciton absorption and luminescence, as well as for other optical processes in solids.

The principal concepts of vibration theory can be introduced by considering the simplest model of a linear chain, consisting of alternating atoms of two types, which differ in mass. In such a one-dimensional lattice, an elementary cell will consist of two atoms—a light one and a heavy one. If both atoms are moving synchronously to the right or to the left of the equilibrium position, such vibrations are referred to as longitudinal acoustic vibrations, since this motion is characteristic of sonic waves. They are designated as LA.

If the atoms of one cell are moving in opposite directions to meet or depart from each other, while the center of mass remains stationary, one speaks of longitudinal optical (LO) vibrations. This name is derived from the fact that electric polarization can occur in LO vibrations and a variable dipole moment can arise, which leads to the absorption and emission of infrared electromagnetic waves.

Apart from longitudinal vibrations there are also transverse acoustic (TA) and transverse optical (TO) vibrations. In the first case, both atoms of the cell move up and down synchronously, whereas in second case they move in antiphase. If one denotes the wavevector of longitudinal waves by $\mathbf{q}$, the frequencies of optical (ω_{LO}) and acoustic (ω_{LA}) vibrations can be given by the equation (Anselm, 1962; Fan, 1967).

$$\omega_{\mathrm{LO,LA}}^2 = \tfrac{1}{2}\omega_0^2\{1 \pm [1 - \beta^2 \sin^2(aq/2)]^{1/2}\} \tag{53}$$

where $\beta \ll 1$ and a are parameters that depend on the mass of the atoms and the elastic lattice constants. According to Eq. (53), $\omega_{LO} > \omega_{LA}$ for all values of $\mathbf{q}$. At point $\mathbf{q} = 0$, $\omega_{LO} = \omega_0$ and $\omega_{LA} = 0$.

In studying vibrations of a three-dimensional lattice in a harmonic approximation, it is assumed that vibrations of atoms for all degrees of freedom occur independently of one another. The number of degrees of freedom equals $3gN$, where N is the number of elementary cells in the crystal, and g is the number of atoms in one cell. The motion of atoms is described by a system of $3gN$ linear differential equations, which are easily reduced to a system of $3g$ homogeneous algebraic equations for the vibration frequency, since solutions are sought in the form of plane waves. Setting the determinant of the system equal to zero, one arrives at the equation for characteristic frequencies, or the so-called secular equation (this term arose in celestial mechanics, where an analogous equation describes the movement of the planets). Altogether $3g$ solutions (dispersion laws) are obtained in the form of functions $\omega_j = \omega_j(\mathbf{q})$ where $j = 1, 2, 3, \ldots, 3g$. Account of the crystal symmetry shows that if some vibration branches $\omega_j(\mathbf{q})$ can be degenerate, then their total number is less than $3g$. Of the $3g$ branches three always belong to acoustic vibrations: one longitudinal and two (or one doubly degenerate) transverse branches.

Semiconductor crystals of Group-IV elements (diamond, silicon, germanium, gray tin) have a diamond-type crystal lattice. The crystal lattices of III–V and some II–VI compounds are of the zinc-blende type. In both cases, each elementary cell has two atoms. Consequently, such crystals will have three acoustic and three optical vibration branches.

Introducing normal coordinates, one can represent the lattice vibrations as a superposition of vibrations of harmonic oscillators. In quantum mechanics, the harmonic oscillator energy has a discrete set of values (see Section 4.1) and never vanishes. By analogy with light quanta (photons) the concept of quasi-particles of vibrational motion phonons is introduced. The energy and the quasi-momentum of one $\mathbf{q}$-type phonon are, respectively, $\hbar\omega_q$ and $\hbar\mathbf{q}$, where $\mathbf{q}$ is the wavevector of a plane wave with frequency ω_q. According to quantum mechanics, it is impossible to "freeze" the crystal atoms in the lattice to stillness even at absolute zero. The action of lattice vibrations on the motion of electrons can be regarded as collisions of electrons with phonons, or as an electron–phonon interaction, and the anharmonicity of vibrations can be seen as a phonon interaction. Phonons play a dominant role in some mechanisms of luminescence.

3.4. Impurity Levels and Bands

The appearance of foreign atoms in a crystal leads to local disturbances in the periodicity of the potential field. If one denotes the potential function of an ideal crystal by $V(r)$, then, in the presence of defects, the potential will be equal to $V(r) = V_0(r) + V'(r)$. The second term differs from zero in a particular volume in the vicinity of the defect.

If we assume that $V'(r)$ is determined by the Coulomb interaction of the electron with the point charge of the defect, i.e., $V'(r) = -e^2/\varepsilon r$ where ε is the dielectric constant of matter, the Schrödinger equation for such a model is reduced to the equation for the hydrogen atom. The energy of bound states will be given by

$$E_n = -\frac{m_e(e_e^*)^4}{2\hbar^2 n^2} = -\frac{me^4}{2\hbar^2\varepsilon^2 n^2}\left(\frac{m_e}{m}\right) = -\frac{13.6}{\varepsilon^2 n^2}\left(\frac{m_e}{m}\right) \quad \text{eV} \tag{54}$$

where e^* is the effective charge of the electron, m is the electron mass in vacuum, and n is the principal quantum number. Since in deriving this relation it was assumed that the energy of the conduction band bottom E_{c_0} was equal to zero, it follows from Eq. (54) that the electron near impurity atoms has levels lying in the forbidden band. Such impurities are called hydrogen-like impurities. The depth of the unexcited level is usually equal to a few or dozens of meV. For example, 'for the GaAs monocrystal at $\varepsilon = 12.5$, $m_e/m = 0.072$, we find $E_l = 6$ meV.

Expression (54) is also valid for the energy of holes near the impurity if one replaces m_e by the effective mass m_h of the hole, sets the energy of the valence band ceiling E_{v0} equal to zero, and assumes that the energy increases with increasing depth of the valence band. Since in GaAs $m_h = 0.5$m, for an unexcited acceptor level a considerably greater $E_l = 43$ meV is obtained.

At small concentrations of impurities the electrons and holes are bound to one another and cannot move in the crystal. Such crystals are referred to as weakly doped. In moderately doped solids, the distances between impurity centers become comparable to the Bohr radius of a hydrogen-like system a_B, the wavefunctions of electrons belonging to different atoms partially overlap, impurity bands rather than local levels are formed, and there is impurity conduction. In strongly doped crystals the impurity bands merge with the main bands. In n-type GaAs an impurity band is formed at donor concentrations $N_d = 5 \times 10^{15}$ cm^{-3}, and at $N_d > 4 \times 10^{16}$ cm^{-3} the impurity ionization energy becomes zero. At the same time for p-type GaAs the activation energy tends to zero only at acceptor concentrations $N_a > 5 \times 10^{18}$ cm^{-3} (Emelyanenko *et al.*, 1965).

It has been shown on the basis of comparison of the luminescence band profile in GaAs with calculated data that the density of states in impurity bands can be described by a Gaussian curve (Lucovsky, 1966):

$$g(E) = g^0 \exp[-(E - E^0)^2/\gamma^2] \tag{55}$$

The quantity γ^2 determines the band halfwidth and can be calculated approximately by the expression (Kane, 1963)

$$\gamma = e^2(4\pi N/K_D)^{1/2}/\varepsilon \tag{56}$$

where N is the impurity concentration and

$$K_D = 2(2N/\pi)^{1/6} a_B \tag{57}$$

is the inverse length of Debye screening. A more detailed discussion of the screening problems was also carried out (Stern, 1966).

In contrast to the regular arrangement of parent atoms in the crystal, impurities are chaotically distributed in space, and within macroscopic volumes there are significant fluctuations in their concentrations. This leads to local distortions of main bands and to the formation of tails of bands. If the tails are taken into account, the density of the state will be determined not by Eq. (45), but rather by a more complicated expression (Kane, 1963):

$$g(E) = \frac{V(2m/\hbar^2)^{3/2}y(E)}{2\pi^2} \tag{58}$$

where

$$y(E) = (\gamma/\pi)^{1/2} \int_{-\infty}^{E/\gamma} (E/\eta - \zeta)\exp(-\zeta^2)d\zeta \tag{59}$$

If $E > \gamma$, then $y(E) \simeq E^{1/2}$ and, consequently, Eq. (58) reduces to Eq. (45). For energies lying below the γ level, it follows from Eqs. (58) and (59) that

$$g(E) = V\left(\frac{\gamma}{\pi}\right)^2 \left(\frac{m}{\hbar^2}|E|\right)^{3/2} \frac{\exp(-E^2/\gamma^2)}{2} \tag{60}$$

i.e., in the forbidden band the density function of states moves to the wing of the Gaussian curve.

The formation of band tails in strongly doped crystals was also considered in a number of other works (Hwang, 1970; Efros, 1973). Apart from shallow donor and acceptor levels, introduction of impurities into the crystal can be followed by the appearance of local levels at a considerable depth in the forbidden band. The electrons and holes that get into them can remain for a fairly long time. Therefore, in the theory of crystal phosphors such levels are referred to as traps. Construction of a theory of deep centers is more complicated than in the case of shallow impurities (Lucovsky, 1965*a*; Bonch-Bruevich, 1971).

3.5. Excitons and Biexcitons

The single-electron approximation turned out to be inadequate to explain all the experimental results on absorption and luminescence of crystals. It has been found by experiment, in particular, that solids without a noticeable content of impurities are able to absorb light quanta with an energy smaller than the bandgap $h\nu_{ex} < E_g$. In this case, conductivity does not increase and free carriers do not arise. The crystal lattice is unable to absorb such a large portion of the energy immediately, as the energy of photons is higher by two to three orders of magnitude than that of phonons. Luminescence lines have been observed in the same spectral region.

To explain the observed phenomena, the concept of special particles, excitons, was introduced. Frenkel's exciton model is used for molecular crystals and the large-radius exciton or the Wannier–Mott model is used for semiconductors. According to Frenkel, the exciton can be represented as an excited state of the molecule, which is able to leap to another molecule and thus migrate over the whole crystal (Frenkel, 1931*a,b*). In Frenkel's exciton the electron and the hole are always localized in one site of the lattice, which is why it is referred to as the small-radius exciton.

The Wannier exciton can be represented as a hydrogen-like atom consisting of a positively charged hole bound to an electron (Wannier, 1937).

The average distance between the electron and the hole exceeds the value of the lattice constant. If there are free electrons and holes in the crystal, excitons are formed by their binding with one another. In doing so, the electron and the hole lose part of their kinetic energies. If there are no free carriers, an energy less than E_g should be expended in order for an exciton to be formed. This energy takes the electron away from the hole but not far enough away for it to become a free particle. Excitons can also be formed with the participation of charges of impurity atoms localized at certain points in the crystal. Such excitons are referred to as bound excitons because they cannot travel in the crystal when the distance between the impurity centers is considerable.

In contrast to the hydrogen atom, excitons can disappear (or annihilate). In this respect they are similar to positronium, which is a hydrogen-like atom consisting of an electron and a positively charged positron.

If the electron and the hole forming an exciton are at a distance considerably exceeding the value of the lattice constant, one can, as an approximation, assume that a force acts between them that is a Coulomb attraction weakened by ε multiplied by the dielectric permeability of the crystal. In this approximation the wavefunction (which is the solution of the Schrödinger equation for the exciton) can be sought in the form of the product of two functions, one of which describes the translational motion of the exciton as a single neutral particle and the second that describes its internal motion. The Schrödinger equation is broken down into two equations, namely, one for the plane wave and one for the hydrogen-like particle. The energy of the exciton as a whole will be given by

$$E'_n = E_{c0} - E^0_{ex}/n^2 + p^2/2m_{ex} \tag{61}$$

where $p = \hbar k$ is the value of the momentum of the exciton gravity center, and E_{ex} is the energy of the unexcited state of the exciton. The second term in Eq. (61) gives a hydrogen-like set of discrete levels, while the third term is equal to the energy of translational motion of the exciton. It will be recalled that shallow impurities in the crystal are also characterized by a hydrogen-like energy spectrum.

The exciton as a system of two interacting particles is beyond the framework of single-electron band theory, yet its energy spectrum can conditionally be combined with the diagrams of bands for one electron. To do this, neglecting the third term in Eq. (61), one uses the first level of the exciton under the bottom of the conduction band at a depth E^0_{ex}.

As excitons move in the crystal and there is a statistical spread in velocities, their energy levels cannot be considered to be sharp. The higher the temperature of the crystal the greater the width of the levels, which is a clear indication of the temperature dependence of absorption and luminescence bandwidths.

Similar to the Bohr radius for the hydrogen atom, the exciton radius can be calculated by

$$r_{ex} = (\hbar^2/me^2)(m/m_{ex})\varepsilon = 0.53(m/m_{ex})\varepsilon \text{ Å} \tag{62}$$

where $m_{ex} = m_e m_h/(m_e + m_h)$ is the reduced exciton mass, and m_e and m_h are the effective masses of the electron and hole, respectively. For $m_{ex} = 0.5m$ and $\varepsilon = 10$ it follows from Eq. (62) that $r_{ex} = 10.6$ Å.

It has been shown theoretically (Lambert, 1958) that two excitons can form a stable complex—an exciton molecule or a biexciton. The binding energy of the excitons in the molecule is equal to 5.5 meV for germanium (Zubov *et al.*, 1972) and 15 meV for silicon (Haynes, 1966). For comparison note that kT at 80 and 300 K are equal to 6.9 and 25.9 meV, respectively. The kinetics and collective properties of excitons in semiconductors are explained on the basis of the ideas about the exciton molecules.

3.6. Probabilities of Spontaneous and Induced Transitions

In the quantum theory of radiation (Heitler, 1954), as in quantum mechanics, transition probabilities are calculated by perturbation theory methods. It was shown that the differential Einstein coefficients for absorption $b_{ji}^{\alpha}(\Omega)$, for stimulated emission $b_{ij}^{\alpha}(\Omega)$, and for spontaneous emission a_{ij}^{α} in the dipole approximation are given by

$$b_{ji}^{\alpha}(\Omega) = b_{ij}^{\alpha}(\Omega) = (8\pi^3/h^2)\,|(\mathcal{D}_{ji}\mathbf{e}_{k\alpha})|^2 \tag{63}$$

$$a_{ij}^{\alpha}(\Omega) = (8\pi^3 \nu_{ij}^3/hc^3)|(\mathcal{D}_{ij}\mathbf{e}_{k\alpha})|^2 \tag{64}$$

where $\mathbf{e}_{k\alpha}$ is the unit vector of photon polarization, $\alpha = 1, 2$, and Ω is solid angle. Coefficients given by Eqs. (63) and (64) are related to the Einstein coefficients for isotropic radiation by the relations (Stepanov and Gribkovskii, 1963)

$$B_{ji} = \left[\sum_{\alpha} \int b_{ji}^{\alpha}(\Omega_1, \Omega_2)\mathrm{d}\Omega_2\right] \Big/ 8\pi \tag{65}$$

$$B_{ij} = \left[\sum_{\alpha} \int b_{ij}^{\alpha}(\Omega_1, \Omega_2)d\Omega_2\right] \Big/ 8\pi \tag{66}$$

$$A_{ij} = \left[\sum_{\alpha} \int a_{ij}^{\alpha}(\Omega_1, \Omega_2)\mathrm{d}\Omega_2\right] \Big/ 8\pi \tag{67}$$

Here the summation is carried out over the two independent polarizations, with the solid angle Ω_2 specifying the direction of radiation propagation. The angle Ω_1 determines the orientation of the dipole moment $\mathcal{D}_{ij}$. In deriving Eqs. (65)–(67), it was taken into account that in the case of isotropic radiation $u^{\alpha}(\nu_{ij}, \Omega) = u(\nu_{ij})/8\pi$. If the levels i and j are characterized by the degrees of degeneracy g_i and g_j, the equality

$$g_j B_{ji} = g_i B_{ij} \tag{68}$$

is valid.

Quantities analogous to Einstein coefficients can also be introduced in considering transitions between wide energy bands (see Section 4.3). Transitions of a quantum-mechanical system from one energy state to another that are accompanied by absorption or emission of photons are called optical transitions.

3.7. Nonradiative Transitions

Apart from optical transitions in atoms, molecules, and crystals, there are also nonoptical transitions from lower to higher states and vice versa. In particular, all kinds of luminescence, apart from photoluminescence, are due to nonoptical excitation of matter. Downward transitions can be accompanied by energy transfer to the crystal lattice, to impurity centers, or to free carriers such as those arising from Auger collisional recombination (Landsberg and Beattie, 1959; Beattie and Landsberg, 1960). One of the channels of excitation energy loss is surface recombination. At the boundaries of a solid body the valence bonds are broken, the potential periodicity is disturbed, and so a large number of energy levels belonging to the surface states appear in the forbidden band. A free electron can pass through these states to the valence band with complete or partial transformation of its energy into heat. Surface nonradiative recombination can be eliminated by making heterostructures in which the luminescing layer is surrounded by more wideband layers transparent for luminescence. For example, in the AlGaAs/GaAs/AlGaAs heterostructure, the internal quantum yield of photoluminescence is equal to 0.997, which is indicative of the absence of nonradiative recombination (Schnitzer *et al.*, 1993).

From the detailed balancing principle it follows that the probabilities of nonradiative transitions between two levels, downward d_{ij} and upward d_{ji}, are related by

$$d_{ji} = d_{ij}\exp(-E_{ij}/kT) \tag{69}$$

For room and lower temperatures in the visible range of frequencies, d_{ji} is practically equal to zero, i.e., there is no nonoptical excitation of a luminescent substance through its interaction with its medium. Luminescence arises owing to the energy of the external sources of excitation or to chemical reactions in the matter itself. At the same time, neither the temperature nor the spectral range of luminescence imposes any restrictions on the value of d_{ij}.

4. THEORETICAL MODELS AND MECHANISMS OF LUMINESCENCE

4.1. The Harmonic Oscillator

The harmonic oscillator is the basic model of matter in classical electrodynamics. In the quantum theory of radiation it is used both as a model of matter and as a model of the electromagnetic field, which is represented by a set of oscillators. Within the framework of linear optics, a pair of levels of a complicated system can be replaced by a harmonic oscillator with an adequate frequency and strength of the oscillator. The Schrödinger equation for the harmonic oscillator has an exact solution; the wavefunctions are expressed in terms of Chebyshev–Hermite polynomials and the energy of the Vth level is given by

$$E_v = (v + \tfrac{1}{2})h\nu_0 \tag{70}$$

where $\nu_0 = (k/4\pi^2 m)^{1/2}$ is the natural frequency of vibrations, and k and m represent the quasi-elastic constant and mass of the oscillator, respectively. The zero-state energy is not equal to zero as in classical mechanics when the oscillator is at rest, but constitutes the quantity $h\nu_0/2$. All the other levels are at an equal distance $h\nu_0$ from one another.

The absorption coefficient of the harmonic oscillator, or, to be more precise, the integral of the absorption coefficient along the line profile $k = W_a/cu$, depends neither on the exciting light density nor on the temperature but coincides with the Kravets integral obtained in classical electrodynamics (Stepanov and Gribkovskii, 1958, 1963). The luminescence power W_l can be given by a linear differential equation (Gribkovskii, 1960*a*):

$$dW_l/dt + A_0 W_l = nA_0 B_0 u(t) h\nu_0 \tag{71}$$

In accordance with the classical theory, from Eq. (71) exponential laws of luminescence rise and decay follow:

$$\begin{aligned} W_l^{\text{in}}(t) &= W_l^0[1 - \exp(-A_0 t)] \\ W_l^d(t) &= W_l^0 \exp(-A_0 t) \end{aligned} \tag{72}$$

where $W_l^0 = nB_0 uh\nu_0$. Calculations show that the natural width of the spectral line of radiation on the scale of energies at spontaneous transitions $i \to j$ is

$$\Delta E_{ij} = \Delta E_i + \Delta E_j \tag{73}$$

where ΔE_i and ΔE_j are the natural widths of the levels i and j.

Since in the harmonic oscillator, the natural width of the level increases proportionally with the level number; on the basis of Eq. (73) one could expect radiation line-broadening at oscillator excitation to higher energy levels. However, a rigorous quantum-electrodynamic calculation, in view of the indistinguishability of quanta emitted at transitions between different pairs of levels $v + 1 \to v$, has shown that whatever the initial store of oscillator energy, its natural energy contour is always given by the familiar classical expression (Gribkovskii, 1960*b*; Stepanov and Gribkovskii, 1963):

$$W_{\text{sp}}(\omega) = \gamma_0/\pi[(\omega - \omega_0)^2 + \gamma_0^2] \tag{74}$$

where $\gamma_0 = A_0/2$ is the line halfwidth. An analogous result was obtained earlier for the special case where at the initial instant of time the oscillator was at the second excited level (Weisskopf and Wigner, 1930).

Detailed analyses (Gribkovskii, 1960*a*; Stepanov and Gribkovskii, 1963) have shown that, in the dipole approximation, all the results of the classical and quantum theories concerning the optical properties of the harmonic oscillator are not only in agreement with one another, but are fully congruent.

4.2. Two-, Three-, and N-Level Systems

Although real objects have a large number of energy levels and solids as well as complex molecules have wide bands of allowed energies, yet in the theory of luminescence relatively simple systems with two to four energy levels are frequently used. By employing such models it is possible to elucidate many characteristic features of the processes taking place in real objects. This is attributed to the fact that all of the energy states do not participate in the excitation of matter and in subsequent luminescence, and wide bands frequently play the part of single reservoirs of free particles.

As the number of levels in a model is increased, the complexity of the calculations increases sharply. It has been shown, however, that with a monomolecular character of transitions between the levels, when the number of transitions $i \rightarrow j$ is directly proportional to the product of the number of particles in the initial state and the transition probability, one can obtain a general solution of the rate equations for a system of particles with any number of energy levels. This enables one to represent the dependence of level populations, absorption and luminescence powers, and other characteristics on the excitation intensity by the same expressions for all systems, independent of the number of levels (Gribkovskii, 1960*c*, Stepanov and Gribkovskii, 1963).

Calculations of the absorption coefficient of isotropic k^{is}, linearly polarized k^{lin}, and natural k^{nat} (the degree of polarization $P = 0$) lead to

$$k^{\text{is}} = k^0(1 + \alpha u)^{-1} \tag{75}$$

$$k^{\text{lin}} = k^0[(\alpha u)^{-1} + (3\alpha u)^{-3/2} \tan^{-1}(3\alpha u)^{1/2}] \tag{76}$$

$$k^{\text{nat}} = k^0(\alpha u)^{-1}[1 - (a^{-1/2} - a^{1/2})\tanh^{-1} a^{1/2}] \tag{77}$$

where α is the parameter of nonlinearity, $a = 3u(2 - 3u)^{-1}$.

Although the absorption coefficient for each of the three variants of excitation is given by totally different relations [Eqs. (75), (76) and (77)] the plots of their dependence on αu practically coincide (Fig. 5).

Under the action of external radiation of frequency ν_{ml}, luminescence can also arise at another frequency ν_{ij}. Its polarization depends not only on the excitation intensity, but first of all on the angle between the matrix elements of the dipole moment $\mathscr{D}_{ml}$ and $\mathscr{D}_{ij}$. With the same geometry of observation and at excitation by linearly polarized light the degree of polarization will be equal to (Gribkovskii, 1960*d*)

$$P^{\text{lin}}(\nu_{ij}) = \frac{P_0^{\text{lin}}(\nu_{ij})(2 - 3\sin^2\xi)}{2 - P_0^{\text{lin}}(\nu_{ij})\sin^2\xi} \tag{78}$$

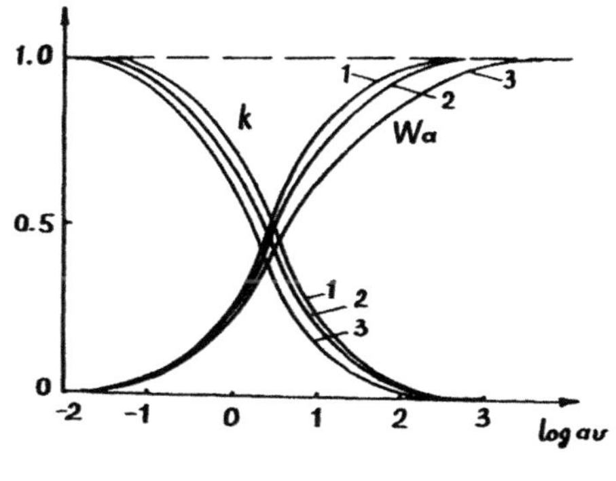

Figure 5. Dependence of k and W_a on the intensity of incident radiation in the case of (1) isotropic, (2) natural, and (3) plane-polarized radiation (Gribkovskii, 1960*a*).

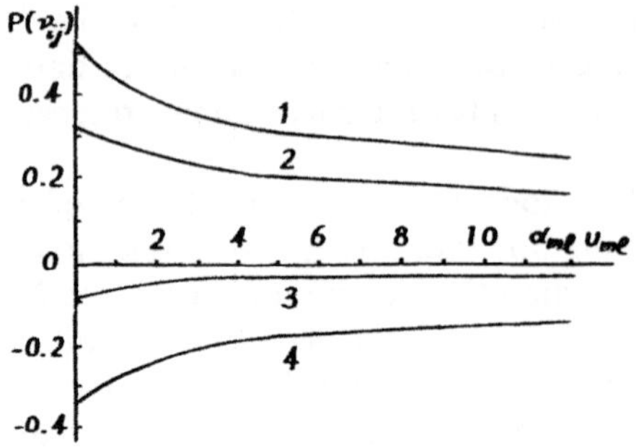

Figure 6. Dependence of the polarization of luminescence on $\alpha_{ml}u_{ml}$ for $\xi = 0°$ (curve 1), 30° (curve 2), 60° (curve 3), and 90° (curve 4) (Gribkovskii, 1960*d*).

where the quantity $P_0^{\text{lin}}(\nu_{ij})$ gives the value of the degree of polarization when the directions of the vectors $\mathscr{D}_{ml}$ and $\mathscr{D}_{ij}$ coincide ($\zeta = 0$). If we substitute the value of $P^{\text{lin}}(\nu_{ij}) = \frac{1}{2}$, which holds at $\alpha_{ml}u_{ml} \ll 1$, into Eq. (78) it becomes the relation obtained within the framework of classical electrodynamics, which is valid in the absence of saturation effects (Levshin, 1951). In this case the degree of polarization varies from $\frac{1}{2}$ to $\frac{1}{3}$ as the angle ζ is increased from 0° to 90° (Fig. 6).

The reader can find the expressions for the degree of polarization of luminescence at excitation by linearly polarized and natural light in a monograph by Stepanov and Gribkovskii (1963).

4.3. *Band-to-Band Absorption and Luminescence*

Optical transitions in undoped crystals are bimolecular in nature, which means that, for a transition to be realized, not only is an electron needed in the initial state but a vacancy (hole) is required in the ultimate state. A quantum-mechanical system with two energy levels is an elementary model of a crystal, which makes it possible to determine, in particular, the dependence of the absorption coefficient on the exciting light intensity, which is also characteristic of more complicated systems (Gribkovskii, 1969, 1975).

One of the basic models of a solid body in both the theory of luminescence and in laser physics is the model of two parabolic bands (Stern, 1963; Lasher and Stern, 1964; McLean, 1960). In its simplest version, where the effective masses m_e and m_h for the electron and for the hole, respectively are isotropic and in optical transitions the rules of selection in relation to the wavevector $\mathbf{k} = \mathbf{k}'$ (where $\mathbf{k}$ and $\mathbf{k}'$ are the wavevectors of the initial and final states) are satisfied (Fig. 7). The absorption coefficient and the power of spontaneous radiation can be given by

$$k(\omega) = \varkappa(\omega)[f_e(E_v) - f_e(E_c)] \tag{79}$$

$$W_{\text{sp}}(\omega) = A_{cv}g(\omega)f_e(E_c)[1 - f_e(E_c)]\hbar\omega \tag{80}$$

where $\varkappa(\omega) = B_{cv}g(\omega)\hbar\omega/v_g$ is the value of the absorption coefficient provided all the states of the valence band are filled with electrons and all the states of the conduction band are vacant, A_{cv} and B_{cv} are integral Einstein coefficients for spontaneous and stimulated transitions proportional to the square of the matrix element $|\mathscr{D}_{cv}|^2$.

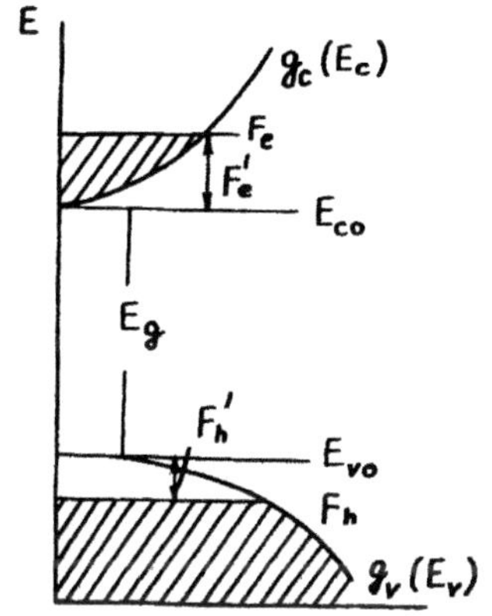

Figure 7. Model of two parabolic bands (Gribkovskii, 1975).

According to Eqs. (79)–(80), the ratio $W_{sp}(\omega)/k(\omega)$ is independent of the kind of distribution function and is universal

$$\frac{W_{sp}(\omega)}{k(\omega)} = \frac{\hbar\omega^3}{\pi^2 v^2}\left\{\exp\left[\frac{\hbar\omega - \Delta F}{kT}\right] - 1\right\}^{-1} \tag{81}$$

where $\Delta F = F_e - F_h$ is the difference between quasi-Fermi levels for the electrons and holes, respectively, and it is taken into account that on the frequency scale ω instead of (41) one can have $A_{cv}/B_{cv} = \hbar\omega^3/\pi^2 v^2 v_g$. The last expression differs from the previously obtained relation between the spontaneous recombination rate and the induced emission function only by the coefficient in front of the brackets (Lasher and Stern, 1964).

The quantity ΔF determines the degree of excitation of the system. If $\Delta F > \hbar\omega$ the right-hand side in Eq. (81) becomes negative. Since $W_{sp}(\omega)$ is always positive, then, consequently, the absorption coefficient takes a negative value. The value $\hbar\omega_{\text{inv}} = \Delta F$ serves as the boundary between negative ($E_g \leqslant \hbar\omega < \Delta F$) and positive absorption (ω_{inv} is the frequency of inversion). At thermodynamic equilibrium $\Delta F = 0$, and Eq. (81) is reduced to the Kirchhoff law: $W_{sp}(\omega)/k^0(\omega) = v_g u^0(\omega)$.

On the basis of Eq. (81) the spontaneous recombination rate summed over the frequency can be expressed in terms of the integral

$$R_{sp} = \int_0^\infty \left(\frac{W_{sp}(\omega)}{\hbar\omega}\right) d\omega = \left(\frac{1}{\pi^2}\right)\int_0^\infty \frac{k(\omega)\omega^2 d\omega}{v^2\{\exp[(\hbar\omega - \Delta F)/kT] - 1\}} \tag{82}$$

In the special case when $\Delta F = 0$, this integral becomes the Roosbroeck–Shockley relation (Van Roosbroeck and Shockley, 1954) and permits calculating the spontaneous recombination rate by the measured value of the absorption coefficient.

If the electrons and holes in the bands are nondegenerate and approximate equalities,

$$f_e(E_c) \approx \exp[(F_e - E_c)/kT] \qquad \text{and} \qquad f_e(E_v) \approx \exp[(F_h - E_v)/kT]$$

are satisfied; then

$$R = Bnp \tag{83}$$

where B is the recombination constant and

$$n = \int_{E_{c_0}}^{\infty} g_e(E_c)f_e(E_c)dE_c \quad \text{and} \quad p = \int_{-\infty}^{E_{c_0}} g_h(E_v)f_h(E_v)dE_v \tag{84}$$

are the total number of electrons and holes in the bands. For germanium monocrystals at $T = 300\,\text{K}$, $R_{sp} = 1.57 \times 10^{13}\,\text{cm}^{-3}\cdot\text{s}^{-1}$, $B = 5.25 \times 10^{-14}\,\text{cm}^3\cdot\text{s}^{-1}$ (Van Roosbroeck and Shockley, 1954). In direct-gap compounds, such as GaAs, GaSb, InP, InAs, and InSb, the recombination constant is larger by a few orders of magnitude than in indirect-gap compounds (Varshni, 1967).

With the definition of luminescence as the excess emission over and above the thermal emission background [Eq. (4)] one finds

$$W_l(\omega) = W_{sp}(\omega)[1 - \exp(-\Delta F/kT)][1 - \exp(-\hbar\omega/kT)]^{-1} \tag{85}$$

At thermodynamic equilibrium $\Delta F = 0$, and from Eq. (85) it follows that $W_l(\omega) = 0$. On the other hand, if the equilibrium is strongly disturbed so that $\Delta F \gg kT$ and $\hbar\omega \gg kT$, then Eq. (85) becomes Eq. (81). In this case, practically all the spontaneous radiation consists of luminescence.

With the aid of Eqs. (81) and (85) one can obtain a universal relation between the spectra of luminescence and absorption (Gribkovskii and Stepanov, 1968; Gribkovskii, 1975)

$$\frac{W_l(\omega)}{k(\omega)} = \frac{\hbar\omega^3}{\pi^2 v^2} \frac{1 - \exp(-\Delta F/kT)}{1 - \exp(-\hbar\omega/kT)} \frac{1}{\exp[(\hbar\omega - \Delta F)/kT] - 1} \tag{86}$$

This equation holds for any optical transition mechanism (bimolecular or monomolecular) and does not depend on the density of state function or on the type of expression (Fermi-Dirac or Boltzmann) that describes the quasi-equilibrium distribution of electrons in the bands. The only requirement for it to be applicable is a separate quasi-equilibrium distribution of excited and unexcited electrons. It has also been shown to be valid for complex molecules (Stepanov and Gribkovskii, 1969, 1970). The universal relation Eq. (86) significantly widens the possibilities for experimental studies of solids. It relates the quantities $W_l(\omega)$, $k(\omega)$, and ΔF to one another and makes it possible to find any one of these quantities by measuring the other two. This is particularly important in studying microscopic volumes of matter when direct measurement of the absorption coefficient is very complicated and sometimes impossible. Methods have been developed, in particular, for calculating the values of ΔF, the amplification spectrum, and luminescence power in absolute units in injection lasers by the measured luminescence spectrum in relative units and the generation frequency (Gribkovskii and Samoilykovich, 1969).

4.4. Luminescence in Impurity Solids

The significant changes in the energy spectrum of a crystal resulting from the appearance within it of impurities and defects are accompanied by changes in the absorption and luminescence spectra, owing to the appearance of new channels of optical and nonoptical transitions. For example, in InSb crystals, upon doping with donors up to concentrations $5 \times 10^{18}\ \mathrm{cm}^{-3}$, a displacement of the absorption band edge from the region of 7.2 μm to that of 3.2 μm was observed (Burnstein, 1954). This is explained by the filling of the lower states of the conduction band with electrons from ionized donors (Moss, 1954). The Burnstein–Moss effect in indirect-gap semiconductors can be calculated approximately on the basis of the solution of the electroneutrality equation

$$n + N_a^- = p + N_d^+ \tag{87}$$

where N_a^- and N_d^+ are the concentrations of ionized acceptors and donors, and n and p are given by the Fermi–Dirac integrals of Eq. (84). On finding the Fermi level the absorption coefficient is calculated for the model of parabolic bands with the k-selection rule (Lasher and Stern, 1964):

$$k(\omega) = (A/E) \int_{E_{co}}^{\infty} g_v(E_v) g_c(E_c)[f_e(E_v) - f_e(E_c)]dE_c \tag{88}$$

where A is the constant proportional to the square of the band-to-band transition matrix element, and $F_e = F_h = F_0$.

Optical transitions between the impurity states and the nearest allowed band (photoionization of impurities) manifest themselves as absorption spectra in the far-infrared region (Pankove, 1971). Similar to band-to-band transitions, probabilities of transitions involving impurity levels are usually calculated by the perturbation theory method. In doing so, shallow impurities are usually described by hydrogen-like wavefunctions and for deep impurities the wavefunctions are found under the assumption of a δ-like (Lucovsky, 1965*a*) or a more complicated (Bonch-Bruevich, 1971) potential of the center.

From the detailed balancing principle it follows that the probability of electron emission from the donor level to the level E_c in the conduction band $P_{dc}(E_c)$ and the probability of electron capture from the level E_c to the donor level $P_{cd}(E_c)$ are bound by the relation

$$P_{dc}(E_c) = P_{cd}(E_c) \exp[-(E_c - E_d)/kT]$$

For the probabilities of transitions between the acceptor level and the valence band we have, respectively,

$$P_{av}(E_v) = P_{va}(E_v) \exp[-(E_a - E_v)/kT]$$

The recombination of electrons and holes in semiconductors with a small concentration of isolated impurity centers has been studied in detail (Shockley and Read, 1952). The outline of the luminescence band and the position of its maximum in strongly doped concentrated semiconductors has been calculated with regard to the nonequilibrium distribution of carriers over the impurity states and the nearest band (Levanjuk and Osipov, 1973*a,b,c*). It was shown (Dumke, 1963; Eagles, 1960) that valence-band–donor optical transitions are more probable by one to two orders of magnitude than acceptor–conduction-band transitions.

Although shallow impurities and Wannier–Mott excitons are characterized by a hydrogen-like energy spectrum (see Sections 3.4 and 3.5), wide absorption and luminescence bands correspond to the band–impurity transitions and the absorption and radiation bands of excitons consist, as a rule, of narrower lines. Optical transitions in solids are frequently accompanied by absorption and emission of phonons, which leads to the appearance of additional maxima in absorption and radiation spectra. In moderately and strongly doped crystals, the impurity states form continuous bands in which the density of states is given by the Gaussian function

$$g_d(E_d) = g_{d0}\exp[-(E_d - E_{d0})^2/\gamma_d^2], \qquad g_a(E_a) = g_{a0}\exp[-(E_a - E_{a0})^2/\gamma_a^2]$$

where E_{d0}, E_{a0}, γ_d, and γ_a are the positions of centers and the values of the donor and acceptor bandwidths (Lucovsky, 1965*b*, 1966). For this model the dependence of the absorption coefficient on the excitation intensity has been studied in detail (Kononenko and Gribkovskii, 1968*a*,*b*; Kononenko, 1970). The luminescence band profile has also been calculated for other mechanisms of recombination, in particular, for the parabolic-band discrete level and parabolic-band–Gaussian-band model (Gribkovskii and Kononenko, 1970).

It should be noted that sometimes in interpreting experimental results, and in formulating rate equations, instead of spontaneous A and nonradiative d transition probabilities, radiative τ_R and nonradiative τ_{NR} lifetimes are used. In the case of the monomolecular mechanism of recombination $\tau_R = 1/A$ and $\tau_{NR} = 1/d$. If the concentration of majority carriers in the crystal is higher by a few orders of magnitude than the concentration of minority carriers and depends weakly on the excitation intensity, then on the basis of the equality $R = Bnp$ one finds the radiative lifetime of electrons $\tau_R^n = 1/Bp$ in a p-type crystal and the radiative lifetime of holes $\tau_R^p = 1/Bn$ in an n-type crystal. For Auger recombination $\tau_0^n = 1/q_p p^2$ and $\tau_0^p = 1/q_n n$. In carbon-doped GaAs the radiative lifetime is given by a more complicated expression, $1/\tau_{NR} = q_p p^2 + Cp^3$, where C is a constant (Calderon *et al.*, 1992).

4.5. Donor–Acceptor Pair Luminescence

Some peculiarities of light absorption and luminescence in solids cannot be explained if the interactions of electrons and holes, which are at donor and acceptor levels, are not taken into account. If one confines oneself only to the Coulomb and van der Waals interactions and designates the distance between the donor and the acceptor of the nth pair as r_n, the energy of the pair can be given by (Hopfield *et al.*, 1963; Thomas *et al.*, 1964; Dean, 1973):

$$E_n(r_n) = E_g - (E_d' - E_a') + e^2/\varepsilon r_n - (e^2/\varepsilon)(a/r_n)^6 \tag{89}$$

where $E_d' = E_{c0} - E_d$, $E_a' = E_a - E_{v0}$ are the ionization energies of the donors and acceptors, and a is the constant of the van der Waals interaction.

Since the quantities r_n and $E_n(r_n)$ assume discrete values, the absorption and luminescence spectra of donor–acceptor pairs (DAPs) should, in principle, consist of individual lines. The distance between the lines decreases as the number n is increased. Empirically, however, the line structure of absorption and luminescence spectra is observed (and not always) only on short-wave wings of corresponding bands (Williams, 1968; Junovich, 1972; Radomsky *et al.*, 1994). This is one of the differences between the spectroscopic properties of DAPs and excitons in whose absorption and radiation spectra fine structure appear on the opposite, long-wave section band (see Section 4.6).

For shallow hydrogen-like impurity centers the dependence of the recombination time on the distance between the donor and the acceptor can be given by (Dohler, 1971; Thomas *et al.*, 1965)

$$1/\tau(r_n) = [1/\tau(0)] \exp(-2r/a_B) \qquad (90)$$

where a_B is the Bohr radius of the impurity. Owing to the dependence of transition probabilities on r, the DAP luminescence spectrum is deformed and shifted toward short wavelengths with increasing impurity concentration (Dobrego and Shlimak, 1967) and excitation intensity (Zacks and Halperin, 1972; Freitas *et al.*, 1994). It changes in the afterglow process as well. The dependence of the luminescence spectrum on the excitation intensity is due to the fact that pairs with a large value of r_n (small $\hbar\omega$) are saturated faster and their radiation ceases to increase with the growth of pumping. Therefore, the specific weight of radiation of the pairs, to which there correspond larger values of $\hbar\omega$, increases. In the decay process, first pairs with a small value of the lifetime (small values of r_n and large values of $\hbar\omega$) and then long-lived ones luminesce. Therefore, the luminescence spectrum is shifted toward longer wavelengths (Heim, 1969).

Excitation of samples by polarized light enables one to study optical anisotropy of electron–hole pairs (Ostapenko and Hoffmann, 1990). It has been shown theoretically that in doped superlattices the DAP luminescence band will have two maxima instead of one just as in bulk samples (Zene and Juarez, 1989). If the decay times of various well-resolved lines are determined experimentally, the expression given by Eq. (90) permits calculation of the value of τ_0 ($1/\tau_0 = W_0$ is the recombination constant) and the Bohr radii of impurities (Thomas *et al.*, 1965; Ohishi, 1986; Yi and Neumark, 1994).

4.6. Luminescence of Free and Bound Excitons

Although an exciton with a large radius as a quantum-mechanical system looks like a hydrogen atom and can be described in the first approximation by hydrogen-like wavefunction and by the energy spectrum (see Section 3.5), there are radical differences in the mechanism of optical transitions in the exciton and in the hydrogen atom. The hydrogen atom, absorbing and emitting light, always remains a system specified in advance and can exist in the ground state as long as desired. At the same time the exciton absorption process is connected with the generation of the exciton itself and exciton luminescence arises at its annihilation.

When there is thermal motion of hydrogen atoms, which do not interact with one another, the profile of their radiation line is broadened owing to the Doppler effect and is given by

$$I(\omega) = I_0 \exp[-(\omega - \omega_0)^2/\gamma^2] \tag{91}$$

where ω_0 is the radiation frequency of the atom at rest, $\gamma = \omega_0(2kT/M)^{1/2}/c$ is the Doppler linewidth, and M is the mass of the atom. According to Eq. (91), the radiation line profile of a moving atom is symmetric in shape and is described by the Gauss function. An absolutely different regularity is observed at exciton annihilation. If it does not interact with phonons, impurity centers, or defects of the crystal, all of the exciton's energy, both internal and kinetic, is fully transferred to the emitted photon (Gross *et al.*, 1971). Therefore, in contrast to the Doppler effect, exciton line-broadening is determined not by the projections of velocities on the observation axis, but by the kinetic energy of the exciton. The zero-phonon lineshape of exciton luminescence reproduces the Maxwell distribution of excitons over the energies

$$n_{ex}(E) = 2\pi^{-1/2} n_{ex}(kT)^{-3/2} E^{1/2} \exp(-E/kT) \tag{92}$$

where n_{ex} is the exciton concentration.

At exciton annihilation with the emission of one, e.g., longitudinal optical phonon with energy $\hbar\omega_{LO}$, a line appears in the luminescence spectrum that is shifted relative to the zero-phonon line by the value $\hbar\omega_{LO}$. It was shown theoretically (Anselm and Firsov, 1956) and experimentally (Gross *et al.*, 1971) that the profile of this line is described by the Maxwell function given in Eq. (92) multiplied by E. At the same time, the line profile of two-phonon annihilation of the exciton is better described by the Maxwell function. The study of exciton–phonon luminescence yields valuable information on the mechanism of thermalization of a system of excitons and its temperature (Permogorov, 1985).

Traveling in the crystal, the exciton not only interacts with the phonons, but is also scattered by the defects, impurity centers, and fluctuations of the chemical composition. This leads to the broadening of luminescence bands. In quantum-size structures the linewidth is additionally increased owing to the fluctuations in the width of the quantum well (Zimmermann and Ringe, 1994). In indirect-bandgap semiconductors the quantum restriction sharply increases the probability of optical transitions (Gaponenko *et al.*, 1994). It has also been found that the luminescence linewidth increases with decreasing mobility of the excitons (Bacher *et al.*, 1992).

If the exciton is formed with the participation of a charge localized on a donor or acceptor impurity, it is called a bound exciton. It cannot move in the crystal and the kinetic energy of its translational motion is close to zero. Therefore, the radiation linewidth of bound excitons is less than that of free excitons. The ground level of a bound exciton lies lower than the level ($n = 1$) of a free exciton, and the luminescence lines are shifted in the spectrum toward longer wavelengths relative to the radiation lines of free excitons. In the radiation of some exciton complexes, a fine structure analogous to the rotational structure of the hydrogen atom can appear (Halsted, 1967; Hopfield, 1964).

In real crystals, the dependence of allowed band energies on the wavevector $E(\mathbf{k})$ is complicated in form. The energy bands in **k**-space are characterized by several maxima and minima, which can be degenerate and split. If free and various kinds of bound excitons are formed in the crystal, the luminescence spectrum will contain a large number of lines. In particular, if the valence band consists of three subbands, there will be three hydrogen-like series in the radiation spectrum: series *A* is the longest wavelength one and belongs to the upper subband of the valence band, series *B* belongs to the middle subband, and series *C* belongs to the lowest subband. In the CdSe crystal 12 lines of series *A*, 12 lines of series *B*, and 2 lines of series *C* were observed (Sobolev, 1966).

The account of the possibility of the formation of excitons made it possible to obtain more exact expressions for edge absorption at both direct and indirect optical transitions (Elliott, 1957, 1963; McLean, 1960). From these equations follows, in particular, a hydrogen-like series of lines in the region of $E_g - E_{ex}^0 < \hbar\omega < E_g$ and a continuous absorption spectrum for $\hbar\omega - E_g > E_{ex}^0$. The line spectrum of absorption of excitons was discovered first in copper protoxide (Cu_2O) crystals (Gross and Karryev, 1952) and then in many other solids (Gross, 1962; Nikitine, 1969).

Theory also explains the Urbach rule—exponential decrease of the absorption coefficient in the region $\hbar\omega < E_g$, which was first discovered in alkali-halide crystals (Urbach, 1953). Theoretical studies of the kinetics of excitons are usually based on the solution of rate equations for both monocrystals (Gribkovskii, 1975) and quantum-size structures (Ping, 1993).

4.7. Streamer Luminescence

Discussion of even principal theoretical and experimental results on streamer luminescence is possible only within the scope of a separate chapter, so the review here has been limited to a presentation of the most general ideas on the subject, and references to the reviews or to a small number of original works have been noted for interested readers.

Streamer luminescence is excited in high-resistivity semiconductors by streamer discharges, which are a variety of incomplete electric breakdown, when current-conducting channels move to the depth of the crystal but there are no short-circuited electrodes. To bring about an incomplete electric breakdown of a solid body, it is necessary to subject it to short pulses of a high-voltage field or to charge it with an electron beam. The specific resistance of the samples should be high enough. Streamer luminescence was investigated in monocrystals of CdS (Nicoll, 1973), CdSe (Gladyshchuk *et al.*, 1986*a*), ZnSe (Baltramejunas *et al.*, 1978), GaAs (Gribkovskii *et al.*, 1977), ZnS (Basov *et al.*, 1974; Baltramejunas *et al.*, 1979), ZnTe (Gribkovskii *et al.*, 1980), CdTe (Gladyshchuk *et al.*, 1982), ZnO (Gladyshchuk *et al.*, 1988), in triple compounds CdSSe, $AgGaS_2$, $ZnIn_2S_4$, and in solid solutions of $CuGaS_2Se_2$.

In appearance (near-field pattern), the following kinds of discharges are distinguished: rectilinear crystallographically oriented streamer discharges (Fig. 8), cord-like bent discharges, corona discharges, diffuse discharges as in amorphous bodies, surface-branched and herringbone discharges (Gribkovskii, 1984, 1988, 1990). In hexagonal CdS, linear streamers propagate in 36 directions: in 18 directions at luminescence of solids. This will allow us to abandon too rough simplifications of the mechanism of luminescence, which are made in order to obtain a final result in analytical form. At excitation of luminescence by ultrashort laser pulses, one cannot assume a priori that quasi-equilibrium distribution of carriers is established even within each energy band singly. Therefore, the theory of luminescence in ultrafast processes should become an independent scientific direction in the near future. Finally, the development of technologies for obtaining quantum-size structures—quantum layers, wires, and dots as well as heterostructures, superlattices, and other previously unknown artificially created luminescent substances—opens up a rich field of activity for theorists.

ACKNOWLEDGMENT

The author would like to thank Dr. S. V. Gaponenko and Dr. G. P. Yablonskii for helpful discussions.

REFERENCES

Alentsev, M. N., Antonov-Romanovskii, V. V., Stepanov, B. I., and Fok, M. B. (1955) [in Russian], *Zh. Eksp. Teor. Fiz.* **28**, 253.

Anselm, A. I. (1962) [in Russian], *Introduction to the Theory of Semiconductors* (Fizmatgiz, Moscow-Leningrad).

Anselm, A. I., and Firsov, Yu. A. (1956) [in Russian], *Zh. Eksp. Theor. Fiz.* **30**, 719.

Antonov-Romanovskii, V. V. (1966) [in Russian], *Photoluminescence Kinetics of Crystal Phosphors* (Nauka, Moscow).

Antonov-Romanovskii, V. V., Stepanov, B. I., Fok, M. B., and Khapaljuk, A. P. (1955) [in Russian], *Dokl. Akad. Nauk SSSR.* **105**, 50.

Apanasevich, P. A. (1977) [in Russian], *Foundations of the Theory of Light with Matter Interaction* (Nauka i Technika, Minsk).

Auzel, F., Chen, Y., and Meichenin, D. (1994), *J. Lumines.* **60/61**, 692.

Bacher, G., Schweizer, H., Kovas, J., Nickel, H., Schlapp, W., and Losch, R. (1992), *Appl. Phys. Lett.* **61**, 702.

Baltramejunas, R. A., Gribkovskii, V. P., Ivanov, V. A., Kuokshtis, E. P., Parashchuk, V. V., and Yablonskii, G. P. (1978) [in Russian], *Fiz. Tech. Poluprov.* **12**, 497.

Baltramejunas, R. A., Gribkovskii, V. P., Ivanov, V. A., Kuokshtis, E. P., Parashchuk, V. V., and Yablonskii, G. P. (1979) [in Russian], *Zh. Prikl. Spektr.* **30**, 161.

Basov, N. G., Molchanov, A. G., Nasibov, A. S., Obidin, A. Z., Pechenov, A. N., and Popov, Yu. M. (1974), *J. Quant. Electron.* **QE-10**, 794.

Basov, N. G., Molchanov, A. G., Nasibov, A. S., Obidin, A. Z., Pechenov, A. N., and Popov, Yu. M. (1977), *J. Quant. Electron.* **QE-13**, 699.

Bassani, F., and Parravicini, G. P. (1975), *Electronic States and Optical Transitions in Solids* (Pergamon, Oxford).

Beattie, A. R., and Landsberg, P. T. (1960), *Proc. Roy. Soc.* **A258**, 486.

Bergh, A. A., and Dean, P. J. (1972), *Proc. IEEE* **60**, 156.

Bonch-Bruevich, V. L. (1971) [in Russian], *Vestnik MGU, Fiz, Astronomy* **12**, 586.

Borisevich, N. A. (1967), *Excited States of Complicated Molecules in Gas Phase [in Russian]*, (Nauka i Technika, Minsk).

Boukaert, L., Smolychowski, R., and Wigner, E. (1936), *Phys. Rev.* **50**, 58.

Burnstein, E. (1954), *Phys. Rev.* **93**, 632.

Calderon, L., Lu, Y., Shen, H. Pamulapati, J., Dutta, M., Chang, W. H., Yang, L. W., and Wright, P. D. (1992), *Appl. Phys. Lett.* **60**, 1597.
Chukova, Ju. P. (1980), *Anti-Stokes Luminescence and New Possibilities for Its Application [in Russian]*, (Sovetskoe Radio, Moscow).
Chernozatonskii, L. A. (1983) [in Russian], *Pisma ZhETF* **38**, 225.
Curie, D. (1960), *Crystalline Luminescence* [in French] (Dunod, Paris).
Dean, P. J. (1973), in: *Progress in Solid State Chemistry, Vol. 8* (J. Q. McCaldin and G. Somorgai Eds.) (Pergamon, Oxford), pp. 1–126.
Dobrego, V. P., and Shlimak, I. S. (1967) [in Russian], *Fiz. Tech. Poluprov.* **1**, 1478.
Dohler, G. H. (1971), *Phys. Stat. Sol. (b)* **45**, 705.
Dumke, W. P. (1963), *Phys. Rev.* **132**, 1998.
Eagles, D. M. (1960), *J. Phys. Chem. Solids* **16**, 76.
Efros, A. L. (1973) [in Russian], *Uspekhi Fiz. Nauk* **111**, 451.
Elliott, R. J. (1957), *Phys. Rev.* **108**, 1387.
Elliott, R. J. (1963), *Theory of Excitons, in: Polarons and Excitons* (C. G. Kuper and G. D. Whitfield, Eds.) (Oliver and Boyd, Edinburgh and London), pp. 269–293.
Emelyanenko, O. V., Lagunov, T. S., Nasledov, D. N., and Talalakin, G. K. (1965) [in Russian], *Fiz. Tverd. Tela* **7**, 1315.
Fan, H. Y. (1967), *Phonon-Electron Interaction: Crystals without Fields* (Springer-Verlag, Berlin).
Feofilov, P. P. (1959), *Polarized Luminescence of Atoms, Molecules and Crystals* [in Russian], (Moscow). Translation: (1962), *The Physical Basis of Polarized Emission* (Heywood, London).
Fok, M. B. (1964), *Introduction to Luminescence Kinetics of Crystal Phosphors* [in Russian] (Fizmatgiz, Moscow).
Freitas, J. A. Jr., Klein, P. B., and Collins, A. T. (1994), *Appl. Phys. Lett.* **64**, 2136.
Frenkel, J. I. (1931*a*), *Phys. Rev.* **37**, 17.
Frenkel, J. I. (1931*b*), *Phys. Rev.* **37**, 1276.
Galginaitis, S. V. (1971), *Met. Trans.* **2**, 757.
Gaponenko, S. V., Germanenko, I. N., Petrov, E. P., Stupak, A. P., Bondarenko, V. P., and Dorofeev, A. M. (1994), *Appl. Phys. Lett.* **64**, 85.
Gladyshchuk, A. A., Gribkovskii, V. P., and Yablonskii, G. P. (1982) [in Russian], *Zh. Prikl. Spektr.* **36**, 97. Translation: *J. Appl. Spectr.* **36**, 85.
Gladyshchuk, A. A., Gurskii, A. L., Parashchuk, V. V., and Yablonskii, G. P. (1986*a*) [in Russian], *Zh. Prikl. Spektr.* **44**, 978. Translation: *J. Appl. Spectr.* **44**, 639.
Gladyshchuk, A. A., Gribkovskii, V. P., Parashchuk, V. V., and Yablonskii, G. P. (1986*b*) [in Russian], USSR Inventor's Certificate N 1491271, IPC H01L21/66 (unpublished).
Gladyshchuk, A. A., Gurskii, A. L., Parashchuk, V. V., Nikitenko, V. A., and Yablonskii, G. P. (1988), *J. Lumines.* **42**, 49.
Gribkovskii, V. P. (1960*a*), Nonlinear Optical Phenomena and Implementation Limits of Harmonic Oscillator Classical Theory [in Russian], Thesis for a Canidates Degree. Institute Fiz. AN BSSR, Minsk.
Gribkovskii, V. P. (1960*b*) [in Russian], *Vestsi AN BSSR, Ser. Fiz.-Tech. Navuk* **1**, 43.
Gribkovskii, V. P. (1960*c*) [in Russian], *Dokl. AN BSSR* **4**, 284.
Gribkovskii, V. P. (1960*d*) [in Russian], *Dokl. AN BSSR* **4**, 199.
Gribkovskii, V. P. (1966) [in Russian], *Izv. Akad. Nauk BSSR, Ser. Fiz.* **4**, 95.
Gribkovskii, V.P. (1969) [in Russian], *Fiz. Tekh. Poluprov.* **3**, 944.
Gribkovskii, V. P. (1975), *Theory of Light Absorption and Emission in Semiconductors* [in Russian] (Nauka i Technika, Minsk).
Gribkovskii, V. P. (1984), (Review) [in Russian], *Zh. Prikl. Spektr.* **40**, 709. Translation: *J. Appl. Spectr.* **40**, 495.
Gribkovskii, V. P. (1988), *Semiconductor Lasers* [in Russian] (Universitetskoe, Minsk).
Gribkovskii, V. P. (1990), Abst. 3rd CECRI Research Conference on Luminescence, Karaikudi, India, p. 12.
Gribkovskii, V. P., and Kononenko, V. K. (1970) [in Russian], *Zh. Prikl. Spektr.* **12**, 45.
Gribkovskii, V. P., and Samoilyukovich, V. A. (1969) [in Russian], *Zh. Prikl. Spektr.* **11**, 170.
Gribkovskii, V. P., and Stepanov, B. I. (1968) [in Russian], *Dokl. Akad. Nauk SSSR* **183**, 67.
Gribkovskii, V. P., Parashchuk, V. V., Rjabtzev, G. I., and Yablonskii, G. P. (1977) [in Russian], *Zh. Prikl. Spektr.* **26**, 551. Translation: *J. Appl. Spectr.* **26**, 404.

Gribkovskii, V. P., Zubritskii, V. V., Ivanov, V. A., Parashchuk, V. V., and Yablonskii, G. P. (1980) [in Russian], *Fiz. Tech. Poluprov.* **14**, 2047.
Gribkovskii, V. P., Parashchuk, V. V., and Yablonskii, G. P. (1982) [in Russian], USSR Inventor's Certificate N 574011, IPC C01R31/26. *Byulleten Izobretenii* **19**, 296.
Gribkovskii, V. P., Gladyshchuk, A. A., Zubritskii, V. V., Parashchuk, V. V., and Yablonskii, G. P. (1983), *Phys. Stat. Sol.* (*a*) **77**, 765.
Gribkovskii, V. P., Zubritskii, V. V., and Yablonskii, G. P. (1985*a*) [in Russian], USSR Inventor's Certificate N 1045785, IPC H01L21/66. *Otkrytija, Izobretenija* **30**, 248.
Gribkovskii, V. P., Gladyshchuk, A. A., Gurskii, A. L., Parashchuk, V. V., Yablonskii, G. P., Pendjur, S. A. and Talenskii, O. N. (1985*b*) [in Russian], USSR Inventor's Certificate N 1268015, IPC H01L21/66 (unpublished).
Gribkovskii, V. P., Zubritskii, V. V., Beljaeva, A. K., and Satsunkevich, V. D. (1986) [in Russian], USSR Inventor's Certificate N 1241943, IPC H01L21/66. *Byulleten Izobretenii* **48**, 293.
Gribkovskii, V. P., Prokopenja, A. N., Rusakov, K. I., and Parashchuk, V. V. (1994) [in Russian], *Zh. Prikl. Spektr.* **60**, 362.
Gross, E. F. (1962) [in Russian], *Uspekhi Fiz. Nauk* **76**, 433.
Gross, E. F., and Karryev, N. A. (1952) [in Russian], *Dokl. Akad. Nauk SSSR* **84**, 471.
Gross, E. F., Permogorov, S. A., and Razbirin, B. S. (1971) [in Russian], *Uspekhi Fiz. Nauk* **103**, 431.
Halsted, R. E. (1967), in: *Physics and Chemistry of II–VI Compounds* (M. Aven, and J. S. Prener, Eds.) (North-Holland, Amsterdam), pp. 385–431.
Harvey, E. N. (1957), *A History of Luminescence from Earliest Times until 1900* (American Philosophical Society, Philadelphia).
Haynes, J. R. (1966), *Phys. Rev. Lett.* **17**, 860.
Heim, U. (1969), *Sol. St. Comm.* **7**, 445.
Heitler, W. (1954), *The Quantum Theory of Radiation* (Clarendon, Oxford).
Henisch, H. K. (1962), *Electroluminescence* (Pergamon, Oxford).
Hopfield, J. J. (1964), *Phys. Semicond. Paris*, p. 725.
Hopfield, J. J., Thomas, D. G., and Gershenzov, M. (1963), *Phys. Rev. Lett.* **10**, 162.
Hwang, C. J. (1970), *Phys. Rev. B: Sol. St.* **2**, 4117.
Jones, H. (1960), *The Theory of Brillouin Zones and Electronic States in Crystals* (North-Holland, Amsterdam).
Junovich, A. E. (1972), in: *Radiative Recombination in Semiconductors* [in Russian] (Nauka, Moscow), pp. 224–304.
Kane, E. O. (1963), *Phys. Rev.* **131**, 79.
Kononenko, V. K. (1970) [in Russian], *Vesti Akad. Nauk BSSR, Ser. Fiz-Mat. Navuk* **4**, 106.
Kononenko, V. K., and Gribkovskii, V. P. (1968*a*) [in Russian], *Zh. Prikl. Spektr.* **9**, 789.
Kononenko, V. K., and Gribkovskii, V. P. (1968*b*) [in Russian], *Vesti AN BSSR, Ser. Fiz-Matem. Navuk* **3**, 98.
Lampert, M. A. (1958), *Phys. Rev. Lett.* **1**, 450.
Landsberg, P. T., and Beattie, A. R. (1959), *J. Phys. Chem. Sol.* **8**, 73.
Lasher, G., and Stern, F. (1964), *Phys. Rev.* **133** A, 553.
Levanjuk, A. P., and Osipov, V. V. (1973*a*) [in Russian], *Fiz. Tech. Poluprov.* **7**, 1058.
Levanjuk, A. P., and Osipov, V. V. (1973*b*) [in Russian], *Fiz. Tech. Poluprov.* **7**, 1069.
Levanjuk, A. P., and Osipov, V. V. (1973*c*) [in Russian], *Fiz. Tech. Poluprov.* **7**, 1575.
Levshin, V. L. (1951), *Photoluminescence of Liquid and Solid Substances* [in Russian] (GITTL, Moscow, Leningrad).
Lucovsky, G. (1965*a*), *Sol. St. Comm.* **3**, 299.
Lucovsky, G. (1965*b*), *Sol. St. Comm.* **3**, 105.
Lucovsky, G. (1966), in: *Physics of Quantum Electronics* (McGraw-Hill, New York), pp. 467–477.
Malyutenko, V. K., Pipa, V. I., Yablonovskii, E. I., and Kolesnikov, I. V. (1990) [in Russian], *Fiz. Tech. Poluprov.* **24**, 866.
McLean, T. P., (1960), *Prog. Semicond.* **5**, 53.
Moss, T. S. (1954), *Proc. Phys. Soc.* **B67**, 775.
Nicoll, F. N. (1973), *Appl. Phys. Lett.* **23**, 465.
Nikitine, S. (1969), in: *Optical Properties of Solids* (S. Nudelman and S. S. Mitra, Eds.) (Plenum, New York), pp. 197–273.

Ohishi, M. (1986), *Jpn. J. Appl. Phys.* **25**, 1545.
Ostapenko, S. S., and Hoffmann, G. (1990). *Sol. St. Comm.* **74**, 447.
Pankove, J. I. (1971), *Optical Processes in Semiconductors* (Prentice Hall, Englewood Cliffs, N.J.).
Permogorov, S. A. (1985) [in Russian], in: *Excitons* (Nauka, Moscow), pp. 130–147.
Ping, Er-X. (1993), *Appl. Phys. Lett.* **62**, 493.
Radomsky, L., Yi G.-J., and Neumark, G. F. (1994), *Appl. Phys. Lett.* **64**, 1027.
Rebane, K. K., Saari, P. M., and Mauring, T. Kh. (1973) [in Russian], *Izv. Akad. Nauk SSSR, Ser. Fiz.* **37**, 848.
Saari, P., and Rebane, K. (1969), *Sol. St. Comm.* **7**, 887.
Schnitzer, I., Yablonovitch, E., Caneau, C., and Gmitter, I.J. (1993), *Appl. Phys. Lett.* **62**, 131.
Shockley, W., and Read, W. T. (1952), *Phys. Rev.* **87**, 835.
Sobolev, V. V. (1966) [in Russian], *Opt. i Spektr.* **20**, 673.
Stepanov, B. I. (1961), *Foundations of the Spectroscopy of Negative Light Flows* [in Russian] (BGU, Minsk).
Stepanov, B. I. (1991), *Introduction to Modern Optics. Light Absorption and Emission by Quantum Systems* [in Russian] (Navuka i Technika, Minsk).
Stepanov, B. I., and Gribkovskii, V. P. (1958), *Dokl. Akad. Nauk SSSR* **121**, 446.
Stepanov, B. I., and Gribkovskii, V. P. (1963), *Introduction to the Theory of Luminescence* [in Russian] (AN BSSR). Translation: *Theory of Luminescence* (1968) (ILIFFE Books, London).
Stepanov, B. I., and Gribkovskii, V. P. (1969) [in Russian], *Opt. i Spektr.* **27**, 583.
Stepanov, B. I., and Gribkovskii, V. P. (1970), *Izv. Akad. Nauk SSSR* **34**, 513.
Stern, F. (1963), *Sol. State Phys.* **15**, 299.
Stokes, G. (1852), *Phil. Trans.* **143**, 463.
Thomas, D. G., Gershenzon, M., and Trumbore, F. A. (1964), *Phys. Rev. A* **133**, 269.
Thomas, D. G., Hopfield, J. J., and Augustyniak, W. M. (1965), *Phys. Rev. A* **140**, 202.
Urbach, F. (1953), *Phys. Rev.* **92**, 1324.
Van Roosbroeck, W., and Shockley, W. (1954), *Phys. Rev.* **94**, 1558.
Varshni, Y. P. (1967), *I. Phys. Stat. Sol.* **19**, 459; *II. Phys. Stat. Sol.* **20**, 9.
Vavilov, S. I. (1924), *Z. Phys.* **22**, 266.
Vavilov, S. I. (1945), *J. Phys. USSR* **9**, 68.
Vavilov, S. I. (1951), in: P. Pringsheim, *Fluorescence and Phosphorescence* [in Russian] (IIL, Moscow).
Vavilov, S. I. (1952) [in Russian], in: *Collected Works. Vol. 2* (AN SSSR, Moscow), pp. 315–333.
Voitikov, S. V., and Gribkovskii, V. P. (1984), *Phys. Stat. Sol.* (*b*) **126**, 415.
Voitikov, S. V., and Gribkovskii, V. P. (1986), *Phys. Stat. Sol.* (*b*) **137**, 655.
Voronin, V. F., Gribkovskii, V. P., and Samoilyukovich, V. A. (1971) [in Russian], *Zh. Prikl. Spektr.* **14**, 531. Translation: *J. Appl. Spectr.* **14**, 399.
Wannier, G. N. (1937), *Phys. Rev.* **52**, 191.
Weinstein, M. A. (1960), *JOSA* **50**, 597.
Weisskopf, V., and Wigner, E. (1930), *Zeit. Phys.* **65**, 18.
Wiedemann, G. (1888), *Ann. Phys. Chem.* **34**, 446.
Williams, F. (1968), *Phys. Stat. Sol.* **25**, 493.
Yablonskii, G. P., and Lutsenko, E. V. (1994) [in Russian], Stepanov Institute of Physics. Print N 693, Minsk.
Yi, G.-J., and Neumark, G. F. (1994), *J. Lumines.* **60/61**, 29.
Zacks, E., and Halperin, A. (1972), *Phys. Rev. B: Sol. St.* **6**, 3072.
Zene, A., and Juarez, A. (1989), *Phys. Stat. Sol.* (*b*) **156**, K89.
Zimin, L. G., Gaponenko, S. V., Malinovskii, I. E., Lebed, V. Yu., Kuznetsov, P. I., Yakusheva, G. G., and Kuznetsov, A. V. (1990), *Phys. Stat. Sol.* (*b*) **159**, 449.
Zimmermann, R., and Runge, E. (1994), *J. Lumines.* **60/61**, 320.
Zubov, B. V., Kalinushkin, V. A., Murina, T. M., Prokhorov, A. M., and Rogachov, A. A. (1972) [in Russian], in: *Nonlinear Processes in Optics* (AN SSSR, Novosibirsk), pp. 277–278.
Zubritskii, V. V. (1993), *Phys. Stat. Sol.* (*b*) **180**, 349.
Zubritskii, V. V., and Purlys, R. P. (1990), *Phys. Stat. Sol.* (*b*) **158**, 91.

2

Experimental Techniques

A. Meijerink

1. INTRODUCTION

Experiments in optical spectroscopy require three basic elements: a light source, a light-dispersing element, and a light detector. The character of these elements has changed over the years, but many of the experimental principles have remained the same. After a historical introduction to optical spectroscopy, this chapter will treat conventional and modern spectroscopic techniques and the different types of light sources, light-dispersing elements, and detectors that are used nowadays in luminescence spectroscopy of solids.

The great advances in the understanding of the nature and properties of light in the 17th and 18th centuries resulted in rapid development of spectroscopic instruments in the 19th century (Baly, 1912). The connection between the dark lines in the solar spectrum and the presence of certain elements in the atmosphere of the sun was made by Kirchhoff and Bunsen in 1859. Some of the dark lines were reproduced when light from an electric arc was passed through a flame tinted yellow by Na salt. Kirchhoff realized that absorption of light of certain wavelengths by the Na atoms in the flame was responsible for the dark lines. Each element was shown to have characteristic emission and absorption lines. Soon this technique was a powerful tool for the qualitative analysis of elements.

The success of optical spectroscopy would not have been possible without the rapid technological development. To replace the sun as a light source, different types of arc lamps were developed and the lamps devised in the last decades of the 19th century provided the basic models for the lamplight sources used in luminescence spectroscopy today.

The theory of diffraction inspired Fraunhofer to invent a new light-dispersing element: the grating. With the new gratings a more accurate determination of the spectral positions of the dark lines in the solar spectrum was possible. Even greater accuracy in the determination of absolute wavelengths became possible with the

A. Meijerink • Department of Condensed Matter, Universität Utrecht, 3508 TA, Utrecht, The Netherlands.

Luminescence of Solids, edited by D. R. Vij. Plenum Press, New York, 1998.

interferometer designed by Michelson in 1894 and the (improved) version developed a few years later by Fabry and Perot. The estimated wavelength accuracy was estimated as $\Delta\lambda/\lambda \approx 10^{-7}$.

The main focus of spectroscopy research in the 19th century was an accurate determination of lines in the solar spectrum and chemical analyses. The progress made with spectroscopic techniques also promoted investigations on luminescent materials. Many materials that glowed with extraordinary brilliance under cathode ray or ultraviolet excitation were investigated.

In the first half of the 20th century the exciting new field of quantum theory developed and fruitfully interacted with the field of optical spectroscopy (Herzberg, 1944). The physical/mathematical foundation for quantum theory eventually developed by scientists such as de Broglie, Schrödinger, Heisenberg, and Sommerfeld successfully explained the positions of the lines in spectra of elements, the observed fine structure, and the Stark and Zeeman effects.

However, this period did not yield many new developments in experimental techniques. Spectra were recorded with the same types of lamps, dispersing elements (interferometers, gratings, prisms), and detector (photographic film), but some progress was made by obtaining a higher resolution, greater sensitivity, and a broader wavelength range. One noteworthy invention was the photomultiplier tube (PMT) at the end of the 1930s. By combining a photoemissive cell with a number of electron multiplier steps a sensitive detector for the ultraviolet, visible, and near-infrared ranges was obtained.

After World War II the field of luminescence spectroscopy expanded rapidly, owing partly to the application of luminescent materials in such devices as fluorescent tubes, color television screens, night viewers, x-ray photography, and scintillators. These applications promoted research on new efficient luminescent materials. A major breakthrough in optical spectroscopy was Maiman's (1960) invention of the laser. This intense monochromatic and coherent light source offered many new possibilities and has changed the field of luminescence spectroscopy dramatically. The remainder of this chapter will give an overview of the various techniques that are used at present to extract information from nature by luminescence spectroscopy

2. ABSORPTION AND LUMINESCENCE SPECTROSCOPY

2.1. Absorption

An absorption spectrum provides information on the wavelengths of electromagnetic radiation that can be absorbed by the samples under study. This is determined by varying the wavelength and recording the intensity of the transmitted beam. When light (photons) is absorbed, the energy of the absorbed photons is used to excite a transition between electronic levels, usually starting from the electronic ground state. The positions of higher energy levels (excited states) can be derived from the absorption spectrum. A typical setup for recording absorption spectra is

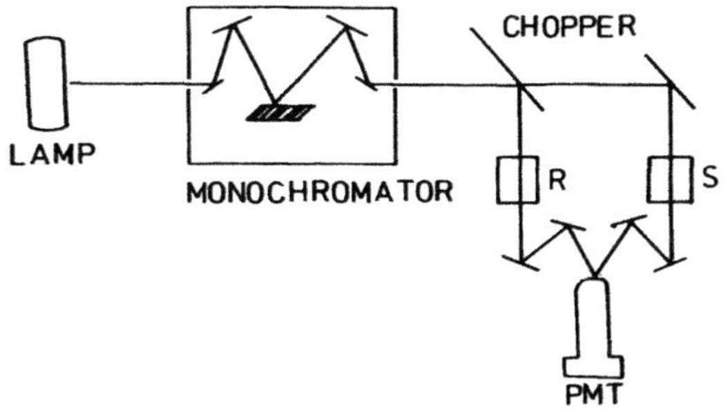

Figure 1. Dual-beam setup for the measurement of absorption spectra.

shown in Fig. 1. Light from a source is fed into a monochromator and the intensity of the transmitted beam is measured. In the commonly used dual-beam setups, the light is passed alternately through a reference and the sample by using a chopper. The transmission T is defined as the intensity of the sample beam divided by the intensity of the reference beam.

The light source often consists of a deuterium lamp for the ultraviolet (180–350 nm) and a tungsten lamp for the near ultraviolet, visible, and infrared spectral regions (350–3000 nm). Xenon lamps are also used. For wavelength selection a grating is used in combination with the appropriate filters to eliminate higher-order contributions from the grating. The light is detected by a PMT (180–800 nm) or a PbS detector (800–3000 nm). The dual-beam setup offers the advantage of easy correction for the instrumental response and a higher sensitivity. The spectral resolution of commercial spectrophotometers is typically not much better than 1 nm. For high-resolution absorption spectra, a collimated beam from an intense stable tungsten lamp is passed through a sample and the transmitted beam is focused on the slit of a long focal length (e.g., 1 m) monochromator and detected by a PMT. An absorption spectrum is recorded by scanning the monochromator. If low-temperature absorption spectra are recorded for a sample in a cryostat, there can be periodic fluctuations in the light intensity owing to etaloning at the windows. If this occurs, a dual-beam setup has to be used, and the resolution of this setup for a 1-m monochromator can be as high as 0.01 nm.

Extremely high-resolution absorption spectra have been reported for a Fourier transform infrared spectrometer. The resolution of commercial setups can be as high as $0.0004\ \mathrm{cm}^{-1}$. A beautiful illustration of high resolution is presented in Fig. 2. The absorption spectra of ${}^5I_8 \rightarrow {}^5I_7$ transitions on Ho^{3+} in $LiYF_4$ are shown for three different ratios of the two isotopes of Li (^{6}Li and ^{7}Li) (Agladze *et al.*, 1993). The very small isotope shift of the eight hyperfine components of the ${}^5I_8(\Gamma_2^1) \rightarrow {}^5I_7(\Gamma_{34}^1)$ transition on Ho^{3+} is clearly resolved. This was the first observation of an isotope shift owing to isotopes in the surroundings of the optically active impurity ion ever reported.

Information on the oscillator strengths or radiative transition probabilities can be derived from absorption spectra if the concentration of the absorbing species is

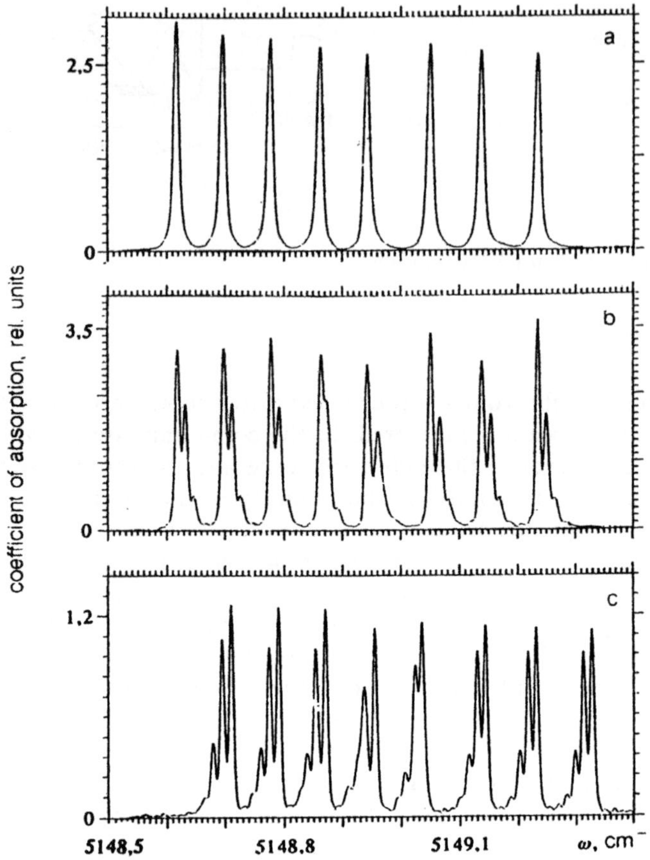

Figure 2. High-resolution absorption spectrum of $LiYF_4$–Ho^{3+} at 4.2 K. The spectrum shows the hyperfine components of the $\Gamma_2(^5I_8) \rightarrow \Gamma_{3,4}(^5I_7)$ transition on Ho^{3+} in: (a) $^7LiYF_4$, (b) $^6Li_{0.0742}{}^7Li_{0.9258}YF_4$, (c) $^6Li_{0.9}{}^7Li_{0.1}YF_4$ (Agladze *et al.*, 1993; with permission from the American Institute of Physics).

known. The attenuation of a light beam traveling through a light-absorbing medium is described by the Lambert–Beer law:

$$I(\nu) = I_0(\nu) \times e^{-[\alpha(\nu)l]} \tag{1}$$

where $I_0(\nu)$ is the intensity of a transmitted beam in the absence of the absorbing species, $\alpha(\nu)$ is the absorption coefficient of the medium for light of frequency ν, and l is the length of the light path in the sample. If the absorbing species are ions or molecules in solution the Lambert-Beer law is often expressed as

$$I(\nu) = I_0(\nu) \times 10^{-[\varepsilon(\nu)cl]} \tag{2}$$

where $\varepsilon(\nu)$ is the molar extinction coefficient and c is the concentration of absorbing species in moles/liter.

The integrated absorption coefficient is related to the cross section for a transition by

$$\int \alpha(\nu)d\nu = N\sigma \tag{3}$$

where N is the number of absorbing species and σ is the absorption cross section. The absorption cross section is related to the Einstein coefficient for spontaneous emission A_{ba} by

$$\sigma(\nu) = A_{ba}\frac{g_b c^2 g(\nu - \nu_0)}{g_a 8\pi n^2 \nu^2} \tag{4}$$

where g_b and g_a are the degeneracies of levels a and b and the function $g(\nu - \nu_0)$ gives the lineshape. For an inhomogeneously broadened line one has a Gaussian lineshape, and $g(\nu - \nu_0)$ equals

$$g(\nu - \nu_0) = \frac{2\sqrt{\ln 2}}{\sqrt{\pi}\,\Delta\nu}\exp\frac{-4\ln 2(\nu - \nu_0)^2}{(\Delta\nu)^2} \tag{5}$$

At the absorption band maximum ($\nu = \nu_0$), $g(\nu)$ is $0.939/\Delta\nu$, where $\Delta\nu$ is the full width at half-maximum (FWHM). With Eqs. (3) to (5) it is possible to calculate the absorption cross section and radiative decay rates from absorption spectra. The oscillator strength f_{ab} for a transition is related to the integrated absorption coefficient via

$$\int \alpha(\nu)d\nu = N\frac{e^2}{4\pi\varepsilon_0 mc}\left[\left(\frac{E_{\text{loc}}}{E}\right)\frac{1}{n}\right] f_{ab}(\text{ED}) \tag{6}$$

for electric dipole (ED) transitions, and via

$$\int \alpha(\nu)d\nu = N\frac{e^2 n}{4\pi\varepsilon_0 mc} f_{ab}(\text{MD}) \tag{7}$$

for magnetic dipole (MD) transitions. Often, local field corrections and the refractive index are ignored and the oscillator strength is estimated from the absorption spectrum using Smakula's relation:

$$Nf_{ab} = 0.87 \times 10^{17}\alpha(\nu_0)\Delta\nu \tag{8}$$

where $\alpha(\nu_0)$ is the maximum absorption coefficient in cm^{-1} and $\Delta\nu$ is the FWHM in electron volts. For a Lorentzian lineshape the constant 0.87×10^{17} is replaced by 1.29×10^{17}.

Equations (3) to (8) are very useful for relating absorption spectra to transition probabilities and microscopic parameters such as the cross section for absorption. To convey an idea of typical values of f_{ab}, A_{ba}, and ε for different types of transitions, Table 1 gives numbers for common optical impurities in the solid state (F-centers, transition metal ions, and rare-earth ions). The numbers are given as a rough estimate as they depend strongly on the host lattice.

Table 1. Typical Values for Oscillator Strengths f_{ab}, Transition Probabilities A_{bn}, and Maximum Absorption Coefficients ε_{max} for Various Transitions on Common Optical Impurities

Center	Type	Transition	f_{ab}	A_{ba} (s^{-1})	ε_{max} ($M^{-1}cm^{-1}$)
F-center	ED allowed	$s \rightarrow p$	1	10^{-7}	10^4
Cr^{3+}	forced ED, spin all.	${}^4A_2 \rightarrow {}^4T_2$	10^{-2}	10^{-5}	10^2
Mn^{2+}	forced ED, spin forb.	${}^6A_1 \rightarrow {}^4T_2$	10^{-8}	10^{-1}	10^{-2}
Eu^{3+}	forced ED	${}^7F_0 \rightarrow {}^5D_2$	10^{-6}	10^{-3}	1
Eu^{3+}	MD	${}^7F_0 \rightarrow {}^5D_1$	10^{-8}	10^{-1}	10^{-2}

Absorption spectra cannot be recorded for polycrystalline samples. The typical particle size of powdered luminescent materials is 1–10 μm. The light is strongly scattered and the intensity of the transmitted beam will be close to zero, even if the light is not absorbed. Diffuse reflection spectra can be measured for these samples. The setup is very similar to the dual-beam spectrometer, except that the reflected light is collected by an integrating sphere and the intensity is measured by a PMT at the focal point of the sphere. Polycrystalline MgO or $BaSO_4$ powder is used as a reference. Both MgO and $BaSO_4$ are perfectly white in the wavelength range 200–3000 nm and the light will be scattered and not absorbed. If the sample absorbs part of the light, the intensity of the reflected scattered light will be reduced and will be lower than the intensity reflected by the reference. Reflection spectra can be used for determination of the position of absorption bands. The technique cannot be used for a quantitative analysis of oscillator strengths or absorption coefficients because the optical path l through the sample is not known.

The intensity of the reflected light depends on the scattering coefficient S and the absorption coefficient K. A useful relationship between the reflectivity $R(=I_{sam}/I_{ref})$ and S and K has been derived by Kubelka and Munk (1969):

$$F(R) = \frac{(1-R)^2}{2R} = \frac{K}{S} = \frac{\varepsilon \cdot c}{S} \tag{9}$$

where R is the reflection by an infinitely thick layer.

With the Kubelka–Munk function the concentrations of impurities can be estimated if R is measured for standard samples with known concentrations of impurities. Also the ratio of molar extinction coefficients for different optical impurities (with known concentrations) in samples can be estimated.

2.2. Luminescence

Two types of luminescence spectra can be distinguished: excitation and emission. In the case of an excitation spectrum the wavelength of the exciting light is varied and the intensity of the emitted light at a fixed (well-chosen) emission

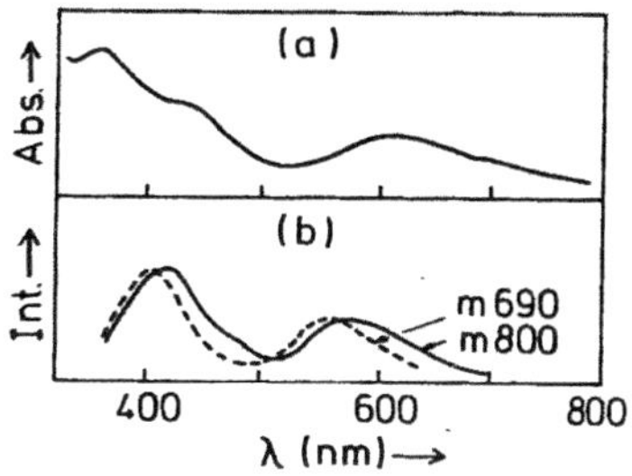

Figure 3. Absorption spectrum (a) and excitation spectra (b) for emission of Cr^{3+} at 690 and 800 nm in borate glass at 300 K (van Die, 1978).

wavelength is measured as a function of the excitation wavelength. The excitation spectrum gives information on the positions of excited states, just as the absorption spectrum does, except that the former reveals only absorption bands that result in the emission of light. The observed differences between the absorption and excitation spectra can yield useful information.

The excitation and absorption spectra of Cr^{3+} in borate glass are shown in Fig. 3 (van Die, 1987). Comparison of the two spectra shows that absorption of wavelengths longer than 700 nm does not result in Cr^{3+} emission as the excitation spectrum stops at 700 nm, while the absorption spectrum shows that there are still Cr^{3+} ions that absorb beyond 700 nm. This is explained by nonradiative relaxation to the ground state for the Cr^{3+} ions that absorb at wavelengths longer than 700 nm. The nonluminescent Cr^{3+} ions in this glass are called "dark sites."

The intensity of the emission is given by

$$I_{em}(\nu) = I_0(\nu) \times q \times (1 - 10^{-\varepsilon(\nu)cl}) \tag{10}$$

where q is the quantum efficiency for the luminescence process (see Section 6) and the other symbols have been defined before. If the product of $\varepsilon(\nu)cl$ is small the emission intensity varies linearly with the concentration of luminescent centers. For higher values of $\varepsilon(\nu)cl$ [i.e., when a substantial amount of the incident light $I_0(\nu)$ is absorbed] this dependence is no longer linear and care should be taken in the interpretation of the excitation spectra (de Mello Donegá *et al.*, 1994). When $\varepsilon(\nu)c$ is very large (i.e., a high concentration of strongly absorbing centers) most of the light will be absorbed in a thin surface layer. The concentration of defects centers can be substantially higher in a surface layer leading to nonradiative decay channels and stronger luminescence quenching in the surface layer than in the bulk. As a result it is often observed that the emission intensity decreases when the absorption strength becomes very great. After an initial steep rise at the position of the absorption edge, the excitation spectrum drops when the absorption strength increases further.

The role of surface defects is illustrated in Fig. 4, where the excitation spectra for three different $PbMoO_4$ samples are shown (Groenink and Blasse, 1980). The cleaved crystal has a very low concentration of surface defects and the excitation spectrum resembles the absorption spectrum. In the excitation spectra of the other two samples, the relative quantum output decreases for high absorption strengths, owing to surface layer defects.

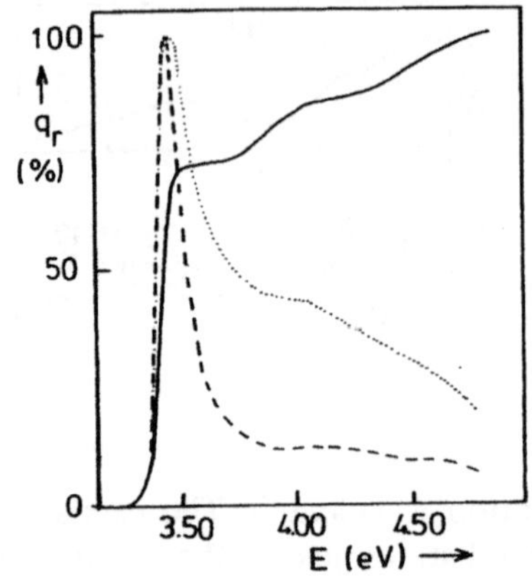

Figure 4. Excitation spectra of the luminescence of $PbMoO_4$ samples at 77 K; q_r gives the relative quantum output: solid line—cleaved crystals; dotted line—powdered crystal; broken line—polished crystal (Groenink and Blasse, 1980; with permission from Academic Press).

An emission spectrum provides information on the spectral distribution of the light emitted by a sample. To record an emission spectrum the sample is excited (usually by visible light or ultraviolet radiation, but other high-energy excitation sources can also be used), the emission is fed into a monochromator, and the emission intensity is recorded as a function of wavelength. A typical setup that is able to measure both excitation and emission spectra (a spectrofluorometer) is shown in Fig. 5. The setup consists of an excitation source, an excitation monochromator, a sample compartment, an emission monochromator, and a detector. The different possibilities and characteristics of all these components will be discussed in the next section.

The choice of components depends on the type of experiment. For low-resolution luminescence measurements in the ultraviolet, visible, and near-infrared spectral regions, a typical setup uses a xenon lamp as excitation source, 0.2-m monochromators, and a PMT for detection of the light. In the sample compartment a cryostat can be built to allow cooling to 77 K (liquid N_2) or 4.2 K (liquid He).

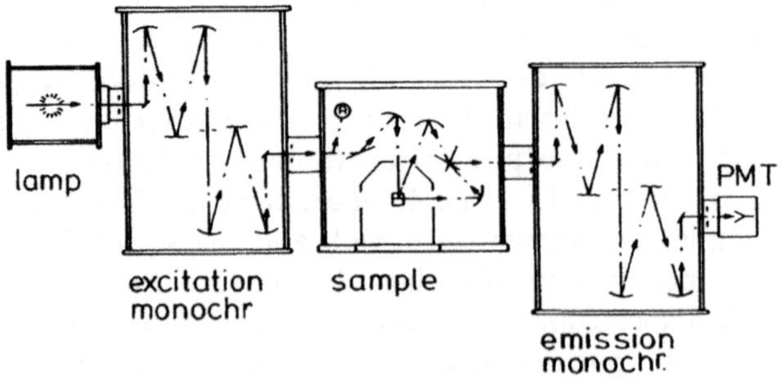

Figure 5. Schematic setup of a spectrofluorometer for the measurement of excitation and emission spectra (courtesy of Spex Industries).

3. SPECTROSCOPIC COMPONENTS

3.1. Excitation Sources

3.1.1. Lamps

There are two types of lamps: incandescent and arc. Incandescent lamps act as blackbody radiators. The highest temperatures are reached by using a tungsten filament in combination with halogen in the bulb to regenerate the slowly evaporating tungsten wire. Incandescent lamps are used as excitation sources in the visible and infrared. The highest efficiencies that can be obtained are 15–30 lumens/W. Incandescent lamps fall short in the ultraviolet range and arc lamps are used to bridge this gap. Arc lamps create radiation by discharging energy through ionized gases. In low-pressure discharge lamps the characteristic lines of the atoms in the discharge are emitted (e.g., for a low-pressure mercury discharge at 185, 254, and 365 nm).

At high pressures and temperatures a continuous background comes up owing to broadening of the spectral lines. The most widely used arc lamps are those of high-pressure xenon, mercury, and deuterium (Kruithof, 1967). The spectral distributions of the light output for typical lamps are given in Fig. 6. Xenon and mercury lamps give a continuous background between 220 and 2000 nm. On top of the background, the line spectrum is still visible, which is undesirable for a continuous excitation source. For mercury lamps the spectral lines are more pronounced and this is the reason that xenon lamps are more widely used in spectrofluorometers as a continuous excitation source. Deuterium lamps are not as bright as those of xenon and mercury, but they produce a good continuous spectrum at short wavelengths and are used for measurements below 240 nm.

3.1.2. Lasers

Lasers have all the characteristics of a perfect excitation source: high power, small divergence, coherence, monochromatism and the capability of very short pulses; they are also (sometimes) inexpensive. The success of lasers is clear from the

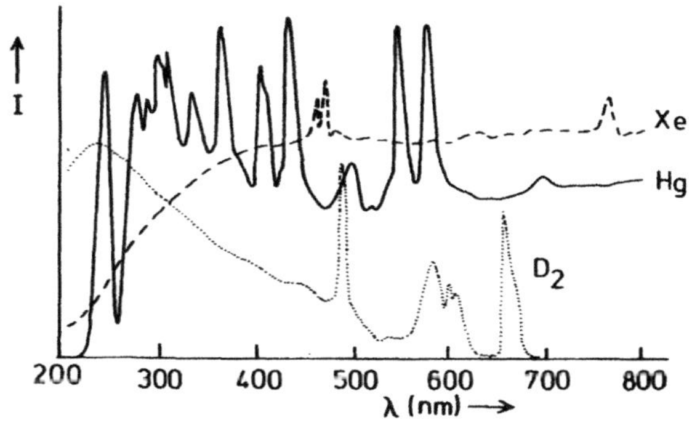

Figure 6. Typical spectral distribution of the output of three widely used arc lamps (high-pressure xenon, mercury, and deuterium).

enormous range of applications in both research and for use in consumer products such as compact disc players (Siegman, 1986; Wright, 1990).

(a) Gas Lasers. Gas lasers use a gas as the lasing medium (Wright and Wirth, 1980). The most widely used gas lasers are helium–neon, argon, krypton, nitrogen, and excimer lasers, while for specific applications (high repetition rate) metal vapor lasers are still used.

Helium–neon, argon, and krypton lasers are usually continuous-wave lasers. A discharge is passed through the laser tube filled with a noble gas under low-pressure (e.g., 1 torr) to obtain laser action. Of all the available types, helium–neon lasers come closest to being the ideal monochromatic source. They have excellent intensity and frequency stability and very good temporal and spatial coherence. Low-power (1 mW) helium–neon lasers are frequently used for optical alignment and as a secondary frequency standard. Their common output wavelength is 632.8 nm, but they can also be operated at 543, 1150, or 3390 nm.

Argon and krypton lasers are powerful continuous-wave lasers in the visible and near-ultraviolet. Transitions on the singly or doubly ionized ions are responsible for laser action. The discharge is concentrated in the center of the laser tube by an external magnetic field. An initial spike of several thousands of volts breaks down the gas and then the voltage drops to 100–400 V while the current jumps to 10–70 A. Owing to the high current density, extensive cooling is required. The output power can be very high—up to 100 W for commercial lasers. Laser action takes place far above the ground level and is followed by hard ultraviolet (74 nm) radiation. The strongest laser line of Ar^+ in argon lasers is at 514.5 nm, but other wavelengths can also be made to lase, e.g., 528.7, 488.0, and 457.9 nm (Ar^+) and 351 and 300 nm (Ar^{2+}). The power generated for krypton is lower (only 10–30% of argon) but adds to the wavelength range. The strongest Kr^+ line is at 647.1 nm. Argon and krypton lasers are both used directly as intense excitation sources and as pumps for dye and solid state lasers.

In excimer lasers a high-pressure mixture of gases such as xenon, hydrogen chloride, and neon (buffer) or krypton, fluorine and helium (buffer) is the lasing medium. A pulsed discharge over tens of kilovolts in the laser tube produces excited complexes such as XeCl* or KrF*. The return to the ground state gives rise to an ultraviolet laser pulse of 5–20 ns duration. In the ground state the complex immediately dissociates, since noble gas compounds are not stable. In this way population inversion is maintained. Excimers are very high-gain lasers and in a single passage most complexes return to the ground state. The output window is therefore 100% transparent. The output beam is a relatively big rectangular beam (e.g., 1 × 3 cm) and the wavelength stability is not very high. Mostly excimer lasers are used to pump dye lasers or as powerful ultraviolet excitation sources. The wavelength depends on the particular combination of noble gas and halogen.

Molecular nitrogen lasers produce pulsed ultraviolet radiation at 337 nm. The characteristics are similar to those of excimer lasers. As the output power is lower than for excimer lasers, nitrogen lasers are less popular than they used to be. Still, in many laboratories one will find nitrogen-laser-pumped dye laser setups.

Of the metal vapor lasers, the copper vapor laser is the most widely used. In a high-temperature gas mixture of copper and neon or helium, laser pulses can be generated at a high repetition rate (up to 20 kHz) and of several millijoules per pulse at 510.6 or 578.2 nm. Copper vapor lasers are often used to pump dye lasers in experiments that require high repetition rates.

(b) Dye Lasers. Dye lasers rely on optical excitation of dye molecules dissolved in a liquid solvent (Schäfer, 1977). The dye molecules are organic with conjugated π-bonds. The emission wavelength varies depending on the structure. Dyes are available for emission from 300 to 1000 nm, and any one typically covers a wavelength range of 10–30 nm. Wavelength selection is achieved by a diffraction grating at the end of the cavity, and the wavelength can be tuned by changing the angle of the grating.

Efficiencies for dye lasers are typically 5–30%. For pulsed dye lasers the most commonly used pump sources are nitrogen, excimer, and frequency-doubled Nd–YAG lasers. Continuous-wave dye lasers are often pumped with Ar^+ lasers. A special design of a continuous-wave dye laser is the ring dye laser, in which the light travels unidirectionally around in the cavity. The primary advantage of the ring laser is that traveling-wave operation as opposed to standing-wave operation prevents spatial hole-burning. This provides a more homogeneous laser medium and prevents mode competition between adjacent axial modes. Very narrow linewidths, down to 3×10^{-5} cm^{-1} (1 MHz), have been reported for ring dye lasers.

In spite of the great versatility of dye lasers (capability of short pulses, narrow linewidths, and large tunable wavelength region) there has recently been a tendency to replace them with solid state lasers having the same or better characteristics, as the latter do not require the inconvenient frequent replacement of (carcinogenic) dye solutions.

(c) Solid State Lasers. The first laser was a solid state device, the well-known ruby laser, built by Maiman (1960). The lasing medium is a ruby crystal (Al_2O_3–Cr^{3+}). The ruby laser is a classic example of a three-level laser. Excitation takes place by flashlamp pumping in the $^4T_2$ and $^4T_1$ levels of the Cr^{3+} ions, followed by a fast relaxation to the ^2E metastable level. The lasing is due to the transition between the metastable ^2E state and the ground state $^4A_2$. The lasing transition gives a narrow line at 694.3 nm. Ruby lasers are operated in the pulsed configuration at low repetition rates (several hertz). High-output powers can be obtained (up to 400 J/pulse) for long pulse durations (2.5 ms).

Ruby lasers are not much used today. A more popular type of solid state laser is the Nd–YAG laser (Henderson and Imbusch, 1989*a*), in which the optically active ion is Nd^{3+}, incorporated in yttrium aluminum garnet (YAG) crystals. The Nd–YAG laser is a four-level laser. Pumping by, e.g., xenon flashlamps, excites the Nd^{3+} ions into a large number of different levels of the $4f^3$ configuration of Nd^{3+}. The lasing transition on Nd^{3+} is the $^4F_{3/2} \rightarrow {}^4I_{11/2}$ transition, which gives a narrow laser line at 1064 nm. Nd–YAG lasers can be operated in a continuous wave or a pulsed configuration. A typical repetition rate is 10 Hz but lasers with higher repetition rates (up to 500 Hz) are available.

The continuous-wave power can be as high as 1000 W. Short, high peak power pulses are produced by Q-switching. The pulse duration is typically 5–10 ns. Mode locked Nd–YAG lasers are capable of giving picosecond pulses. The 1064-nm output of a YAG laser can be frequency-doubled to 532 nm or tripled to 355 nm. The 532 or 355 nm can be used directly for excitation in the visible or ultraviolet or it can be used to pump a dye laser if a tunable excitation source in the visible range is required. The high peak powers of YAG lasers allow efficient frequency-doubling by, e.g., KDP frequency-doubling. Typical efficiencies for frequency-doubling are 30%. Besides Nd–YAG lasers, Nd-glass lasers are also popular. The lasing transition is the same as for Nd–YAG ($^4F_{3/2} \rightarrow {}^4I_{11/2}$). The Nd^{3+} ions are incorporated in a glass matrix. Nd^{3+}–glass lasers are most widely used as amplifiers, as they can generate very high power densities. A third type of solid state laser is the color center laser (Henderson and Imbusch, 1989*a*), which is a tunable continuous-wave laser that operates in the infrared. The optically active species is a lattice defect.

A small revolution has taken place in the last 5 years in the development and successful marketing of two new types of solid state lasers: the Ti-sapphire laser and optical parametric oscillator (OPOs) (Padgett and Dunn, 1994). The former is a vibronic laser producing femtosecond or picosecond pulses over a wide tuning range (650–1100 nm). OPOs are a convenient alternative to dye lasers. They are tunable from 220 to 3000 nm without the need of replacing the dyes. The linewidth and efficiency are comparable or better than for dye lasers.

The Ti-sapphire laser belongs to the class of vibronic devices in which broad (vibronic) emission bands of transition metal ions are used to create a tunable laser. The heart of a Ti-sapphire laser is formed by an Al_2O_3–Ti^{3+} crystal. Ti^{3+} has the $3d^1$ configuration and the transition involved in the lasing is the spin-allowed parity-forbidden $^2T_{2g} \leftarrow {}^2E_g$ transition between the two crystal field components for the $3d$ electron. Other vibronic lasers in this class are the alexandrite laser and the fosterite laser, with Cr^{3+} and Cr^{4+}, respectively, as optically active centers. Different types of solid state lasers and their characteristics are shown in Fig. 7. Of the tunable (vibronic) solid state lasers, the Ti-sapphire laser is the most widely used at present. Ti-sapphire lasers can be used as well in continuous-wave as in pulsed modes. The broad emission bands make it possible to generate very short femtosecond pulses by mode-locking (Kafka *et al.*, 1989, 1992).

A second type of tunable laser that might possibly replace the dye laser, is the OPO. Strictly speaking, an OPO is not a laser since the generation of its coherent and monochromatic light is not based on stimulated emission radiation. The important element in an OPO is a nonlinear crystal that splits the (laser) pump beam into two beams of lower energy (Baumgartner and Byer, 1979). The first OPOs and the theoretical basis for the device were developed in the 1960s, but it is only recently that OPOs have matured into an important type of commercial laser. High-quality laser sources and new (improved) nonlinear crystals have improved the performance of OPOs.

The theory behind the OPOs is based on second-order nonlinear interaction of light and matter. The efficiency of an OPO is high (30% or more) and flat over a broad wavelength range. The spectral linewidth is determined by the quality of the

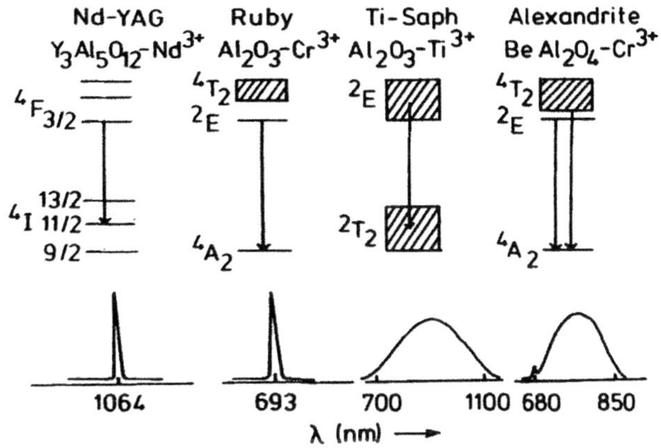

Figure 7. Energy-level schemes and gain profiles for four popular solid state lasers.

pump laser. Commercial OPOs, pumped by a single-frequency seeded TEM_{00} mode Nd–YAG laser beam, have linewidths as narrow as $0.02\,cm^{-1}$. Pumping with Ti-sapphire lasers can be used to generate ultrashort (femtosecond) pulses in the infrared (1100–2600 nm) from OPOs. At present, commercially available OPOs use pulsed pump lasers since the nonlinear process requires high peak powers for high efficiencies. Recently continuous-wave operation has also been achieved for OPOs pumped by an Ar^+ laser.

A third class of solid state lasers with which impressive progress has been made in the last 5 years is the semiconductor diode laser (Zory, 1993). The concept for laser action is simple and was demonstrated as early as 1962. A semiconductor diode laser is shown in Fig. 8. Laser action takes place in a thin layer of undoped $Ga_{1-x}Al_xAs$ that is sandwiched between layers of wider-bandgap undoped $Ga_{1-y}Al_yAs$ ($y > x$) (the confinement layers). The confinement layers border on

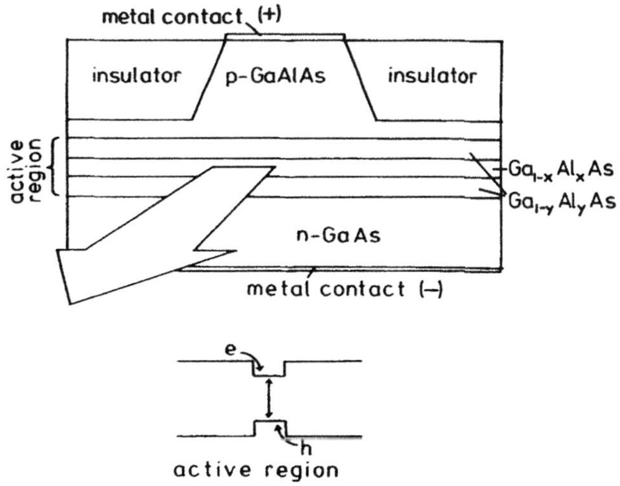

Figure 8. Schematic representation of a GaAlAs semiconductor laser.

p-type GaAs on one side and *n*-type GaAs on the other side (the cladding layers). A voltage applied in the forward bias direction (negative to *n*-type and positive to *p*-type) causes charge carriers to flow toward the active region, where they are trapped in the very thin layer between the confinement layers. There the electrons and holes recombine. In the case of a direct bandgap semiconductor, the recombination results in the emission of photons with energies that are about equal to the bandgap.

This process occurs in light-emitting diodes as well as in semiconductor lasers. To obtain laser action, reflective facets on both sides of the active region reflect the light back and forth and stimulated emission is generated if the current through the junction is high enough to sustain the laser action. The bandgap of $Ga_{1-x}Al_xAs$ depends on x, and increases with the Al content. Laser action has been demonstrated at wavelengths between 620 ($x = 0.45$) and 895 nm ($x = 0$). Diode lasers with different types of semiconductors operating from the infrared to the blue have been produced.

$Ga_{1-x}Al_xAs$ lasers are commonly used for information storage in compact disc players. The amount of information that can be stored depends critically on the wavelength of the laser: a shorter wavelength will significantly increase the information density. The search for a shorter-wavelength semiconductor laser resulted in the development of the ZnSe/ZnMgSSe laser, emitting at 447 nm. High-power and tunable solid state lasers at most wavelengths between 630 and 2000 nm are now available.

One of the important applications of semiconductor lasers in laser spectroscopy is pumping of solid state lasers. Instead of flashlamps, semiconductor lasers can be used to pump, e.g., Nd–YAG lasers. The 810 nm of an $Ga_{1-x}Al_xAs$ laser matches an excitation wavelength of the Nd^{3+} ions in YAG. The high efficiency and simple operating scheme of semiconductor lasers results in compact and energy-efficient solid state laser systems. A remaining problem is the limited output power of diode lasers and the higher price in comparison to flashlamp pumping. However, enormous progress has been made since 1990, and the price has gone down from \$1000/W in 1990 to \$100/W in 1993. The average power per centimeter of laser diode array has steadily increased from 20 W/cm in 1990 to 100 W/cm in 1993. By stacking ten diode arrays per centimeter, powers of 1 kW/cm^3 can be reached. One of the important limitations for obtaining higher powers is the necessity for cooling to prevent self-destruction of the laser arrays. For the highest powers water cooling is required. The semiconductor diode laser is still being improved and its potential will increase as time goes on.

3.2. Light-Dispersing Elements

Light-dispersing elements are used for two purposes in luminescence spectroscopy:

1. To produce monochromatic light from a continuous light source.
2. To analyze the wavelength distribution of the emitted light.

There are various types of light-dispersing elements, namely monochromators based

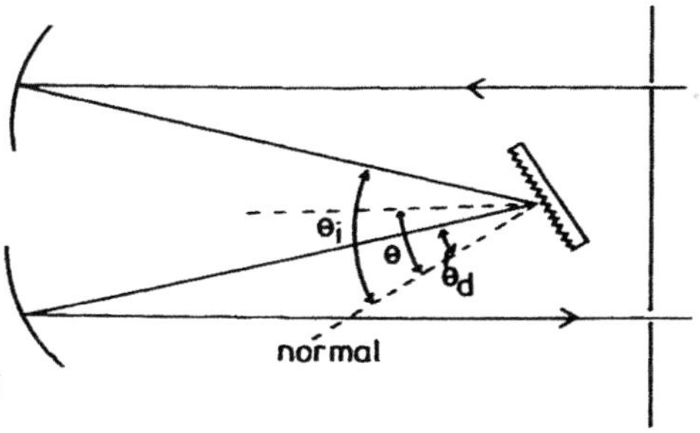

Figure 9. Czerny–Turner configuration of a grating monochromator (Wright, 1990).

on gratings or prisms and interferometers (Moore *et al.*, 1989). The choice of a certain type of light-dispersing element depends on the required performance. Important parameters are the resolving power, the light throughput, and the free spectral range.

The most commonly used light-dispersing element in luminescence spectroscopy is the grating monochromator. Of the various designs for monochromators with flat gratings (Ebert, Czerny–Turner, and Littrow) or curved gratings (Rowland, Paschen–Runge, and Wadsworth), the most popular design is the Czerny–Turner configuration, shown in Fig. 9. The principles of operation and the performance characteristics of a grating monochromator will be illustrated using this design. The monochromator consists of narrow entrance and exit slits, a grating, and focusing mirrors. Gratings are made by ruling or etching a series of lines onto a polished reflecting surface. The number of lines varies typically from 300 to 3600 lines/mm. The light is diffracted at the grating rulings in all possible directions. For constructive interference the difference in pathlength for the light diffracted from two neighboring grooves has to be a multiple of λ:

$$n\lambda = d(\sin\theta_i + \sin\theta_d) \tag{11}$$

where n is the order of diffraction and d is the spacing between two rulings.

An important parameter for the performance of a grating is the angular dispersion:

$$\frac{d\theta_d}{d\lambda} = \frac{n}{d\cos\theta_d} \qquad \text{[rad/nm]} \tag{12}$$

To obtain a large angular dispersion one has to use a grating with a small d (i.e., many lines per millimeter) or work in high orders (large n). In practice it is often not the angular dispersion that is important, but rather the linear distance between two diffracted wavelengths at the exit slit. This linear dispersion is strongly dependent on the focal length of the monochromator: the longer the focal length, the better the linear dispersion:

$$dx/d\lambda = nf/d\cos\theta_d \qquad \text{[mm/nm]} \tag{13}$$

where $dx/d\lambda$ is the linear dispersion at the exit slit and f is the focal length of the monochromator.

The maximum resolution is determined by the value of $\Delta\lambda$ for which there is constructive interference at λ and destructive interference at $\lambda + \Delta\lambda$. It is not hard to derive that

$$\lambda/\Delta\lambda = nN \tag{14}$$

where N is the total number of diffracting grooves on the grating. So, for a grating that is 10 cm wide, with 1200 lines/mm the resolution is 120,000 in first-order. At 500 nm the best possible resolution with this grating in first-order can be calculated using Eq. (14) to be 0.004 nm. To obtain this resolution a long monochromator with narrow slits has to be used. The slitwidth [dx in Eq. (13)] required to obtain maximum resolution can be calculated to be 5 μm for a 1-m focal length monochromator used in first-order at small values of θ_d, using Eq. (13).

The performance of a grating can be affected by the presence of imperfections, which can give rise to ghosts or stray light. Periodic imperfections produce satellite peaks. In the old days, periodic imperfections were caused by the joint in the leather belt that was used to connect the motor to the grating ruling engine. The satellite lines caused by these periodic errors, called Rowland ghosts, are very close to the main line and have a 10^3 to 10^4 times lower intensity. Other, unrelated periodic errors cause spurious lines, called Lyman ghosts, that show up anywhere in the spectrum. The small nonperiodic variations in the spacings between the rulings cause background noise called grass, which is found around the main line. Finally, all other random imperfections give rise to stray light. One can find all these imperfections in a classically ruled grating. A new technique for the production of gratings is based on etching a photosensitive coating on a flat surface by two interfering laser beams. Ghosts and grass can be eliminated in these gratings because the interference pattern of the two laser beams will result in a grating without periodic errors. Stray light can be effectively reduced by using double or even triple monochromators.

In experiments where very weak signals are detected, the efficiency, i.e., the light throughput in a monochromator, is important. To obtain high efficiency in a selected wavelength range, gratings are ruled with the grooves at a specific angle with the grating surface, θ_b, known as the blaze angle. The intensity of the diffracted light is the highest in the direction for which the groove acts as a mirror (angle of reflection from the groove surface). The blaze angle is related to the blaze wavelength (for a Littrow configuration monochromator) by

$$n\lambda_b = 2d \sin \theta_b \tag{15}$$

In a wavelength range of several hundred nanometers around the blaze wavelength the efficiency is high. The efficiency also depends strongly on the polarization of the light.

A last important aspect is the f/number of a monochromator. The highest collection efficiency and resolution are achieved if the f/number of the light collection optics (lenses) matches that of the monochromator. The f/number of a monochromator is defined as f/W, the focal length divided by the diameter of the focusing optics in the monochromator. The f/number of a lens is defined as the focal length divided by the diameter of the lens. If the f/numbers match, the light exactly fills the grating. However, if the f/number of the focusing lens is too small, the light overfills the grating and some is lost (lower efficiency) and if the f/number is too large, the grating will only be partly filled and the number of rulings used is smaller (smaller resolution).

A second type of light-dispersing element is the interferometers, of which there are different types (Fabry–Perot, Michelson, Mach, and Zehnder). Here attention will be focused on the Fabry–Perot interferometers and etalons. An etalon consists of a pair of flat parallel highly reflecting plates at a fixed spacing. The plates form a cavity in which the light is trapped. The light travels back and forth between the two plates and for each successive round trip the pathlength difference δ is given by

$$\delta = 2d \cos \theta \tag{16}$$

If δ is an integral number of wavelengths ($n\lambda$) there is constructive interference. Because the reflectivity of the plates is less than 100%, the amplitudes of the interfering rays are not equal and if $\delta = (n + \frac{1}{2})\lambda$ the destructive interference will not be complete. The closer the reflectivity R is to 100%, the better the wavelength separation will be.

The characteristics of a Fabry–Perot etalon are illustrated in Fig. 10 for different values of R. The resolution of an etalon is given by

$$\frac{\lambda}{\Delta\lambda} = 1.49 \frac{2d \cos \theta}{\lambda} \frac{2\sqrt{R}}{1 - R} = 1.49 \frac{2d \cos \theta}{\lambda} \sqrt{F} \tag{17}$$

where F is called the finesse of the etalon. Equation 17 shows that the resolution increases with d, the distance between the plates, and with R, the reflectivity. The increase in resolution comes at the expense of the intensity (for higher R) or the free spectral range (for larger d). The free spectral range is determined by the next wavelength for which δ is an integral number of wavelengths. For a large d the next wavelength showing constructive interference is very close.

A Fabry–Perot interferometer is an etalon for which the spacing d can be adjusted or the effective optical pathlength can be varied by changing the gas pressure in the space between the plates. The resolution of a Fabry–Perot interferometer can be very high. For example, with a spacing d of 1 cm and with $R = 99\%$

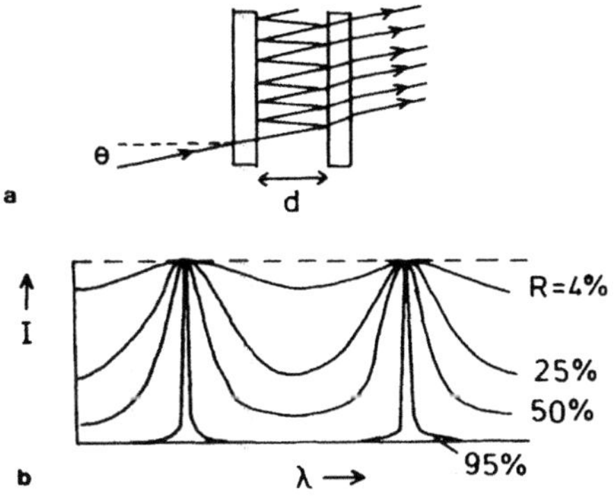

Figure 10. (a) Paths of the transmitted and reflected beams in a Fabry–Perot etalon. (b) Transmission characteristics of a Fabry–Perot etalon for different reflectivities R of the partial reflecting mirror (Wright, 1990).

it will have a resolution of 0.000084 nm at 500 nm ($F = 313$). The light throughput is still 64%.

The main disadvantage of interferometers is the small free spectral range, which imposes a serious limit on their use. One of the important applications is in lasers. By using an etalon one of the many axial modes can be separated out and the laser linewidth can be reduced to very small values.

3.3. Light Detection

The success of an experiment in luminescence spectroscopy often depends on the ability to detect weak signals. Among the detectors for photons the PMT is the one most widely used. The sensitivity of PMTs is very high; single photons can be detected. The human eye comes close to this sensitivity: a dark-adapted eye can detect bunches of ten photons. Other detectors are photodiodes and charge-coupled devices (CCDs). For the CCD detector great progress has been made in the past years and their sensitivity comes close to that of PMTs. Various detection techniques can be used to improve the signal-to-noise ratio such as lock-in amplification and boxcar averaging. These will be discussed at the end of this section.

3.3.1. Photomultipliers

A PMT consists of a photoemissive layer (the photo cathode), focusing electrodes, an electron multiplier, and an electron collector (the anode) (Moore *et al.*, 1989; Hamamatsu, 1994). All these stages are sealed in a vacuum tube (see Fig. 11). Photons entering the PMT will hit the photoemissive layer and can eject an electron if the energy of the photon is higher than the work function ϕ of the photoemissive material. For most metals ϕ is rather large (4 eV) but for alkali metals ϕ is lower (e.g., 1.8 eV for Cs). Mostly photoemissive materials in PMTs consist of an alloy of metals including alkali metals, e.g., $CsNa_2KSb$, Cs_3Sb, or KCsSb. Semiconductors are also used as photoemissive materials, e.g., GaAs or (In, Ga)As.

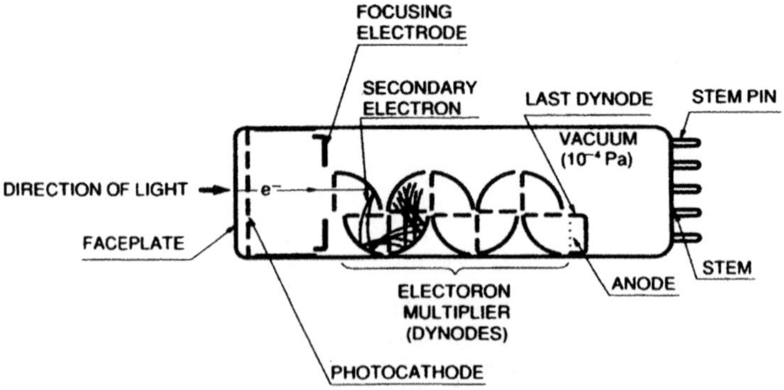

Figure 11. General design of a photomultiplier tube.

The ejected electrons are focused by electrodes and accelerated to the first dynode, which is at a higher voltage. There, a number δ of secondary electrons (typically 3–10) are ejected. At each following dynode every electron ejects δ secondary electrons, resulting in a total amplification (number of electrons at the anode per incident photon) of $G = \delta^n$, where n is the number of dynodes. Typically the overall gain is 10^5–10^8. For dynode materials Be–Cu alloys and GaP are frequently used. The electrons that are collected at the anode give the output signal, a small current in the picoampere to microampere range.

The PMT can have a head-on or side-on design. In the head-on design a semitransparent photocathode is deposited on the inner surface of the entry window. The side-on design uses an opaque photocathode and a circular electron multiplier structure. The side-on type is less expensive and nowadays its performance is comparable to that of head-on tubes. The more expensive head-on design is still preferred for some special applications where spatial uniformity or a large photosensitive area is required.

The choice of a specific PMT is (apart from the price) determined by the spectral response, the sensitivity, the dark current, and the response time. The spectral response depends on the choice of the photoemissive material, and the sensitivity of the photocathode varies with the wavelength of the incident light. On the long-wavelength side the response is determined by the work function of the photoemissive material, whereas on the short-wavelength side the window material often determines the cut-off wavelength of the spectral range. In Table 2 the characteristics for some popular PMTs are summarized. The quantum efficiencies listed in this table correspond to the percentage of incident photons that lead to emission of an electron from the cathode.

The spectral response characteristics are not easy to measure. In practice the cathode or anode luminous sensitivities are given for specific PMTs and the spectral response curve typical of a certain type of PMT is used. The cathode luminous sensitivity is the photoelectric current at the photocathode per incident light flux of 10^{-5} to 10^{-2} lumens from a tungsten lamp operated at a temperature of 2856 K. By

Table 2. Characteristics of Photomultiplier Tubes with Different Photoemissive Materials[a]

Cathode	Maximum radiant sensitivity (mA/W)	Range (nm)	λ_{peak} (nm)	Peak Q.E. (%)	Dark current (A/cm²)
Cs–O–Ag (*S*1)	2	400–1200	800	0.08	9×10^{-13}
Cs_3Sb (*S*11)	20	300–650	440	13	10^{-16}
$CsNa_2KSb$ (*S*20)	53	300–850	470	18	3×10^{-16}
Bialkali (*S*22)	70	160–650	400	25	3×10^{-18}
GaAs–Cs	62	185–930	300–800	25	10^{-16}
Multialkali	74	185–900	400	25	3×10^{-16}

[a]The type of tube is indicated by the *S*-number and/or the photocathode materials. The numbers are given for operation at room temperature.

analogy, the anode luminous sensitivity is the anode output current (after amplification in the electron multiplier) per incident light flux on the photocathode of 10^{-10} to 10^{-5} lumens. The luminous sensitivities are given in amperes per lumen.

Even in the dark there is a small current in PMTs, and for a good signal-to-noise ratio, the dark current, which is made up of contributions from different sources must be small. Thermionic emission of electrons from the photoemissive cathode material is an important source of dark current, which is especially a problem for infrared-sensitive PMTs (small work function) at room temperatures (see Table 2). To reduce the dark current from this source the PMT can be cooled, e.g., to −30°C by dry ice or thermoelectric cooler. If a high sensitivity in the infrared is not required, the best choice of PMT is one with a low sensitivity in the infrared and a small dark current. Other sources of dark current are ionization of residual gases (e.g., Cs), glass scintillation, and γ-emission from radioactive isotopes (e.g., ^{40}K) in the glass (all three causing luminescence and thus current), field emission, and leakage currents. The last can be a serious problem for cooled PMTs. Condensation of water in the PMT socket or at the pin connectors can give rise to high leakage currents. Condensation of water inside the cooled PMT housing is due to imperfect insulation of the cooled compartment.

A third important characteristic of PMTs is the response time, which depends on the design and the applied voltage. One can distinguish between the transit time (the average time it takes an electron to travel from cathode to anode) and the rise and fall times, which are determined by the spread in transit time owing to differences in pathlengths for electrons in the PMT. The time response becomes faster for higher voltages over the PMT.

A good way to measure low-level light intensities with PMTs is photon counting. At low intensities, the incident photons create pulses that are well separated in time, and the number of pulses per second can be counted. The signal-to-noise ratio is improved by discriminating against noise pulses. The amplitude of a noise pulse is generally lower than for pulses from photons as most of the noise pulses originate from a later stage in the PMT and so are amplified by fewer dynodes. By discriminating against the lower pulses a significant reduction in noise is obtained. In photon-counting experiments fast PMTs are generally used to allow for high pulse rates. If the pulse rate becomes too high and the pulses overlap in time the photon-counting signal will no longer be proportional to the light intensity.

3.3.2. Photodiodes

Beyond 900 nm the sensitivity of PMTs is low. In this region photodiodes, which are photovoltaic detectors are used. Light is detected by photoexcitation of electron–hole pairs near a junction. Absorption of photons with an energy greater than the bandgap leads to the formation of electron–hole pairs. The electron and hole are separated and a small potential difference across the junction is created. Apart from *p–n* junctions, other types of junctions can be used: *p–i–n*, Schottky barriers (a metal deposited on a semiconductor surface), and heterojunctions (a junction between two different semiconductors).

Table 3. Characteristics of Different Photodiode Detectors

Semiconductor	T(K)	Range (nm)	D^*_{max} (cm $Hz^{1/2}$ W^{-1})	Rise time (ns)
Si	300	200–1100	2×10^{13}	$\geqslant 0.006$
Ge	300	400–1800	10^{11}	0.3
Ge	77	400–1700	10^{13}	0.3
InAs	77	1000–3200	4×10^{11}	700
PbSnTe	77	2000–18,000	$\simeq 10^{11}$	20

The voltage that is generated when a photodiode is illuminated can be measured with a simple circuit. The highest sensitivity is obtained when the signal from the photodiode is amplified with an operational amplifier. Table 3 gives an overview of commonly used photodiodes for different wavelength regions together with some of their characteristics. D^* is a measure of the signal-to-noise ratio and is defined as $\sqrt{A}$/NEP (the square root of the area divided by the noise equivalent power: the higher the value of D^*, the better the signal-to-noise ratio). As a comparison, D^* for a PMT is about 10^{16} cm $Hz^{1/2}/W^1$, much higher than that of photodiodes. For a photodiode material with a small bandgap (e.g., Ge) the D^* value at 300 K is low owing to the presence of thermally generated charge carriers. Cooling to 77 K by liquid nitrogen reduces the number of thermally generated free charge carriers, and the value of D^* increases.

Photodiodes and other infrared detectors (e.g., photoconductive detectors such as PbS and thermal detectors) suffer from a relatively large amount of noise. The signal-to-noise ratio can be improved by the use of lock-in amplification (Moore *et al.*, 1989), which effectively eliminates the low-frequency noise. The signal is modulated at a selected frequency ω_L, which is chosen to be far from power-line frequencies and their harmonics or other known noise frequencies. Modulation can be achieved, e.g., by a chopper in the light beam. The detection system is set to detect signals only at the frequency of the modulation.

3.3.3. Charge-Coupled Devices

A detector that has gained in popularity owing to an impressive improvement of its performance in the past 5 years is the charge-coupled device (CCD) (Gallagher, 1994). The revolution in semiconductor manufacturing and computation has fueled the rapid improvement of this solid state detector, which has replaced the image-tube detector. CCDs are not only used for scientific purposes but also have applications in consumer products such as camcorders, fax machines, and cameras.

The basis for the CCD image detection technique dates back to 1970. A CCD consists of arrays of metal-oxide–semiconductor (MOS) capacitors, which are made from p-type silicon grown on an insulating layer of SiO_2. On top of the SiO_2 layer a thin metal structure (called the gate) is deposited. Electric contacts are made on

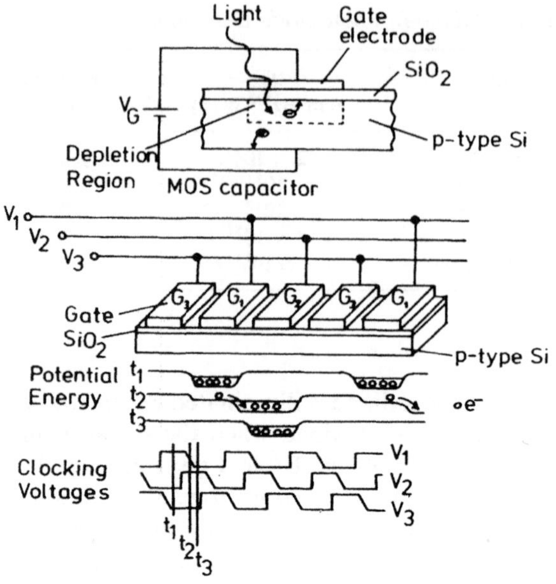

Figure 12. Illustration of the operation of a CCD. On top the trapping electrons under the gate of a MOS capacitor under illumination is depicted. The transfer of the collected charge in a three-phase CCD line is illustrated below. One pixel is formed by three gates (G_1, G_2, and G_3). The time sequence of the voltage applied to lines V_1, V_2, and V_3 and the potential energy diagrams with the moving trapped electrons are shown in the two lower figures (Gallagher, 1994; with permission from Pen Well Publishing Co.).

the gates and the Si substrate. When a positive voltage is applied to the gate, the holes are pushed away from it (see Fig. 12). Illumination of the MOS capacitor results in the creation of free electrons and holes. The electrons are collected at the Si–SiO_2 interface under the gate as long as the potential is applied (the integration time). The number of electrons trapped under the gate is proportional to the number of incident photons (up till a certain maximum of about 10^4–10^6 per gate, at which point the potential well is saturated).

Each pixel in a CCD consists of three gates. During illumination a voltage is applied to the middle gate (G_1) via voltage line V_1 and electrons are collected under all the middle gates G_1. After illumination the image on the CCD plate has to be read out. This is done by transferring the charge collected in each pixel to the detector, point by point and array by array. The transfer of the electrons takes place over long distances without the loss of electrons. The transfer is controlled by applying voltages to G_1, G_2, and G_3 of each pixel in the sequence shown in Fig. 12. The accumulated electrons are shifted by first applying an equal voltage V_2 to all G_2 gates and the electrons are smeared out over the area under gates G_1 and G_2. Then the voltage V_1 is reduced and all the electrons move to gate G_2. The same voltage sequence is then applied to gates G_2 and G_3 to move the electrons to gate G_3. This voltage sequence is repeated many times (controlled by the CCD clock).

The charge is moved array after array to the horizontal array and from the horizontal array to the sense detector. There, the number of stored electrons is measured for every pixel. The number of electrons (proportional to the incident light intensity) for every point on the CCD plate is known after this readout procedure, and the image can be made visible on a screen and be processed with a computer. The sensitivity of a CCD is very high, and thermal electrons are the main source of noise. By cooling the CCD to 77 K this noise can be reduced to less than one electron per pixel per hour. The sense detector is capable of detecting a charge package of five electrons or less. In combination with the high quantum efficiency for electron generation by light in Si, this makes a CCD capable of measuring very low light levels.

One problem in the readout scheme described above (called line address transfer) occurs when illumination continues during readout, which causes image-smearing. To prevent this, a mechanical shutter can protect the CCD from illumination during the readout period. Another solution is interline transfer, where the charges are moved from the imaging pixels to neighboring (not illuminated) shift registers.

The resolution of a CCD is determined by the number of pixels. Typical 2D CCDs have 256 × 256 up to 4096 × 4096 pixels. For a large number of pixels the long readout time can become a problem. The spectral response of a CCD is determined by the light absorption in the Si substrate and this is similar to that of the Si photodiode (200–1100 nm). CCDs are also used as x-ray detectors. For many applications in luminescence spectroscopy a 1D CCD array is sufficient to record a spectrum. The CCD array can measure a complete spectrum instantaneously without the need to scan a monochromator. In place of a slit and a PMT, a CCD array is placed at the exit slit where the light that is diffracted from the grating is focused. The dispersed light is incident on the CCD and every pixel on the array will detect light of a slightly different wavelength. The wavelength resolution is determined by the number of pixels on the CCD array and the linear dispersion of the monochromator at the position of the array.

The great advantage of a CCD array is that all wavelengths in a spectrum can be recorded at the same time. This can be an important advantage if the intensity is expected to change during the time required for a scan. Also the high sensitivity of a CCD camera can be used to measure very weak signals, using long integration times. With a 2D CCD plate, different spectra can be recorded with very short time intervals between them. This can be achieved by focusing the signal at different heights of the entrance slit for different times. In this way the temporal evolution of a spectrum can be followed in time.

4. *TIME-RESOLVED SPECTROSCOPY*

Important information can be obtained from the lifetime of an excited state on the nature of the state and the environment of the ion or molecule. Many different techniques to measure the lifetime of an excited state have been developed over the years, capable of measuring ever shorter lifetimes, down to the femtosecond regime. Before discussing these technique it is important to realize that there are two types of lifetimes, T_1 and T_2 (Henderson and Imbusch, 1989*b*; Allen and Eberly, 1975). The lifetime T_1 is the population lifetime and is determined by the radiative and nonradiative transition rates to the ground state or other excited states. T_2 is the dephasing time and is determined by the rate of dephasing interactions that affect the coherence of an excited state. Only T_1 will be considered for the lifetime measurements discussed in this section. The next section will describe techniques that can be used to study T_2 (photon echo, fluorescence line-narrowing, and hole-burning).

In a simple lifetime experiment the rate at which the population in the excited state changes is measured. This is done by applying a short, intense excitation pulse at $t = 0$, which creates a population $N(0)$ in the excited state. The intensity of the emission from the excited state is measured as a function of time. As the luminescence intensity is proportional to $N(t)$, it decreases exponentially with time:

$$I(t) = I(0) \exp(-t/\tau_R) \tag{18}$$

where τ_R is the radiative decay time, which is the inverse of the transition probability. The values for τ_R can vary from 10^{-8} to 1 s, depending on the nature of the transition involved. For an electric dipole allowed transition, τ_R can be as short as 10 ns, while for a strongly spin-forbidden transition in organic molecules decay times as long as seconds have been observed. In the case of short decay times (spin-allowed transitions) one speaks of fluorescence, and in the case of long decay times (spin-forbidden transitions) one speaks of phosphorescence. To measure the lifetimes of excited states one needs a pulsed excitation source (to give a sharp intense pulse at $t = 0$) and a detection system capable of recording the temporal evolution of the emission intensity in time.

Pulsed excitation can be obtained by a continuous light source in combination with a mechanical shutter. This type of pulsed excitation allows the measurement of long decay times (1 ms or longer). Faster decay times (microsecond regime) can be measured using a continuous light source in combination with a chopper wheel. The most commonly used pulsed excitation sources nowadays are pulsed laser systems and flashlamps, the latter being available with pulsewidths of less than 1 ns and repetition rates up to 50 kHz. Flashlamps can be operated with different gas fillings, e.g. deuterium, hydrogen, xenon, and nitrogen for excitation in different spectral regions in the ultraviolet and visible ranges. A flashlamp is an inexpensive pulsed excitation source and is very adequate for decay time measurements. Still, the most widely used pulsed excitation sources are pulsed lasers. The pulsewidths vary from about 10 ns (nitrogen, excimer, and Q-switched Nd-YAG) down to the femtosecond regime (mode-locked lasers).

The luminescence signal is detected with a PMT or photodiode. To record the temporal evolution of the light intensity the detector is connected to an oscilloscope, a multichannel analyzer, or a device with a more complicated detection scheme such as pulse-height analysis. An oscilloscope or multichannel analyzer can measure the luminescence intensity as a function of time after a single excitation pulse. Since the signal from a single excitation is usually too weak, the signal is averaged over a large number of pulses. Oscilloscopes can measure very fast decaying signals (down to the nanosecond regime). Multichannel analyzers usually have a time resolution that is longer than 10 μs. In the case of weak signals with short decay times pulse height analysis has proven to be useful for decay time measurements.

In pulse height analyses the interval between the excitation pulse and the arrival of the first photon at the detector is measured for a large number of excitation pulses.

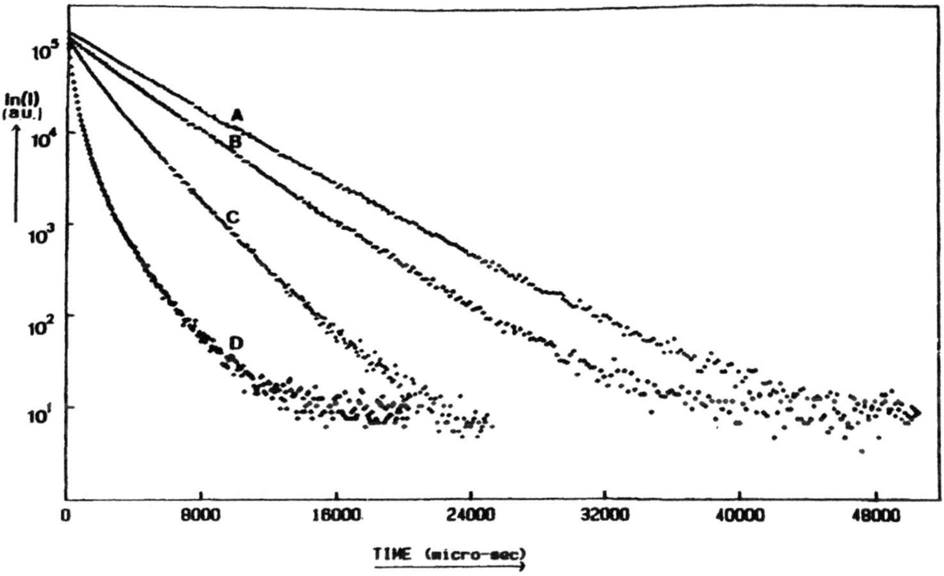

Figure 13. Decay curves of the $^6P_{7/2}$ emission of Gd^{3+} for excitation in the same level at room temperature: (A) $L_{0.4}Gd_{0.6}MgB_5O_{10}$; (B) $La_{0.4}Gd_{0.6}MgB_5O_{10}$–$Fe^{2+}$0.01%; (C) $La_{0.4}Gd_{0.6}MgB_5O_{10}$–$Fe^{2+}$0.1%; (D) $La_{0.4}Gd_{0.6}MgB_5O_{10}$–$Fe^{2+}$1% (van Schaik *et al.*, 1994; with permission from the Electrochemical Society).

To do this, a ramp voltage rises linearly in time, starting at the moment of the excitation pulse (the start pulse). As soon as the first photon is detected (stop pulse) the voltage reached is read. The height of the voltage is a measure of the time elapsed between the start and stop pulses. The distribution of the times between the start and stop pulses gives a good decay curve, provided that the ratio between them is 0.2 or lower to assure good statistics. Since only one time is obtained for every excitation pulse, high repetition rates are required to have an acceptable total measuring time.

The limiting factor in the short-time regime (subnanosecond) is the PMT. The fastest PMTs still have a rise time of about 200 ps. Si photodiodes can have considerably shorter rise times (6 ps) and can be used for decay time measurements in the picosecond regime. To measure on even shorter timescales (femtoseconds), pump-probe experiments can be used. This technique will be discussed in the next section.

Decay curves are not always single exponential. The deviation from exponentiality can provide important information. As an example, Fig. 13 shows the decay curves for the Gd^{3+} emission in $La_{0.4}Gd_{0.6}MgB_5O_{10}$ for samples with different concentrations of Fe^{2+} (van Schaik *et al.*, 1994). If no Fe^{2+} is present, a single exponential decay curve is observed with a 4.3 ms decay time. The long decay time reflects the parity-forbidden character of the transition within the $4f^7$ configuration of Gd^{3+}. When Fe^{2+} is added the decay time becomes shorter and nonexponential owing to energy transfer from Gd^{3+} to Fe^{2+}.

The energy transfer process provides an extra decay channel for the Gd^{3+} ions in the excited state and shortens the decay time. The rate for energy transfer depends strongly on the distance between a Gd^{3+} ion and neighboring Fe^{2+} ions. The distribution of Fe^{2+} ions around Gd^{3+} ions will vary and thus the decay rates will be different for different Gd^{3+} ions, resulting in a nonexponential decay curve. Analysis of the decay curves provides information on the energy transfer rates and the transfer mechanism. A good review of how analysis of decay curves can be used to obtain information on energy transfer processes can be found in Di Bartolo (1983).

The difference in lifetime of luminescence can be used to separate overlapping luminescence spectra by measuring them for different delay times after an excitation pulse. For this type of time-resolved spectroscopy a boxcar averager is used. The boxcar averager requires two input signals: a gate and the signal from the detector. The signal is measured only during the time a voltage is applied to the gate. With a function generator a voltage can be applied to the gate starting at a certain time after the excitation pulse (delay time) for a certain period of time (gatewidth). The delay time and gatewidth can be varied. To detect a rapidly decaying luminescence signal, short times are chosen, and for slowly decaying signals a long delay time and gatewidth are selected. The signal from the detector is applied to the input of the boxcar averager, measured during the selected time interval, and averaged over a number of excitation pulses to improve the signal-to-noise ratio.

The best known example of the ability to separate a rapidly and a slowly decaying luminescence signal is found in the field of luminescing organic molecules. The spin-forbidden triplet–singlet transition (phosphorescence) and the spin-allowed singlet–singlet transition (fluorescence) occur simultaneously under continuous excitation. The use of a boxcar averager makes it possible to separate the two emission bands. In addition, the time evolution of luminescence spectra can be followed by recording spectra for different delay times.

One nice example is time-resolved emission from Pr^{3+} in $La_{0.8}Pr_{0.2}F_3$ (Fig. 14) (Huber *et al.*, 1977). A narrow-band laser is used to excite a small fraction of the Pr^{3+} ions in the inhomogeneously broadened line, which is chosen to be some $12\,cm^{-1}$ from the maximum. For zero delay time the emission originates from the excited subset of Pr^{3+} ions. For longer delay times nonresonant energy transfer between neighboring Pr^{3+} ions leads to spectral diffusion, and a broad background emission that is very similar to the inhomogeneously broadened line starts to rise. From the time evolution of the emission spectra, information on the energy transfer rates and mechanism for nonresonant energy transfer between Pr^{3+} ions can be obtained. Similar experiments have been done for transition-metal ions and rare-earth ions in glasses and to study energy transfer to traps. The technique is similar to fluorescence line-narrowing, which is discussed in the next section.

An important additional advantage of the boxcar averager is the noise reduction. Since the detector noise has an important component that is time-independent, the signal-to-noise ratio is improved by measuring only during the time that the signal is high. The boxcar averaging is especially useful when the fraction of the time during which the signal is high is small (a low duty cycle). In the case of a low duty

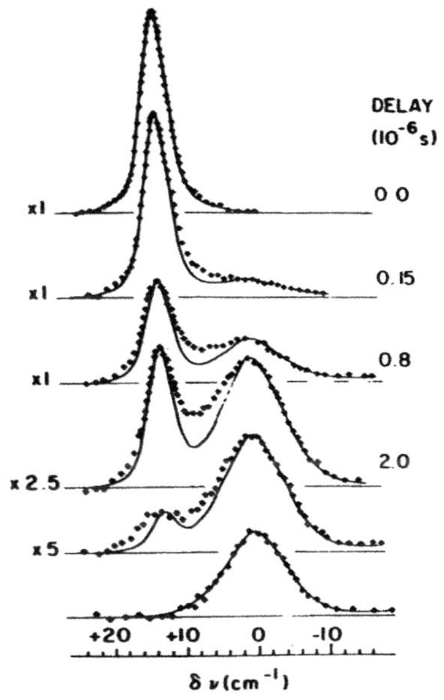

Figure 14. Time-resolved spectra for the $^3P_0 \rightarrow {}^3H_6$ emission of Pr^{3+} in $La_{0.8}Pr_{0.2}F_3$ at 14 K. Excitation is at $+12\,cm^{-1}$ from the maximum of the inhomogeneously broadened emission line (i.e., at $\delta\nu = +12\,cm^{-1}$). The solid line is a theoretical fit (Huber *et al.*, 1977; with permission from the American Physical Society).

cycle, the use of a boxcar averager is a better method for noise reduction than lock-in amplification.

5. HOLE-BURNING, FLUORESCENCE LINE-NARROWING, AND PHOTON ECHO

In the previous section different techniques for measuring the population decay time T_1 were discussed. The measurement of the dephasing time T_2 is more complicated. Various techniques to measure the dephasing time in the time or in the frequency domain will be discussed in this section. Techniques in the frequency domain are hole-burning and fluorescence line-narrowing (Macfarlane and Shelby, 1987). In the time domain photon echo and optical free induction decay experiments can provide information on T_2 (Allen and Eberly, 1975; Macfarlane and Shelby, 1987).

The homogeneous linewidth of a spectral transition between a ground state and an excited state contains information on the dephasing time T_2. The homogeneous linewidth Γ_{hom} is given by

$$\Gamma_{\text{hom}} = 1/\pi T_1 + 1/2\pi T_2 \tag{19}$$

The physical background for this equation lies in Heisenberg's uncertainty principle: the shorter the coherence lifetime of an excited state (determined by population decay and dephasing interactions), the greater the uncertainty in the energy of the transition between the ground state and excited state. If T_1 is long in comparison to T_2, the homogeneous linewidth provides information on T_2 directly. The problem is to measure homogeneous linewidths.

Zero-phonon lines in excitation and emission spectra of ions or molecules in solids usually show inhomogeneous broadening that is several orders of magnitude larger than the homogeneous linewidth. The inhomogeneous linewidth varies typically from 0.1 cm^{-1} (for a dilute solution of ions or molecules in high-quality crystal) to a few hundred cm^{-1} (in glasses). The inhomogeneous broadening is caused by imperfections in a crystal and the disorder in a glass. The resulting small variations in the environment of the luminescing species give rise to small differences in the energy of the transition, resulting in an inhomogeneously broadened zero-phonon line, consisting of many homogeneously broadened lines of individual ions or molecules (Moerner and Basche, 1993). With narrow-band lasers it is possible to eliminate the inhomogeneous broadening and measure the homogeneous linewidth via optical hole-burning or fluorescence line-narrowing.

5.1. Optical Hole-Burning

In the case of optical hole-burning (OHB) a narrow-band laser with a spectral width Γ_L and frequency v_1 excites a resonant subset of ions to an excited state, thus depleting the ground state for those ions for which v_1 is within the homogeneous linewidth Γ_{hom} (Macfarlane and Shelby, 1987). This creates a hole in the original (inhomogeneously broadened) absorption band at frequency v_1. The hole can be probed by scanning with a second laser at a low enough intensity to prevent additional hole-burning. The absorption or excitation spectrum recorded with the second laser beam (also of spectral width Γ_L) will show a dip (or hole) at the position of v_1. The width of the hole represents the homogeneous linewidth if $\Gamma_L < \Gamma_{\text{hom}}$. OHB seems to be a straightforward technique for measuring homogeneous linewidths, but extreme care has to be taken to avoid contributions of residual broadening such as power-broadening, dynamic inhomogeneous broadening, or broadening owing to local heating.

There are three different types of hole-burning: (1) photochemical hole-burning, (2) nonphotochemical hole-burning, and (3) transient hole-burning. In the case of the first a photochemical reaction takes place in or around the ion or molecule after excitation. As a result, the ion or molecule experiences a large shift in the absorption frequency. The lifetime of holes burned this way can be very long (permanent) at low temperatures. The second is typical for amorphous systems. Excitation promotes a reorientation in the close environment of the excited species, causing a small shift in the aborption frequency. The reorientation is pictured as a two-level system consisting of two conformational minima separated by an energy barrier V. In the excited state the two-level system can tunnel from one minimum to the other. At low temperatures (kT small in comparison to V) these holes are permanent. Finally, there are transient holes which occur because population from the ground state is transferred to a long-lived excited state. The lifetime of these holes is short and is determined by the radiative decay rate of the metastable excited state (typically 1 μs to 1 s). In view of the short lifetime, the burning and probing laser beams are needed simultaneously.

Essential for a hole-burning experiment is a narrow-band laser for which $\Gamma_L < \Gamma_{\mathrm{hom}}$. Continuous-wave ring dye lasers can produce linewidths down to 1 MHz and are commonly used. The power of the laser beam to burn a hole can be 5–100 mW/cm^2. The same laser that is used to burn the hole is used to scan it at a reduced intensity (10–1000 times attenuated). The luminescence from the sample is detected by a PMT while scanning the ring dye laser over the wavelength region in which the hole was burned. For transient hole-burning two continuous-wave ring dye lasers are used simultaneously, one at a fixed frequency to burn the hole and one that is scanned to record the excitation spectrum around the hole.

5.2. Fluorescence Line-Narrowing

Fluorescence line-narrowing (FLN) is closely related to hole-burning. Again, a narrow-band laser is used to excite a subset of ions or molecules within the inhomogeneously broadened line. The resonant luminescence from the ions or molecules is measured. In the absence of spectral diffusion the linewidth is given by

$$\Gamma = \Gamma_L + 2\Gamma_{\mathrm{hom}} \tag{20}$$

The homogeneous linewidth enters twice, since it contributes in both excitation and emission. The FLN experiment is more complicated than OHB. The resonant emission has to be detected with a high resolution. For the narrow linewidths (of the order of several megahertz) an interferometer with a high finesse, which has a small free spectral range, has to be used. If there are several hyperfine lines in the FLN emission spectrum the small free spectral range can become a problem as different lines may be transmitted in different orders. If this is the case, two or more tandem interferometers have to be used.

A second complication is the presence of scattered laser light that is resonant with the emission that is being detected. For weak luminescence signals the scattered laser light intensity may be much higher than the emission intensity. To discriminate against the scattered laser light one can use choppers or electronic shutters, which block the laser during signal detection and block the detector when the laser excites the sample. This method only works if the lifetime of the excited state is not too short. An advantage of FLN is that the spectral dynamics can be followed by measuring the time evolution of the emission spectrum, as was discussed in the previous section.

5.3. Photon Echo

A different type of experiment to measure T_2 dephasing times is coherent transient spectroscopy. As an example of this type of technique, the photon echo experiment will be discussed. Other techniques, such as optical free induction decay and optical nutation will not be treated here. Detailed information can be found in

Allen and Eberly (1975), Macfarlane and Shelby (1987), and Yen and Selzer (1981). In the case of coherent transient spectroscopy the dephasing time T_2 is measured directly and not via the homogeneous linewidth. With a laser, a coherent state of the system is created and the decay of the coherence owing to dephasing interactions follows in time. Only a qualitative picture of the photon echo experiment is given here and for a thorough theoretical understanding of photon echo spectroscopy the reader is referred to Allen and Eberly (1975) and Macfarlane and Shelby (1987).

To understand the photon echo experiment, the interaction between a driving electric field (of the electromagnetic radiation produced by the laser) and a two-level system will be considered first. The driving field will promote the system to the excited state $|2\rangle$ (absorption) and back to the ground stae $|1\rangle$ (stimulated emission) with a frequency known as the Rabi frequency. A complete cycle of absorption and stimulated emission corresponds to a 2π-laser pulse: the larger the amplitude of the driving field (intensity of the laser light), the higher the Rabi frequency and thus the shorter a 2π-pulse. If the driving field is exactly in resonance with the energy difference between states $|1\rangle$ and $|2\rangle$, the polarization of the system will continue to oscillate at the frequency of the light when the laser is turned off. If the driving field is slightly out of resonance with the system, the polarization of the system will start to oscillate at its own frequency when the driving field is turned off.

These processes are indicated in a Bloch diagram (Fig. 15). The w-axis represents the population in the ground state and an excited state. A down-pointing vector corresponds only to population in the ground state and an upward-pointing vector only to population in the excited state. If $w = 0$ the ground state and excited

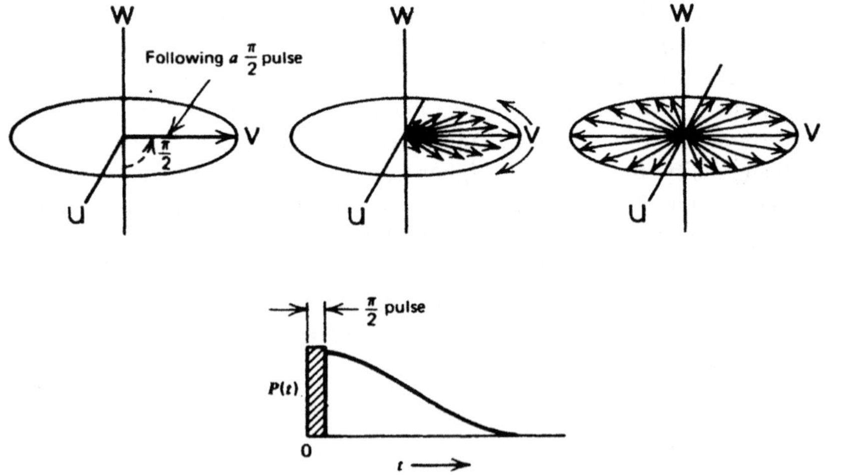

Figure 15. Bloch diagram showing optical free induction decay after a $\pi/2$-pulse for an inhomogeneously broadened line. The lower sketch shows the decrease of the polarization density P as a result of the dephasing of the Bloch vectors of different two-level systems oscillating at slightly different resonance frequencies (Allen and Eberly, 1975; with permission from Dover Publications, Inc.).

state populations are equal. The v-axis corresponds to the component of the oscillating polarization in phase with the driving field and the u-axis gives the out-of-phase component. In a photon echo experiment first a $\pi/2$-pulse (one-quarter of the Rabi cycle) is applied, which brings the two-level systems to $w = 0$, all in phase with the driving field.

When the driving field ends, the polarization of those systems that are in resonance with the driving field will continue to oscillate at their natural frequency and stay at $v = 0$. However, as many systems in the inhomogeneously broadened line were coherently excited to $w = 0$, the polarization of other systems with a resonance frequency slightly different from the exciting laser will oscillate at a slightly different frequency. The out-of-phase component u will increase in time for the off-resonance systems. The larger the difference with the frequency of the driving field is, the faster the out-of-phase component u will grow. In the diagram of Fig. 15 this is seen as a fanning out of the vectors. The wave emitted by the ensemble of two-level systems rapidly decays away as a result of the disappearing constructive interference of the in-phase components of the oscillating polarizations of the different two-level systems.

This behavior is called free induction decay (FID). The FID time becomes shorter for greater inhomogeneous broadening. The trick of a photon echo experiment is to circumvent the influence of FID owing to inhomogeneous broadening by applying a second pulse of the driving field at a variable delay time (called t_{12}, $t_2 - t_1$ in Fig. 16) after the first $\pi/2$-pulse. The second pulse is a π-pulse that shifts the out-of-phase component of every system by 180°. As a result, those systems that were ahead in phase with respect to the driving field are now behind by the same amount and vice versa. The systems will continue to oscillate at their natural frequencies and the systems that are behind in phase will have caught up again at a time t_{12} after the π-pulse (t_4 in Fig. 16). At time t_4 all systems will have caught up and coherence will have been restored.

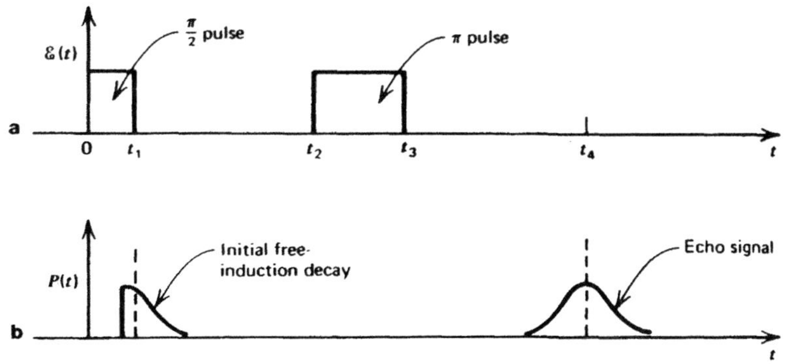

Figure 16. (a) Pulse sequence for a photon echo experiment. (b) The macroscopic polarization $P(t)$ resulting from the $\pi/2$- and the π-pulse. The first pulse is the free induction decay after the $\pi/2$-pulse. The second output pulse is the echo signal as a result of rephasing at a time $t_{12} = t_2 - t_1$ after the π-pulse (Allen and Eberly, 1975; with permission from Dover Publications, Inc.).

The rephased forward directed polarization will emit an echo pulse at that time. The intensity of the echo pulse depends quadratically on the number of rephased oscillators. The echo intensity decreases if t_{12} becomes longer owing to dephasing interactions, which destroy the coherence of individual systems. If the coherence of a two-level system is destroyed (dephasing) at any time between the $\pi/2$-pulse and the echo signal at $2t_{12}$, the system will not be rephased at $t = 2t_{12}$ and it will not contribute to the echo signal. The dephasing time T_2 can be found by measuring the intensity of the photon echo signal as a function of the delay time t_{12}:

$$I_{\text{echo}} = I_0 \exp(-4t_{12}/T_2) \tag{21}$$

The first photon echo experiment was performed by Kurnit *et al.* (1964) in ruby, using a Q-switched ruby laser and a variable optical delay line to produce a second pulse at a variable delay time t_{12}. In later experiments it was found that it is advantageous to use continuous-wave lasers and Stark switching or laser frequency-switching techniques in combination with heterodyne detection to do photon echo experiments. A good review of the possibilities can be found in Macfarlane and Shelby (1987).

One example of a photon echo experiment is given here to illustrate the technique. In Fig. 17 the intensity of the photon echo signal is shown for the $^7F_6 \rightarrow {}^5D_4$ transition on Tb^{3+} in $LiYF_4$ (Lin and Cone, 1990). The echo intensity decreases exponentially with the delay time t_{12}, as expected from Eq. (21). From the decay time the linewidth can be calculated using Eq. (20) to be 285 kHz. From

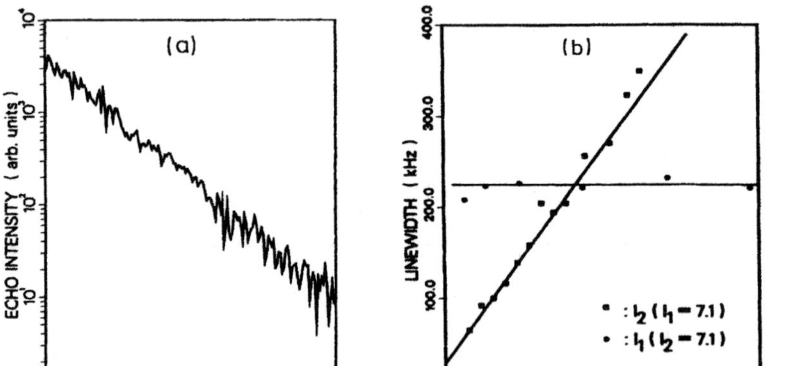

Figure 17. (a) Typical photon echo decay curve of $LiYF_4$–Tb^{3+} at 1.3 K. The apparent linewidth of 285 kHz is greater than the true homogeneous linewidth. (b) Dependence of the linewidths determined from the photon echo decay curves for different intensities of the first laser pulse I_1 and the second laser pulse I_2. The dephasing rate is observed to be dependent only on I_2. The true homogeneous linewidth derived from the intersection point for $I_2 = 0$ is 26 kHz (Lin and Cone, 1990; with permission from the American Physical Society).

Fig. 17b it is clear that this is not the true homogeneous linewidth. The linewidth derived from the decay time of the photon echo signal is plotted in the figure as a function of the intensity of the $\pi/2$-pulse, I_1, and of the π-pulse, I_2, showing that the linewidth derived from the decay time of the photon echo signal increases with the intensity of the π-pulse.

The increase in the dephasing rate for higher intensities of the π-pulse is ascribed to instantaneous spectral diffusion owing to interactions among Tb^{3+} ions in the excited state. This experiment shows that one has to be careful in deriving the true homogeneous linewidth from the decay of the photon echo signal. Extrapolation of the I_2 line in Fig. 17b to $I_2 = 0$ gives the true homogeneous linewidth of 28 kHz.

6. QUANTUM EFFICIENCY AND NONRADIATIVE PROCESSES

Up until now the focus has been on radiative transitions between an excited state and the ground state, and the experimental techniques that have been discussed are based on the emission of a photon when the system returns from the former to the latter state. This is not always the case. There is also the possibility of nonradiative decay (Di Bartolo, 1991), in which the system relaxes from an excited state to the ground state without the emission of a photon. In this case, the energy is dissipated by the emission of phonons, i.e., lattice vibrations or heat. Nonradiative relaxation is sometimes desirable and sometimes it is not. In applications of phosphor materials where a high light output is required (e.g., phosphors in fluorescent lamps, color television, and x-ray photography), nonradiative relaxation from the emitting level should be avoided. In other cases fast nonradiative relaxation is required to feed emitting levels. For example, in lasers fast nonradiative relaxation to the starting level for laser action is necessary. Also, many experimental techniques are based on nonradiative relaxation.

The most commonly encountered models for nonradiative relaxation are the configuration coordinate diagrams (Blasse and Grabmaier, 1994; see also Section 5.4 in Chapter 3). The possibility of nonradiative relaxation via the conduction band is described by the model of Struck and Fonger (1991).

The relative contribution of nonradiative decay processes can be determined from the quantum efficiency q of a phosphor, which is defined as

$$q = \frac{\text{Number of photons emitted}}{\text{Number of photons absorbed}} \tag{22}$$

An alternative measure is the energy efficiency η, which is defined as the ratio between the amount of energy emitted and the amount absorbed and can never exceed 100%. An efficient phosphor has a high quantum efficiency. The present generation of phosphors in fluorescent lamps have quantum efficiencies close to 100%. The quantum efficiencies of some well-known standard luminescent materials are tabulated in Table 4.

The measurement of absolute quantum efficiencies is difficult. The number of photons being absorbed has to be determined from diffuse reflection spectra of the

Table 4. Quantum Efficiencies for Some Standard Luminescent Materials at 300 K for Excitation 254-nm

Phosphor	λ_{em}^{max} (nm)	q (%)
Zn_2SiO_4: Mn^{2+}	530	70
$MgWO_4$	475	84
Y_2O_3-Eu^{3+}	615	93
$CaWO_4$	430	75
SrP_2O_7-Sn^{2+}	460	86

sample in comparison to $BaSO_4$ or another totally reflecting material. The number of photons emitted has to be determined by collecting them all with an integrating sphere and detecting them with a PMT at the focal point. The process is rather laborious and there are many pitfalls in the calibration procedures. As a result, different values have been reported for absolute quantum efficiencies of the same material (Bril and de Jager-Veenis, 1976). The determination of absolute quantum efficiencies is best left to experts.

A good indication of a quantum efficiency can be obtained by measuring relative quantum efficiencies. The light outputs of the luminescent material and of a reference material with a known quantum efficiency are measured under identical conditions. The signals measured for the two samples have to be corrected for the differences in response of the PMT in the different spectral regions. By using a PMT with a flat response (e.g., one with a GaAs photocathode) this correction procedure is easy in the spectral region from 300 to 800 nm. Also the amount of light that is absorbed at the excitation wavelength is determined for the sample and the reference material by diffuse reflection spectroscopy. With this information one can get a good idea of the quantum efficiencies of new luminescent materials.

An alternative method for obtaining a good estimate of quantum efficiencies is the measurement of the temperature dependence of the decay time and/or the light output of a luminescent material. At low temperatures the quantum efficiency is near unity. Increasing the temperature leads to thermal quenching of the luminescence, and the thermal quenching results in a reduction in the light output. Also, the decay time decreases owing to the increased (nonradiative) transition probability from the excited state to the ground state:

$$1/\tau = A_R + A_{NR} \tag{23}$$

Here A_R is the radiative decay rate (which is assumed to be independent of temperature) and A_{NR} is the nonradiative decay rate (which is assumed to be zero at the lowest temperatures and to increase with temperature).

This method works well if the nonradiative decay processes can be neglected at low temperatures and if the radiative decay time is not a function of temperature. Muñoz *et al.* (1988) have observed the decay time and the light output as a function of temperature for Eu^{2+} in KCl. For Eu^{2+} the radiative decay time has a small temperature dependence, which can be neglected for a rough estimate of the

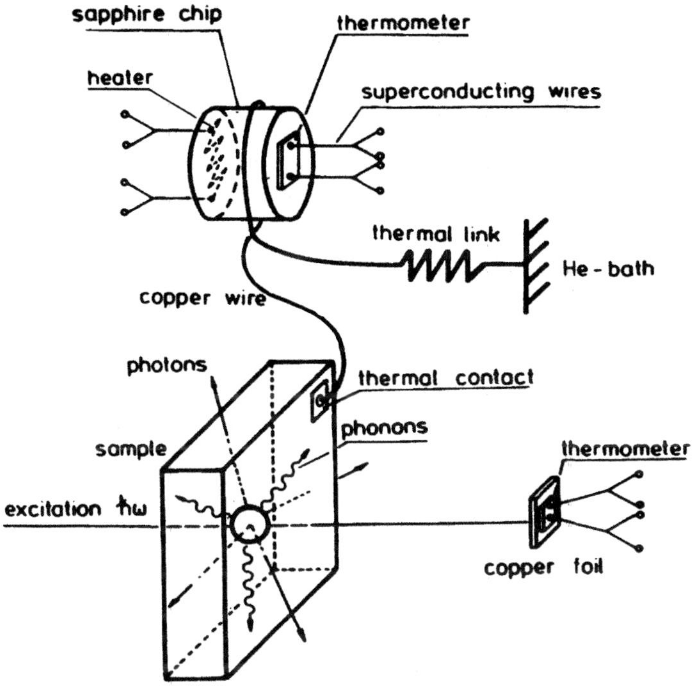

Figure 18. Sample thermometer arrangement for CAS and CTS (Bimberg *et al.*, 1991).

quantum efficiency. From the temperature dependence of the light output and the decay time the quantum efficiency at room temperature can be estimated to be close to 100%, but it starts decreasing above 600 K.

There are now also newer techniques for measuring quantum efficiencies (Bimberg *et al.*, 1991), which are related to the photothermal detection techniques that were developed to measure absorption spectra of nonluminescent or weakly luminescent samples: the CAS (calorimetric absorption spectroscopy), CTS (calorimetric transmission spectroscopy) and PAS (photoacoustic spectroscopy).

A schematic setup for CAS and CTS is shown in Fig. 18 (Bimberg *et al.*, 1991). Monochromatic light (from a laser or a lamp and monochromator) is directed toward a sample, which is attached to insulating wires that hold it in vacuum in a ^3He or ^4He cryostat. Absorption of the light is followed by nonradiative losses, which will cause the sample temperature to increase. The temperature of the sample is measured by a thermometer that is placed at a distance (to prevent heating owing to light absorption by the thermometer) but in good thermal contact with the sample via superconducting wires (e.g., NbTi).

For temperatures between 1.3 and 4.2 K an Allen–Bradley carbon resistor serves as a low-temperature thermometer. Below 1.3 K down to 0.4 K (the minimum temperature that can be obtained in a ^3He cryostat) a Ge resistance thermometer is the better choice. The thermometer is attached to a small sapphire chip (high heat conductivity, small heat capacity). The temperature of the sample under illumination is measured as a function of the excitation wavelength: the more light absorbed, the higher the temperature. The power dissipated in the sample can be related to the sample temperature by calibrating the setup with a small heater connected to the sapphire chip.

The sensitivity of CAS is high, especially at low sample temperatures. For weakly luminescent or nonluminescent samples CAS is a good alternative for excitation spectra. In combination with CTS absolute quantum efficiencies can be determined. The total power absorbed by the sample is measured by determining the power of the transmitted excitation beam using a second low-temperature thermometer attached to a blackened copper foil, which absorbs all the transmitted light (see Fig. 18). The power absorbed by the sample is the power of the excitation beam minus the power of the transmitted beam. The total power absorbed by the sample is the sum of the power emitted radiatively and the nonradiative losses. The power owing to nonradiative losses is measured in the CAS experiment. The energy efficiency of the luminescent material can now be calculated by

$$\eta = 1 - \frac{\Delta P_{\mathrm{NR}}}{\Delta P_{\mathrm{NR+R}}} \tag{24}$$

where ΔP_{NR} is the part of the power that is dissipated nonradiatively (CAS experiment) and $\Delta P_{\mathrm{NR+R}}$ is the total absorbed power (CTS experiment). The quantum efficiency can be calculated from the energy efficiency by multiplying by ν_a/ν_e, where ν_a is the frequency of photons absorbed and ν_e the frequency of photons emitted. With this technique the quantum efficiency for the Cr^{3+} emission in ruby was calculated to be 71% at room temperature (Li *et al.*, 1991).

An alternative calorimetric method is compensation photocalorimetry. In this experiment the sample is connected to a thermal bath at a slightly lower temperature. The heat flow from the sample to the thermal bath is compensated for by a small heater that keeps the sample at constant temperature. During illumination, the current through the compensating heater will be lower owing to heating of the sample by nonradiative relaxation processes. Quantum efficiencies can also be determined with this technique. For ruby a value of 73% was found, in good agreement with the value noted previously (Seelert and Strauss, 1987).

Photoacoustic spectroscopy (PAS) measures light absorption in condensed matter via excitation of sound waves in an adjacent gas. The sample is sealed in a small chamber with a gas. Nonradiative relaxation after light absorption produces heat. Under pulsed excitation the heat is produced at the repetition rate of the excitation source. The heat that is produced will cause density fluctuations in the adjacent gas, and sound waves are produced in the gas at the frequency of the modulation of the excitation source. The sound waves are detected with a sensitive microphone. With PAS absorption spectra can be recorded for nonluminescent samples and also quantum efficiencies can be determined. The sensitivity of PAS is lower than for CAS.

A third technique that uses heat production owing to nonradiative processes is photothermal deflection spectroscopy (PDS) (Strauss, 1990). In this technique two laser beams are used. The pump beam is scanned over the wavelength region of interest. If light is absorbed the refractive index in the area of the pump beam will change. There are two processes that contribute to the change in refractive index:

1. A thermal contribution, where the heat released owing to nonradiative relaxation will result in (thermal) expansion.
2. An expansion of the material owing to the expansion of ions in the excited state with respect to the ground state.

The change in refractive index is probed by a second laser beam. The probe beam travels at an angle through the zone excited by the pump beam. If the refractive index changes owing to light absorption, the probe beam is deflected. To detect absorption of light, the position of the probe beam after deflection is measured as function of excitation wavelength with a position-sensitive detector. The photothermal deflection technique has been used not only to measure absorption spectra, but also to obtain an estimated value of the expansion of ions in the excited state with respect to the ground state (Strauss, 1990). For the $^4T_2$ excited state of Cr^{3+} in ruby this value was estimated to be 0.12 Å.

7. POLARIZATION AND PERTURBATION SPECTROSCOPY

7.1. Polarization Spectroscopy

In the past half-century a number of techniques have been developed to obtain extra information on the symmetry of the ground and excited states and on the site symmetry of the luminescing species. Pioneering work in this field has been done by the teams of Feofilov (1961) and Kaplyanskii (1964). In polarization spectroscopy, the polarization of the excitation light is varied and/or the polarization of the emitted light is analyzed. Perturbation spectroscopy is based on applying an external perturbation (e.g., stress or an electric or magnetic field) along a well-defined crystallographic axis and measuring changes in the luminescence spectra. Understanding these experiments requires a basic knowledge of group theory. A good introduction to the subject can be found in Cotton (1963).

The polarization of light can be classified as linear, elliptical, or circular. For linearly polarized light the electric field vector always points in the same direction when the light travels through a medium. Circularly polarized light (a special case of elliptically polarized light) is characterized by an electric field vector that rotates clockwise (right-handed circularly polarized light) or counterclockwise (left-handed circularly polarized light). First, techniques to make polarized light will be discussed and then an illustrative example of polarization spectroscopy will be described.

Linear polarizers change unpolarized light into linearly polarized light. The simplest type of linear polarizer uses a dichroic material that selectively absorbs one polarization direction and transmits the other. Dichroic polarizers are made from long polymer molecules that have been embedded in a plastic film. Stretching the film aligns the molecules. Then the film is dyed with a pigment and the dye molecules get attached to the aligned polymer molecules. As a result, absorption is strong in one plane and weak in the other. The most common of these polarizers is the well-known Polaroid. These polarizers are inexpensive, but the damage threshold is low and the K value (a measure of the quality, defined as $^{10}\log(T_0/T_{90})$, where T_0 is the transmittance for parallel polarizers and T_{90} is the transmittance for crossed polarizers) is not very high ($\sim 10^4$).

High-quality linear polarizers are made using birefringent materials such as calcite, quartz, or magnesium fluoride. A birefringent crystal will split an incoming monochromatic light beam into two beams with orthogonal polarization. One beam is called the extraordinary beam because it violates the law of reflection and the refractive index is a function of the direction. The other beam is called the ordinary beam. There are various designs for polarizers based on birefringent materials (Moore *et al.*, 1989). One popular design is the Glan–Taylor model. The quality of this type of polarizer is high (typically $K = 10^6$). Circularly polarized light can be made out of linearly polarized light with a $\frac{1}{4}\lambda$ plate, usually a thin quartz plate in which a phase change of $(2n + 1)\pi/2$ is produced between the ordinary and extraordinary wave, resulting in circularly polarized light.

A $\frac{1}{2}\lambda$ plate is similar, but now a phase difference of $(2n + 1)\pi$ is introduced, resulting in a rotation of the polarization direction. In all experiments in which a high degree of polarization is important care has to be taken to prevent depolarization. Usually the polarizing optics are the last optical components before the sample, apart from cryostat windows. Depolarization or polarization rotation at cryostat windows can be a serious problem, but it can be reduced by tuning the stress in the windows with the screws in the window mounts to minimize depolarization.

As an example of polarization spectroscopy a uniaxial system will be considered. The z-axis is the unique axis and the x- and y-axis are equivalent. Three types of linearly polarized light can now be distinguished, σ, π, and α (Henderson and Imbusch, 1989*c*), which are illustrated in Fig. 19. For π-polarization the light travels in a direction perpendicular to the z-axis and the electric field vector is in the z-direction. The magnetic field vector is in the x, y-plane, perpendicular to z. In the

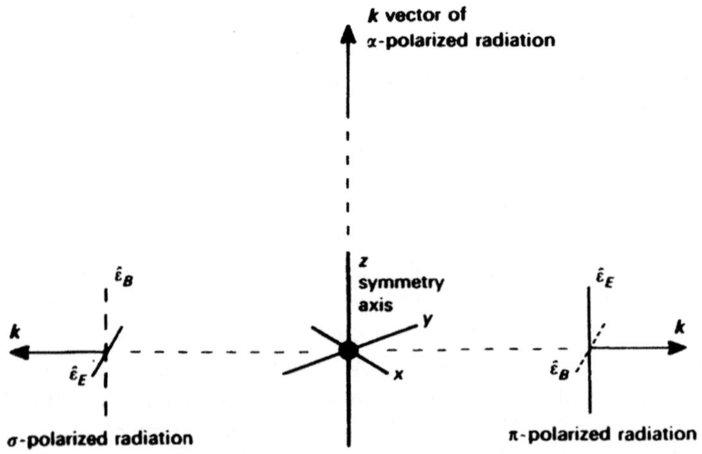

Figure 19. Illustration of the three types of linearly polarized light (σ, α, and π) in the presence of a symmetry axis (Henderson and Imbusch, 1989*a*; by permission of Oxford University Press).

case of α-polarization the light travels in the z-direction and the electric and magnetic field vectors are in the x, y-plane. In the case of σ-polarization the light travels perpendicularly to the z-axis and the magnetic field vector is in the z-direction while the electric field vector is in the x, y-plane. The selection rules for transitions between levels of specific symmetries for α-, σ-, and π-polarized light can be derived with group theory and have been tabulated for magnetic dipole (MD) and electric dipole (ED) transitions.

As an example, the absorption spectra for the ${}^7F_0 \rightarrow {}^5D_1$ transition on Eu^{3+} in $LiYF_4$ crystal shows that the Eu^{3+} ion is in S_4 site symmetry (Görller–Walrand *et al.*, 1993). The experiment has been performed in a commercial spectrophotometer in which polarization was obtained with a Glan–Thompson polarizer. The spectra were recorded for single crystals that were oriented with Laue photography with two of the faces perpendicular to the c-axis (for α-spectra) or parallel to the c-axis (for σ- and π-spectra).

In Table 5 the selection rules for MD and ED transitions for S_4 site symmetry are shown. From group theory it follows that the ${}^7F_0$ ground state transforms as Γ_1 and the ${}^5D_1$ excited state is split into Γ_1 and $\Gamma_{3,4}$. The fact that two transitions are observed (one in π, α-polarization and the other in σ-polarization) shows that the transition has a MD character. For an ED transition only the $\Gamma_1 \rightarrow \Gamma_{3,4}$ transition is expected. The $\Gamma_1 \rightarrow \Gamma_1$ is forbidden as an ED transition for all types of polarization, and the $\Gamma_1 \rightarrow \Gamma_{3,4}$ transition should show up for σ- and α-polarization if an ED transition was involved.

The spectra show that the $\Gamma_1 \rightarrow \Gamma_{3,4}$ transition is observed in π- and α-polarization, as is expected for a MD transition. In addition, the symmetry of the two crystal field components of the ${}^5D_1$ level can be identified. The higher-energy crystal field component at 19,040 cm^{-1} has Γ_1 symmetry and the lower-energy component at 19,018 cm^{-1} has $\Gamma_{3,4}$ symmetry. This example illustrates how extra information on the nature of transitions and the symmetry of excited states can be obtained if the site symmetry of the luminescing species is known. If the site symmetry is not known, a careful examination of the polarized absorption and emission spectra can often provide information on the site symmetry of the luminescing species.

A second beautiful example was illustrated by Berg and McClure (1989). In $NaF:Cu^+$ the Cu^+ ion is in octahedral site symmetry. Normal (one-photon)

Table 5. Selection Rules for Induced Electric Dipole and Magnetic Dipole Transitions in S_4 Symmetry

	Electric dipole			Magnetic dipole		
S_4	Γ_1	Γ_2	$\Gamma_{3,4}$	Γ_1	Γ_2	$\Gamma_{3,4}$
Γ_1	—	π	σ, α	σ	—	π, α
Γ_2	π	—	σ, α	—	σ	π, α
$\Gamma_{3,4}$	σ, α	σ, α	π	π, α	π, α	σ

polarized absorption spectra cannot provide extra information as all three axes are equivalent in octahedral symmetry. With two-photon absorption (or excitation), experiments in polarization spectroscopy can be done. In the two-photon excitation spectra for $NaF:Cu^+$ the polarization directions of the two laser beams are different for the two spectra. If both beams have ⟨001⟩ polarization, transitions from the A_{1g} ground state to E_g excited states are symmetry-allowed. However, if the two laser beams have different polarization directions (⟨010⟩ and ⟨001⟩) the selection rules change and only the $A_{1g} \rightarrow T_{2g}$ transitions are observed. With this technique the T_{2g} and E_g excited levels can be observed and the symmetries can be identified.

7.2. Perturbation Spectroscopy

Apart from polarization spectroscopy, there are more tricks available to help the spectroscopists solve the problem of symmetry of excited states and the environment of luminescing species. The changes in the (polarized) luminescence spectra that result from applying an external perturbation along a well-defined crystallographic axis provide useful information. Three types of perturbations are commonly used: stress, magnetic field, and electric field.

7.2.1. Stress Perturbation

If stress is used as a perturbation one speaks of piezospectroscopy. Stress can be applied to a crystal by hydraulic pressing or a system of weights and levers. Many interesting experiments were performed in the late 1960s and early 1970s by the Kaplyanskii (1964) and Runciman (1965) groups. Nowadays, the technique is not often used. An area in which stress experiments have been used is resolving the symmetry of anisotropic centers in cubic crystals. In the crystal without stress the differently oriented centers are equivalent and one absorption line is observed. The degeneracy is removed by applying stress. The stress-induced splitting patterns, the number of components, and the polarization dependence for both MD and ED transitions have been calculated and are tabulated for different types of symmetry

Table 6. Stress-Induced Splitting (Number of Components)

Centers	Stress ⟨100⟩	Stress ⟨111⟩	Stress ⟨110⟩
Tetragonal	2	1	2
Trigonal	1	2	2
Rhombic I	2	2	3
Rhombic II	3	1	3
Monoclinic I	2	3	4
Monoclinic II	3	2	4
Triclinic	3	4	6

centers. As an example, Table 6 shows the splitting predicted for seven types of centers if stress is applied along the ⟨100⟩, ⟨111⟩, or ⟨110⟩ axis (Kaplyanskii, 1964). The symmetry of an anisotropic center can now be determined uniquely.

The splitting for a tetragonal center can be easily understood. Three types of tetragonal centers are possible. The anisotropy axis is directed along the ⟨100⟩, ⟨010⟩, or ⟨001⟩ direction. All three centers are equivalent if no stress is applied. Under stress along the ⟨100⟩ direction, the ⟨100⟩ tetragonal center has the stress parallel to the anisotropy axis, while the stress is perpendicular to the anisotropy axis for the ⟨010⟩ and ⟨001⟩ centers. As a result, there is splitting into two lines, one corresponding to the ⟨100⟩ centers and the other to the ⟨010⟩ and ⟨001⟩ centers. If the stress is applied along the ⟨111⟩ direction, the stress field is at the same angle for all three tetragonal centers and no splitting occurs. Only a stress-induced shift of the line can be observed. Further information can be obtained from the polarization behavior of the different lines. Examples of piezospectroscopy experiments to solve the structure of various defects in alkali halides can be found in Hayes and Stoneham (1985).

The pressure experiments described above use uniaxial pressure. It is interesting to note here that hydrostatic pressure (isotropic pressure) does not cause splitting lines, but it can cause a shift in the resonance frequency. The pressure-induced shift of the *R*-lines in ruby is a famous example, and this shift is used to calibrate the pressure in diamond anvil cells. Small chips of ruby are included in the high-pressure volume in a diamond anvil cell. Since diamonds are transparent in the visible range, the ruby chips can be excited through the diamond anvils with a laser and the *R*-line emission can be collected and fed into a monochromator to determine the position of the *R*-lines from which the pressure in the cell can be derived. Because of the transparency of diamonds over a wide spectral range and the high pressure that can be obtained in diamond anvil cells (up to tens of GPa), diamond anvil cells have become increasingly popular for spectroscopic studies of luminescent materials under high pressures (Jayaraman, 1983).

7.2.2. Magnetic Field Perturbation

The second type of perturbation spectroscopy uses an external magnetic field. The splitting of energy levels by a magnetic field has been known since the end of the last century, and is commonly referred to as the Zeeman effect. Fields up to 2.5 T can be achieved with conventional electromagnets. For higher fields (up to 10 T), superconducting magnets are used. When an ion or molecule is placed in a magnetic field the (spin) degeneracy of an excited state and the ground state can be raised owing to interaction between the magnetic field and the magnetic moments of the electrons. The interaction is given by

$$H_Z = \mu_B B(L + 2S) \tag{25}$$

where H_Z is the contribution to the total Hamiltonian from the Zeeman effect, μ_B is the Bohr magneton ($\mu_B = 4.666 \times 10^5\ \mathrm{cm}^{-1}$/gauss), B is the applied magnetic field, and L and S are the orbital angular and spin operators, respectively.

Just as for the stress experiments described above, the magnetic field is applied along one of the crystallographic axes. This provides a unique axis in an isotropic (cubic) system. Polarized absorption and emission spectra (α, π, and σ) in the presence of a unique axis provide information on the nature of the transition (MD or ED). Also, in analogy with stress experiments, the site symmetry for differently oriented centers can be resolved by applying a magnetic field along the $\langle 100 \rangle$, $\langle 110 \rangle$, and $\langle 111 \rangle$ axes. These techniques were developed at the same time as the stress experiments. A nice illustration from this period is the work on KCl-Sm^{2+} by the Fong group (Fong and Wong, 1967).

Zeeman splitting is a technique that is still used frequently. As an example, the spectroscopy of trivalent rare-earth ions in MF_2 (M = Ca^{2+}, Sr^{2+}, Ba^{2+}) will be discussed. The trivalent rare-earth ions replace divalent alkaline earth ions. The extra positive charge can be compensated for by an interstitial F^- ion, which can be distant (resulting in a cubic site) or local (resulting in C_{3v} or C_{4v} site symmetry, depending on the position of the interstitial F^- ion with respect to the rare-earth ion).

Zeeman experiments have been performed to identify sites with different types of charge compensation. For rare-earth ions with an odd number of electrons the Kramers degeneracy of Stark levels is lifted in a magnetic field, and splitting into $-M_J$ and $+M_J$ levels occurs. The splitting factors for the magnetic field parallel ($s_\parallel$) and perpendicular ($s_\perp$) to the symmetry axis of the system (introduced by the charge compensating F^- ion) are

$$s_\perp = g(J + \tfrac{1}{2})\mu_B B \qquad s_\parallel = 2gJ_Z\mu_B B \tag{26}$$

Here g is the Landé factor:

$$g = 1 + \frac{J(J+1) + S(S+1) - L(L+1)}{2J(J+1)} \tag{27}$$

The effective splitting factor is given by $s_{\text{eff}} = s_\parallel^2 \cos^2\theta + s_\perp^2 \sin^2\theta$, where θ is the angle between the magnetic field and the symmetry axis. The symmetry of a site can be determined by studying the Zeeman splitting patterns for the magnetic field along different crystallographic axes (Tissue, 1989).

Other applications of Zeeman splitting in luminescence spectroscopy make use of the feature that the energy difference between levels can be continuously tuned by varying the magnetic field strength. This has been used, e.g., to study the dependence of the relaxation rate on the energy difference for relaxation between the two Zeeman levels ($\bar{E}$ and $2\bar{A}$) of the ^2E level of Cr^{3+} in ruby (Fokker, 1991). In another study the energy transfer rate between inequivalent Er^{3+} ions was measured as a function of the energy mismatch. The energy mismatch could be varied continuously by changing the strength of the applied magnetic field (Tissue, 1989).

7.2.3. Electric Field Perturbation

A third type of perturbation spectroscopy uses an electric field as a perturbation. The splitting of energy levels owing to an electric field is called the Stark effect. Just as for the previously discussed perturbation techniques, the Stark effect has been an important technique for studying the symmetry of excited states and site symmetries of ions and for determining the nature of transitions. Several types of Stark effects are distinguished. If the electric field removes the degeneracy of an electronic level in a single ion one speaks of the Stark effect, and if it removes the degeneracy of two equivalent ions one speaks of the pseudo-Stark effect. In the case of a linear Stark effect, levels shift owing to a first-order perturbation and the shift is proportional to the electric field. In the case of a quadratic Stark effect owing to a second-order perturbation, the shift is proportional to E^2. It is not easy to observe the Stark effect. Very high electric fields are required to produce shifts that are comparable to the inhomogeneous linewidths. The direct observation of the Stark effect is therefore only possible in systems with a very small inhomogeneous broadening and then only when very high electric fields are used.

Hole-burning techniques have been used to measure Stark effects that cannot be observed otherwise. A nice example is the work by Shelby and Macfarlane (1978), who observed a splitting into four of a hole burned in the $^3H_4 \rightarrow {}^1D_2$ absorption line of Pr^{3+} in LaF_3 when an electric field was applied parallel to the C_2-axis. The Stark splitting was only 0.15 MHz/V cm^{-1}. In the time regime an even higher resolution can be obtained. The Stark splitting results in a modulation of the photon echo intensity. Many more complicated techniques have been developed. For more information the reader is referred to Kaplyanskii and Macfarlane (1987).

8. NEW DEVELOPMENTS

In the previous sections the historical background and the techniques that are presently used in luminescence spectroscopy were discussed. This section will discuss some recent developments in experimental techniques and the expected influence on the future. The trends in luminescence spectroscopy can be characterized by "faster" and "smaller." Great progress has been made in the production of shorter laser pulses (femtosecond) and fast detection schemes (pump-probe). The trend toward smaller structures is exemplified by single-molecule detection and near-field imaging. In the following sections femtosecond spectroscopy, single-molecule spectroscopy and near-field imaging will be discussed in more detail.

8.1. Ultrafast Spectroscopy

The generation of femtosecond laser pulses has progressed from a job for highly qualified laser engineers ten years ago to rather easily operated commercially available laser systems today (Krausz and Winter, 1994). This development is largely due to vibronic solid state lasers, discussed in Section 3.1.2. The shortest pulses that can be produced are about 6 fs, and this was achieved 10 years ago (Shank, 1988).

The major breakthrough in femtosecond spectroscopy is very recent and is due to the availability of commercial femtosecond laser systems such as the Ti-sapphire laser. The field is still developing rapidly. The main effort is not directed so much at making the pulses shorter, but more at extending the wavelength range in which femtosecond pulses can be produced. With the combination of a Ti-sapphire laser and an OPO, femtosecond pulses can be produced in the infrared range. Frequency-doubling of the femtosecond output of a Ti-sapphire laser provides femtosecond pulses in the visible and near-ultraviolet. Great progress is also being made in obtaining higher peak powers using femtosecond amplifier rods of Ti-sapphire (high damage threshhold).

In this section the use of femtosecond laser pulses in luminescence spectroscopy will be discussed using some illustrative examples. The detection is complicated for femtosecond pulses. Conventional detectors (PMTs and photodiodes) are too slow to respond on a femtosecond timescale, and this presents a problem even before the start of the experiment. To measure the width of the pulses produced by the laser one needs a special technique, such as autocorrelation.

In the case of second harmonic autocorrelation a part of the ultrafast pulse (usually 50%) is split off and mixed with the remainder of the original pulse with a variable time delay. Since the second harmonic generation (SHG) response is quadratic with the intensity of the laser pulses, that signal will strongly depend on the overlap of the two laser pulses in time. The larger the overlap, the higher the SHG signal will be. By measuring it as a function of the delay time information on the pulse width is obtained. Another type of autocorrelator is the interferometric model. In this case the interference maxima and minima for the original pulse and the delayed pulse are measured as functions of the delay time.

In ultrafast spectroscopy experiments pump-probe techniques are generally used to obtain a femtosecond time resolution. A first ultrashort light pulse excites the system and a second delayed pulse probes the changes as a function of time. A special type of pump-probe experiment is the photon echo technique discussed in Section 5. By using femtosecond pulses and femtosecond delay times, dephasing times in the femtosecond regime can be measured. Femtosecond dephasing processes have been studied by photon echo in, e.g., semiconductors and conjugated polymers.

Another type of pump-probe experiment makes use of the broad spectral range of the Fourier transform limited femtosecond pulses. The shorter the pulsewidth, the larger the spectral width will be, according to the uncertainty principle ($\Delta\nu\Delta\tau \approx 1$). The time evolution of the absorption spectrum after an initial pump pulse can be measured by recording the transmitted light intensity as a function of wavelength for various delay times after the pump pulse. An example is shown in Fig. 20. An 80-fs pump pulse ($\Delta\lambda \approx 20$ nm) is resonant with the exciton absorption band of a CdZnTe/ZnTe quantum-well system (Becker *et al.*, 1992). A second (probe) pulse of 14 fs ($\Delta\lambda \approx 100$ nm) is used to measure the transmission spectrum at different times after the pump pulse.

The transmission spectra are resolved with a multichannel analyzer system. The results are shown in Fig. 20b. The pump pulse causes a difference in the transmission spectrum owing to bleaching of the exciton band. An initial rapid decay of the bleaching (time constant 11 fs) is ascribed to fast exciton ionization. This type of pump-probe technique, where the absorption of a spectrally broadened ultrashort

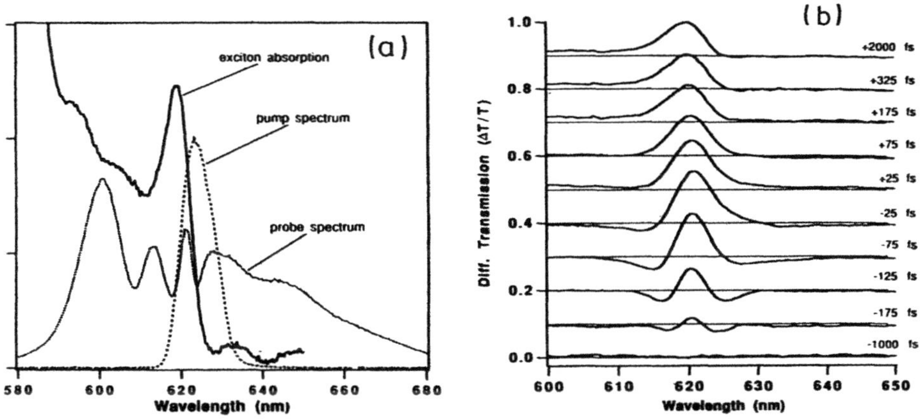

Figure 20. (a) Room-temperature absorption spectrum of the CdZnTe/ZnTe quantum well (solid line) and the spectra of the 80-fs pump pulse and the 14-fs probe pulse. (b) Spectrally resolved pump-probe response of the CdZnTe/ZnTe quantum well in the vicinity of the room temperature excitonic resonance, for various delays between pump and probe, plotted as $\Delta T/T$ (Becker *et al.*, 1992; with permission from the American Physical Society).

probe pulse is measured as a function of time, is frequently used to study femtosecond processes in semiconductors and charge transfer processes in photosynthesis model systems. Electron transfer rates can be extracted from the time evolution of the absorption spectra after an initial pump pulse.

8.2. Single-Molecule Spectroscopy

In Section 5 various techniques for the measurement of homogeneous linewidths in an inhomogeneously broadened line were discussed. The most elegant way to observe a homogeneous linewidth is, of course, to measure the linewidth for a single luminescing species. The first experiments that showed that it is possible to detect the luminescence of one pentacene molecule in a *p*-terphenyl single crystal were reported just over 5 years ago (Moerner and Kador, 1989). This observation opened up an exciting new field in luminescence spectroscopy. By studying the luminescence of a single molecule, phenomena such as spectral diffusion, spectral hole-burning, and interaction with the local environment can be probed directly for one molecule without having to extract this information from the average behavior of an ensemble of molecules (Moerner and Basché, 1993).

The observation of the luminescence for a single molecule is not easy. Naturally, the collection efficiency and sensitivity of the light-detection system have to be high in order to measure the weak signal. A second requirement is that a small volume of the matrix (glass or crystal) with a very low dopant concentration be investigated, so that it will contain a sufficiently small number of molecules that individual molecules can be observed. Also, a low volume reduces background scattering, which contributes to undesired noise.

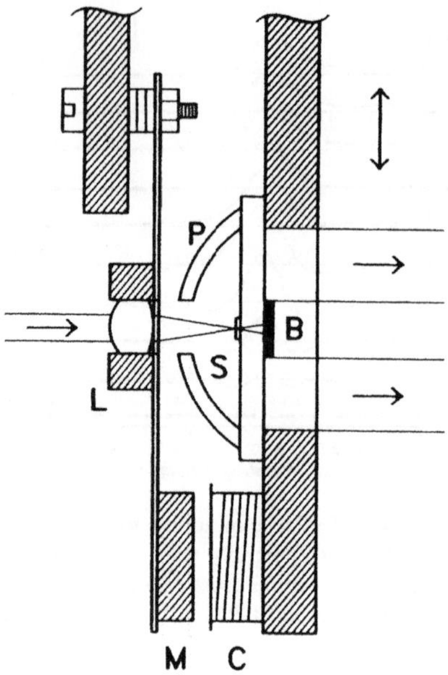

Figure 21. Experimental setup for single-molecule fluorescence excitation spectroscopy in a solid at liquid helium temperatures. The excitation laser beam is focused by a lens L in the cryostat on the sample S and then blocked by black tape B. The fluorescence is collected by a parabolic mirror P. The lens can be positioned with a magnet M and coil C (Moerner and Basché, 1993; with permission from the American Physical Society).

The third condition is that a molecule has to be excited many times to obtain a detectable number of photons per second. As a result, single-molecule spectroscopy has only been successful for a luminescing species in which allowed transitions are involved. In the case of a forbidden transition or efficient relaxation to a metastable level (e.g., the triplet level in an organic molecule) the number of excitation/emission cycles per second is too small to detect the luminescence from a single molecule.

The ideal system so far has been pentacene in *p*-terphenyl single crystals. There are four possible sites for pentacene (O_1–O_4) with singlet–singlet transitions around 17,000 cm^{-1}. For two of the sites the intersystem crossing efficiency to the metastable triplet level is low (0.2%) and single-molecule spectroscopy is possible.

The experimental setup used is depicted in Fig. 21. A thin layer (e.g., 10 μm) of *p*-terphenyl doped with 10^{-9} to 10^{-6} moles/liter of pentacene is excited with a laser beam that is focused by a lens in the cryostat to the diffraction-limited beam waist diameter of about 5 μm. The excitation beam is blocked directly behind the sample. The emitted light is collected by a small parabolic mirror around the sample and focused on a PMT. The Stokes-shifted luminescence is separated from scattered laser light by filters and detected by the PMT. The emission intensity is measured as a function of the excitation wavelength. A narrow-band laser is used as an excitation source, e.g., a single-mode ring dye laser with a resolution of 1 MHz.

Moerner and Basché (1993) have observed fluorescence excitation spectra of single pentacene molecules in a *p*-terphenyl crystal at 1.8 K in which the linewidth of about 8 MHz corresponds to the homogeneous linewidth. Many elegant experiments have proved that the line corresponds to the luminescence of one pentacene molecule. One of these experiments is the observation of antibunching on a short timescale (owing to the lifetime of the excited state) and bunching on a long timescale. The observation of bunches of photons followed by dark periods is explained by occasional intersystem crossing to the triplet excited states. Once the molecule is trapped in the triplet excited state it takes a while before it returns to the ground state and is again available for an excitation/emission cycle.

The observation of spectral diffusion for a single molecule has stimulated theoretical work in this field. Spectral diffusion is the change in resonance frequency owing to changes in the local environment of a luminescing species. Even at 1.5 K spectral diffusion is observed for pentacene in *p*-terphenyl. By studying spectral diffusion for a single molecule, instead of extracting information from an ensemble average, more insight is obtained. Indeed, different types of spectral diffusion were observed for different molecules. Also other previously discussed phenomena such as the Stark effect, polarization studies, pressure effects, hole-burning, and lifetime measurements have been demonstrated for single molecules. Most of the experiments have been performed on pentacene in *p*-terphenyl. In the future single-molecule spectroscopy can be expected to provide insight in interactions between a single luminescing species and the host for many glassy and crystalline systems. More information on single-molecule spectroscopy can be found in Moerner and Basché (1993).

8.3. Near-Field Optical Spectroscopy

In recent years various new optical detection techniques have been developed with a subwavelength resolution (Betzig and Trautman, 1992). All these techniques come from near-field optics. With conventional far-field optics (optical microscopes) the resolution is limited to $\lambda/2$, which imposes 200-nm limit on the resolution using visible light. In the case of near-field optical spectroscopy or microscopy a submicron-sized aperture is scanned over the surface to be investigated from a very short distance. The resolution that can be reached depends critically on the size of the aperture. Resolutions down to 10 nm have been reported. To obtain the subwavelength resolution a tapered optical fiber tip is used. The tip, called the probe, is placed very close to the surface, at a distance less than the wavelength of light.

At this distance the light is still highly collimated with a dimension determined by the radius of the probe and not the wavelength of the light. This region is called the near-field. The probe can be used in two ways: it can pick up light from the surface or transfer light to the surface. In both cases the transfer of light occurs by tunneling through the evanescent light field (exponential distance dependence) or through energy transfer.

An experimental setup is shown in Fig. 22. The probe is the most crucial part in the setup. Two types of probes can be used: active and passive. Passive (or semiactive) probes consist of a coated tapered optical fiber tip. Light is coupled into the fiber and the evanescent light field can be used for near-field imaging at the fiber tip (etched to a nanometer-size radius). In the case of active probes an active light source of submicron dimensions (e.g., a nanometer-size luminescent

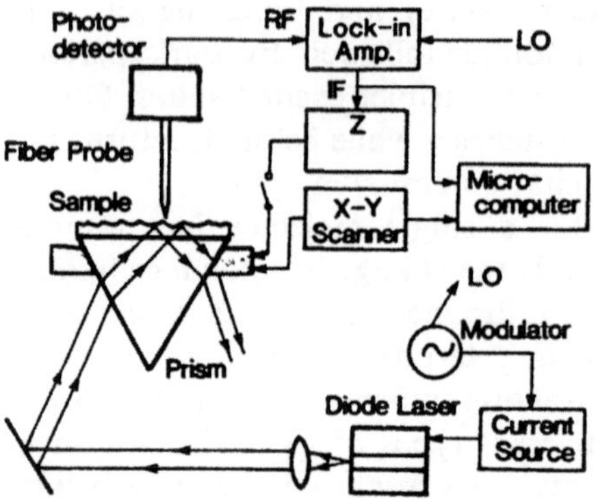

Figure 22. Setup for near-field optical microscopy. The evanescent light field from the sample is picked up by the probe very close to the sample and measured while scanning the probe over the sample (Jiang *et al.*, 1991; with permission from the *Japanese Journal of Applied Physics*).

crystal) is attached to the optical fiber tip. The light coupled into the fiber excites the luminescent material on the top of the tip. The luminescent material then emits light that can be used for near-field imaging. The probe scans the surface of the sample and the light from the sample is collected and analyzed with conventional techniques.

The amount of light that reaches the detector is influenced by structures present on the sample surface. For example, a contrast is obtained for Al on a glass substrate. The opacity of Al provides a good contrast and very small Al structures can be resolved. Since the evanescent light field decreases exponentially with the distance between the probe and the sample, it is essential that this distance remain constant during a scan in order to obtain the best possible lateral resolution (down to $\lambda/40$). If the surface that is scanned is rough, a surface force feedback mechanism comparable to atomic force microscopy can be used to keep the distance between the probe and the surface constant. The exponential dependence of the evanescent field can also be used to obtain a high vertical resolution, down to $\lambda/100$ (Paesler *et al.*, 1990).

There are many examples in the literature of the various applications of near-field scanning optical microscopy. In addition, spectroscopy can be done with very high spatial resolution. The light from the probe can be used to excite luminescent ions or molecules in the sample. The light that is emitted can be dispersed in a spectrometer and analyzed. With this technique variations in the luminescence spectra can be studied with very high spatial resolution. With this technique, called near-field scanning optical spectroscopy, spectra of nanostructures can be selectively recorded. It is thus a promising technique for detecting small variations in luminescence that are due to nanometer-size inhomogeneities (Kopelman *et al.*, 1994).

The province of near-field optical techniques is young and it can be expected that it will be exploited in the future in many areas where subwavelength light sources can be useful, e.g., for measuring local photoconductivity, detecting dopant distributions in semiconductors, studying individual quantum-confined semiconductor particles, and high-resolution imaging of living cells (*in vivo*).

ACKNOWLEDGMENTS

The author is very grateful to Mrs. C. Derks for typing the manuscript and to Mr. R. T. Wegh for proofreading it.

REFERENCES

Agladze, N. I., Popova, M. N., Koreiba, M. A., Malkin, B. Z., and Pekurovskii, V. R. (1993), *Zh.Eksp.Theor.Fiz.* **104**, 4171.

Allen, L., and Eberly, J. H. (1975), *Optical Resonance and Two-Level Atoms* (Dover, New York).

Baly, E. C. C. (1912), *Spectroscopy* (Longmans, Green, London) Ch. 1 and 2.

Baumgartner, R. A., and Byer, R. L. (1979), *J. Quantum Elect.* **15**, 432.

Becker, P. C., Lee, D., Johnson, A. M., Prosser, A. G., Feldman, R. D., Austin, R. F., and Behringer, R. E. (1992), *Phys. Rev. Lett.* **68**, 1876.

Berg, J. M., and McClure, D. S. (1989), *J. Chem. Phys.* **90**, 3915.

Betzig, E., and Trautman, J. K. (1992), *Science* **257**, 189.

Bimberg, D., Wolf, T., and Böhrer, J. (1991), in: *Advances in Nonradiative Processes in Solids* (B. Di Bartolo, Ed.) Plenum, New York, p. 577.

Blasse, G., and Grabmaier, B. C. (1991), *Luminescent Materials* (Springer-Verlag, Berlin), Ch. 4.

Bril, A., and de Jager-Veenis, A. W. (1976), *J. Electrochem. Soc.* **123**, 396.

Cotton, F. A. (1963), *Chemical Applications of Group Theory* (Wiley, New York).

de Mello Donegá, C. M., Meijerink, A., and Blasse, G. (1994), *J. Lumines.* **62**, 189.

Di Bartolo, B. (Ed.) (1983), *Energy Transfer Processes in Condensed Matter* (Plenum, New York).

Di Bartolo, B. (Ed.) (1991), *Advances in Nonradiative Processes in Solids* (Plenum, New York).

Feofilov, P. (1961). *The Physical Basis of Polarized Emission* (Consultants Bureau, New York).

Fokker, P. A. (1991), Ph.D. Thesis, University of Utrecht, The Netherlands.

Fong, F. K., and Wong, E. Y. (1967), *Phys. Rev.* **162**, 348.

Gallagher, P. K. (1994), *Laser Focus World* **30**, 57.

Görller-Walrand, C., Binnemans, K., and Fluyt, L. (1993), *J. Phys: Cond. Mat.* **5**, 8359.

Groenink, J. A., and Blasse, G. (1980), *J. Sol. St. Chem.* **32**, 9.

Hamamatsu Catalog (1994), *Photomultiplier Tubes* (Hamamatsu Photonics, Japan).

Hayes, W., and Stoneham, A. M. (1985). in: *Crystals with Fluorite Structure* (W. Hayes, Ed.) (Clarendon, Oxford).

Henderson, B., and Imbusch, G. F. (1989*a*), *Optical Spectroscopy of Inorganic Solids* (Clarendon, Oxford), Ch. 11.

Henderson, B., and Imbusch, G. F. (1989*b*), *Optical Spectroscopy of Inorganic Solids* (Clarendon, Oxford), Ch. 4, 5.

Henderson, B. and Imbusch, G. F. (1989*c*), *Optical Spectroscopy of Inorganic Solids* (Clarendon, Oxford), p. 175.

Herzberg, G. (1944), *Atomic Spectra and Atomic Structure* (Dover, New York).

Huber, D. L., Hamilton, D. S., and Barnett, B. (1977), *Phys. Rev. B* **16**, 4642.

Jayaraman, A. (1983), *Rev. Mod. Phys.* **55**, 65.

Jiang, S., Tomita, H., Ohsawa, H., and Ohtsu, M. (1991), *Jpn. J. Appl. Phys.* **30**, 2107.

Kafka, J. D., Watts, M. L., Roach, D. J., Keirstead, M. S., Schaaf, H. W., and Bear, T. (1989), in: *Ultrafast Phenomena VII* (C. B. Harris, E. P. Ippen, G. A. Mourou and A. Z. Zewail, Eds.) (Springer-Verlag, Berlin), p. 66.

Kafka, J. D., Watts, M. L., and Pieterse, J. W. J. (1992), *J. Quant. Electr.* **28**, 2151.

Kaplyanskii, A. A. (1964), *Optics Spectrosc.*, USSR **16**, 329, 357.

Kaplyanskii, A. A., and Macfarlane, R. M. (Eds.) (1987), *Spectroscopy of Solids Containing Rare Earth Ions* (North-Holland, Amsterdam).

Kopelman, R., Tan, W., and Birbaum, D. (1994), *J. Lumines.* **54**, 380.

Krausz, F., and Winter, E. (Eds.) (1994), *Appl. Phys. B* **58** (Special Issue on Ultrashort Pulse Generation).
Kruithof, A. A. (1967), Modern Light Sources, in: *Advanced Optical Techniques* (A. C. S. van Heel, Ed.) (North-Holland, Amsterdam).
Kubelka and Munk (1969), Quoted in: *Reflectance Spectroscopy: Principles, Methods, Applications* (G. Kortüm, Ed.) (Springer-Verlag, Berlin), p. 111.
Kurnit, N. A., Abella, I. D., and Hartman, S. R. (1964), *Phys. Rev. Lett.* **13**, 567.
Li, Y., Duncan, I., and Morrow, T. (1991), *J. Lumines.* **50**, 69.
Lin, G. K., and Cone, R. L. (1990), *Phys. Rev. Lett. B* **41**, 6193.
Macfarlane, R. M., and Shelby, R. M. (1987), in *Spectroscopy of Solids Containing Rare Earth Ions* (A. A. Kaplyanskii and R. M. Macfarlane, Eds.) (North-Holland, Amsterdam), Ch. 3.
Maiman, T. (1960), *Nature* **187**, 493.
Moerner, W. E., and Kador, L. (1989), *Phys. Lett.* **62**, 2535.
Moerner, W. E., and Basche, T. (1993), *Ange. Chem.* **32**, 457.
Moore, J. H., Davis, C. C., and Coplan, M. A. (1989), *Building Scientific Apparatus* (Addison-Wesley, Menlo Park, Calif.).
Muñoz, G. H., de la Cruz, C. L., Muñoz, A. F., and Rubio, J. O. (1988), *J. Mat. Sci. Lett.* **7**, 1310.
Padgett, M. J., and Dunn, M. H. (1994), *Laser Focus World* **30**, 69.
Paesler, M. A., Moyer, P. J., Jahneke, C. J., Johnson, C. E., Reddick, R. C., Warmack, R. J., and Ferrell, T. L. (1990), *Phys. Rev. B* **42**, 6750.
Runciman, W. A. (1965), *Proc. Phys. Soc.* **86**, 625.
Schäfer, F. P. (1977), *Dye Lasers* (Springer-Verlag, Berlin), Vol. **I**.
Seelert, W., and Strauss, E. (1987), *J. Lumines.* **36**, 355.
Shank, C. V. (1988), in: *Ultrasonic Laser Pulses* (W. Kaiser, Ed.) (Springer-Verlag, Berlin).
Shelby, R. M., and Macfarlane, R. M. (1978), *Opt. Comm.* **27**, 399.
Seigman, A. E. (1986), *Lasers* (University Science Books, Mill Valley, Calif.).
Strauss, E. (1990), *Phys. Rev. B* **42**, 1917.
Struck, C. W., and Fonger, W. H. (1991), *Understanding Luminescence Spectra and Efficiency Using W_p and Related Functions* (Springer-Verlag, Berlin).
Tissue, B. M. (1989), Ph.D. Thesis, University of Utrecht, The Netherlands.
van Die, A. (1987), Ph.D. Thesis, University of Utrecht, The Netherlands.
van Schaik, W., Poort, S. H. M., Schlotter, J. J. H., Dorrestein, E., and Blasse, G. (1994), *J. Electrochem. Soc.* **141**, 2201.
Wright, J. C. (1990), Lecture Notes on Experimental Spectroscopy, University of Wisconsin, Madison.
Wright, J. C., and Wirth, M. J. (1980), *Anal. Chem.* **152**, 1087A.
Yen, W. M., and Selzer, P. M. (1981), in: *Laser Spectroscopy of Solids* (W. M. Yen and P. M. Selzer, Eds.) (Springer-Verlag, Berlin), Ch. 5.
Zory, P. S. Jr. (1993), *Quantum Well Lasers* (Academic, San Diego).

3

Photoluminescence

Shigeo Shionoya

1. INTRODUCTION

Luminescence in solids is the phenomenon in which electronic states of solids are excited by some energy from an external source and the excitation energy is released as light. When the energy comes from short-wavelength light, usually ultraviolet light, the phenomenon is called photoluminescence. In this chapter, photoluminescence in solids is classified from the point of view of the nature of the electronic transitions producing photoluminescence, and the spectral characteristics of each kind of photoluminescence are described and discussed.

Photoluminescence is divided into two major types, namely intrinsic and extrinsic luminescence. In the former there are three kinds, band-to-band luminescence, exciton luminescence, and cross-luminescence. The latter is divided into unlocalized and localized types, depending on whether excited electrons and holes of the host lattice participate in luminescence processes or whether the luminescence excitation and emission processes are confined to localized centers. Luminescence materials and phenomena that have been studied by a large number of scientists as early as the late 19th century belong to this category.

In describing the spectral characteristics of luminescence belonging to each category, emphasis is placed on interpreting why such features are brought about as the reflection of the nature of the electronic transition. This chapter is intended to provide a brief textbook on photoluminescence.

As to luminescence materials, primarily inorganic ionic crystals and semiconductors are dealt with. Materials important from the point of view of luminescence applications are pointed out. In the final section, luminescence of low-dimensional systems, which is one of the more recent important and interesting topics in luminescence research, is described briefly.

Shigeo Shionoya • The University of Tokyo, Tokyo, Japan.

Luminescence of Solids, edited by D. R. Vij. Plenum Press, New York, 1998.

2. CLASSIFICATION OF PHOTOLUMINESCENCE IN SOLIDS

Photoluminescence in solids, i.e., in inorganic and organic insulators and semiconductors, is classified in terms of the nature of the electronic transitions producing it. In this section, this classification is explained very briefly. Detailed explanations and discussions of each kind of luminescence will be given in subsequent sections.

2.1. Intrinsic Luminescence

There are three kinds of intrinsic luminescence: (1) band-to-band luminescence, (2) exciton luminescence, and (3) cross-luminescence.

2.1.1. Band-to-Band Luminescence

Luminescence owing to the band-to-band transition, i.e., to the recombination of an electron in the conduction band with a hole in the valence band, can be observed in very pure crystals at relatively high temperatures. This has been observed in Si, Ge, and some IIIb–Vb compounds such as GaAs. At low temperatures, this luminescence is transformed into exciton luminescence, which will be explained in the next subsection. Why this luminescence is observed only in very pure materials will be discussed in Section 3.3. From the point of view of luminescence applications, it should be noted that light emission from very bright type light-emitting diodes (not popular types) and semiconductor lasers is due to the band-to-band transition.

2.1.2. Exciton Luminescence

An exciton is a composite particle of an excited electron and a hole interacting with one another. It moves in a crystal conveying energy and produces luminescence owing to the recombination of the electron and the hole. There are two kinds of excitons: the Wannier (or Wannier–Mott) exciton and the Frenkel exciton.

The Wannier exciton model expresses an exciton as composed of an electron in the conduction band and a hole in the valence band bound together by the Coulomb interaction. Thus, a Wannier exciton is analogous to a hydrogen atom. This model works well for inorganic semiconductors such as IIIb–Vb and IIb–VIb compounds. The Wannier exciton moves in a crystal but does not contribute to electric conduction. It emits luminescence by the recombination of the electron and hole composing it. The expanse of the wavefunctions of the electron and hole in a Wannier exciton is usually much larger than the lattice constant.

Wannier excitons are stable only at relatively low temperatures, where the binding energies of excitons are higher than the thermal energy. Luminescence of Wannier excitons is observed only at such temperatures. At higher temperatures, where the thermal energy is higher, the excitons are no longer stable and band-to-band luminescence appears instead.

The Frenkel exciton model is used in cases where the expanse of the electron and hole wavefunctions is smaller than the lattice constant. Typical examples are organic molecular crystals such as anthracene and tetracene and inorganic complex salts including transition-metal ions such as vanadates (e.g., YVO_4), tungstates ($CaWO_4$), cyanoplatinates [$BaPt(CN)_4 \cdot 4H_2O$], and uranyl salts ($Cs_2UO_2Cl_4$). In these materials, the excited state of an isolated molecule or a complex ion transfers from molecule to molecule or from complex ion to ion usually owing to the dipole–dipole interaction. It is therefore regarded as the exciton state. Luminescence characteristics are similar to those of isolated molecules or complex ions.

2.1.3. Cross-Luminescence

So-called cross-luminescence is produced by the recombination of an electron in the valence band with a hole created in the outermost core band (van Eijk, 1994). This luminescence was first observed in BaF_2 under x-ray pulse excitation in 1982. It has now been observed in a number of alkali and alkaline-earth halides and double halides.

Cross-luminescence can take place only when the energy difference between the top of the valence band and that of the outermost core band is smaller than the bandgap energy, i.e., $E_{c-v} < E_g$. If, on the other hand, $E_{c-v} > E_g$, an Auger process occurs so that the hole in the outermost core band is filled nonradiatively with an electron in the valence band and another electron in the valence band is raised to the conduction band. Therefore, cross-luminescence is observable only in materials satisfying the condition of $E_{c-v} < E_g$. In the case of BaF_2, $E_g = 10.5\,\text{eV}$ and $E_{c-v} = 7.8\,\text{eV}$, and the peak of the luminescence spectrum is 5.6 eV. A notable characteristic of cross-luminescence is that the decay time is very fast, of the order of a nanosecond or less. In BaF_2 it is 0.8 ns. It is this characteristic that enables crystals showing efficient cross-luminescence to be used as scintillators.

The name cross-luminescence was proposed because in the case of BaF_2, e.g., the valence band is composed mostly of the F^- $2p$ orbit, while the outermost core band mostly of the Ba^{2+} $5p$ orbit. In x-ray spectroscopy, transitions between the bands formed by different ions in ionic crystals are called cross-transitions. The naming followed this terminology in x-ray spectroscopy.

2.2. Extrinsic Luminescence

Luminescence caused by intentionally incorporated impurities, in most cases metallic impurities, or defects is classified as extrinsic luminescence as opposed to intrinsic luminescence. Most of the observed types of luminescence that have practical applications belong to this category. Intentionally incorporated impurities are called activators and materials made luminescent in this way are usually called phosphors.

Extrinsic luminescence in ionic crystals and semiconductors is classified into two types: unlocalized and localized. In the unlocalized type, the electrons and holes of the host lattice, i.e., free electrons in the conduction band and free holes in the valence band, participate in the luminescence process, while in the case of the localized type the luminescence excitation and emission processes are confined in a localized luminescence center.

2.2.1. Unlocalized Type

In semiconductors, most important impurities are donors and acceptors that dominate semiconductive properties, and these act as luminescence activators. There are two types of luminescence transitions, i.e., the transition of a free carrier to a bound carrier, and the transition of a bound electron at a donor to a bound hole at an acceptor (which is also called donor–acceptor pair transition). These kinds of luminescence lines and bands are usually observed in typical compound semiconductors such as IIIb–Vb and IIb–VIb compounds.

Donor–acceptor pair luminescence is caused in two ways: The first is that after the band-to-band excitation an electron is trapped at an ionized donor, and a hole at a compensated acceptor, and they recombine. The second is the direct excitation of an electron in a compensated acceptor to an ionized donor level. Therefore, the donor–acceptor pair luminescence is an intermediate type between localized and unlocalized luminescences. The emission energy of this luminescence depends on the spatial separation between the donor and the acceptor in a pair, as will be explained in Section 6.1.1. This luminescence is very important in terms of practical applications.

In addition to those mentioned above, there are also the luminescence of Wannier excitons bound to impurities and the luminescence owing to isoelectronic traps. These will be explained in Sections 4.2 and 6.2.

2.2.2. Localized Type

Various kinds of metallic impurities intentionally incorporated in ionic crystals and semiconductors often create efficient localized luminescence centers. Some kinds of lattice defects also act as localized luminescence centers. Localized-type centers are classified into: (a) allowed-transition type, and (b) forbidden-transition type, in regard to electric dipole transitions.

The electric dipole transition can take place between energy levels with different parities. The selection rule for this transition in atoms is for the azimuthal quantum number $\Delta l = \pm 1$. When atoms and ions are incorporated in crystals, the forbidden character of the dipole transition is altered by the perturbation of the crystal electric field, so that the forbidden transition becomes allowed to some degree.

Typical examples of localized-type activators and luminescent defects will now be described. Many phosphors that are important from a practical point of view are synthesized by incorporating following activators:

(a) Allowed-transition type

(i) $s \leftrightarrow p$ transition—F center (an electron trapped at an anion vacancy), Tl^0

(ii) $s^2 \leftrightarrow sp$ transition—Tl^+, Sn^{2+}, Pb^{2+}, Sb^{3+}, Bi^{3+}

(iii) $f \leftrightarrow d$ transition—Eu^{2+}, Ce^{3+}

Oscillator strengths for this type of transition are of the order of magnitude of 10^{-1}–10^{-3}.

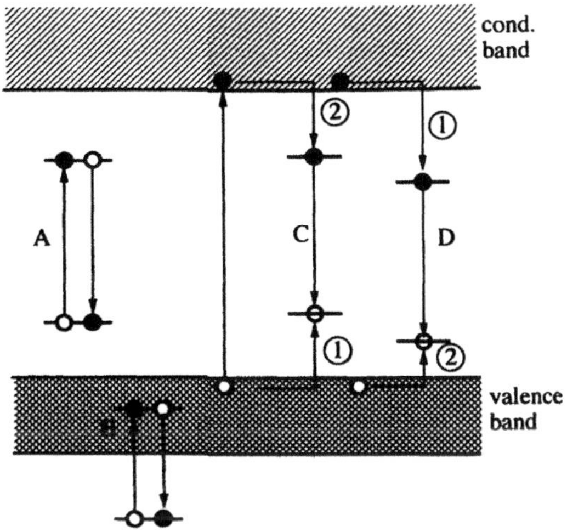

Figure 1. Energy levels of localized centers and luminescence transitions. Relative positions of the levels to the bands are shown.

(b) Forbidden-transition type

(i) $d \leftrightarrow d$ transition—Ti^{3+}, Cr^{3+}, Cr^{4+}, Mn^{2+}, Mn^{4+}, Fe^{2+}, Fe^{3+}

(ii) $f \leftrightarrow f$ transition—Pr^{3+}, Nd^{3+}, Sm^{3+}, Eu^{3+}, Tb^{3+}, Er^{3+}, Tm^{3+} (and other trivalent rare-earth ions)

Oscillator strengths of this type of transition, made allowed, are 10^{-4}–10^{-8}.

Energy levels of localized centers are seen in the energy band scheme in Fig. 1. In the case of A, both the ground and the excited states are located in the forbidden gap. In the case of B, the states are embedded in, or located below, the valence band. There are many intermediate cases between A and B. The relative positions of the energy levels of a localized center to the host energy bands are considered to be essentially determined by their depths from the vacuum level. These positions are not always well understood even for localized centers that are important from a practical point of view.

Localized centers are also excited by the band-to-band transition as shown by C and D in Fig. 1. In the case of C, first a hole is captured by the center and then an electron is captured to produce luminescence. In the case of D, the electron is captured first and then the hole. Further, localized centers are excited in some cases by excitons that move in the crystal. Sometimes excitation of localized centers by nonradiative resonant energy transfer from other kinds of localized centers also takes place and this will be explained in Section 7.1.1.

3. BAND-TO-BAND LUMINESCENCE

3.1. Direct and Indirect Transitions

This section begins with a very brief explanation of the important selection rule of optical transitions in crystals. The quantum state of an electron in an energy band is specified by the wavenumber vector **k** of the electron wave. The energy E of electrons is expressed as a function of **k**. Then $E(\mathbf{k})$ is represented by energy surfaces in the **k**-space composed of k_x, k_y, and k_z. The quantity $\hbar\mathbf{k}$, called crystal momentum, is conserved in a process in which an electron in a crystal interacts with another particle. In optical absorption processes in crystals the **k**-value of an electron making a transition does not change, because the **k**-value of a photon in the optical region is negligible compared with that of an electron. In other words, optical absorption can take place only between states with the same **k**-value. This rule is called the crystal momentum selection rule. Further, it is noted that this absorption is expressed by a vertical arrow in **k**-space.

The band structure at the forbidden energy gaps of ionic crystals and semiconductors are of two types: direct and indirect transition type. In many of IIIb–Vb compounds, such as GaN, GaAs, GaSb, and InP, and almost all of the IIb–VIb compounds, such as ZnO, ZnS, ZnSe, ZnTe, CdS, CdSe, and CdTe, both the bottom of the conduction band and the top of the valence band are located at the (000) point, called the Γ-point in **k**-space, as shown in Fig. 2a. Therefore, the band-to-band transition at the forbidden gap takes place without changing the value of **k**, so that the **k**-selection rule is satisfied. The transition satisfying the **k**-selection rule is called direct transition.

On the other hand, in Si, Ge, and some IIIb–Vb compounds such as AlAs and GaP, the top of the valence band is located at the Γ-point, while the bottom of the conduction band is elsewhere. In the case of Si, AlAs, and GaP, the bottom is located at or near the (100) point, called the X-point, as shown in Fig. 2b. In this kind of band structure, the direct transition to the bottom of the conduction band from the top of the valence band is forbidden from the **k**-selection rule. However, this

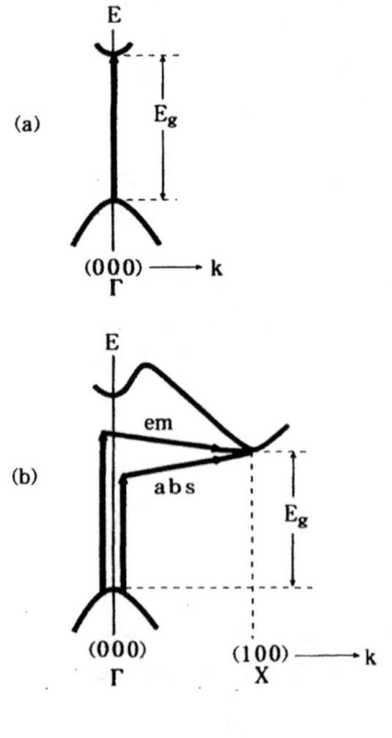

Figure 2. Band structures at the forbidden gap: (a) direct-transition type, (b) indirect-transition type.

transition is made partially allowed by the assistance of phonons. Namely, the transition can take place if it is accompanied by simultaneous emission or absorption of phonons having a phonon crystal momentum **K** in order that the **k**-selection rule is obeyed. This kind of phonon-assisted transition is called an indirect transition, and it has a probability that is three to four orders of magnitude lower than that of a direct transition.

3.2. Luminescence Transition Rates and Spectra

The rate of band-to-band luminescence transition as a function of photon energy, i.e., band-to-band luminescence spectra, can be obtained on the basis of the principle of detailed balance by using experimentally obtained band-to-band absorption spectra (van Roosbroeck and Shockley, 1954).

Under the condition of thermodynamic equilibrium, the number of radiative recombinations of electrons and holes in the frequency interval dv is equal to that of the generation of electrons and holes by thermal radiation. If the number of recombinations per unit volume per unit time is $R(v)dv$, and the number of the generation is $P(v)\rho(v)dv$, where $P(v)$ is the rate of the absorption of a photon and $\rho(v)$dv is the density of photons, then,

$$R(v)dv = P(v)\rho(v)dv \tag{1}$$

where $P(v)$ is given by

$$P(v) = \alpha(v)v_g \tag{2}$$

Here $\alpha(v)$ is an absorption spectrum expressed by the absorption coefficient and v_g is the group velocity of photons expressed as $cdv/d(nv)$, where c is the velocity of light, n is the refractive index assumed to be constant, and $\rho(v)dv$ is given by Planck's law of radiation. Then,

$$R(v)dv = \frac{8\pi n^2 v^2 \alpha(v)dv}{c^2[\exp(hv/kT) - 1]} \tag{3}$$

which gives the band-to-band luminescence spectra.

As an example of the comparison between a band-to-band luminescence spectrum experimentally observed and that calculated by Eq. (3), results for GaAs—a direct-transition-type semiconductor—are shown in Fig. 3 (Sarace *et al.*, 1965). As noted in Section 2.1.1, band-to-band luminescence is observable in very pure materials at relatively high temperatures. The experimental spectrum in the figure is that observed at 77 K. The calculated spectrum was obtained by using an experimentally observed absorption spectrum. As seen, agreement between experiment and calculation is satisfactory. The high-energy tail of the spectrum is determined by the Boltzmann factor, as understood from the spectral shape, reflecting the thermal distribution of electrons and holes in the bands.

The total number of recombinations of electrons and holes per unit volume per unit time is given by

$$R = \int R(v)dv \tag{4}$$

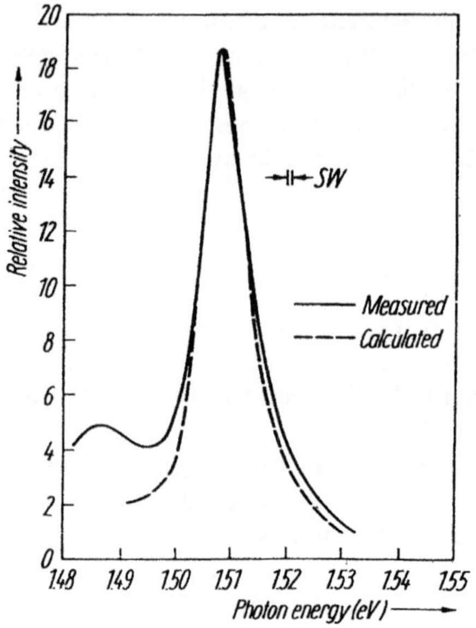

Figure 3. Comparison between measured and calculated luminescence spectra of GaAs at 77 K (Sarace *et al.*, 1965; with kind permission from American Physical Society).

where R is proportional to the product of the number of electrons n and holes p. In intrinsic semiconductors $np = n_i^2$. Then

$$R = Bn_i^2 \tag{5}$$

where B is called the recombination probability.

Values of B are calculated from observed absorption spectra for Si, Ge, and most of the IIIb–Vb compounds (Varshni, 1967). B-values for direct-transition-type semiconductors, such as GaAs and GaSb, are $1–10 \times 10^{-10}\,\mathrm{cm^3\,s^{-1}}$, while those for the indirect-transition-type, e.g., Si, Ge, and GaP, are $0.2–5 \times 10^{-14}\,\mathrm{cm^3\,s^{-1}}$. It is noted that B values for the indirect type are four orders of magnitude smaller than those for the direct type. This is to be seen as being very significant when one considers these semiconductors as candidates for luminescence materials to be used for light-emitting diodes and semiconductor lasers.

3.3. Recombination Cross Sections and Impurities

The radiative recombination cross section σ is given by

$$\sigma = R/n_i^2 v = B/v \tag{6}$$

where v is the velocity of the thermal motion of the electrons and holes and is obtained from $mv^2/2 = kT$. Values of σ for direct-transition-type semiconductors such as GaAs and GaSb are $0.5–10 \times 10^{-17}\,\mathrm{cm^2}$, while those for the indirect-transition-type (Si, Ge, and GaP) are $0.2–5 \times 10^{-21}\,\mathrm{cm^2}$.

Band-to-band luminescence in semiconductors is observable only in very pure materials. If there are some impurities having the impurity levels in the forbidden gap, electrons and/or holes are trapped by them and recombine with each other via such levels either radiatively emitting luminescence or nonradiatively emitting a number of phonons.

Values of capture cross sections of impurities for carriers depend very much on the nature of the impurities, i.e., whether they are neutral, positively charged, or negatively charged. Therefore, it is difficult to discuss the values in a simple way. However, one may assume that the values are given roughtly by the size of the atomic cross sections, i.e., 10^{-15} cm^2. This value is almost two orders of magnitude larger than radiative recombination cross sections in direct-transition-type materials. Thus, one can see why band-to-band luminescence appears only in very pure materials.

4. WANNIER EXCITON LUMINESCENCE

4.1. Energy States of Wannier Excitons

As noted in Section 2.1.2, a Wannier exciton is a composite particle made up of an electron in the conduction band and a hole in the valence band bound together by the Coulomb interaction. The energy of an exciton is given by

$$E_{ex} = E_g - G_{ex} \tag{7}$$

where G_{ex} is the binding energy of the exciton. The Wannier exciton is analogous to the hydrogen atom, and its binding energy is obtained from this analogy. Namely, on the basis of the well-known expression for the energy states of the hydrogen atom the binding energy of an exciton with the quantum number n is given as

$$G_{ex}^n = \mu e^4/(8\varepsilon^2 h^2)(1/n^2), \qquad n = 1, 2, 3, \ldots \tag{8}$$

by making a suitable replacement of $m \rightarrow \mu$ and $\varepsilon_0 \rightarrow \varepsilon(=\varepsilon_r \varepsilon_0)$, where μ is the reduced mass of the exciton given by $1/\mu = 1/m_e + 1/m_h$ and ε is the static dielectric constant. The Bohr radius of an exciton is

$$a_{ex}^n = \varepsilon h^2/(\pi \mu e^2) n^2 \tag{9}$$

For typical semiconductors, $\mu \doteqdot 0.1$ and $\varepsilon_r \doteqdot 10$, so that $G_{ex}(n = 1) \doteqdot 10^{-3} \times 13.6$ eV (Rydberg constant) $\doteqdot$ 10 MeV. Similarly, $a_{ex} \doteqdot 10^2 \times 0.53$ Å (Bohr radius of the hydrogen atom) $\doteqdot$ 50 Å. G_{ex} values (in MeV) for important semiconductors are as follows: Si: 14.7, Ge: 4.15, GaP: 21.5, GaAs: 4.2, GaSb: 1.6, ZnO: 59, ZnS: 40, ZnSe: 17, ZnTe: 12, CdS: 28.

Absorption spectra of excitons are composed of a series of sharp lines as seen from Eqs. (7) and (8). As an example, a spectrum of a GaAs crystal is shown in Fig. 4 (Ulbeich and Weisbuch, 1976). Exciton absorption lines of $n = 1$, 2, and 3 are clearly seen; $n = \infty$ gives the bandgap.

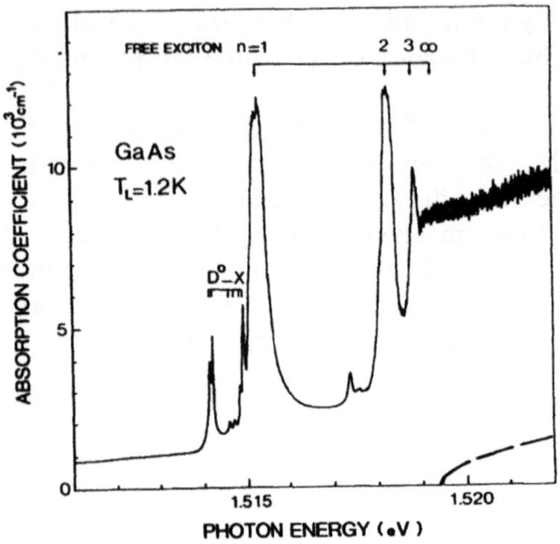

Figure 4. Absorption spectrum of free excitons in GaAs at 1.2 K. $D^0 - X$ lines are due to excitons bound to neutral donors. The dashed line shows calculated absorption of the band-to-band transition (Ulbeich and Weisbuch 1976; with kind permission from C. Weisbuch and also from North-Holland Publishing Co.).

4.2. Luminescence of Free and Bound Excitons

Excitons move freely in a crystal and are annihilated radiatively, emitting a sharp line luminescence. The exciton binding energy of ZnO is as high as 59 MeV as noted above, so that the exciton luminescence at about 3.3 eV is observed distinctly even at room temperature. Oscillator strengths of excitons per molecule are obtained for IIb–VIb compounds to be of the order of magnitude of 10^{-3}.

If excitons collide with impurities such as donors or acceptors, they are bound to them and afterward are annihilated emitting luminescence. As noted above, the Wannier exciton is analogous to the hydrogen atom. Therefore, when an exciton is bound to a neutral or an ionized donor or acceptor, a bound exciton complex of the H_2 or the H_2^+ type, respectively, is created.

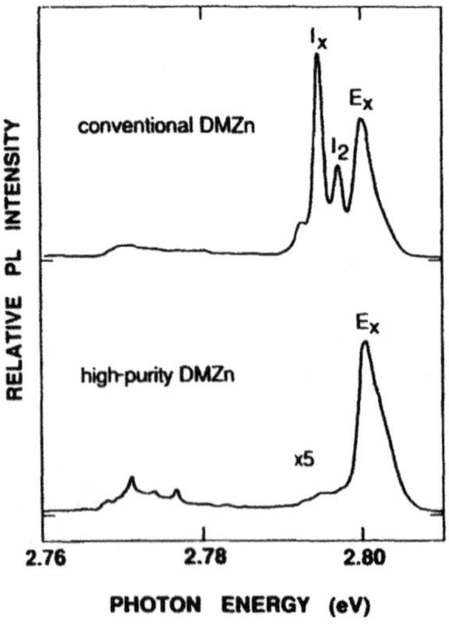

Figure 5. Luminescence spectra of free and bound excitons in ZnSe at 10 K. See the text for details (Kukimoto, 1990; with kind permission from Elsevier Science Publishers, V.B.).

Emission spectra of free and bound excitons in ZnSe crystals are shown in Fig. 5 (Kukimoto, 1990) as an example. E_x is the free exciton line, and I_2 and I_x are lines of excitons bound to neutral donors. Bound exciton lines are shifted a little toward lower energies by an amount corresponding to the binding energy of the exciton to that impurity.

ZnSe crystals used for Fig. 5 were grown by the MOVPE (metal organic vapor phase epitaxy) method using dimethyl zinc (DMZn) and dimethyl selenide as the raw materials. The upper and lower spectra are those of crystals grown by using conventional DMZn including chlorine impurities and purified DMZn, respectively. It is seen that in the lower spectrum two bound exciton lines disappear completely.

The high-energy side of the spectral shape of the free exciton line is approximated by the Boltzmann factor. This is a reflection of the thermal distribution of excitons on the exciton dispersion curve expressed by $\hbar^2k^2/2M$, where $M = m_e + m_h$ is the mass of exciton.

The bound exciton is characterized by very large oscillator strengths, known as "giant oscillator strengths," so the lifetime of the bound exciton is very short. The oscillator strength of a free exciton, f_{ex}, is given per molecule of the host. For the bound exciton the oscillator strength f_b is given per impurity that binds the exciton. Then, f_b/f_{ex} is the ratio of the volume surrounding the impurity in which the bound exciton moves around to the volume of the unit cell: f_b/f_{ex} can reach a value of about 10^3.

In CdS, f_{ex} for a free exciton is 3×10^{-3}; f_b for a bound exciton to a neutral donor is 9, i.e., 3×10^3 times enhanced. The calculated lifetime of this bound exciton is 0.4 ns, which is close to the experimental value.

4.3. Luminescence of Self-Trapped Excitons

It was found in the 1960s that pure alkali halide crystals emit intrinsic luminescence of broad bands in the near-ultraviolet and visible regions. Later on it was concluded that this luminescence is due to self-trapped Wannier excitons (Song and Williams, 1993).

In ionic crystals such as alkali halides in which the ionicity is strong, electron–phonon interactions are strong, so that electrons and holes cause large distortions in the lattice surrounding them. If the lattice distortion energy is very high, holes or electrons are made immovable and are self-trapped, because the self-trapped state is more stable than the one in which they would move dragging the distortions. It is known that holes are self-trapped in many ionic crystals, but there are no known examples of self-trapped electrons.

In alkali halides holes are self-trapped by two adjacent halogen ions approaching one another. This molecular ion type center X_2^-, called a V_K center, captures an electron to become $(V_K + e)$ and emits intrinsic luminescence. When an exciton is directly created, it is self-trapped relaxing to the $(V_K + e)$ state and emits intrinsic luminescence.

In the case of NaCl, the $n = 1$ exciton absorption is at 7.96 eV, and intrinsic luminescence consists of two fairly broad bands with peaks at 5.47 and 3.47 eV. The high-energy band, called the σ-band, is polarized parallel to the molecular axis of the V_K center. The decay of the σ-band is very rapid with a time constant of the order of 10^{-8} s. The low-energy band, called the π-band, is polarized perpendicular to the molecular axis. The decay is not rapid, with a time constant of the other of 10^{-4} s. The σ-band is emitted from the singlet excited state of the $(V_K + e)$ molecular center, and the π-band from the triplet excited state. Intrinsic luminescence of most of the alkali halides consists of σ- and π-bands, but in some of them, e.g., KCl and RbCl, only the π-band is observed.

4.4. Luminescence of Excitonic Molecules

When Wannier excitons are created in high densities by high-intensity excitation, the collision of excitons becomes very frequent, and as a result excitonic molecules (or biexcitons) are formed owing to the attractive covalent interaction between two excitons, analogous to the formation of a hydrogen molecule from two hydrogen atoms. Excitonic molecules emit luminescence with high efficiencies, as described below. Excitonic molecule luminescence was first observed in CuCl in 1968. Since then it has been observed in CuBr, CdS, most of the other IIb–VIb compounds, and GaAs (Hanamura, 1976; Klingshirn and Haug, 1981).

The energy of the excitonic molecule E_m is given by

$$E_m = 2E_{ex} - G_m \tag{10}$$

where G_m is the binding energy of the molecule. The luminescence of the excitonic molecule is the process in which a molecule emits a photon of energy $E_{ex} - G_m$, leaving a single exciton.

Figure 6 shows spectra of excitonic molecule luminescence in CdS observed by nitrogen laser nanosecond pulse excitation (Shionoya *et al.*, 1973). Under weak mercury lamp excitation only luminescence lines of A free excitons $[A(\Gamma_6)]$ and bound excitons (I_1, I_2) are seen. Under strong nanosecond pulse excitation the M line of excitonic molecules grows superlinearly.

The low-energy side of M line spectral shapes is approximated by the Boltzmann factor, contrary to the case of the exciton emission line, where the high-energy side is approximated. The energies of an exciton and an excitonic molecule as functions of **k** are

$$E_{ex}(\mathbf{k}) = E_{ex} + \hbar^2k^2/2M \tag{11}$$

and

$$E_m(\mathbf{k}) = 2E_{ex} - G_m + \hbar^2k^2/4M \tag{12}$$

The photon energy of the M line is given by

$$E = E_m(\mathbf{k}) - E_{ex}(\mathbf{k}) = E_{ex} - G_m - \hbar^2k^2/4M \tag{13}$$

Since the third term of Eq. (13) has a negative sign, the low-energy side, but not the high-energy side, of the M lineshape is expressed by the Boltzmann factor.

The high-energy side of M lineshapes is somewhat broadened, and it is suggested that this is due to elastic collisions between excitonic molecules and also those between a molecule and a single exciton. Taking account of this effect enabled

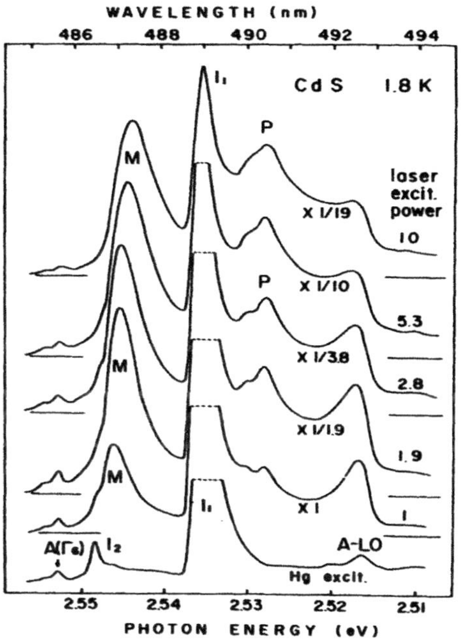

Figure 6. Luminescence spectra of excitonic molecules in CdS at 1.8 K. See the text for details (Shionoya *et al.*, 1973; with kind permission from Elsevier Science Ltd.).

a theoretical analysis of observed *M* lineshapes, and from this the position corresponding to $E_{ex}(2.5524\,\text{eV}) - G_m$ was determined, and $G_m = 5.4\,\text{meV}$ was obtained.

The term G_m/G_{ex} is a function of the electron-to-hole effective-mass ratio $\sigma = m_e/m_h$. For $\sigma \to 0$, G_m/G_{ex} should be 0.35, similar to that of the hydrogen molecule. For $\sigma \to 1$, on the other hand, G_m/G_{ex} should be 0.027, similar to that of positronium. The dependence of G_m/G_{ex} on σ for $0 < \sigma < 1$ has been investigated theoretically.

The transition of excitonic molecule luminescence has a giant oscillator strength f_m for exactly the same reason as in the case of bound excitons. The ratio of f_m/f_{ex} is given by the ratio of the volumes occupied by an excitonic molecule and the unit cell, and can be as high as 10^2–10^3.

Excitonic molecules are directly created by two-photon absorption with a photon energy of $E_{ex} - G_m/2$. The probability of this two-photon absorption is very high because of two positive effects: the first that the exciton state acts as the intermediate virtual state in the two-photon resonant condition, and the second that the transition from the exciton state to the molecular state has a giant oscillator strength. This two-photon absorption line is very sharp, so that observing it is useful for an accurate determination of the value of G_m.

The dynamic processes of the formation and radiative annihilation of excitonic molecules were studied in CuBr by means of ultrafast spectroscopy (Unuma *et al.*, 1982). Time-resolved luminescence spectra of excitonic molecules were observed under both band-to-band and two-photon direct excitation. The observed time dependence of luminescence intensity was analyzed on the basis of population dynamics, and the molecular lifetime was found to be 60 ps. The oscillator strength of excitons in CuBr is known to be 0.014. The ratio f_m/f_{ex} was theoretically determined to be 3,300, giving a calculated lifetime 24 ps, which is in good agreement with the experimental value.

5. CHARACTERISTICS OF LOCALIZED-CENTER LUMINESCENCE

In this section we discuss the characteristics, especially the spectral shapes, of luminescence in which localized centers participate. Various types of spectral shapes are observed, the most common being broad bell-shaped bands. Luminescence spectra of the allowed-transition type [(a) in Section 2.2.2] are always of this shape. Sometimes the spectra are accompanied by a phonon structure at low temperatures. Spectra of trivalent rare-earth ions that are due to the $f \rightarrow f$ transitions [(b–ii] are composed of rather sharp lines. Spectra of transition-metal ions that are due to $d \rightarrow d$ transitions [(b–i)] are sometimes made up of broad bands and sometimes composed of lines. Donor–acceptor pair (DAP) luminescence sometimes shows somewhat peculiar shapes.

5.1. Configuration-Coordinate Model

These spectral characteristics can be interpreted in a unified way on the basis of the configuration-coordinate model of a localized center as shown in Fig. 7. The configuration coordinate represents changes of the nuclear coordinates of all the lattice ions constituting the luminescence center. If the motion of electrons is compared to that of nuclei it is seen that electrons move 10^3 times or more faster. Therefore, it is considered that an electron always occupies a stationary energy state corresponding to each position of the nuclei, and that the latter are vibrating within the potential determined by the electron.

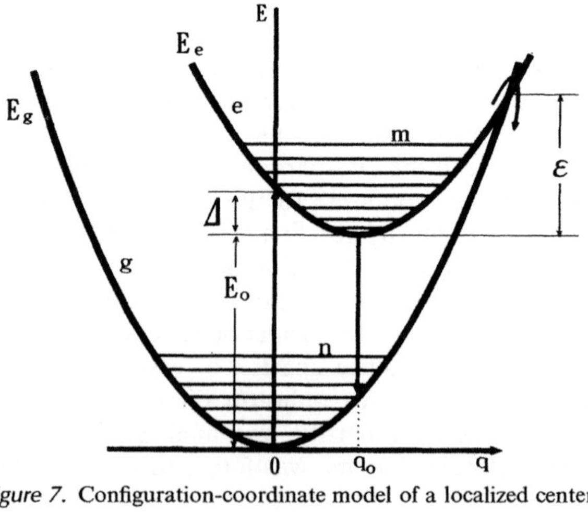

Figure 7. Configuration-coordinate model of a localized center.

The ordinate of the configuration-coordinate model represents this potential energy for the ground and excited states. The configuration coordinate is essentially of $3N$ dimensions, where N is the number of ions constituting the luminescence center. The local mode of lattice vibration that influences the electronic energy most strongly is thought to be the one in which ions surrounding the activator ion in the center vibrate maintaining total symmetry. If only this mode of vibration is taken into account, the configuration coordinate is approximated as being one-dimensional.

Absorption and emission processes are dominated by the Franck–Condon principle, which is interpreted as follows: The time for a transition between two electronic states is of the order of 10^{-16} s, which is shorter by far than 10^{-13} s, the vibrational period of nuclei. Thus during the time an electron makes a transition the nuclei are regarded as being stationary, so in Fig. 7 absorption and emission transitions are shown by vertical arrows.

Wavefunctions of excited states are considered to be more spread out generally than those of ground states, so that the energy minimum point of the excited state q_0 is assumed to be shifted toward the outside. The terminal state of the absorption transition is at a higher-energy point than the minimum in the excited state. The lifetime of the excited state is of the order of 10^{-8} s in the case of allowed transitions and is much longer than the lattice vibration period. Therefore, just after the absorption has terminated, relaxation toward the minimum energy point in the excited state takes place accompanied by the emission of phonons. As a result, the emission is produced from the minimum point. Thus it is expected that the emission wavelength will always be shifted toward the longer side than the absorption wavelength. This is well known experimentally and is called Stokes' law.

5.2. Theoretical Treatments of Spectral Shapes

Absorption and emission processes in localized centers are treated quantum-mechanically by the method of adiabatic approximation (also called the Born–Oppenheimer method), and spectral shapes are obtained. A brief description of this method follows.

The total Hamiltonian $H(\mathbf{r}, \mathbf{R})$ (where $\mathbf{r}$ is the electron coordinate, and $\mathbf{R}$ is the nucleus coordinate) representing the electronic state of a localized center is divided into three parts:

$$H(\mathbf{r}, \mathbf{R}) = H_e(\mathbf{r}) + H_l(\mathbf{R}) + H_{el}(\mathbf{r}, \mathbf{R}) \tag{14}$$

H_e, H_l, and H_{e-l} represents the energies of the electrons, the lattice and the electron–lattice interaction, respectively. As noted in Section 5.1, an electron occupies a stationary state corresponding to each position of the nuclei. Then, assuming that the electron wavefunction ψ depends on both $\mathbf{r}$ and $\mathbf{R}$, while the wavefunction of the vibration of the nucleus, ϕ, only on $\mathbf{R}$, Eq. (14) breaks down as follows:

$$[H_e(\mathbf{r}) + H_{el}(\mathbf{r}, \mathbf{R})]\psi_j(\mathbf{r}, \mathbf{R}) = E_j(\mathbf{R})\psi_j(\mathbf{r}, \mathbf{R}) \tag{15a}$$

and

$$[H_l(\mathbf{R}) + E_j(\mathbf{R})]\phi_{j,n}(\mathbf{R}) = E_{j,n}\phi_{j,n}(\mathbf{R}) \tag{15b}$$

where the subscripts j and n represent the electronic state, and the quantum number of the vibration, respectively. The total wavefunction Ψ for Eq. (14) is approximated as

$$\Psi_{j,n}(\mathbf{r}, \mathbf{R}) = \psi_j(\mathbf{r}, \mathbf{R})\phi_{j,n}(\mathbf{R}) \tag{16}$$

Equation (15a) gives the energy of an electronic state $E_j(\mathbf{R})$ without the vibration of the nuclei, and this $E_j(R)$ gives the potential for the vibration of the nuclei in Eq. (15b). This adiabatic approximation—treating electronic states and vibrational states of nuclei by separating them in this way—can only be used if the energy difference between the two electronic states is large enough compared with lattice vibration energy. Therefore, absorption and emission processes in the ultraviolet and visible regions can reasonably be treated by this approximation.

The case in which the vibrational frequencies of the ground and excited states are equal and only the minimum energy point of the excited states is shifted, as shown in Fig. 7, will now be described. If harmonic vibration is assumed, and vibrational frequency is denoted by Ω, the mass of atoms by M, and the configuration coordinate by q, then, the potential energies of the ground and excited states E_g and E_e, respectively, are expressed as

$$E_g = \tfrac{1}{2}M\Omega^2 q^2 \tag{17a}$$

and

$$E_e = \tfrac{1}{2}M\Omega^2(q - q_0)^2 + E_0 \tag{17b}$$

The number of phonons S emitted just after the absorption transition at $q = 0$ is expressed as

$$S = \tfrac{1}{2}M\Omega^2 q_0^2/h\Omega = \Delta/h\Omega \tag{18}$$

S is regarded as a constant expressing the strength of the electron–phonon interaction.

The transition probability P is proportional to the square of the matrix element of the dipole moment. If the wavefunctions of the ground and excited states are expressed on the basis of the adiabatic approximation, P is given by

$$P = |\langle\psi_e\phi_{e,m}|\mathbf{r}|\psi_g\phi_{g,n}\rangle|^2 \tag{19}$$

If it is assumed that the dipole moment is constant and independent of q (Condon approximation), the matrix element becomes $\langle\psi_e|\mathbf{r}|\psi_g\rangle\langle\phi_{e,m}|\phi_{g,n}\rangle$. Then, one observes that the q-dependence of the transition probability, and hence the spectral shape, is determined by the overlap integral of the wavefunctions $\langle\phi_{e,m}|\phi_{g,n}\rangle$.

The overlap integral in oscillators in which the origin is shifted is given as

$$\langle\phi_{e,m}|\phi_{g,n}\rangle = e^{-S/2}(n!/m!)^{1/2}S^{(m-n)/2}L_n^{m-n} \tag{20}$$

where L_n^{m-n} is the Laguerre polynomial. In the vibrational levels of the ground state at $T = 0$ only the $n = 0$ level is occupied, and $L_0^m(S) = 1$. In this case the transition probability P_{0n} from the $n = 0$ level to the $m = n$ level in the excited state, i.e., the

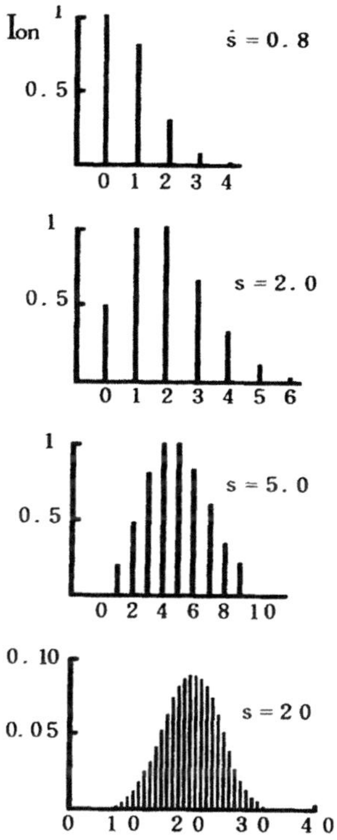

Figure 8. Spectral shapes given by the Poisson distribution.

intensity of the absorption line, is given by squaring Eq. (20):

$$P_{0n} \propto I_{0n} \propto e^{-S}S^n/n! \qquad n = 0, 1, 2, \ldots \tag{21}$$

One notes that this is the expression for the Poisson distribution.

In Fig. 8, the shapes of the Poisson distribution for $S = 0.8$, 2.5, 5.0, and 20 are shown. These give the shapes of absorption spectra at very low temperatures approximated as 0 K. I_{0n} becomes a maximum at $n = S - \frac{1}{2}$. The line of the transition from $n = 0$ to $m = 0$ is called the zero-phonon line. Its intensity I_{00} is proportional to e^{-S}, and the stronger the electron–phonon interaction is the weaker it becomes. At $S \doteqdot 10$, the zero-phonon line becomes invisible. At $S \geqq 20$, the envelope of I_{0n} is approximated by Gaussian curves.

Exactly the same argument holds true for emission spectra, and the spectral shapes of the transitions from the $m = 0$ level in the excited state are given by the Poisson distribution. Therefore, absorption and emission spectra have a common zero–phonon line, and display a mirror image relationship with bilateral symmetry on their high- and low-energy sides.

5.3. Examples of Typical Spectra

Shapes of absorption and emission spectra exhibited by various types of localized luminescence centers are reasonably well explained by the theory described

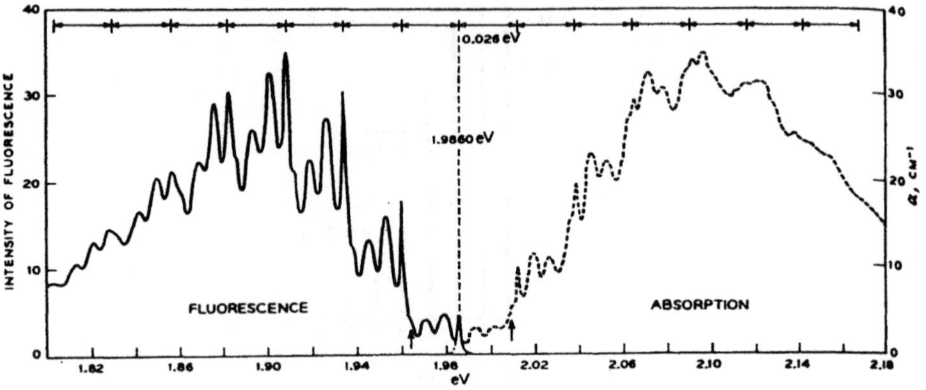

Figure 9. Absorption and fluorescence spectra of ZnTe : O at 20 K (Merz, 1968; with kind permission from the American Physical Society).

above. The spectra of ZnTe : O at low temperatures are shown in Fig. 9 (Merz, 1968), as typical of a case in which both absorption and emission spectra are represented by a Poisson distribution and the mirror image relationship is present. An O atom in ZnTe behaves as a typical isoelectronic trap, which will be explained in Section 6.2, and creates a localized center by trapping an electron and a hole. As seen in the figure, there is a distinct zero-phonon line common to the absorption and emission spectra at 1.9860 eV. Both the spectra are represented by the Poisson distribution $S \doteqdot 3$ with a coupled phonon energy of 0.026 eV.

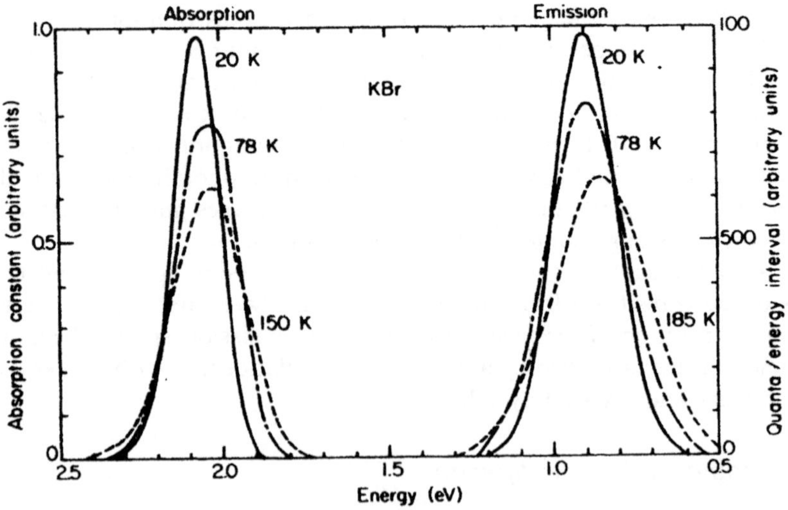

Figure 10. Absorption and emission spectra of *F*-centers in KBr at various temperatures (Gebhardt and Kühnert, 1964; with kind permission from Elsevier Science Publishers, V.B.).

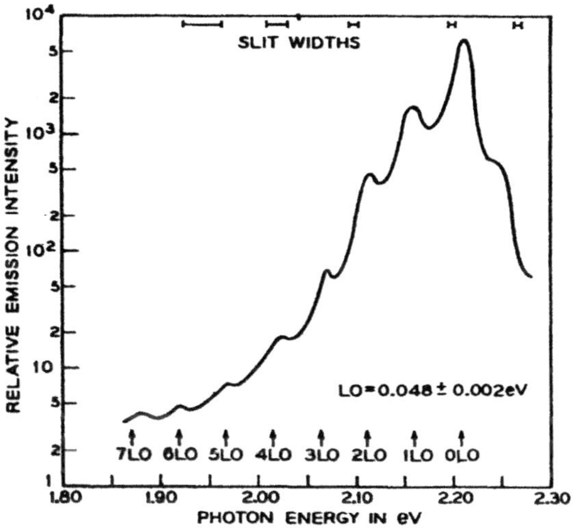

Figure 11. Spectrum of edge emission (shallow DAP luminescence) in GaP:Si,S at 20 K (LO stands for longitudinal optical phonon energy). (Thomas *et al.*, 1964; with kind permission from the American Physical Society).

Figure 10 shows absorption and emission spectra of the *F*-center in KBr at various temperatures (Gebhardt and Kühnert, 1964). It is seen that the shapes of both spectra are Gaussian, corresponding to the case of $S \doteqdot 20$, and the invisible zero-phonon line is expected to be at about 1.5 eV. These kinds of spectra are typical of localized luminescence centers of the allowed-transition type.

Figure 11 shows a spectrum of the so-called edge emission in GaP:Si, S at 20 K (Thomas *et al.*, 1964), which will be explained in Section 6.1.2. The spectrum is represented by the Poisson distribution of $S = 0.87$ with a coupled longitudinal optical (LO) phonon energy of 0.048 eV.

5.4. Temperature Dependence of Luminescence Efficiency

The temperature dependence of luminescence efficiency in localized centers is determined as follows. In the configuration-coordinate model in Fig. 7, the minimum point of the excited-state potential curve is shifted outward, so that the curves of the ground and excited states cross at some point. The energy of the crossing point is written as ε, as shown in the figure. With increasing temperature, vibrational levels in the excited state are excited, so that the system reaches the crossing point with a probability proportional to $\exp(-\varepsilon/kT)$. Then it transfers to the ground state and falls to the minimum of the ground state, emitting a number of phonons. In this way the nonradiative transition is caused with the probability $R = R_0 \exp(-\varepsilon/kT)$.

Thus, if A is the probability of radiative transition, the temperature dependence of luminescence efficiency is given by

$$\eta = A/(A + R) = A/[A + R_0 \exp(-\varepsilon/kT)] \tag{22}$$

The efficiency always decreases with increasing temperature, starting from some T where $R_0 \exp(-\varepsilon/kT)$ becomes appreciable as compared with A, and the luminescence eventually disappears. This phenomenon is called temperature quenching of luminescence.

The luminescence decay of localized centers should be a simple exponential, i.e., $I = I_0 \exp(-At)$. If temperature quenching begins, A in this equation is replaced by $A + R$. Therefore, decay time becomes faster with increasing temperature with a simultaneous decrease in luminescence intensity.

6. EXTRINSIC LUMINESCENCE OF UNLOCALIZED TYPE

Donor–acceptor pair luminescence is a very important example of the unlocalized extrinsic type from the viewpoint of both basics and applications, and is described in detail in Section 6.1. Another important luminescence of the unlocalized type is that due to isoelectronic traps. This luminescence transition plays an important role in semiconductors of the indirect-transition type, and is described in Section 6.2.

6.1. Donor–Acceptor Pair Luminescence

6.1.1. Theoretical Considerations

As noted in Section 2.2.1, the emission energy of donor–acceptor pair (DAP) luminescence depends on the spatial separation between the donor and acceptor in a pair. In the initial state from which the emission starts, an electron is located at the donor level, with a hole at the acceptor level, as shown in Fig. 12. The energy of this state E_i is given by

$$E_i = E_g - (E_d + E_a) \tag{23}$$

if the origin of the energy scale is taken to be the position of the acceptor level. In the final state, there is no electron in the donor level so that an effective charge of $+1$ is left there, while in the acceptor level an effective charge of -1 is left. As a result the Coulomb attractive force works between the donor and acceptor, so that

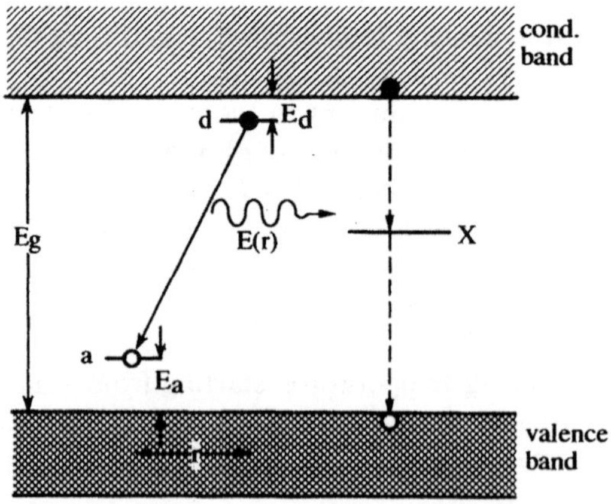

Figure 12. Transition of donor (d)–acceptor(a) pair luminescence and nonradiative recombination through the X-center.

the energy of the final state E_f is given by

$$E_f = -e^2/4\pi\varepsilon r \tag{24}$$

where ε is the dielectric constant for the static field, and r is the intrapair separation. The emission energy as a function of r is thus expressed as

$$E(r) = E_i - E_f = E_g - (E_d + E_a) + e^2/4\pi\varepsilon r \tag{25}$$

From this equation it can be seen that $E(r)$ shifts toward lower energies with increasing r. The values of r in crystals are discrete, so that spectra of DAP luminescence should be composed of numerous sharp lines, if lines from each pair are separated.

The transition probability of DAP luminescence also depends on r. The probability is considered to be proportional to the square of the overlap of the donor and acceptor wavefunctions. The donor and acceptor are hydrogen-like, so that their wavefunctions decay exponentially with distance. The spread of the wavefunction of a donor is generally much larger than that of an acceptor, so that it is enough to consider only the spread of the donor wavefunction. Then the transition probability as a function of r, using the Bohr radius r_B of the donor, is given as

$$W(r) = W_0 \exp(-2r/r_B) \tag{26}$$

6.1.2. Shallow Donor–Acceptor Pair Luminescence

There are two types of DAP luminescence: one owing to shallow DAPs and the other to deep DAPs. The spectral shapes of the two types are quite different. A typical example for the former is the so-called "edge emission," which is usually observed in relatively pure crystals of IIb–VIb and IIIb–Vb compounds such as CdS, ZnSe, and GaP and appear just below the band edge.

Figure 11, in the previous section, showed a spectrum of the edge emission in GaP:Si, S, in which Si is the acceptor and S is the donor, both occupying the P site. The spectral shape is well represented by the Poisson distribution, as was described in Section 5.2 with $S = 0.87$. This spectral shape is characteristic of shallow DAP luminescence. In DAP luminescence one pair is regarded as one localized center to which the configuration-coordinate model is applied. The strongest peak, at 2.22 eV, is the assembly of numerous unresolved zero-phonon lines of DAPs with different intrapair separations. The interval of peaks toward the low-energy side is 0.048 eV, equal to the LO phonon energy.

If the high-energy side of the zero-phonon peak is observed in detail with high spectral resolution, there are a large number of very sharp lines. The energy positions of these lines are well represented by Eq. (25), as shown in Fig. 13 (Thomas *et al.*, 1964). In the figure, Type I is the case of GaP:Si, S shown in Fig. 11. Type II is the case of GaP:Zn, S in which Zn acceptors occupy the Ga site, while S donors occupy the P site; i.e., in Type I the acceptor and the donor occupy the same kind of site, while in Type II they occupy different kinds of sites. Discrete values of r are different in the two types. In Fig. 13 the solid lines follow Eq. (25), and the dashed lines correspond to the following equation, in which the van der Waals interaction energy

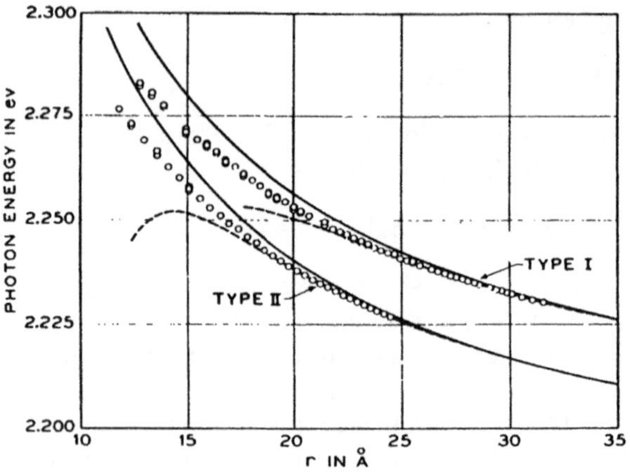

Figure 13. Energies of very sharp lines (open circles) in spectra of shallow DAP luminescence in GaP:Si,S (Type I) and GaP:Zn,S (Type II) at 20 K are plotted against the interapair separation r. Solid lines are calculated values following Eq. (25), and dashed lines follow Eq. (27) (Thomas *et al.*, 1964; with kind permission from the American Physical Society).

is taken into account as the fourth term,

$$E(r) = E_g - (E_d + E_a) + e^2/4\pi\varepsilon r - (e^2/\varepsilon)(a/r)^6 \tag{27}$$

It is seen that adding the van der Waals term somewhat improves agreement with calculation in the intermediate region of r.

In DAP luminescence, the emission energy shifts toward higher energies with a decrease in the intrapair separation r and simultaneously the transition probability becomes larger, i.e. the luminescence decay becomes more rapid, as was explained above. Therefore, the peak of a narrow emission band composed of many unresolved pair lines should shift toward lower energies during decay. Figure 14 (Thomas *et al.*, 1965) shows time-resolvd emission spectra of GaP:Si,S after pulse excitation. It is clearly seen that the zero-phonon peak located at 2.23 eV just after pulse excitation, shifts with the lapse of time toward lower energies and reaches 2.20 eV at a decay time of 1 s.

A similar kind of spectral shift in DAP luminescence should be observed with changing excitation intensity. With increasing excitation intensity, intensities of emission lines located at longer wavelengths with larger r should be saturated because of smaller transition probabilities, but intensities of lines at shorter wavelengths are not saturated. Therefore, the peak of a band composed of unresolved pair lines should shift to the short-wavelength side. This spectral shift was also observed. These two kinds of spectral shifts during decay and with changing excitation intensity are important features resulting from the mechanism of DAP luminescence.

6.1.3. Deep Donor–Acceptor Pair Luminescence

Typical examples of deep DAP luminescence are the green emission of ZnS:Cu, Al(Cl) phosphors and the blue one of ZnS:Ag,Al(Cl) phosphors. These two kinds of luminescence have been well known since the very early days of luminescence research in the 1920s. Spectral shapes of both of these examples are broad Gaussian

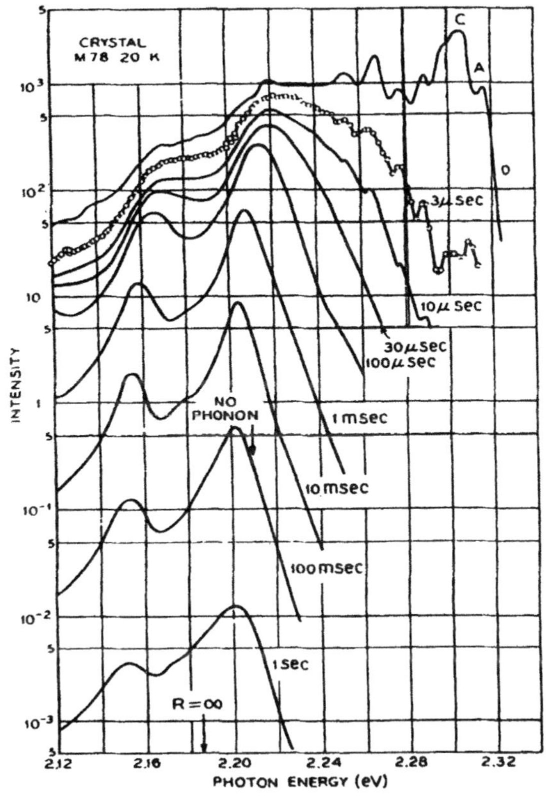

Figure 14. Time-resolved spectra of shallow DAP luminescence in GaP:Si,S at 20 K after pulse excitation (Thomas *et al.*, 1965; with kind permission from the American Physical Society).

bands. The peak and halfwidth at room temperature of the ZnS:Cu, Al spectrum are 2.35 eV and 0.32 eV, respectively, and for ZnS:Ag,Cl are 2.76 eV and 0.32 eV, respectively. The bandgap energy of cubic ZnS is 3.7 eV at room temperature. These two kinds of phosphors, which are currently used, e.g., for television tubes, are very important in terms of applications.

The mechanism of these kinds of luminescence and the nature of the luminescence centers were discussed intensively by many luminescence scientists for some 40 years—from the 1930s to the end of 1960s. These subjects represented important problems to be solved in the luminescence research of this era. In 1950s, it became clear that Cl^- or Al^{3+} ions provide charge compensation for Cu^+ and Ag^+ activators. Then, Cl^- and Al^{3+} were named coactivators because they are indispensable for the formation of luminescence centers, but the luminescence wavelength is dominated by the activator species. Further, on the basis of the concepts of semiconductor physics, activators were recognized as acceptors and coactivators as donors. Various models for the transitions that produce the luminescence under discussion were proposed in the 1950s and early 1960s.

Time-resolved emission spectra of these phosphors were observed (Era *et al.*, 1968*a,b*) in the latter half of 1960s. Figure 15 shows those of ZnS:Cu, Al at 4.2 K. It is clearly seen that after pulse excitation the peaks of the emission bands display

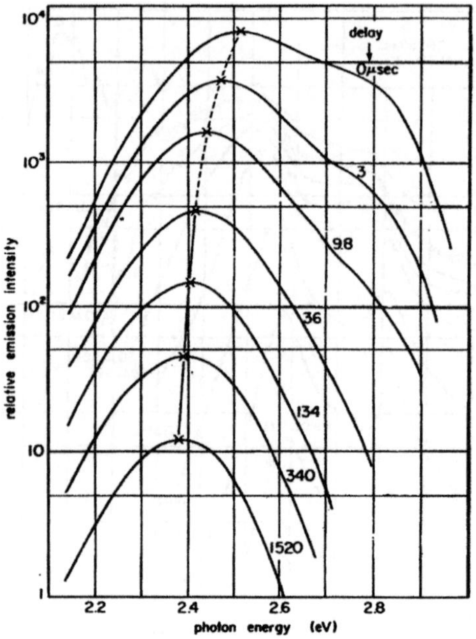

Figure 15. Time-resolved spectra of deep DAP luminescence in ZnS:Cu, Al at 4.2 K (Era *et al.*, 1968*a*; with kind permission from Elsevier Science Ltd.)

a significant shift toward lower energies with time. Further, it was observed that with increasing excitation intensity the emission peaks shift toward higher energies. Also, for ZnS:Ag, Al, similar spectral shifts were observed quite clearly both during decay and with increasing excitation intensity.

These observations of the two kinds of spectral shifts are reasonably well explained by the mechanism of deep DAP luminescence. As noted in Section 6.1.2, the configuration-coordinate model is applied to a DAP regarded as a localized center. Emission spectra of these phosphors are Gaussian bands and no zero-phonon lines are observed. These characteristics correspond to the case of the Poisson distribution with $S \geqq 20$, which is expected for stronger electron–phonon interactions.

Cu and Ag acceptor levels in ZnS are pretty deep, i.e., roughly 1.2 and 0.8 eV, respectively, and the depth of the Al donor is about 0.1 eV. It is known that in semiconductors the electron–phonon interaction associated with impurity levels such as donor and acceptor levels becomes stronger with an increase in the depth of the levels. Therefore, deep Cu–Al and Ag–Al pairs produce broadband spectra as expected from the configuration-coordinate model. The whole spectrum is the pile of numerous Gaussian bands with different zero–phonon lines corresponding to different intrapair separations. Thus, it was revealed that both the green luminescence in ZnS:Cu, Al(Cl) and the blue luminescence in ZnS:Ag, Al(Cl) are dominated by the deep DAP mechanism.

Further investigations were performed for ZnS:Cu, Al on the luminescence mechanism and the nature of the luminescence center. The polarization of the green luminescence was measured using melt-grown single crystals (Shionoya *et al.*, 1965; Suzuki and Shionoya, 1971a). It was found that in hexagonal ZnS:Cu, Al the luminescence is polarized preferentially perpendicular to the *c*-axis, independent of whether the excitation light is polarized or unpolarized and whether its wavelength belongs to the band-to-band transition or to the transition in the pair center. This fact indicates that Cu acceptors and Al donors are randomly distributed in a crystal and not associated. The polarization perpendicular to the *c*-axis is just the effect of the uniaxial crystal field.

Direct experimental evidence for the pair luminescence mechanism was obtained in the following way (Suzuki and Shionoya, 1971*b*). In ZnS : Cu, Al two kinds of photoinduced absorption in the infrared region are observable. One is due to holes created at Cu acceptors and the other to electrons trapped at Al donors. The relationship between the decay rates of the green luminescence and of two kinds of photoinduced absorption was studied. It was found that the two absorption decay rates are equal and are about one-half of that of the luminescence. This fact clearly indicates that the green luminescence is caused by the recombination of electrons trapped at Al donors and holes located at Cu acceptors according to the bimolecular kinetics. In this way, the luminescence mechanism and the nature of the luminescence center in ZnS : Cu, Al were unambiguously elucidated. Thus, one of the most important problems in luminescence research since the 1930s was perfectly solved, and the discussions were brought to an end.

The luminescence in these ZnS phosphors is highly efficient, making them important from practical points of view. The efficiency of unlocalized-type extrinsic luminescence in semiconductors is determined as follows.

In semiconductors there are generally deep levels of some kind that act as nonradiative recombination centers, as shown by the *X*-center in Fig. 12. Such levels are created by a very small quantity of impurities, lattice defects, or dislocations. Even in very pure materials such levels exist, and it is impossible to remove them completely using the present material preparation methods. Nonradiative recombination centers capture electrons or holes first by nonradiative transitions accompanied by multiphonon emission. Then they are charged and capture the other carriers nonradiatively.

In the deep DAP luminescence process, electrons and holes are created by the band-to-band transition. These either recombine radiatively through the DAP route or nonradiatively through the deep-level route. The luminescence efficiency is determined by the competition of these two processes, and is given by the ratio of the probabilities of the two processes occurring. It is considered that deep acceptors capture holes promptly and prevent their reliberation to the band. It is reasonable to assume that the capture cross section for holes is larger for deep acceptors than for nonradiative recombination centers, and this is the main reason for the high efficiency of deep DAP luminescence.

6.2. Luminescence Owing to Isoelectronic Traps

If in a semiconductor an atom of an element belonging to the same column of the periodic table as that of a constituent atom of that semiconductor is introduced and replaces the constituent atom, it attracts an electron or a hole because of the difference in electron affinity. Such an atom is called an isoelectronic trap. If the electron affinity of the introduced atom is larger (smaller) than that of the constituent atom, that atom becomes an electron (hole) trap. If an electron (hole) is trapped, a hole (electron) is attracted through the Coulomb force. In this way a bound exciton is formed and luminescence is produced.

ZnTe:O, whose absorption and emission spectra were shown in Fig. 9, is a typical example of an isoelectronic trap. The electron affinity of O is much greater than that of Te, so that an O atom acts as an electron trap. Another typical example is GaP:N. A N atom acts as an electron trap, and a bound exciton is formed, which shows efficient green luminescence at 2.37 eV (Trumbore *et al.*, 1965). On the other hand, a Bi atom in GaP acts as a hole trap, and produces red luminescence.

It was noted in Section 3.1 that GaP is an indirect-transition type semiconductor, and its band structure was shown in Fig. 2b. Luminescence owing to isoelectronic traps is especially important in indirect-type semiconductors. Donors or acceptors attract electrons or holes through a long-range Coulomb force. On the other hand, isoelectronic traps attract electrons or holes by the short-range force that is a result of the difference in electron affinities. The wavefunction of trapped electrons or holes is thus extremely localized in real space, but is very widely spread out in **k**-space. This is very different from the wavefunction of electrons at donors or of holes at acceptors.

In GaP:N, electrons trapped at N isoelectronic traps create an energy level a little below the bottom of the conduction band at point X in Fig. 2b. The wavefunction of trapped electrons is widely spread out in **k**-space and has a fairly large amplitude at point Γ, although the maximum amplitude is at point X. Hence, recombination owing to direct transition of trapped electrons with holes at point Γ becomes possible, and has a high probability, with an oscillator strength of 0.1. On the other hand, the oscillator strength of bound excitons at S donors is much smaller (0.001).

It should be noted that isoelectronic traps play an important role in creating high-efficiency luminescence centers in indirect-type semiconductors such as GaP. Green-emitting GaP:N and red-emitting GaP:Zn–O (Zn–O is isoelectronic with Ga–P) are important luminescent materials for light-emitting diodes from the point of view of applications.

7. EXTRINSIC LUMINESCENCE OF LOCALIZED TYPE

7.1. Luminescence of Allowed-Transition-Type Ions

As noted in Section 2.2.2(a), $s \rightarrow p$, $s^2 \rightarrow sp$ and $f \rightarrow d$ transitions are typical allowed transitions producing efficient broadband luminescence. Among them the $s^2 \rightarrow sp$ and $f \rightarrow d$ transitions are the ones that are important in terms of practical phosphors.

7.1.1. The $s^2 \rightarrow sp$ Transition

Among metallic ions showing luminescence owing to the $s^2 \rightarrow sp$ transition, Sn^{2+}, Pb^{2+}, and Sb^{3+} ions are important activators in various kinds of practical phosphors. The most representative one is the blue-emitting $Ca_5(PO_4)_3(F,Cl):Sb^{3+}$ phosphor for fluorescent lamps. If Mn^{2+} is also incorporated, there is an energy

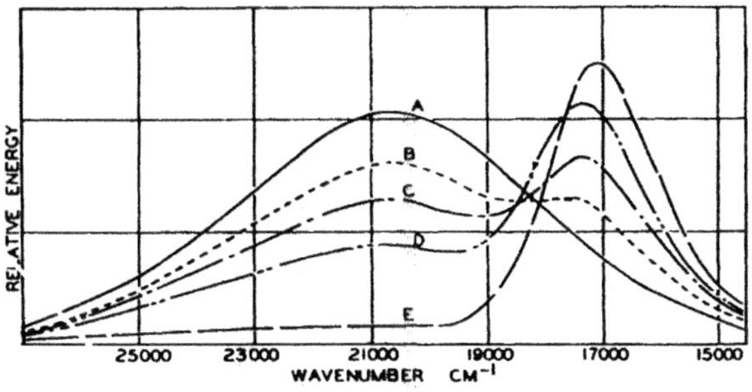

Figure 16. Emission spectra of $Ca_5(PO_4)_3(F,Cl):Sb^{3+},Mn^{2+}$ at room temperature. Sb^{3+} concentration: fixed at 0.01 mole/mole Ca, Mn^{2+} concentration: A:0, B:0.005, C:0.010, D:0.020, E:0.080 mole/mole Ca (Butler and Jerome, 1950; with kind permission from the Electrochemical Society).

transfer from Sb^{3+} to Mn^{2+}, which produces orange luminescence, and, as a result, white light-emitting phosphors suitable for fluorescent lamps are obtained.

The ground state of the s^2 electron configuration is $^1S_0$. In the excited state of the sp configuration, there are four electronic states, $^1P_1$ and $^3P_{0,1,2}$, with $^3P_0$ being the lowest. Therefore, luminescence is due to the $^3P_0 \rightarrow {}^1S_0$ transition. Figure 16 shows emission spectra of five kinds of $Ca_5(PO_4)_3(F,Cl):Sb^{3+},Mn^{2+}$ phosphors, in which the Sb^{3+} concentration is fixed while the Mn^{2+} concentration is varied (Butler and Jerome, 1950). It is seen that an increase in Mn^{2+} concentration leads to an increase in the Mn^{2+} emission intensity and an accompanying decrease in the Sb^{3+} emission intensity. The Sb^{3+} emission spectrum is a very broad Gaussian band with a halfwidth of 6300 cm^{-1} (0.65 eV). The excited-state wavefunction formed by the p-orbit is much more spread out than the ground-state wavefunction of the s-orbit. Therefore, according to the configuration-coordinate model shown in Fig. 7, q_0 is very large in the Sb^{3+} center, bringing about a very broad band.

Figure 17 shows excitation spectra for Sb^{3+} and Mn^{2+} emission (Johnson, 1961). Excitation is made by the $^1S_0 \rightarrow {}^1P_1$ transition of Sb^{3+}. It is understood in the figure that Sb^{3+} is efficiently excited by the 253.7 nm excitation line produced by the Hg discharge in fluorescent lamps, which is a necessary condition of fluorescent

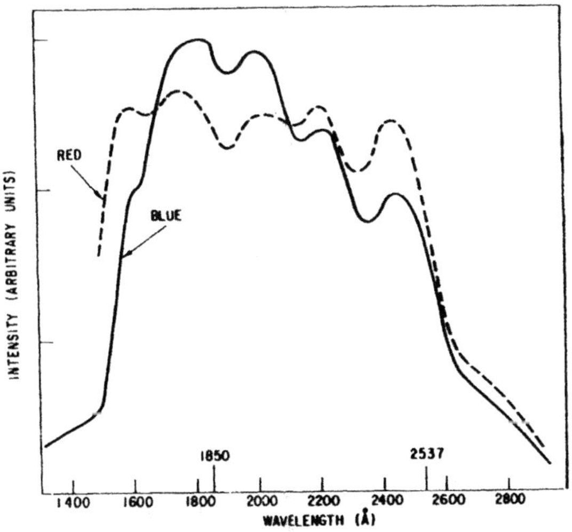

Figure 17. Excitation spectra of $Ca_5(PO_4)_3:Sb^{3+},Mn^{2+}$ at room temperature. Solid line for Sb^{3+} blue emission, and dashed line for Mn^{2+} orange emission (Johnson, 1961; with kind permission from the Electrochemical Society).

lamp phosphors. It is further seen that the Mn^{2+} excitation spectrum is almost the same as that of Sb^{3+}. This is clear experimental evidence that energy transfer takes place from Sb^{3+} to Mn^{2+}, and that Sb^{3+} acts as the sensitizer of the Mn^{2+} luminescence for the 253.7 nm excitation. This will be discussed further in Section 7.2.2.

The decay time of the Sb^{3+} luminescence is 7.7 μs. The luminescence transition of $^3P_0 \rightarrow {}^1S_0$ noted just above is forbidden by the spin selection rule $\Delta S = 0$. Thus the decay time is fairly long, in spite of allowed-transition luminescence.

7.1.2. The f → d Transition

The Eu^{2+} ion with the $4f^7$ electron configuration and the Ce^{3+} ion with the $4f^1$ configuration show efficient luminescence owing to the $4f \rightarrow 5d$ transition, and are both important activators for various kinds of practical phosphors. The luminescence colors or wavelengths of these two ions change widely from near-ultraviolet to red regions depending on the nature of host lattices. As far as luminescence characteristics are concerned, those of the Eu^{2+} ion are more complicated and interesting, and so will be the case dealt with here.

The $4f^7$ configuration of the Eu^{2+} ion is the same as that of the Gd^{3+} ion. Energy-level structure of trivalent rare-earth ions will be explained in Section 7.2.3. As will be seen later in Fig. 22, the ground state of the $4f^7$ configuration is ^8S, while the lowest excited state is ^6P_J. In most Eu^{2+}-activated phosphors, $5d$ levels are located below the ^6P_J state, causing broadband luminescence owing to the allowed $4f^6 5d \rightarrow 4f^7$ transition. However, in some phosphors the ^6P_J state is lower, and in such cases the observed luminescence is due to the $4f^{7*} \rightarrow 4f^7$ transition.

The *d*-orbit is fivefold degenerate in free space. In crystals of cubic symmetry it is split into the threefold *dε*- and twofold *dγ*-orbits, as will be explained in Section 7.2.2. The split interval is proportional to the crystal field strength as shown schematically in Fig. 18 (Blasse, 1978).

For weaker crystal fields the ^6P_J state of the $4f^{7*}$ configuration is lower, such as in the case of $SrFCl:Eu^{2+}$ (Sommerdijk *et al.*, 1974), which shows luminescence

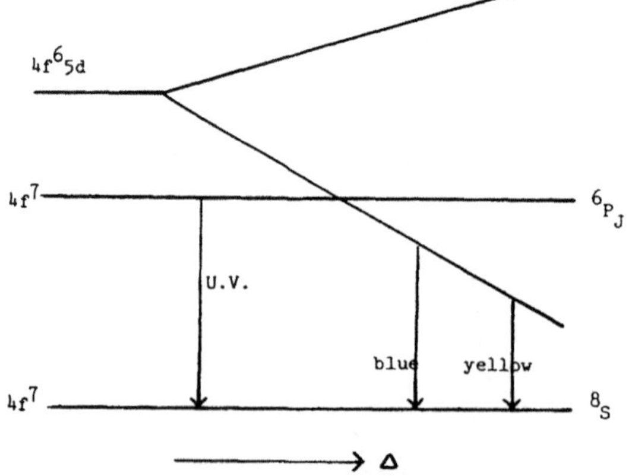

Figure 18. Schematic energy level diagram of Eu^{2+} as a function of the crystal field strength Δ (Blasse, 1978).

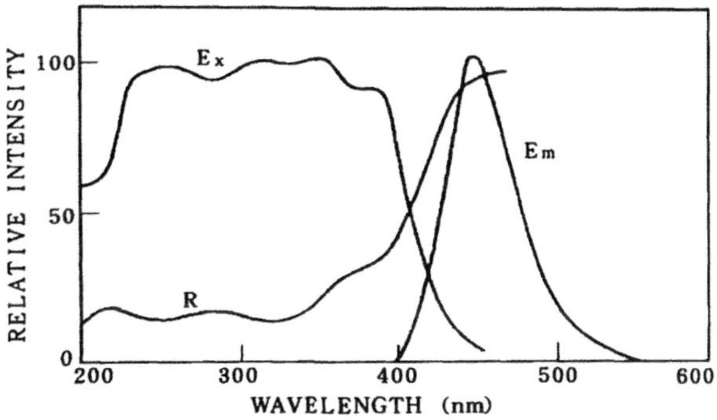

Figure 19. Emission (E_m), excitation (E_x), and reflection (R) spectra of $BaMgAl_{10}O_{17}:Eu^{2+}$ at room temperature (Verstegen, 1974; with kind permission from the Electrochemical Society).

of two sharp lines at about 365 nm at 77 K. Sharp luminescence lines are characteristic of the $f \rightarrow f$ transition, as will be explained in Section 7.2.3. At room temperature, the excited electrons are raised to the 5*d* state, so that broadband luminescence at 390 nm owing to the $5d \rightarrow 4f$ transition is observed.

With increasing crystal field strength, the 5*d* state becomes lower than the $4f^{7*}$ state, so that luminescence owing to the $5d \rightarrow 4f$ transition is produced. The stronger the crystal field is, the longer the luminescence wavelength becomes. In this way, the luminescence color of Eu^{2+} ions varies through a very wide range. Typical examples of Eu^{2+}-activated phosphors and their luminescence colors and peak wavelengths are as follows: SrB_4O_7F–near-ultraviolet, 360 nm; BaFBr–near-ultraviolet, 380 nm; $BaMgAl_{10}O_{17}$–blue, 450 nm; $(Ba, Ca, Mg)_5(PO_4)_3Cl$–blue-green, 480 nm; $SrAl_2O_4$–green, 530 nm; CaS–red, 650 nm.

Figure 19 shows emission, excitation, and reflection spectra of the $BaMgAl_{10}O_{17}:Eu^{2+}$ phosphor (Verstegen, 1974), which is an important blue phosphor for fluorescent lamps. The halfwidth of the emission spectrum is 0.32 eV, and is much smaller than that for the Sb^{3+} emission noted in Section 7.1.1. This is because the wavefunction of the excited 5*d* state is not much spread out compared to that of the ground 4*f* state, so that q_0 is small in the configuration-coordinate model shown in Fig. 7. The decay time of the Eu^{2+} ion is of the order of 10^{-7}–10^{-6} s, which is not as fast as that of allowed-transition luminescence. It is suggested that some kind of spin-forbidden nature is involved in the $4f^6 5d \rightarrow 4f^7$ transition of Eu^{2+}. On the other hand, the decay time of Ce^{3+} is fast, of the order of 10^{-8}–10^{-7} s, because the transition is spin-allowed.

In Section 2.2.2 it was pointed out that localized centers are excited by the band-to-band excitation in two ways, as shown by C and D in Fig. 1. One finds these two types of localized centers in Eu^{2+}-activated phosphors. $BaFBr:Eu^{2+}$ belongs to type C. By vacuum-ultraviolet excitation, Eu^{2+} ions are photoionized, Eu^{2+} changing to Eu^{3+}, and photoconductivity owing to electrons is observed (Iwabuchi *et al.*, 1994). On the other hand, $SrAl_2O_4:Eu^{2+}$ belongs to type D. After the Eu^{2+} ions are directly excited by ultraviolet light, holes are liberated to the valence band, Eu^{2+} changing to Eu^{1+}, and photoconductivity owing to holes appears (Matsuzawa *et al.*, 1996).

7.2. Luminescence of Forbidden-Transition-Type Ions

7.2.1. Effects of Crystal Fields

As noted in Section 2.2.2b, the forbidden character of the electric dipole transition is partially removed in crystals by the perturbation of crystal electric fields. In a case in which the crystal structure does not have inversion symmetry, the crystal field includes the odd component. By the perturbation of this component the dipole transition forbidden by the parity selection rule is made somewhat allowed. In this way oscillator strengths of the order of magnitude of 10^{-4}–10^{-6} are realized. Even in the case in which the crystal structure has inversion symmetry, the same kind of argument is possible by taking account of dynamic crystal fields with odd parity. However, oscillator strengths in this case are smaller than those for no inversion symmetry.

7.2.2. The $d \rightarrow d$ Transition

Iron group ions show luminescence owing to the $3d^{n*} \rightarrow 3d^{n}$ transition in the visible to infrared region. Among them, Cr^{3+} and Mn^{4+} ions with the $3d^3$ configuration and the Mn^{2+} ion with the $3d^5$ configuration are important for phosphor applications. In most host lattices luminescence spectra of Cr^{3+} and Mn^{4+} are made up of lines in the red to near-infrared region, with $Al_2O_3:Cr^{3+}$ (ruby) being a representative case. However, in some host lattices Cr^{3+} luminescence spectra are bands, and those of Mn^{2+} are always bands, which vary over a wide visible range from blue to red. These spectral characteristics are interpreted based on crystal field theory (ligand field theory), which will now be described.

The d-orbit is fivefold degenerate in the free ion state. If a crystal field of cubic symmetry is exerted, the orbit is split into a threefold degenerate $d\varepsilon$-orbit [composed of $d(xy)$-, $d(yz)$-, and $d(xz)$-orbits] and twofold degenerate $d\gamma$-orbits [$d(x^2 - y^2)$ and $d(z^2)$]. In the case of octahedral symmetry, $d\varepsilon$ is lower, while in the case of tetrahedral symmetry $d\gamma$ is lower. The energy difference between split $d\varepsilon$ and $d\gamma$ levels is expressed by $10\,Dq$, a parameter showing the strength of the crystal field, which is given approximately by

$$10Dq \doteqdot 5eqa^4/3r^5 \tag{28}$$

where q is the charge of ligand, a is the radius of the d-orbit and r is the distance between the metallic ion and the ligand.

The case of the d^3 configuration will now be used to explain the method of obtaining energy levels. In the free ion state ($Dq = 0$), there are eight energy terms, ^4F, ^4P, ^2G, $^2F \ldots$, from the lowest, which is due to the Coulomb interaction among electrons. In the presence of a crystal field, each term is split. In the case of octahedral symmetry, the lowest term ^4F is split into $^4A_2$, $^4T_2$, and $^4T_1$. These notations express irreducible representations of the O_h group (octahedral group), and mean that the symmetries of the wavefunctions of these states are represented by these irreducible representations. The energy of the Coulomb interaction expressed by a parameter B changes only slightly with the change of $10Dq$. It has been derived theoretically that E/B, where E is energy of each level measured from the

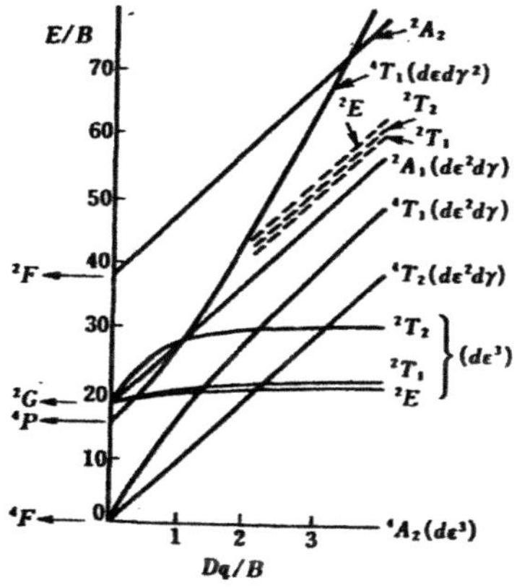

Figure 20. Energy-level diagram for the d^3 configuration in O_h symmetry derived from crystal field theory (Tanabe and Sugano, 1954; with kind permission from the Physical Society of Japan).

lowest level, is a function only of Dq/B. In this way energy-level diagrams expressed by E/B vs. Dq/B are obtained for all the d^n ($n = 2$–8) configurations (Tanabe and Sugano, 1954).

The diagram for the d^3 configuration in O_h symmetry is shown in Fig. 20. Electron configurations based on $d\varepsilon$- and $d\gamma$-orbits are shown on the right edge. The ground state is $^4A_2(^4F)$. For weaker Dq the lowest excited state is $^4T_2(^4F)$, whose energy increases with increasing Dq. For stronger Dq, on the other hand, the lowest excited state changes to $^2E(^2G)$, whose energy is almost constant with changing Dq.

Absorption spectra of $Al_2O_3:Cr^{3+}$ are shown in Fig. 21 (A. Misu, unpublished). The site symmetry of Cr^{3+} is C_3, which has no inversion symmetry. Therefore, dipole transitions are allowed to some degree by the crystal field, as noted above. In absorption spectra one observes two groups of lines named R (R_1 and R_2) and B and two bands U and Y. These are due to $d \rightarrow d$ transitions. Absorption bands below $35{,}000\,\mathrm{cm}^{-1}$ are due to charge transfer from O^{2-} to Al^{3+}. Luminescence is emitted as the reverse transition of R_1 and R_2 absorption.

The absorption transitions of R, U, B, and Y are well interpreted on the basis of Fig. 20 by assuming that Dq/B in the present case is a little less than 3. The terminal state of the R absorption is ^2E represented by a horizontal energy curve. In a horizontal curve level, the energy is not affected by the vibration of ions surrounding the center ion, because such a vibration corresponds to the periodic change in the crystal field. Therefore, a transition to a horizontal level is expected to

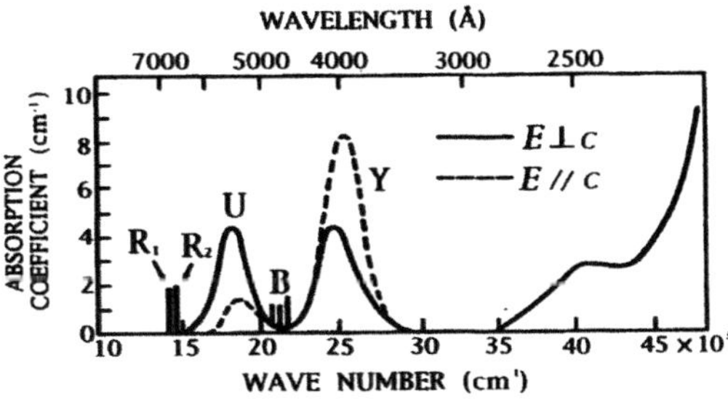

Figure 21. Absorption spectra of $Al_2O_3:Cr^{3+}$ (ruby) at room temperature. $E \parallel C$, $E \perp c$: light polarized parallel and perpendicular, respectively, to the c-axis (with courtesy Akira Misu).

become a sharp line. On the other hand, a transition to a level represented by an oblique energy curve becomes a band. Thus, the assignment of observed absorption transitions is made as follows:

$$R: {}^4A_2(d\varepsilon^3) \rightarrow {}^2E(d\varepsilon^3), \qquad U: {}^4A_2(d\varepsilon^3) \rightarrow {}^4T_2(d\varepsilon^2 d\gamma)$$

$$B: A_2(d\varepsilon^3) \rightarrow {}^4T_2(d\varepsilon^3), \qquad Y: {}^4A_2(d\varepsilon^3) \rightarrow {}^4T_1(d\varepsilon^2 d\gamma)$$

Both R and B sharp lines are due to the $d\varepsilon^3 \rightarrow d\varepsilon^3$ transition, in which only the spin state, and not the orbital state, is changed. This is considered to be the origin of the sharp lines. The ^2E state is doubly degenerate, and the degeneracy is lifted by the C_3 crystal field of low symmetry. Therefore, two closely situated R lines are observed. Oscillator strengths of these transitions are $f(R) \doteqdot 10^{-6}$ and $f(U \text{ and } Y) \doteqdot 2 \times 10^{-4}$.

R_1 and R_2 luminescence lines are located at 693.4 and 691.9 nm, respectively, at 77 K (Schawlow, 1961). They are very sharp with a linewidth of 0.3 cm^{-1}. At 60°C the R_1 and R_2 lines shift to 694.3 and 692.9 nm, repectively, and the linewidth is broadened to 12 cm^{-1} owing to the scattering of lattice phonons by the Cr^{3+} ions. The R-line luminescence is efficiently produced by the R, U, B, and Y absorption, and the quantum efficiency reaches 0.8 at low temperatures. The decay time is 4.3 ms at 77 K, but decreases to 3.0 ms at room temperature.

$LiSrAlF_6:Cr^{3+}$ emits near-infrared broadband luminescence with a peak at 840 nm and a halfwidth of 0.34 eV (Payne *et al.*, 1989). It is known that in fluorides the crystal field is weaker than in oxides, so that the lowest excited state, i.e., the emitting state, is $^4T_2$, thus producing broadband luminescence.

The Mn^{4+} ion shows red luminescence of spectra composed of several lines including phonon lines owing to the $^2E \rightarrow {}^4A_2$ transition; $3.5MgO \cdot 0.5MgF_2 \cdot GeO_2:Mn^{4+}$ is a typical Mn^{4+}-activated phosphor.

As noted at the beginning of this section, the Mn^{2+} ion of the $3d^5$ configuration produces efficient broadband luminescence over a wide range from blue to red, and is a very important activator for practical phosphors. In the energy-level diagram of the $3d^5$ configuration, the ground state is $^6A_1(^6S)(d\varepsilon^3 d\gamma^2)$, and the lowest excited state is $^4T_1(^4G)(d\varepsilon^4 d\gamma)$ for weaker Dq. The energy curve of $^4T_1$ is oblique, falling toward the right. Therefore, the emission peak shifs toward longer wavelengths with increasing Dq. In this way the luminescence color of the Mn^{2+} ion changes over the wide range. The halfwidth of emission spectra is about 0.3 eV. Typical examples of Mn^{2+}-activated phosphors and their luminescence colors and peak wavelengths are as follows: $SrSb_2O_6$–blue, 470 nm; $ZnSi_2O_4$–green, 530 nm; ZnS–orange, 585 nm; $Ca_5(PO_4)_3(F,Cl)$–orange, 587 nm; β-$Zn_3(PO_4)_2$–red, 630 nm.

$ZnS:Mn^{2+}$ is an important electroluminescence phosphor used as thin films. It is known that the position of the Mn^{2+} ground state is below the top of the valence band corresponding to case B in Fig. 1. Experiments in photoelectron spectroscopy carried out for MnSe revealed that the Mn^{2+} ground state is located 3.5 eV below the top of the valence band (Bressler *et al.*, 1993); the situation will also be almost the same for $ZnS:Mn^{2+}$.

As noted in Section 7.1.1, in $Ca_5(PO_4)_3(F,Cl):Sb^{3+},Mn^{2+}$ energy transfer from Sb^{3+} to Mn^{2+} takes place efficiently for the 253.7-nm Hg-line excitation. Mn^{2+} has no appreciable absorption for the 253.7-nm line, so that the energy transfer is nonradiative. Such an energy transfer between two localized centers is caused by quantum-mechanical resonance that is due to the Coulomb interaction between the two centers. It is pointed out that this kind of energy transfer is very effective for exciting luminescence of forbidden-transition-type centers such as Mn^{2+} (Dexter, 1953).

7.2.3. The $f \rightarrow f$ Transition

Most trivalent rare-earth (lanthanoid) ions of Ce^{3+}–Yb^{3+} produce efficient luminescence of line spectra owing to the $4f^{n*} \rightarrow 4f^{n}$ transition in the visible to near-infrared region. Among them the Eu^{3+} ion producing red luminescence and the Tb^{3+} ion producing green luminescence are especially important, acting as useful activators for various kinds of practical phosphors.

Trivalent rare-earth ions of Ce^{3+}–Yb^{3+} have the electron configuration $4f^{n}5s^{2}5p^{6}(n = 1\text{–}13)$. The configuration of divalent ions includes one more $4f$ electron. The $4f$ electron shell is located inside the $5s5p$ shell, so the influence of the crystal field on the $4f$ energy levels is weak, much weaker than on the $3d$ levels. As a result, the $4f$ levels in solids are not very different from those of the free ion state, and further do not change very much when the host lattice is changed. Absorption and emission spectra of trivalent rare-earth ions owing to $4f \rightarrow 4f$ transitions are thus composed of a number of lines. In the consideration based on the configuration-coordinate model shown in Fig. 7, the situation corresponds to $q_0 = 0$. The extent of $4f$ wavefunctions is not changed very much by excitation, so that $q_0 = 0$ is reasonable.

The $4f$ energy-level structure was very precisely investigated by using $LaCl_3$ as the host lattice, and the energy-level diagrams are provided for 13 ions of $Ce^{3+}(4f^1)$ to $Yb^{3+}(4f^{13})$ (Crosswhite and Moos, 1967; AIP Handbook, 1972). These diagrams are very useful in studies of luminescence characteristics of trivalent rare-earth ions. Figure 22 shows the diagrams of Eu^{3+}, Gd^{3+}, and Tb^{3+}, whose luminescence is discussed in this chapter. Energy levels prescribed by the inner quantum number J are further split into several sublevels in solids by the Stark effect of the crystal field. The number of split sublevels depends on the symmetry of the crystal field, and is, at the maximum $(2J + 1)$ for J of integer values and $(J + \frac{1}{2})$ for J of half-integer values. The widths of the levels in the figure indicate the extent of Stark splitting. A semicircle under a level indicates that that level emits luminescence.

Oscillator strengths of $f \rightarrow f$ transitions that are made electric-dipole-allowed by crystal fields are 10^{-5}–10^{-8}, depending on whether the field has inversion symmetry. Magnetic dipole transitions are sometimes observed, and their oscillator strengths are in the range 10^{-7}–10^{-8}. Electric quadrupole transitions are presumed to have oscillator strengths of 10^{-9}, but no examples have been found. Selection rules for $f \rightarrow f$ transitions are as follows: For rare-earth ions the inner quantum number J is regarded as a good quantum number because the spin–orbit interaction

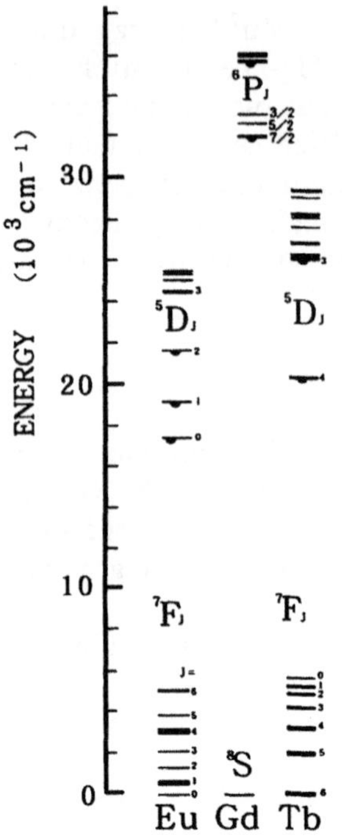

Figure 22. Energy levels of Eu^{3+}, Gd^{3+}, and Tb^{3+} in $LaCl_3$ (Crosswhite and Moos, 1967; with kind permission from Interscience Publishers).

is strong. The selection rules for the electric dipole transition made allowed by the crystal field with no inversion symmetry are $\Delta J \leqq 6$. If either of the two levels is $J = 0$, then $\Delta J = 2, 4.6$. For the magnetic dipole transition, $\Delta J = 0, \pm 1$, but $0 \leftrightarrow 0$ is forbidden.

The Eu^{3+} ion produces intense red luminescence owing to the $^5D_0 \rightarrow {}^7F_J$ transitions. Typical examples of emission spectra are shown in Figs. 23 (Pallila, 1968) and 24 (Blasse and Bril, 1970) for $YVO_4:Eu^{3+}$ and $Na(Lu, Eu)O_2$, respectively. The two spectra are fairly different, which is a reflection of the difference in crystal structure and is understood on the basis of the selection rules noted above. The

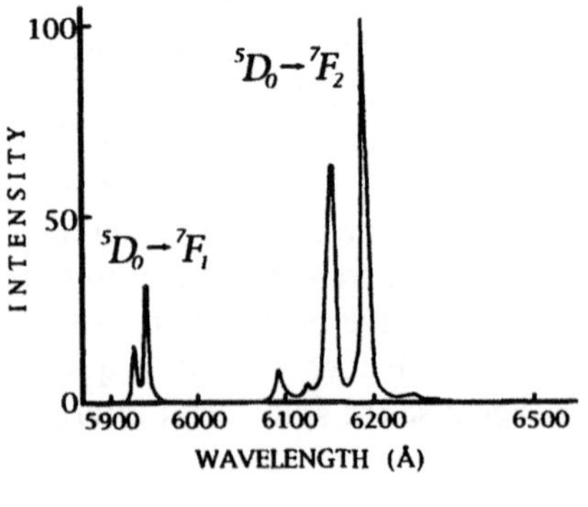

Figure 23. Emission spectrum of $YVO_4:Eu^{3+}$ at room temperature (Pallila, 1968; with kind permission from the Electrochemical Society).

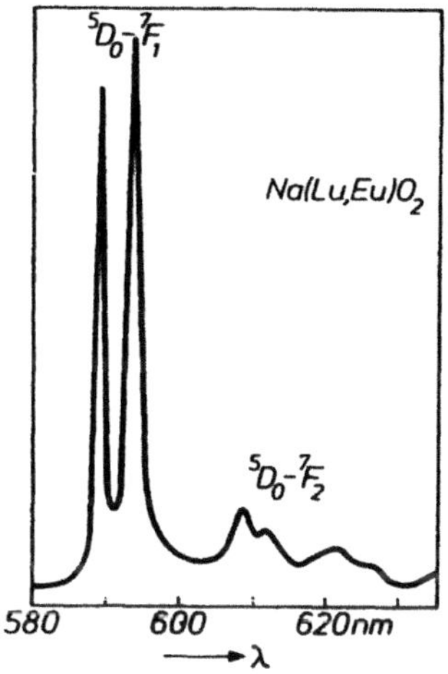

Figure 24. Emission spectrum of Na(Lu, Eu)O_2 at room temperature (Blasse and Bril, 1970; with kind permission from N. V. Philips' Gloeilampenfabrieken).

crystal structure of YVO_4, shown in Fig. 23, has no inversion symmetry, so that the emission lines of the $^5D_0 \rightarrow {}^7F_2$ transition, which is electric-dipole-allowed, is very strong. On the other hand, the structure of Na(Lu, Eu)O_2 in Fig. 24 does have inversion symmetry, so that the lines of the $^5D_0 \rightarrow {}^7F_1$ transition, which is magnetic-dipole-allowed and independent of the crystal structure, become very strong. From the point of view of applications the important Eu^{3+}-activated red phosphors are, in addition to $YVO_4{:}Eu^{3+}$, $Y_2O_2S{:}Eu^{3+}$ and $Y_2O_3{:}Eu^{3+}$, and their emission spectra are similar to that of $YVO_4{:}Eu^{3+}$.

The Tb^{3+} ion produces intense green luminescence owing to $^5D_4 \rightarrow {}^7F_J$ transitions, which are electric-dipole-allowed. Figure 25 (Nakazawa and Shionoya, 1967) shows an emission spectrum of Tb^{3+} in $Ca(PO_3)_2$ glass. The lines of $^5D_4 \rightarrow {}^7F_5$ at 18,400 cm^{-1} (543 nm) are the strongest, as is true of all the Tb^{3+}-activated phosphors, so that efficient green phosphors are available. This transition is allowed by both the electric and magnetic dipoles, which is considered to be a reason for the fact that it has the greatest strength. Practically important Tb^{3+}-activated phosphors are $LaPO_4{:}Ce^{3+}, Tb^{3+}$ (in which Ce^{3+} acts as the sensitizer) and $Y_3Al_5O_{12}{:}Tb^{3+}$.

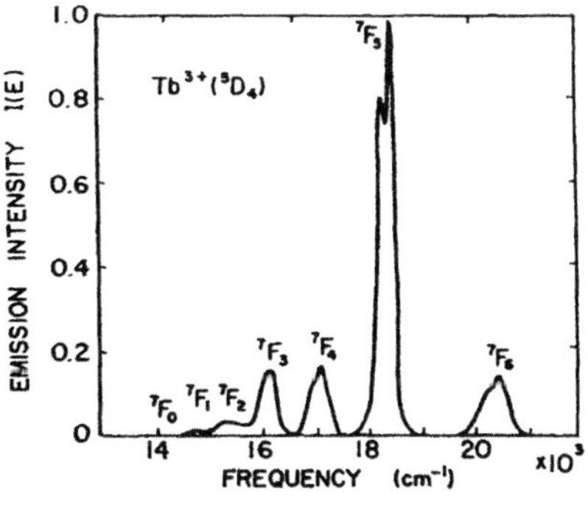

Figure 25. Emission spectrum of Tb^{3+} in $Ca(PO_3)_2$ glass at room temperature (Nakazawa and Shionoya, 1967; with kind permission from the American Institute of Physics).

8. LUMINESCENCE OF LOW-DIMENSIONAL SYSTEMS

Luminescence of low-dimensional systems (Yoffe, 1993), i.e., two-dimensional, one-dimensional, and zero-dimensional systems, is one of the recent important and interesting topics in luminescence research. Two-dimensional systems include quantum wells and layered materials, quantum wires and linear chainlike materials constitute one-dimensional systems, and quantum dots and microcrystallites are zero-dimensional. In low-dimensional systems, electrons and holes are spatially confined causing quantum confinement effects. As a result energy-level structures, and hence optical properties, become essentially different from those in three-dimensional bulk systems.

Let us consider a very thin semiconductor layer that has a thickness less than the extent of the wavefunction of electrons and is regarded as a two-dimensional system. The situation is treated in quantum mechanics as an electron in a well with an infinite barrier height. The motion of the electron in the direction perpendicular to the layer is quantized, and the energy levels are given by

$$E_n = (h^2/8\pi^2 m_e)(n\pi/L_z)^2, \qquad n = 1, 2, 3, \ldots \tag{29}$$

where L_z is the thickness of the layer. In semiconductor quantum wells both electrons and holes are quantum-confined. Therefore, the absorption corresponding to the band-to-band transition shifts to higher energies with decreasing L_z. Also in one- and zero-dimensional systems, the absorption shows blueshifts as the characteristic size of the system representing the low dimensionality like L_z decreases.

Characteristics of excitons in low-dimensional systems are also substantially different from those in three-dimensional systems. The binding energy and oscillator strength are noticeably enhanced in low-dimensional systems because of the increase in the overlap of electron and hole wavefunctions. The enhancement becomes generally more significant with decreasing dimensionality. Theoretical calculations for the binding energy in two-dimensional systems show that in the two-dimensional limit corresponding to $L_z = 0$ the binding energy reaches a value as high as four times that of three-dimensional systems. Calculations for one- and zero-dimensional systems give a much more marked enhancement of the binding energy than for two-dimensional systems. The enhancement of oscillator strength in low-dimensional systems should be noted from the point of view of luminescence materials for practical applications.

As noted above, two quantum effects work toward the transition energy of excitons in low-dimensional systems. The first is the effect of the quantization of the motion of electrons and holes causing blueshifts with decreasing characteristic size of the system. The second is the enhancement of the exciton binding energy causing redshifts. In most cases the blueshifts exceed the redshifts, so that the net result is that the exciton absorption shows a blueshift with decreasing characteristic size.

Figure 26 shows exciton luminescence spectra of CuCl microcrystallites, whose average radii show a wide variation (Itoh *et al.*, 1988). In No. 1 with a radius of 5.7 nm the peak position is shifted a little to the high-energy side from that in bulk crystals. When the radius decreases to 2.2 nm, the peak position exhibits a marked blueshift.

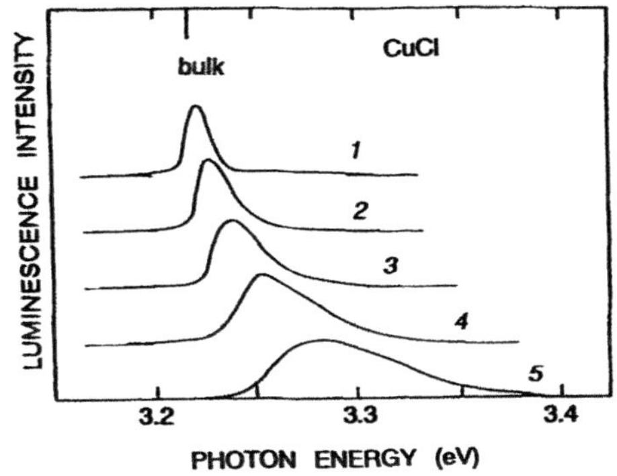

Figure 26. Exciton luminescence spectra of CuCl microcrystallites at 77 K. Average radius (in nanometers) of microcrystallites is (1) 5.7, (2) 4.9, (3) 3.4, (4) 2.7, and (5) 2.2. The peak position for bulk crystals is shown in the figure (Itoh *et al.*, 1988; with kind permission from Akademie Verlag, Berlin).

Recently it was reported that $ZnS:Mn^{2+}$ nanocrystals with sizes varying from 3.5 to 7.5 nm exhibit efficient yellow luminescence at room temperature, with decay times that are extremely shortened (Bhargava *et al.*, 1994). As noted in Section 7.2.2 the Mn^{2+} ion in solids produces efficient luminescence in the visible region owing to the forbidden $d \rightarrow d$ transition ($^4T_1 \rightarrow {}^6A_1$). The quantum efficiency of the yellow luminescence observed for 3.5 nm nanocrystals reaches 18%, which is almost the same as for ordinary $ZnS:Mn^{2+}$ powder (bulk) phosphors.

Figure 27 shows luminescence and excitation spectra of $ZnS:Mn^{2+}$ bulk and nanocrystals. In the luminescence spectrum of nanocrystals the peak is only slightly different from that of the bulk, but in the excitation spectrum it is considerably shifted to the high-energy side. The peak position of 332 nm (3.73 eV) of the bulk corresponds to the bandgap energy of the ZnS host. The observed blueshift to 265 nm (4.68 eV) in nanocrystals is a reflection of the bandgap increase owing to the quantum confinement effect as noted above, which is reasonable.

The luminescence quantum efficiency of nanocrystals decreases with the crystal's increasing size, as shown in Fig. 28. This observation is explained as follows. In $ZnS:Mn^{2+}$ nanocrystals, electrons and holes produced by the band-to-band

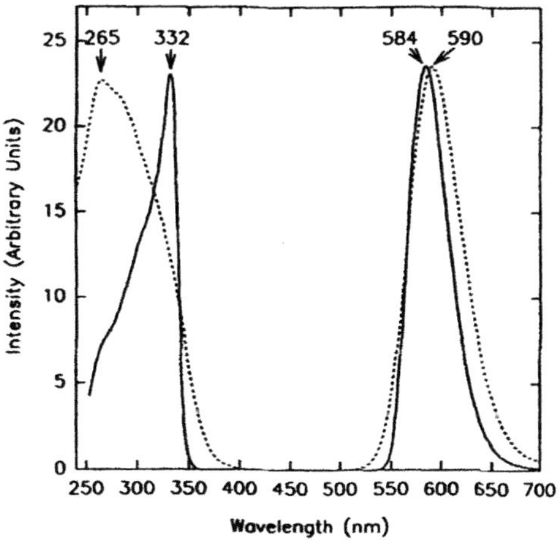

Figure 27. Luminescence and excitation spectra of $ZnS:Mn^{2+}$ at room temperature: Solid line—bulk phosphors; dashed line—nanocrystals (Bhargava *et al.*, 1994; with kind permission from the American Physical Society).

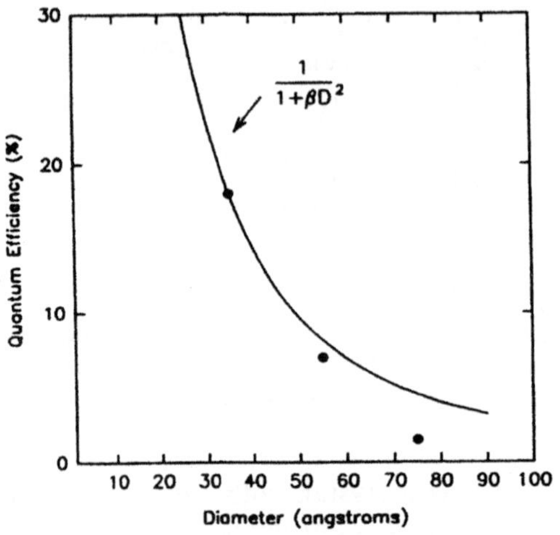

Figure 28. Variation of luminescence quantum efficiency of ZnS : Mn^{2+} nanocrystals as a function of the diameter. The solid line represents the equation $1/(1 + \beta D^2)$ (Bhargava *et al.*, 1994; with kind permission from the American Physical Society).

excitation are either captured by Mn^{2+} centers to produce luminescence or recombine nonradiatively. The quantum efficiency is expressed as $\eta = R_r/(R_r + R_{nr})$, where R_r and R_{nr} are radiative and nonradiative recombination rates, respectively. In nanocrystals nonradiative recombination is considered to take place at the surface. Then R_{nr} should depend on the number of surface atoms per unit volume, which is inversely proportional to the size D of the nanocrystals, i.e., $R_{nr} \propto D^{-1}$. R_r is regarded as equal to the electron–hole capture rate, which is proportional to the number density of Mn^{2+} at the Zn^{2+} site and is, in the case of a single Mn^{2+} ion within a nanocrystal, inversely proportional to the volume of the nanocrystal, i.e., $R_r \propto D^{-3}$. In this way, one obtains the D-dependence of the quantum efficiency as

$$\eta = 1/(1 + \beta D^2) \tag{30}$$

As seen in Fig. 28, observed results are approximated by this equation.

The luminescence decay constant of ZnS : Mn^{2+} bulk phosphors is 1.8 ms at room temperature. It was found that the decay time in nanocrystals is dramatically shortened by five to six orders of magnitude. Decay curves of 3.0 nm size nanocrystals observed by picosecond pulse excitation are double exponential with time constants of 3.7 and 20.5 ns. Because the quantum efficiency is high, almost the same as in bulk phosphors, the observed shortening of decay time has to mean that there is an extremely large enhancement of oscillator strength.

To interpret these observations, it is suggested that the hybridization of the s–p electron state of the host with the d-electron state of Mn^{2+} is caused, to a significant extent, by the spatial overlap of these states owing to the confinement. However, assuming this kind of hybridization does not completely account for the observed extraordinary shortening of the decay time corresponding to the exceedingly large enhancement of the oscillator strength. One concludes that some unknown dominant mechanism must be involved in this surprising phenomenon in luminescence of zero-dimensional systems. In any case, what is certain is that these kinds of nanocrystals doped with some luminescence activators are very promising materials for future applications of luminescence.

REFERENCES

AIP (American Institute of Physics) Handbook, 3rd Edition (1972) (McGraw-Hill, New York), p. 7–25.
Bhargava, R. N., Gallagher, D., Hong, X., and Nurmikko, A. (1994), *Phys. Rev. Lett.* **72**, 416.
Blasse, G. (1978), in: *Luminescence of Inorganic Solids* (B. DiBartolo, Ed.) (Plenum, New York), p. 457.
Blasse, G., and Bril, A. (1970), *Philips Tech. Rev.* **31**, 304.
Bressler, B. R., Klöpper, Th., and Gumlich, H. E. (1993), *J. Vac. Sci. Technol.* **B11**, 1621.
Butler, K. H., and Jerome, C. W. (1950), *J. Electrochem. Soc.* **97**, 265.
Crosswhite, H. M., and Moos, H. W. (Eds.) (1967), *Optical Properties of Ions in Crystals* (Interscience, New York).
Dexter, D. L. (1953), *J. Chem. Phys.* **21**, 836.
Era, K., Shionoya, S., and Washizawa, Y. (1968*a*), *J. Phys. Chem. Solids* **29**, 1827.
Era, K., Shionoya, S., Washizawa, Y., and Ohmatsu, H. (1968*b*), *J. Phys. Chem. Solids* **29**, 1843.
Gebhardt, W., and Kühnert, H. (1964), *Phys. Lett.* **11**, 15.
Hanamura, E. (1976), in: *Optical Properties of Solids: New Developments* (B. O. Seraphin, Ed.) (North-Holland, Amsterdam), Chap. 3.
Itoh, T., Iwabuchi, Y., and Kotaoka, M., (1988), *Phys. Stat. Sol. (b)* **145**, 567.
Iwabuchi, Y., Mori, N., Takahashi, K., Matsuda, T., and Shionoya, S. (1994), *Jpn. J. Appl. Phys.* **33**, 178.
Johnson, P. D. (1961), *J. Electrochem. Soc.* **108**, 159.
Klingshirn, C., and Haug, H. (1981), *Phys. Rep.* **70**, 315.
Kukimoto, H. (1990), *J. Crystal Growth* **101**, 953.
Matsuzawa, T., Aoki, Y., Takeuchi, N., and Murayama, Y. (1996), *J. Electrochem. Soc.* **143**, 2670.
Merz, J. L. (1968), *Phys. Rev.* **176**, 961.
Nakazawa, E., and Shionoya, S. (1967), *J. Chem. Phys.* **47**, 3211.
Pallila, F. C. (1968), *Electrochem. Tech.* **6**, 39.
Payne, S. A., Chase, L. L., Smith, K. K., Kway, W. L., and Newkirk, H. W. (1989), *J. Appl. Phys.* **66**, 1051.
Sarace, J. C., Kaiser, R. H., Whelan, J. M., and Leite, R. C. C. (1965), *Phys. Rev.* **137**, A623.
Schawlow, A. L. (1961), *Advances in Quantum Electronics* (J. R. Singer, Ed.) (Columbia University Press, New York), p. 50.
Shionoya, S., Kobayashi, Y., and Koda, T. (1965), *J. Phys. Soc. Jpn.* **20**, 2946.
Shionoya, S., Saito, H., Hanamura, E., and Akimoto, O. (1973), *Sol. St. Comm.* **12**, 223.
Sommerdijk, J. L., Verstegen, J. M. P. J., and Bril, A. (1974), *J. Lumines.* **8**, 502.
Song, K. S., and Williams, R. T. (1993), *Self-Trapped Excitons, Solid State Science*, Vol. 105 (Springer-Verlag, Heidelberg).
Suzuki, A., and Shionoya, S. (1971*a*), *J. Phys. Soc. Jpn.* **31**, 1462.
Suzuki, A., and Shionoya, S. (1971*b*), *J. Phys. Soc. Jpn.* **31**, 1455.
Tanabe, Y., and Sugano, S. (1954), *J. Phys. Soc. Jpn.* **9**, 753, 756.
Thomas, D. G., Gershenzon, M., and Trumbore, F. A. (1964), *Phys. Rev.* **113**, A269.
Thomas, D. G., Hopfield, J. J., and Augustyniak, W. M. (1965), *Phys. Rev.* **140**, A202.
Trumbore, F. A., Gershenzon, M., and Thomas, D. G. (1965), *Appl. Phys. Lett.* **9**, 4.
Ulbeich, R. G., and Weisbuch, C. (1976), unpublished; reprinted in Weisbuch, C. (1977), Thése de doctorat d'Etat, Univ. Paris 7, (The figure was cited from Sturge, M. D. (1982). In: *Excitons* (E. I. Rashba and M. D. Sturge, Eds.) (North-Holland, Amsterdam).
Unuma, Y., Masumoto, Y., and Shionoya, S. (1982), *J. Phys. Soc. Jpn.* **51**, 1200.
van Eijk, C. W. E. (1994), *J. Lumines.* **60/61**, 936.
van Roosbroeck, W., and Shockley, W. (1954), *Phys. Rev.* **94**, 1558.
Varshni, Y. P. (1967), *Phys. Stat. Sol.* **19**, 459; **20**, 9.
Verstegen, J. M. P. J. (1974), *J. Electrochem. Soc.* **121**, 1623.
Yoffe, A. D. (1993), *Adv. Phys.* **42**, 173.

4

Cathodoluminescence

Stefan Myhajlenko

1. INTRODUCTION

1.1. Historical Perspective

The term cathodoluminescence (CL) derives from the use of high-energy electrons (cathode rays) as the means of excitation. The earliest use of CL was probably by William Crookes in the 1880s, who applied the technique to the characterization of lead glasses. However, it was the development of the analytical (x-ray) electron microprobe in the late 1950s that promoted the current interest in CL. Early investigations were in mineralogy and petrology. Structural and compositional variations in quartzes, dolomites, and feldspars were visible with CL imaging, as well as residual skeletal structures of fossils (Smith and Stenstrom, 1965). During the 1960s, the commercial availability of scanning electron microscopes (SEMs), along with advances in emerging technologies, such as space exploration and microelectronics, brought about a further increase in CL applications. For example, CL was useful in the development of optoelectronic materials and devices (Schiller and Boulou, 1975). Today, there are CL systems commercially available, such as those marketed by Oxford Instruments for SEMs (MonoCL™: Wright, 1991) and Nuclide Corporation for optical microscopes (Luminoscope®: Marshall, 1993).

1.2. Current Outlook

CL has applications in many areas: archaeology, biology, cathode-ray tubes, ceramics, herbicides, medicine, minerals, polymers, semiconductors, and superconductors, to name a few. Although CL instrumentation is well established, developments continue. It is quite often the case that CL attachments are added to imaging instruments, such as the scanning probe microscope (Montelius *et al.*, 1992). Novel investigations also give rise to developments, e.g., the measurement of the lifetimes

---

Stefan Myhajlenko • Center for Solid State Electronics Research, College of Engineering and Applied Sciences, Arizona State University, Tempe, Arizona 85287, United States.

Luminescence of Solids, edited by D. R. Vij. Plenum Press, New York, 1998.

of excited-states by electron–photon correlation (Ahn and Krivanek, 1985). CL modeling and quantitation represent current challenges in the field.

A host apparatus can vary from a simple flood-illumination instrument to an advanced analytical electron microscope. The SEM represents what is probably the most familiar kind of equipment. Modern SEMs are equipped with an array of imaging and analytical capabilities, and CL is one of these modes. The CL mode is convenient to use as is it contactless, relatively nondestructive, and requires little sample preparation. Imaging of panchromatic CL radiation (detection of all wavelengths) is the easiest and most widely used option. The inherent usefulness of SEM–CL lies in visualization. The old cliché "a picture is worth a thousand words" truly emphasizes the simple power of the technique. Image contrast may arise from a number of different mechanisms (often occurring simultaneously), e.g., enhanced nonradiative recombination, emission at undetectable wavelengths, variations in radiative efficiency, optical scattering, and optical absorption. Careful interpretation of CL contrast can reveal the presence of crystallographic defects, impurity inhomogeneties, gettering, stress variations, space-charge regions, and interfaces. More complex equipment arrangements can provide spectrally or temporally resolved images with submicron spatial resolution. The sample "physics" must be optimized in such circumstances. For example, high spatial resolution can be achieved by the confinement of excess carriers, which restricts carrier diffusion. Confinement can be realized via surface recombination effects (Cumberbatch *et al.*, 1981) and with quantum well structures (Warwick, 1991). Spectrally resolved imaging allows the "mapping" of, for instance, specific impurities and crystal compositions, e.g., the effects of enhanced Zn diffusion by dislocations in AlGaAs/GaAs epilayers (Roedel *et al.*, 1989). CL thus complements related techniques that lack spatial resolution, e.g., nonscanning photoluminescence (PL). It should be noted that the high-energy excitation used in CL compared with PL leads to emission from all luminescence processes in the solid.

It is important to appreciate the limitations from the outset. For instance, samples must: (1) be luminescent; (2) be resilient to vacuum conditions (withstand $<10^{-5}$ Torr); (3) not decompose under electron bombardment; and (4) be resistant to atomic displacement since this damage typically has deleterious effects on luminescence efficiency; and finally (5) they should be electrically conductive to avoid sample charging (although optically transparent metal coatings can help). This chapter summarizes basic principles, instrumentation, and recent applications of CL in a wide range of solids. The following list provides support material: Pankove (1971), Lumb (1978), Aspnes (1985), Yacobi and Holt (1986, 1990), Marshall (1988), Holt and Joy (1989), Holt (1992), and Remond *et al.* (1992).

2. FUNDAMENTAL ASPECTS

2.1. Electron-Beam Interactions in Solids

An appreciation of how energetic electrons interact with solids is important to an understanding of CL. The dielectric response of the solid is fundamental to physical interpretation. One can view the incoming fast electrons as an impulse of applied electric field (charge) to the solid. When electrons enter a solid: (1) the host charge can redistribute to oppose the sudden change in electric field; (2) interactions with the crystal lattice can result in scattering (deviations from the original path); (3) inelastic scattering can result in well-defined energy losses of the incident electron; and (4) such characteristic energy losses can involve single-particle and collective (plasmons) excitations of the host electrons (Pines, 1964). It is the subsequent relaxation to the ground state, between characteristic energy levels of the solid, that produces the photons that are detected as CL. These transitions may involve discrete atomic levels (rare-earth metal *d*- and *f*-states), extended bands (conduction band to hydrogen-like shallow impurity states), and *F*-centers. Transitions between tightly bound inner core states result in x-ray photon emissions of very high energies (>1 keV). Excitation of bulk (volume) plasmons in inorganic crystals such as SiC, GaAs, or Al_2O_3 gives rise to energy losses in the range 10–30 eV. Characteristic CL photons fall in the visible part of the electromagnetic spectrum (0.5–6.0 eV). Other electron–solid interactions can also produce light. When a fast electron is slowed down by Coulombic interaction with the host atoms, the deceleration (energy loss) results in bremsstrahlung (braking radiation). This background light is characterized by a continuous spectrum from zero energy to the incident-electron energy. Another source of light is Cerenkov radiation, which occurs when the velocity of the electron in the solid exceeds the speed of light in the solid (dependent on the refractive index). Radiative decay of surface plasmons in metals can also produce visible photons.

The retardation of electrons of given energy is characterized by a depth-dependent energy loss function and the range (the ultimate penetration depth into the solid). The incoming electrons undergo either elastic (large-angle) or inelastic (small-angle) scattering on sensing the lattice potential. These interactions result in zigzag trajectories within the solid. The deviations from the projected primary trajectory broaden the original probe diameter, determined by the initial electron energy and the density of the solid. The cumulative effect of multiple scattering and subsequent diffusion produces a pear-shaped excitation volume in a typical solid (see Fig. 1). This interaction volume defines the source for CL. In semiconductors, the CL generation volume is typically larger than the initial interaction volume (of radius, r_g) because of minority carrier transport (determined by the carrier diffusion length L). In bulk solids, one expects the effective CL probe diameter to be proportional to $(L^2 + r_g^2)^{1/2}$ (Davidson, 1977).

Electrons that undergo large-angle elastic scattering can sometimes exit the solid with no significant energy loss. In the classical Rutherford picture, the

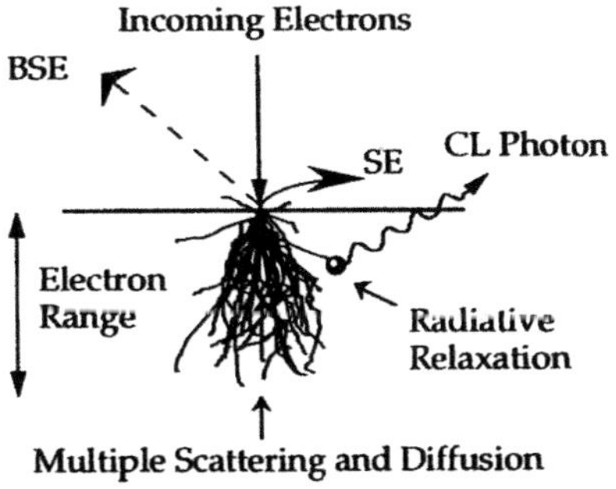

Figure 1. Schematic of electron–solid interaction depicting random trajectories of individual electrons, a pear-shaped excitation volume resulting from multiple scattering and diffusion, and emission of backscattered electrons (BSE), secondary electrons (SE), and photons (CL).

deflection of an electron of given energy by the unscreened positive charge of the nucleus yields a differential cross section $d\sigma/d\Omega$ as a function of the scattering angle that is primarily dependent on the atomic number Z. This cross section is a measure of the probability that a scattering event will occur. Two important parameters derived from this description are: (1) the mean free path (average distance between scattering events), and (2) the scattering probability. These terms are useful in Monte Carlo calculations of electron trajectories. The cumulative effect of elastic scattering results in backscattered electrons (BSE), the basis of SEM (Z-contrast) compositional imaging. Inelastic scattering, where (electron) momentum transfer is small and energy loss is relatively large, leads to the production and emission of secondary electrons, photons (CL), thermal and acoustic phonons, surface and bulk plasmons, characteristic Auger electrons, and x-rays.

Primary electron energy dissipation is a complex cascade process that is transmitted ultimately into single-particle excitations and phonons (heat). Single-particle excitations are more likely with semiconducting and insulating materials because of the low free electron density. Collective excitations (plasmons) occur more readily in higher conductivity solids (metals). Single-particle excitations can decay into electron–hole pairs (Raether, 1980). Radiative recombination of electron–hole pairs can produce CL. In fact, because of the cascade nature of deexcitation, CL generation is complex, e.g., generated x-rays can provide a secondary means of CL production (known as roentgenoluminescence). In particular, the production of secondary electrons or secondaries (SEs) is important to the CL generation process, and they are generated along the entire zigzag path of energy dissipation, which may be many microns in length. Secondaries are produced by the excitation of core and conduction electrons (single-particle) and the decay of plasmons (collective) (Seiler, 1983; Reimer, 1985). Those electrons that escape from the solid into the vacuum are used in SEM (emissive) imaging. They are low in energy (1–50 eV) and must exist within a mean free path of the surface (measured in tens of angstroms) in order to escape. SE yield is influenced by adsorbed layers, such as hydrocarbons and water vapor, substances commonly found in SEM vacuums (Halbritter, 1983).

Secondaries are pertinent to CL since they are derived from processes that also generate electron–hole pairs. Secondary and CL production are related to the rate of energy loss. Bethe (1930) describes the mean energy loss dE per unit pathlength ds (the stopping power in eV/Å) as

$$dE/ds = -785(\rho Z/AE)\ln(bE/J)\ [\text{eV/Å}] \tag{1}$$

where E is the instantaneous energy of the electron (in eV), ρ is the density of the solid (in g/cm^3), Z is the atomic number, A is the atomic weight, and J is the mean ionization potential (in eV) of the solid (in effect, the average energy loss per interaction for all possible processes). The quantum-mechanical value for $b[(e/2)^{1/2} = 1.166]$ is most commonly used in calculations (Reimer, 1985). Values for J are tabulated (ICRU, 1984) or may be approximated (for $E > 10\,\text{keV}$) with an empirical relation developed by Berger and Seltzer (1964)

$$J = \begin{cases} 11.5Z\ [\text{eV}] & \text{for } Z \leqslant 12 \\ 9.76Z + 58.8Z^{-0.19}\ [\text{eV}] & \text{for } Z \geqslant 13 \end{cases}$$

For a multielement solid the individual J terms are added together according to Bragg's rule.

A useful term obtained by numerical integration of Eq. (1) (avoiding the singularity when E approaches J/b) is the Bethe range. This is the overall pathlength an electron would travel in the course of slowing down from energy E to zero energy if the rate of energy loss were equal to the mean rate of energy loss. Kanaya and Okayama (1972) describe an alternative expression for the range that is in excellent agreement with experiment: the electron range (in mass-thickness units, $g\,cm^{-2}$) is found to be proportional to E^n. For example, $n = 1.67$ applies to energies in the range 10–100 keV, while $n = 1.33$ is more appropriate for 1–10 keV (Bongeler *et al.* 1993). This dependence on energy (of ρR) comes from the different energy loss contributions to the stopping power (eV/Å). When the kinetic energy of an electron is less than that required to ionize the inner shells (K, L, M), that process no longer contributes to energy loss, and other mechanisms must then take over. This fact can influence the inelastic mean free path (and range) at low electron energies (<5 keV) (Ashley *et al.*, 1979). This effect is important given the current interest in low-voltage SEM (LVSEM). Joy and Luo (1989) describe a semiempirical modification of Bethe's formulation [Eq. (1)] for lower voltages that uses an energy dependence for J, where J is replaced by the term $J' = J(1 + kJ/E)^{-1}$ and $k \approx 1$. The modified range R is given by [residual ranges values at 50 eV are available (Tung *et al.*, 1979)]

$$R = \text{Residual range at } 50\,\text{eV} + \int_{0.05}^{E} [1/(dE/ds)]dE \qquad (2)$$

A knowledge of dE/ds can help determine the magnitude and spatial distribution of any related emissions, such as x-rays and CL (as the rate of excess carrier generation is proportional to energy loss). However, there is no simple mathematical solution to describe the huge variety of electron interactions in solids. The problem lies in accounting for all possible scattering events that a single energetic electron may experience during energy dissipation. The Monte Carlo calculation lends itself naturally to these problems (Joy, 1988; Newbury, 1989; Joy, 1995). Early electron–solid interaction calculations addressed characteristic x-ray distributions (Green, 1963) and electron-scattering trajectories (Bishop, 1965): the trajectories in Fig. 1 simulate such a calculation. Specifically, each scattering event and pathlength traveled by the electron between events is determined by a probability function based on either experimentally determined (empirical) and/or theoretical expressions. Random numbers are used to predict scattering angles. This approach provides the basis for more detailed CL calculations when combined with established analytical formulations, such as those for excess carrier distributions (Berz and Kuiken, 1976; Donolato, 1978; Jakubowicz, 1986). The three-dimensional (3D) SEM CL simulation model of Phang *et al.* (1992) and the complementary electron-beam-induced current (EBIC)/CL model of Czyzewski and Joy (1990) are based on this approach. The hope is to develop models as widely applicable as those used for x-ray microanalysis, e.g., Gaussian $\Phi(\rho z)$ (Bastin *et al.*, 1986). Presently, there is no unified interpretation for the CL response of a solid.

The last thing to be noted in this section is the charged nature of the excitation. Solids irradiated with low-energy electrons (<5 keV) can sometimes undergo a change in charging status. This is known as the crossover (voltage) phenomenon and is observed with insulators, polymers, metals, and semiconductors (Brunner and Menzel, 1983). Specifically, the total electron yield from a solid (BSE and SE) is found to be dependent upon the applied beam voltage. This phenomenon can influence the relative surface potential of the material. For example, between crossover voltages E_1 and E_2, the sample may charge positively by a few volts because of the SE energy distribution that peaks between 2 and 5 eV. On the other hand, for voltages below E_1 and above E_2, the sample can charge negatively (Joy, 1987; Boyes, 1989). Excessive negative charging (with surface potentials up to the applied beam energy) can occur for beam voltages greater than E_2. The extent of the charging depends on the primary beam current, material conductivity, and the geometrical configuration of the SE detector in a vacuum chamber. A comprehensive description of charging effects in a wide range of materials is given by Cazaux and Le Gressus (1991). Subsequent work suggests that crossover and charging phenomena can influence CL processes: GaAs (Myhajlenko, 1991) and Ce:YAG (Feng and Warde, 1993).

2.2. Luminescence Processes In Solids

Luminescence processes occur in both organic and inorganic solids. In the case of organic materials, luminescence results from molecular excitation. Transitions typically involve the vibrational modes of the solid and emissions typically exhibit Stokes' shifts. Luminescence from inorganic solids is usually associated with impurities, excited states of isolated atoms and ions, and defects. Luminescence from isolated ions/atoms can resemble that from organic solids since transitions may involve localized excited and ground states. Luminescence processes generally compete with nonradiative processes. Competing nonradiative mechanisms are associated with surfaces and interfaces, deep-level impurities and complexes (e.g., metals, point defects, microvoids, and atomic-scale aggregates), spatially extended crystal defects (e.g., dislocations and loops), and Auger-related processes (localized multiple-particle interactions). These competing processes typically exchange charge between closely separated energy levels and may involve strong coupling with phonons (lattice or local modes). The luminescence of solids is generally dominated by extrinsic factors. The luminescence of semiconductors is usually associated with substitutional shallow impurities whose properties can be described by the modified Bohr hydrogen-atom approximation. In minerals, the luminescence is typically associated with activating ions, such as transition metals. The properties of activators are best described by crystal field theory.

2.2.1 Shallow Impurities

Figure 2 shows intrinsic and extrinsic optical transitions in solids with moderate energy gaps, such as Si, GaAs, GaP, ZnSe, and CdTe. At room temperature, the luminescence is dominated by radiative recombination of excess free electrons and free holes across the conduction–valence bands (energy gap, E_g). Discrete shallow

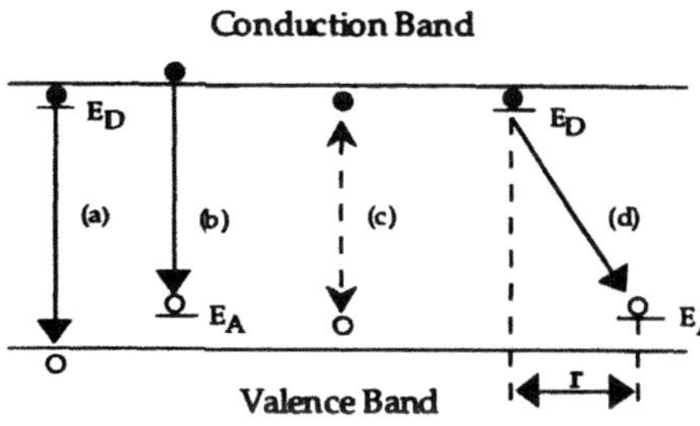

Figure 2. Electronic transitions in bandgap solids that produce optical emissions: (a, b) band-to-impurity, (c) excitonic interaction, and (d) donor–acceptor pair transitions.

impurity luminescence is usually observed only at low temperatures (<30 K) because of thermal ionization. Transitions (a) and (b) are extrinsic and involve energy levels associated with donor (E_D) and acceptor (E_A) impurities. The levels represent the ionization energies of the respective impurities. These transitions are referred to as free-to-bound since they involve both a free and a bound particle. Transition (a) is commonly denoted as (D^0, h). In this transition, an electron bound to a neutral donor D^0 recombines with a free hole h in the valence band. Transition (b) described as e, A^0 involves a free electron e and a hole bound to a neutral acceptor A^0. Transition (d) is known as donor-acceptor pair (DAP) recombination and occurs in compensated crystals. This bound-to-bound transition involves electrons and holes trapped on donors and acceptors. The emission energy hv is given by

$$hv = E_g - (E_A + E_D) + (q^2/\varepsilon r) - (q^2 b^5/\varepsilon r^6) \quad (3)$$

where q is the electronic charge, ε is the static (low-frequency) dielectric constant, b is an adjustable van der Waals parameter, $q^2/\varepsilon r$ represents the Coulomb interaction between a DAP separated by distance r, and $q^2 b^5/\varepsilon r^6$ represents a polarization term (van der Waals energy for interacting dipoles). The last term is usually negligible and is only applicable to very close pairs (Dean, 1968). Since r varies in a discrete manner, DAP emission should exhibit a large number of discrete lines. However, lines usually overlap (for $r > 40$ Å) producing a broad emission band. DAP recombination is also sensitive to excitation level. DAP peaks exhibit blueshifts with increasing excess carrier injection density, and the more widely separated pairs saturate first.

2.2.2. Excitons and Polaritons

Excitonic properties are associated with the periodicity of the host crystal lattice. The intrinsic exciton can be viewed as a virtual and polarized excitation mode of the crystal. From the physical perspective, the exciton is a Coulomb-coupled electron–hole pair (see Fig. 2c). This interaction may be weak (Wannier, 1937) or strong (Frenkel, 1931). The basic properties of Wannier–Mott excitons are described by the effective-mass approximation (modified hydrogen atom), with the exciton ionization energy given by

$$E_x = -m^* q^4 / 2h^2 \varepsilon^2 n^2 \quad (4)$$

where n is an integer $\geqslant 1$ for ground and excited states and $m^* = m_e m_h/(m_e + m_h)$ is the reduced effective mass of the electron, m_e, and hole, m_h. The exciton is characterized by an energy level E_x and wavevector **K** (momentum vector describing the center-of-mass motion for electron and hole). Wannier excitons are free to wander around the crystal because the excitation can be shared with the entire lattice, a coherent superposition of a conduction electron with all of the electrons in the valence band. It is worth noting that it is not really appropriate to represent excitons on a single-electron energy diagram (i.e., Fig. 2) as excitons are bosons (zero-spin particles).

For excitons to form: (1) the free carrier concentration (impurity content) must be low to avoid Coulombic screening of the electron and hole interaction; (2) the prevailing electric fields (e.g., surface depletion region) must be low to prevent field or impact ionization (physical separation of electron–hole pair); and (3) the available thermal energy (i.e., temperature) must be less than E_x to avoid thermal dissociation. Further explanation of the two limiting cases noted above are that Wannier–Mott excitons occur in "narrow-gap" semiconducting crystals (e.g., GaAs and ZnSe), where ε is large and m^* is small, while Frenkel excitons occur in larger-bandgap solids and molecular crystals (e.g., anthracene), where ε is small and m^* is large. Intermediate situations exist with, e.g., Xe, KI, and BaO. Other interesting excitonic attributes include excitonic molecules (two-exciton interaction), electron–hole droplets (Bose condensation of excitons), and polaritons (the interaction between electromagnetic waves and polarization fields). Excitonic phenomena also occur in artificially grown (quantum-mechanical) materials. These low-dimensional solids are grown one atomic layer at a time by techniques such as molecular beam epitaxy (MBE) and metal–organic chemical vapor deposition (MOCVD). Manufactured layers such as single quantum wells (SQW), multiple quantum wells (MQW), and superlattices (SL) form the building blocks of interesting concepts, such as quantum dots, quantum wires, and quantum boxes, as well as practical electronic devices (Petroff *et al.*, 1987). Excitons are strongly bound to these systems because of one-dimensional (1D) or two-dimensional (2D) confinement. Figure 3 depicts the bandgap energy scheme for a GaAs/GaAlAs SQW structure. Carriers are confined in this structure by sandwiching the GaAs quantum-well layer between layers of wider-energy gap GaAlAs material. Quantum confinement occurs when the GaAs layer thickness is less than the de Broglie electron wavelength (for layer thicknesses $L_x < 300$ Å). The problem of calculating the electron energy levels in a SQW is much akin to the Schrödinger "particle in a box" problem (see the dotted lines in Fig. 3). The electron energy levels are dependent primarily on well thickness (proportional to $1/L_x^2$). There is a corresponding relation for excitonic energies. Energy levels in real SQW structures are more complex than those predicted by the box problem (Del Sole *et al.*, 1988).

Bound excitons (BE) of a simpler nature also occur in bulk crystals. Extrinsic BE luminescence is the radiative recombination of free excitons (FE) that are bound to neutral and excited impurities, isoelectronic traps, and defects. Exciton decay from these bound states yields emissions with much narrower spectral linewidths than the corresponding FE because of reduced kinetic broadening: the exciton is spatially localized at an impurity. The BE emission energy is given by $h\nu = E_g - E_x - E_{bx}$, where E_{bx} is the binding energy of the exciton to the impurity. The nomenclature

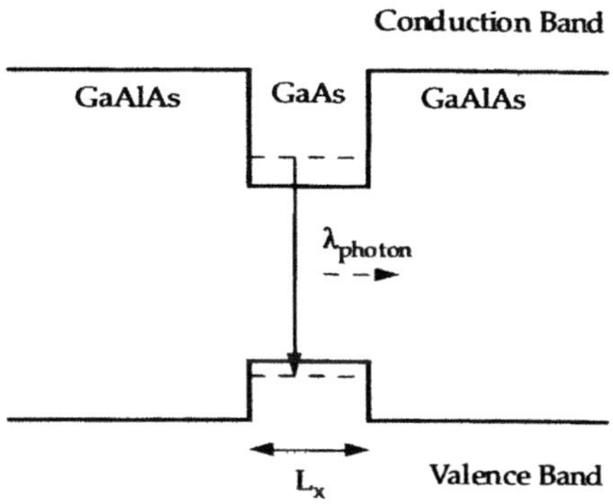

Figure 3. Schematic of a SQW GaAs/GaAlAs structure depicting exciton emission from the well. The exciton emission wavelength λ_{photon} dependence on well thickness (L_x) is a solution to the "particle in a box" problem.

assigned to neutral (and excited) donor and acceptor BE are: $D^0, X(D^+, X)$ and $A^0, X(A^-, X)$, respectively. BE luminescence typically dominates the near-band-edge emissions from semiconductors. The BE transition is particularly relevant to indirect-gap materials (energy minimum at $k \neq 0$) since momentum can be conserved at the impurity. Such a transition can leave a neutral donor impurity in an excited orbital (2*s*, 2*p*) state. This is also known as the two-electron satellite (TES) transition. Alternatively, a donor electron can be excited into the conduction band. Donor impurity identification is possible by measuring the energy difference between the principal D^0X and its TES, especially in wide-gap semiconductors such as ZnSe (Dean *et al.*, 1981).

A complete description of exciton-related luminescence must include polaritons. The polariton is a mixed-mode excitation of the crystal, which is the polarizing interaction between a propagating transverse electromagnetic wave and a harmonic oscillator resonating at the same frequency. This coupling between photons and free excitons is strong in the presence of spatial dispersion (ε dependent on k) (Hopfield, 1958, 1966). This coupling leads to a splitting of the dispersion curve into upper and lower polariton branches. The lower branch is more photon-like in character, while the upper branch is more exciton-like. This branching is manifest in luminescence spectra at or very close to the FE resonance energy. Understanding of the polariton response of a crystal is far from complete because many factors influence the initial distribution of polaritons in space or energy, including the nature of the crystal boundary itself. Polariton propagation must be disrupted by some inelastic event (e.g., impurity scattering) in order to give the resulting photon (when the exciton has truly collapsed) an opportunity of escaping the crystal. Current theoretical descriptions of polariton luminescence are viewed in terms of a transport-related problem (Steiner *et al.*, 1986).

2.2.3. Activators

Another important luminescence mechanism is associated with activated impurities. Well-known examples are the Eu-doped phosphors used for color TV screens. The sharp (red) emission is primarily characteristic of the impurity ion Eu^{3+} itself and not the host crystal Y_2O_3. Radiative transitions take place between excited and ground states of the impurity (the conduction–valence bands need not be involved). The impurity energy levels can exist within the bandgap of a solid. These luminescence-activating impurities are primarily transition metals (partially filled *d*-shells) and rare-earth elements (partially filled *f*-shells).

Crystal field theory provides a description of the energy levels based on symmetry coordination (e.g., octahedral and tetrahedral) within a given crystal environment. The crystal field splitting behavior of Cr^{3+} in Al_2O_3 (ruby) and $Be_3Al_2(SiO_6)_3$ (emerald) is well documented in this respect and known to give the gems their color (Imbusch, 1978). Activated/coactivated impurity luminescence is commonly observed in oxides, phosphors, ceramics, minerals, and wide-gap semiconductors. Coactivators (or codoping) refers to the situation where two different impurities must be present to produce a specific luminescence signature. The enhanced optical activity of Er in the presence of O in Si is a recent technological example (Priolo *et al.*, 1993). In comparison with semiconductors, interpretation of CL from minerals is still in its infancy. The identification of luminescence activator/coactivator/quencher impurities in minerals is a formidable challenge. This is due both to the complex nature of luminescence processes in these solids and the fact that emission is usually associated with trace impurities in concentrations typically below x-ray microprobe detection limits, i.e., <0.01 at. %) (Waychunas, 1988).

2.2.4. Deep Impurities

"Deep" implies that the attributes of the impurity cannot be described by the effective-mass approximation. Substitutional transition-metal impurities in covalent semiconductors fall into this category, e.g., Cr in GaAs. The impurity typically exhibits different charge states, e.g., 3^+ ($3d^3$) can represent the neutral state of Cr and 2^+ ($3d^4$) the first ionized state. The configurational-coordinate (CC) diagram provides a useful means of describing the ion-plus-lattice system. This description was originally formulated for *F*-centers in alkali halides (Bartram and Stoneham, 1975). The CC diagram describes the energy of the ground and first excited state of an impurity as a function of the normalized lattice position. The equilibrium atomic (lattice) configuration is determined by the electronic distribution of the system (including lattice distortion). Transitions between ground and intermediate excited states typically involve atomic displacements (phonon emissions). Relaxation back to the ground state can be partially accommodated by photon emission. Equilibrium is again restored by phonon emission (atomic displacement back to the original lattice coordination).

The difference in energy that is accommodated by lattice displacements is known as the Franck–Condon shift. Also known as the Stokes' shift if the transitions are optical: absorption (excitation) and emission (relaxation). If the ground- and excited-state curves on the CC diagram should overlap, relaxation may be nonradiative. Whether or not an impurity center emits light depends on how closely the two curves overlap, as well as on the strength of local phonon coupling. Thermal activation (temperature) plays a key role and is best expressed through the Huang–Rhys factor (Huang and Rhys, 1950). The impurity is referred to as a killer (or quencher) center if transitions are predominantly nonradiative. For instance, Mn^{2+} is an activator responsible for CL in carbonates. However, emission can be dramatically reduced in the presence of Fe^{2+} or Fe^{3+} quenching ions (Waychunas, 1988).

2.3. Cathodoluminescence Generation

An appreciation of the CL generation process may help answer some basic questions, e.g., how many photons are generated per incoming electron? One begins by assuming that the CL intensity produced within the crystal is proportional to the rate of electron energy loss since this determines the rate of electron–hole pair production. Other material properties determine the subsequent excess carrier distribution and radiative recombination rate. Of the many photons generated, only a small fraction emerge to contribute to the detected CL signal.

2.3.1. Parameter Definitions

The photon detection rate $N(\lambda)$ is given by (Steyn *et al.*, 1976)

$$N(\lambda) = Q(\lambda)[1 - A(\lambda)]D(\lambda) \qquad [\text{counts s}^{-1}] \tag{5}$$

where $Q(\lambda)$ is a material-dependent quantum (photon) yield (a measure of the number of photons of wavelength λ generated per second in the solid); $A(\lambda)$ is the fraction of photons lost internally in the solid through various mechanisms (self-absorption, Fresnel losses, and total internal reflection); and $D(\lambda)$ is the overall system detection efficiency. The $D(\lambda)$ term includes optical collection and coupling efficiencies, spectrometer throughput, and detector quantum efficiency. The system detection efficiency is determined by experiment. An estimate of $N(\lambda)$ can be performed with "back-of-the-envelope" type calculations.

Optical losses $A(\lambda)$ within a solid have been described by Pfefferkorn *et al.* (1980)

$$A(\lambda) \approx 1 - [(1 - R)(1 - \cos\theta_c)(e^{-\alpha z})] \tag{6}$$

where $(1 - \cos\theta_c)$ describes the total internal reflection with a critical angle $\theta_c = \sin^{-1}(n_2/n_1)$; $e^{-\alpha z}$ is optical self-absorption [α is the absorption coefficient (cm^{-1}) and z represents the photon pathlength traveled through the solid to reach the top surface]; $(1 - R)$ accounts for normal reflection losses at the solid–vacuum interface. For normal incidence (Reimer, 1985),

$$R = \left(\frac{n - 1}{n + 1}\right)^2 \tag{7}$$

where the refractive indices are $n_1 = n$ (solid) and $n_2 = 1$ (vacuum). Luminescence generated by reabsorbed radiation (photon recycling) can sometimes be important (Roedel and Keramidas, 1979). Photon recycling primarily affects the external quantum efficiency ($\eta_e = A(\lambda)\eta_i$ described below) and is more evident in heavily doped material (Van Roos, 1983).

$Q(\lambda)$ is expressed in terms of $G\eta_i$, the electron–hole pair generation rate G weighted by the internal radiative quantum efficiency η_i. For solids with energy gaps, η_i is the ratio of the band-edge radiative recombination rate to the total recombination rate. When radiative and nonradiative processes compete in parallel, η_i can be expressed in terms of carrier lifetimes, i.e., $\eta_i = \tau_{nr}/(\tau_r + \tau_{nr})$, where τ_r represents the radiative lifetime and τ_{nr} is a collective (reciprocal) lifetime for all nonradiative processes. The luminescence efficiency of many solids is often dominated by τ_{nr}. In semiconductors, τ_{nr} is typically determined by Hall–Shockley–Read recombination statistics of deep levels and traps (Blakemore, 1987). Surface states and Auger processes also contribute to nonradiative recombination. The surface plays a very significant role in CL because excess carriers are generated within the top few micrometers of the solid.

The electron–hole pair generation rate G (s^{-1}) for semiconductors is (Davidson, 1977)

$$G = EI_b(1 - \eta)/qE_i \tag{8}$$

where I_b is the electron-beam current, E_i is the average energy required to create an electron–hole pair [$E_i = 2.8E_g + E'$, where $0.5\,\text{eV} < E' < 1.0\,\text{eV}$ reflects phonon participation (Klein, 1968)], and η is the backscattered electron coefficient (dependent on the material and beam voltage). Joy (1987) provides a useful estimate for η. Any spatial variation of the parameters in Eqs. (6)–(8) will give rise to CL image contrast. For instance, variations in surface morphology (e.g., hillocks) will influence light extraction via geometrical effects through the $(1 - \cos\theta_c)$ term. The most influential parameter is η_i, as the recombination physics is sensitive to many different factors.

2.3.2. Excess Carrier Distributions

CL emanates from the entire volume of the sample. A more detailed calculation requires a knowledge of the spatial distribution of excited carriers. We describe the case for semiconductors because analytical investigations have been more extensively explored with this group of solids than with any other group. The electron–hole pair generation rate is related to electron energy loss [re: Eq. (1)] and we describe this by a generation function g'. The generated excess minority carriers (Δp, holes in n-type material) are subject to diffusion and recombination processes ($\Delta p/\tau$). The steady-state excess carrier distribution requires the solution of the 3D continuity equation, subject to specific boundary conditions (Van Roosbroeck, 1953). The 1D case can be described by a plane z parallel to the surface ($z = 0$) as

$$D(d^2\Delta p/dz^2) - \Delta p/\tau + g' = 0 \tag{9}$$

with a boundary condition that takes into account surface recombination at $z = 0$,

$$D(\partial\Delta p/\partial z)|_{z=0} = s\Delta p|_{z=0} \tag{10}$$

where D is the diffusion coefficient [the first term in Eq. (9) describes carrier diffusion] and s is the surface recombination velocity (a measure of surface transport). The behavior of s is nontrivial since it is dependent on surface and bulk properties, as well as on the excess carrier concentration (Corrieg *et al.*, 1990). The solution of Eq. (9) for an arbitary generation function is a formidable problem. This function has been described by: (1) a point source located some depth in the solid

(Hackett, 1972); (2) a uniform spherical source (Berz and Kuiken, 1976); and (3) a Gaussian source (Wittry and Kyser, 1967). The last description is more consistent with experiment and Monte Carlo simulations.

If we have a knowledge of the 3D excess carrier distribution $\Delta p(r)$, and assume that the CL intensity is linearly proportional to the steady-state excess carrier density, then the total CL intensity $I_{CL}(r)$ over the sample volume V can be expressed as (see p. 65, Yacobi and Holt, 1990)

$$I_{CL}(r) = A(\lambda)D(\lambda) \int_V [\Delta p(r)/\tau_r]d^3r \tag{11}$$

To get a better feel for the numbers, one can consider the 1D solution to the Van Roosbroeck continuity equation for a point source located at some depth h below the surface $z = 0$ (Hackett, 1972). The term h represents an electron penetration depth: up to this depth the electrons are elastically scattered and then are assumed to disperse by inelastic (energy loss) means. The 1D solution has applications to CL experiments with a defocused electron probe. Defocusing reduces carrier injection densities and beam heating effects. Mathematical integration of Hackett's solution to Eqs. (9) and (10) yields the total steady-state excess carrier density Δp_{ss}:

$$\Delta p_{ss} = G\tau\{1 - [S/(1 + S)] \exp(h/L)\} \tag{12}$$

where L is the minority carrier diffusion length, and S represents the reduced surface recombination velocity ($S = s\tau/L$). One can see that the effect of surface recombination on excess carrier production $G\tau$ is determined by the $S/(1 + S)$ and $\exp(h/L)$ terms. For a fixed beam voltage (generation depth h), the effects of high surface recombination will be reduced in materials with short diffusion lengths. The radiative recombination of excess carriers gives rise to CL, at a rate determined by $\Delta p_{ss}/\tau_r$. The higher Δp_{ss} is for a given excitation, the larger the CL signal (all factors being equal). Solution of the continuity equation with a Gaussian generation function yields better estimates of the excess carrier density (Davidson and Dimitriadis, 1980).

2.4. Cathodoluminescence Modeling

We have reviewed several analytical approaches, primarily applicable to semiconductors that attempt to describe CL intensity behavior. It is more common to express the CL response of a solid in terms of the experimental operating (excitation) conditions, i.e., beam current and accelerating voltage. A well-known example is the CL response of phosphors (Garlick, 1949):

$$I_{CL} = kf(I_b)(E - E_d)^m \tag{13}$$

where k is a constant, f describes the dependence on beam current, E_d is the dead voltage, and m normally lies between 1 and 2. The dead voltage reflects the fact that a minimum beam voltage is required for the production of CL. The effect, well known in ZnS, is viewed as a threshold voltage necessary to penetrate a luminescence dead (skin) layer. The luminescence dead layer is normally attributed to surface space-charge regions. The surface electric field can persist under low-level injection conditions, especially if the Fermi level is pinned by a high density of surface states. The physical separation of electron–hole pairs by this field greatly reduces the probability of radiative recombination in this subsurface region (see Section 5.3). A dead-layer effect can also result from competition between drift and diffusion (Goulding, 1977).

A nonlinear dependence with excitation [i.e., $I_{CL} \propto I_b^n$ in Eq. (13)] means that the CL intensity is no longer linearly proportional either to g' or to the excess carrier density. In actual fact, a variety of behaviors can be observed by varying the beam current at a fixed beam voltage. For instance, at room temperature I_{CL} can exhibit a range of behavior from $n < 1$ (suggesting strong nonradiative deep-level recombination) to $n = 2$ (suggesting space-charge recombination). Other factors that contribute to a nonlinear response include sample temperature, degree of electron probe focus (Casey and Jayson 1971), and sample volume (Myhajlenko *et al.*, 1986; Batstone and Steeds, 1986). By sample volume, we are discriminating between bulk and thinned samples, rather than describing a density-of-states effect, such as quantum confinement. Thinned refers to material that is transparent to high-energy electrons (typically for $E > 100\,\text{keV}$), i.e., suitable for transmission electron microscopy (TEM). A strong dependence on beam current is also observed with optoelectronic devices, where electron-beam pumping can lead to superradiance or even near-lasing characteristics (Steeds, 1989*a*).

For the 1D case (equivalent to flood illumination or a strongly defocused probe), assuming $I_{CL} \propto I_b^n$, Eq. (11) may be expressed in the form (Rao-Sahib and Wittry, 1969)

$$I_{CL} \approx \int_0^\infty [\Delta p(z)]^n dz \tag{14}$$

This integral must be evaluated by numerical means. In Section 2.1 we noted how the Monte Carlo calculation can describe the energy dissipation rate of an electron as it penetrates into a solid (Joy, 1988). This knowledge can be used to calculate electron–hole pair densities. In the Monte Carlo calculation, the amount of energy lost (ΔE) at each scattering event (computed trajectory step) can be converted into an equivalent number of electron–hole pairs [i.e., $\Delta E/E_i$ (Klein, 1968)]. If each trajectory step is treated as a pointlike (generation) source, then, the analytical point source solutions to the continuity equation (Hackett, 1972; Donolato, 1978; Jakubowicz, 1986; Pasemann and Hergert, 1986) can be applied to calculate the excess carrier density. This number can then be used to estimate the CL intensity produced at each step. This approach has been used for CL calculations of dislocation contrast behavior (Czyzewski and Joy, 1990) and signal strengths from bulk material and device structures (Napchan, 1989; Holt and Napchan, 1994).

More extensive CL models using a complete numerical approach have been proposed by Phang *et al.* (1992); Pey *et al.* (1993*a*, 1995*a*, *b*). In these models, the electron–solid interactions are taken into account by the usual Monte Carlo calculation. Carrier transport (continuity equation) and optical losses [$A(\lambda)$] are also solved by numerical methods (Ortega and Poole, 1981). An example of such a calculation is shown in Fig. 4. The plot details CL contrast $C(E)$ behavior versus CL image linewidth $W(E)$ as a function of beam voltage for a defect located at varying

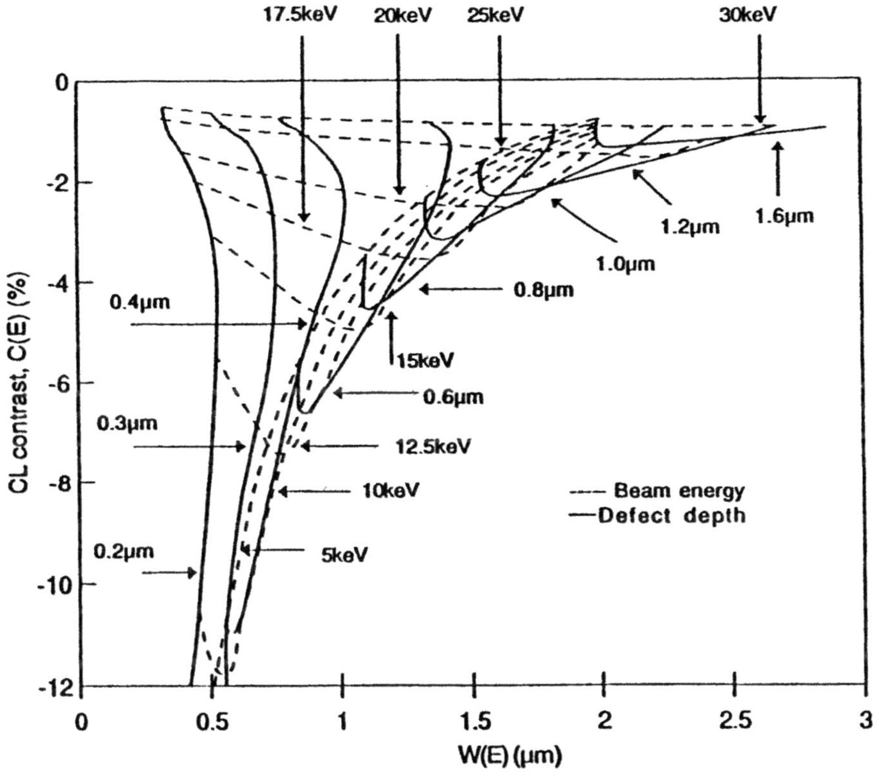

Figure 4. Plot of CL contrast $C(E)$ vs. full-width half-maximum of CL contrast profile (linewidth) $W(E)$ of point defects for different depths: defect size ≈ 0 and defect strength $= 1479\ \mu m^{-2}$ (Pey *et al.*, 1993; with kind permission from IOP Publishing, Ltd., Bristol).

depths (0.2–1.6 μm) (Pey *et al.*, 1993*a*). This is an attempt to calculate defect contrast behavior. The model predicts that E_{min} the beam voltage at which the CL contrast is at a minimum is a function only of the defect depth, i.e., the contrast is not sensitive to defect strength. Subsequent results suggest that defect depths may be estimated from the experimental determination of E_{min} (Pey *et al.*, 1993*b*).

There is still a considerable way to go to describe a unified analytical model. For the moment, one should view the Monte Carlo simulation as a laboratory tool to assist with interpretation when no theory or explicit mathematical solution is available (Holt and Napchan, 1994). To move quantitation one step further requires improvements in spectral analysis. The hope is that impurity identification and concentrations may be determined from spectral signatures in a manner similar to that currently achievable with x-ray microanalysis. One approach suggests that this may be feasible from: (1) a knowledge of the CL spectral intensity; (2) a measurement of the CL decay time constant (an effective measurement of the carrier lifetime); and (3) application of corrections for mixed level injection, absorption, and surface recombination [MAS, analogous to *ZAF* and $\Phi(\rho z)$] (Warwick, 1987). A knowledge of the CL lifetime is important because it provides a relative measure of the competition between radiative and nonradiative processes. Most radiative transitions can be identified through systematic comparative studies. Nonradiative processes usually involve unknown impurities and defect complexes. A lack of any consistent behavior in basic spectral attributes with impurity type, defect content, and excitation conditions represents a significant hurdle to advancement.

The same questions come up with respect to PL quantitation. For example, analysis of PL for the determination of impurity concentrations and compensation ratios in semiconducting crystals have been reported in Si (Tajima 1978), InP (Bugajski and Lewandowski, 1985), and GaAs (Lu *et al.*, 1990). CL examples include the measurement of doping variations in bulk InP ingots (Warwick *et al.*, 1983) and under ohmic contacts on InP epitaxial layers (Graham *et al.*, 1985). These methods rely on a comparison of impurity–exciton peak ratios or the dependence of spectral linewidths on impurity concentrations. However, because of the special and selective nature of these approaches, these methods have yet to gain widespread acceptance.

3. EXPERIMENTAL ASPECTS

3.1. Basic Design and Instrumentation

The most common medium for implementation of a CL system is the SEM. Therefore this section relates primarily to the SEM in descriptions of system design. Section 3.2 describes other CL system manifestations. Panchromatic (all wavelengths) detection at ambient temperatures represents a basic arrangement. One elegant approach makes use of a solid state detector (SSD) mounted onto the objective pole piece of the SEM (Marek *et al.*, 1985). This setup allows detection of emission CL (ECL) at the expense of BSE detection. A more common approach for detection of ECL is the use of a photomultiplier tube (PMT) in conjunction with a simple collection mirror or lens arrangement. Placement of the PMT onto the SEM chamber must minimize the detection of spurious light, e.g., scintillation from the SE detector. As the generated light travels in all directions in the crystal, some of the CL can emerge from the back of the sample. This is known as transmitted CL (TCL). The transmitted light suffers considerable perturbation from bulk optical absorption. TCL is usually detected by SSDs, such as Si and Ge (Vale and Greer, 1977; Chin *et al.*, 1979).

Spectral analysis with a spectrometer is the next capability added to a CL system. Wavelength-selected (monochromatic) CL imaging is readily achieved with optical filters or a grating spectrometer. A less commonly used approach is Fourier transform spectroscopy (FTS) with a Michelson interferometer (Davidson *et al.*, 1981). The interferometer has advantages over the spectrometer in the near-to-mid-infrared because of simultaneous radiation detection, improved signal-to-noise ratio when detector noise is a limiting factor (Fellgett advantage) and higher throughput for a given spectral resolution (Jacquinot advantage). These benefits are important considering the relatively poor performance of SSD infrared detectors in comparison with PMTs. FTS has been implemented onto SEMs (Saba and Holt, 1985) and TEMs (Graham *et al.*, 1987). However, real time FTS imaging has yet to be fully explored on these systems. The next system progression is cryogenic capabilities.

Systems cooled with liquid nitrogen (77 K) are common. However, it is preferable to attain liquid helium temperatures (< 30 K). Lower temperatures aid signal detection: radiative efficiencies are higher; spectral lineshapes are sharper; and beam damage rates are much lower. In addition, low-temperature CL can be compared with a wealth of existing PL data, which aids interpretation.

The next degree of complexity is temporal dispersion. Of the methods available for high-speed electron-beam blanking, e.g., tuning resonant structures (Menzel and Kubalek, 1979) and beam intensity modulation by Wehnelt biasing (Davidson, 1981), beam deflection across an aperture is the most widely used with a rise/fall time of about 10 ps (Fujioka and Ura, 1983). Different approaches have been used to implement time-resolved measurements, including phase modulation (Hwang, 1971); gated single-photon counting (Rasul and Davidson, 1977), analogue sampling with a boxcar averager (Steckenborn, 1980), streak camera (Lohnert *et al.*, 1981), and time-correlated photon counting (delayed coincidence) (Davidson *et al.*, 1981; Myhajlenko and Ke, 1984). The fastest detector systems are based on single-shot streak cameras (time resolution ≈600 fs) and microchannel-plate photomultiplier tubes (MCP-PMT) with rise times of about 100 ps. A combination of the two systems (MCP and streak camera technology) has recently been introduced as a two-dimensional single-photon counter with 5 ps temporal resolution after deconvolution (Watanabe *et al.*, 1994). The two-dimensional aspect refers to simultaneous measurement of time and wavelength information. Time-resolved experiments include CL decay lifetime measurements; generation of 'lifetime-maps' (Steckenborn, 1980); "snapshots" of deexcitation and transport processes, such as tracking "hot" excitons as they cool down and diffuse around the crystal (Christen and Bimberg, 1988).

3.1.1. General Description

The key components of a CL system are depicted in Fig. 5. High collection and detection efficiency are critical to CL measurements. The mirror–window–lens combination represents a basic light collection–transfer–coupling arrangement. Collection components include plane mirrors, parabolic mirrors, ellipsoidal mirrors,

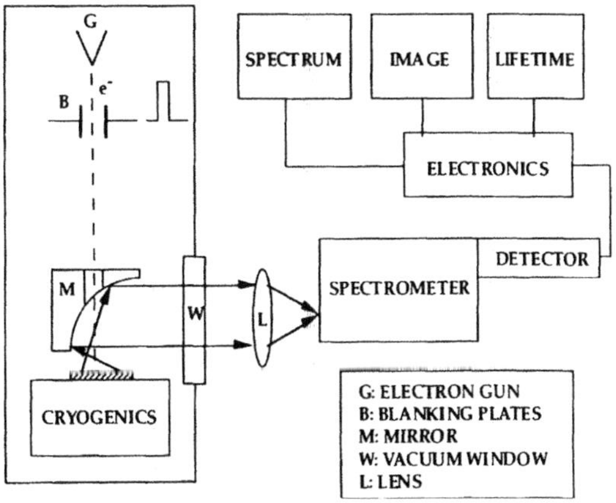

Figure 5. Schematic of an advanced CL system implemented on a SEM.

optical fibers, and hemispherical mirrors. Light transfer and coupling to detectors or spectrometers use one or more of the following: lens, mirror, light guide, fiber optics, and vacuum window. The properties of the window material are important because of spectral absorption and fluorescence (if the window is excited by stray electrons). For example, fused quartz can have significant water (OH^-) absorption bands, which can lead to spectral artifacts in the near-infrared (at 1.24 and 1.39 μm). CaF_2 is an alternative material.

Steyn *et al.* (1976) and Trigg (1985) have reviewed the merits and limitations of several optical design configurations. The ellipsoidal mirror is the most popular collection element used in both custom-built and commercial CL systems, as it affords considerable design flexibility, very high collection efficiency (typically $>80\%$ for 2π-steradians emission), and a second focus that can be located conveniently outside the vacuum system. The sample is placed at the first focus of the mirror in this arrangement. Light at the second focus can either be directed into a spectrometer by a lens or collimated into a circular-to-rectangular fiber bundle. The fiber approach obviates the need for straight optical paths. Mirrors suitable for high-magnification work in SEM or TEM have been designed for efficient light collection from areas as small as 10^2 μm^2 (Roberts, 1981; Pennycook, 1981*a*).

The next component in Fig. 5 is the spectrometer. The spectrometer allows a measurement of the emission spectrum from a sample or the selection of a specific wavelength for monochromatic imaging. The dispersion element may be a prism, diffraction grating, or interferometer. The most common dispersion element is the holographic diffraction grating. By appropriate selection of gratings, spectral analysis is possible from the vacuum ultraviolet (with an evacuated monochromator) to the far-infrared (λ up to 50 μm). Typical systems cover the wavelength range 0.3–5.5 μm. Modern spectrometers of the Czerny–Turner configuration are compact with focal lengths 0.25–0.34 m. They provide adequate throughput and spectral resolution for most CL work (including low-temperature measurements). With respect to the latter, this is because CL cold-stage attachments are normally of the cold-finger type. CL cold-stage performance does not compare favorably with PL cryostats as samples cannot be immersed directly into the liquid helium. CL sample cooling must rely on good thermal contact with the cryogenic heat exchanger. The proximity of the collection mirror to the "warm" objective lens pole piece (especially in TEMs) also makes background blackbody radiation shielding difficult. Custom-designed CL cold stages typically achieve temperatures between 10 and 30 K. Some newer commercial designs can cool down to 5 K (Warwick, 1991).

3.1.2. Detectors

Table 1 lists the detectors most frequently found on spectrally resolved CL systems. The spectral range covers the vacuum ultraviolet to mid-infrared. PMTs provide the most versatile and efficient detection in the visible to near-infrared. PMTs have a high dynamic range: high internal gain (10^5–10^7 current amplification) with fast (nanosecond) response times. The three most commonly used photocathodes are the multialkali S-20 (Na–K–Sb–Cs), the S-1 (Ag–O–Cs), and GaAs:Cs. The end-on single-photon counting GaAs PMT is a popular choice for

Table 1. Spectral Range of Detectors Used in CL Detection Systems[a]

Detector type	Spectral range
PMT (S-20)	300–850 nm
PMT (S-1)	400–1200 nm
PMT (GaAs:Cs)	185–930 nm
CCD-Si	200–1100 nm
CCD-InGaAs	850–1700 nm
SSD-Si	200–1100 nm
SSD-Ge	800–1900 nm
SSD-PbS	1.0–3.5 μm
SSD-InSb	1.0–5.5 μm

[a](PMT: photomultiplier, CCD: linear diode array, SSD: solid state detector).

detection of weak CL signals. However, single-photon counting PMTs can be easily damaged by high photon fluxes and must be operated below a threshold count rate. PMT signal-to-noise performance can be further improved by photocathode cooling and pulse height discrimination. At the opposite extreme, side-on type PMTs provide analogue detection and are preferable in high CL signal environments. The discrete photon pulses overlap and produce an average dc photocurrent proportional to the photon flux.

CCD Si diode array detectors are used for parallel spectral acquisition into an optical multichannel analyzer (OMA). For many CL applications, an intensified diode array detector provides much improved sensitivity. CCD applications have included time-resolved CL spectroscopy of semiconductors (Lohnert *et al.*, 1981), TEM CL investigations of defects (Spellward *et al.*, 1987), and mineralogical studies (Remond *et al.*, 1992). CCD detectors are helpful in CL studies of solids that are susceptible to electron-radiation damage.

CL measurements into the wavelength range 1–5.5 μm requires SSD and appropriate infrared optics (e.g., sapphire and CaF_2). SSDs are typically less sensitive and have slower response times (0.1 μs–0.1 ms) than PMTs. This relatively poor performance status of SSDs has hindered general usage. However, in recent years, improved technology has produced SSDs with near ideal performance. SSDs are either of the photovoltaic (PV) type (like solar cells they generate a photovoltage when subject to infrared radiation) or the photoconductive (PC) type (resistance changes under optical illumination). PV detectors (e.g., Ge, InSb and HgCdTe) are operated with no external bias, whereas PC detectors (e.g., PbS, PbSe, and HgCdTe) are normally operated under reverse bias. In general, cooling of an SSD (typically to 77 K) reduces detector noise and can greatly improve detectivity, albeit at the expense of both spectral and temporal response. Detectivity (D^*) represents the signal-to-noise performance of a detector, and can be further improved with a cold optical stop (filter). The stop is in effect a wavelength bandpass filter that restricts unwanted background blackbody radiation from the detector's field of view.

In CL imaging applications, the weak nature of the infrared signals means that experiments are easily limited by noise (including preamplifier) and dc drift of the detector. There is also a trade-off between speed and responsivity of the detector. In these circumstances, the signal-to-noise ratio can be improved by the use of ac-phase-sensitive detection (a lock-in amplifier). Lock-in detection eliminates unwanted low-frequency, dc, and $1/f$ noise voltage components from the signal. The CL signal must be made periodic by either chopping the light or modulating the electron beam. Since high-speed beam blanking is not required, the necessary modulation signal (Hz–KHz) can be applied to the alignment coils of a SEM or TEM. Signal-to-noise improvement comes from output signal integration determined by the time constant of a low-pass filter. Further signal processing also eliminates nonsynchronous noise frequencies about the reference (modulation) frequency, e.g., light from the hot filament of the electron gun. The number of published infrared CL imaging experiments is small, but growing: high-detectivity cooled Ge-photodiodes used for deep-level imaging of semi-insulating GaAs (Warwick and Brown, 1985); imaging of grain boundaries in $BaTiO_3$ (Koshek and Kubalek, 1985); defect imaging of InP (Myhajlenko *et al.*, 1984); InSb detectors have been used for imaging of HgCdTe (Petrov and Gareeva, 1988; Warwick, 1989; Myhajlenko *et al.*, 1990) and Si (Cumberbatch *et al.*, 1981).

3.1.3. System Calibration

Recall that $D(\lambda)$ in Eq. (5) represents the overall system detection efficiency as a function of wavelength (Steyn *et al.*, 1976). The system transfer function is determined by the combined spectral performance characteristics of each of the different components of the CL system. A knowledge of this function allows emission spectra to be corrected to true relative spectral intensity distributions. The calibration of a CL system in absolute radiance terms also represents a necessary step toward quantitation. Calibrations can also help differentiate between detection and sample artifacts (Hamilton *et al.*, 1978). The radiance calibration process requires a standard tungsten lamp source whose output characteristics can be referenced to some recognized standard. The number of photons emitted by the lamp at a given temperature (and wavelength) can be calculated from Planck's equation. $D(\lambda)$ can then be determined for each detector configuraton by comparing the calculated photon count rate with the detected rate as a function of wavelength. Napchan *et al.* (1993) have described the calibration of a CL system in the range 0.4–1.8 μm, where the corrections are essentially done in real time. It should be remembered that a plot in terms of transition energy must be corrected since the data are acquired in terms of wavelength intervals (light is dispersed by a grating).

3.2. Configurational Permutations

3.2.1. STEM/TEM

Dark-dot features in CL images are usually associated with extended crystal defects, such as dislocations. However, etch-pit studies and/or TEM are required to confirm the origin of this contrast (Titchmarsh *et al.*, 1977; Roedel *et al.*, 1991). In situ TEM CL would obviate matters. Systems have been designed for TEMs equipped with scan coils (Roberts, 1981; Yamamoto *et al.*, 1984; Bailey *et al.*, 1989) and scanning transmission electron microscopes (STEM) (Petroff *et al.*, 1978; Pennycook, 1981*a*). STEM/TEM offers the opportunity of correlating CL with crystallographic diffraction contrast (Hirsch *et al.*, 1965), as well as the promise of high spatial resolution (50–200 Å). Beam spreading b in a thin sample of thickness t owing to elastic scattering is given by (Goldstein *et al.*, 1977)

$$b \propto (Z/E)t^{3/2}(\rho/A)^{1/2} \tag{15}$$

Beam spreading is minimized in thin samples by using high accelerating voltages. However, factors other than carrier diffusion limit the potential CL resolution. For instance, radiation damage (atomic displacement) is deleterious to luminescence efficiency. There is also a minimum requirement of sample thickness. The CL signal is expected to be very weak because only a small fraction of the incoming beam energy is absorbed by the crystal. Electron–hole pair production results from inelastic energy-loss processes. In very thin samples, these losses are primarily by single-electron (cell) excitations and plasmons. The plasmon mean free path has been used to model the CL dependence on thickness (Yuan *et al.*, 1989). However, there are additional difficulties with experimental interpretation, which include crystallographic contrast factors (including anomalous transmission in certain crystal orientations, thin-film optical interference, and modified recombination physics (Myhajlenko *et al.*, 1985; Berger *et al.*, 1985; Steeds, 1989*a*; Yuan *et al.*, 1989).

Sample preparation can also influence the luminescence efficiency. Traditional ion-milling techniques, while appropriate for TEM sample preparation, may leave residual surface damage deleterious to CL. On the other hand, chemical treatments are known to passivate semiconductor surfaces. Jet-thinning of compound semiconductors with solutions of Br_2 or Cl_2 in methanol facilitate TEM CL measurements (Myhajlenko *et al.*, 1984). Given all these mitigating factors, the best reported spatial resolution (aided by surface recombination) is approximately 500 Å (Pennycook, 1981*b*) (see Section 5.2). Most TEM CL investigations to date have focused on defects in semiconducting materials. These studies are not trivial (Steeds, 1989*a*, *b*). There have been some applications in mineralogy, i.e., a study of the CL properties of forsterite found in the Allende (CV3) meteorite (Benstock, 1991; Benstock and Buseck, 1991). The fact that CL signals were detected is encouraging for future TEM CL studies of minerals.

3.2.2. Fiber-Optic CL

Hoenk and Vahala (1989) describe an elegant CL collection system based solely on an optical fiber. One end of the fiber is positioned within 0.5 mm of the sample, while the other end couples the collected light into a spectrometer. This arrangement permits spectral analysis and monochromatic imaging when implemented on a SEM. Fiber position is controlled by a vacuum compatible, three axis motorized translation stage. The fiber collection efficiency is not as high as that for mirrors. The fraction of light collected is determined by the fractional solid angle subtended by the collector at the sample emission point. However, the coupling efficiency to the spectrometer is much better since additional optics are kept to a minimum. The fiber-optic CL system has been used to measure ambipolar diffusion lengths in semiconductor heterostructures (Zarem *et al.*, 1989). The CL intensity as a function of distance is measured by fixing the position of the collection fiber at the edge of a metallized mask and scanning the electron beam across the masked region of the sample. The metal thickness allows electron penetration but prevents light emission. The fiber detects the luminescence from carriers that diffuse and recombine in the unmasked region.

3.2.3. TCL/SRTCL

TCL is the detection of light that exits from the back surface of the sample. Contrast arises from optical absorption and optical scattering. TCL can reveal the presence of dislocations, buried layers, growth striations, clusters, and precipitates (Cocito *et al.*, 1986). The complementary use of TCL with ECL can allow depth-resolved and wavelength-selective imaging (Salviati, 1993). In general, TCL does not reveal anything that is not also visible by ECL (Rowley, 1989). Spectrally resolved

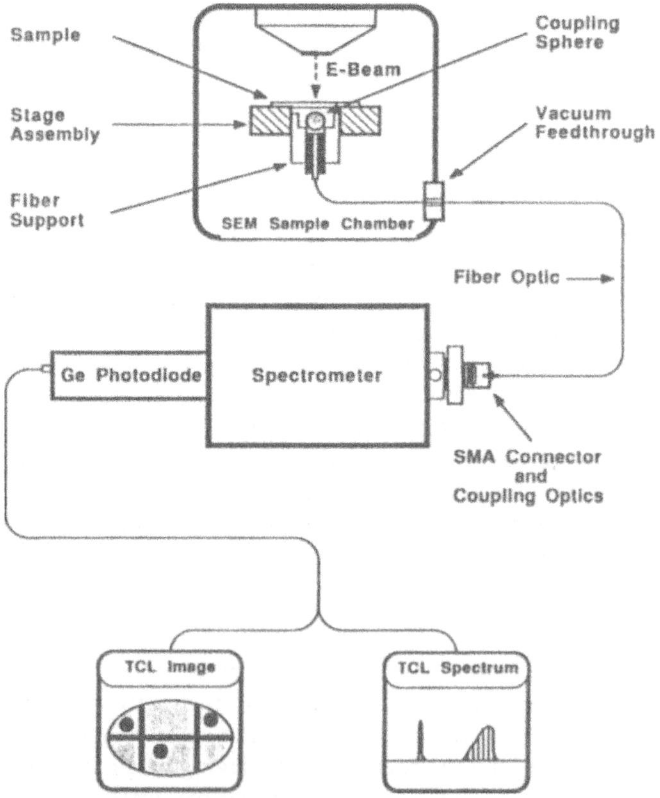

Figure 6. Schematic of a room temperature SRTCL system implemented on a SEM (Myhajlenko *et al.*, 1992; with kind permission from Scanning Microscopy International, Chicago).

TCL (SRTCL) is an unusual extension of the technique. The evaluation of buried InGaAs quantum wells in vertical cavity surface-emitting lasers represents a custom application (Myhajlenko *et al.*, 1992). The device, a 80-Å $In_{0.2}Ga_{0.8}As$ quantum well sandwiched between two distributed Bragg reflector mirrors, emits light through the bottom of the device. Figure 6 shows a SRTCL system based on fiber-optic coupling. A glass coupling sphere (lens) focuses the TCL from the sample into the optical fiber and spectrometer.

3.2.4. Nonscanning Ultrahigh-Vacuum CL

Nonscanning CL systems are typically flood-illumination-type instruments, which do not have any imaging capabilities. A typical system consists of an electron gun attached to a vacuum chamber equipped with optical ports. Nonscanning CL systems provide excellent performance with low-luminescence-efficiency solids, e.g., Si (Vouk and Lightowlers, 1977). One simply adjusts the excitation conditions to generate an ample CL signal, i.e., changes the current and probe diameter. Basic system configurations offer much greater design flexibility than that available with either SEM or TEM, e.g., in the design of an ultrahigh-vacuum (UHV) CL system. Such a system can explore the fundamental properties of surfaces and interfaces. The base pressure of an UHV system is about 10^{-11} Torr, compared with 10^{-6} Torr for SEM. CL of clean surfaces is achieved by cleaving in UHV (Brillson and Viturro, 1988).

3.2.5. Cathodoluminescence Microscope Attachment

The cathodoluminescence microscope attachment (CMA) is a popular tool for studies in earth sciences. The CMA is essentially another flood-illumination-type instrument. Probe diameters are typically 1 mm–2 cm. The typical CMA consists of a cold-cathode electron gun and a low-vacuum chamber (10^{-2} Torr), which attaches to an optical (petrographic) microscope (Sippel, 1965). Electrons are generated by residual gas ionization. The cold-cathode gun plasma-discharge also obviates the need for conductive films with samples that normally charge (Le Gressus *et al.*, 1984). The less common hot-cathode gun version of the CMA requires a conductive film for CL observations of insulating materials. The CL is normally detected through the microscope objective eye piece and recorded on color photographic film/plate (see Fig. 7). The electrons e^- are either deflected or impinge onto the sample at some oblique angle of incidence. The deflected configuration (with magnet) also reduces bombardment damage since negative residual ions do not reach the sample. The sample is usually in the form of a thin section ($\approx 50\,\mu m$ in thickness) mounted on a glass slide. The attachment offers the convenience of being able to switch between CL observations and optical evaluation, such as polarized light microscopy.

The CMA greatly augments the capabilities of the petrographic microscope because of the sensitivity of CL to trace amounts of transition-metal and rare-earth impurities. The CMA readily reveals contrast not visible in the optical microscope. Samples that appear uniform in plane light exhibit rich CL features. For example, CL can reveal chemical zonation (variations in impurity concentration) in many

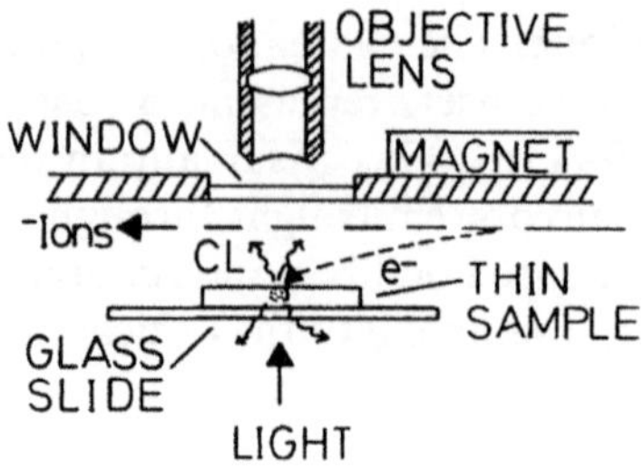

Figure 7. Schematic of a CL microscope attachment configured with deflected beam geometry.

minerals (Marshall, 1988). Developments with CMAs parallel those already described for SEM CL systems. There are commercial CMA systems available, such as the Luminoscope® and Technosyn (Marshall, 1993).

3.2.6. Scanning Tunneling-Induced Luminescence

The relatively new manifestation of CL known as scanning tunneling-induced luminescence (STL) is implemented on a scanning tunneling microscope (STM). The technique offers atomic scale resolution. The STM is a surface profilometer that relies on quantum-mechanical wavefunction overlap between a sharp probe tip and a conducting specimen surface (Binnig *et al.*, 1982). Under an applied voltage bias, a tunneling current is established between the probe tip and the sample surface, the magnitude of which depends exponentially on probe–sample separation. Consequently, the STM has excellent vertical sensitivity (<0.01 nm). The lateral resolution is strongly dependent on the spatial extent of atomic wavefunctions (≈ 0.2 nm). In a nutshell, STM topography reflects the charge density contours of the surface. The interactions between the probe tip and electronic states near the surface Fermi level provide the basis for various spectroscopic techniques (Bonnell, 1993), which can have chemical and electrical sensitivity to individual atoms (Stroscio and Kaiser, 1993). Optical characterization is also feasible with STM. In metals, luminescence can be generated during the tunneling process from the excitation of localized surface plasmon modes (Coombs *et al.*, 1988).

STL has also been observed in semiconductors (Montelius *et al.*, 1992), and is included in a description of CL as the STM tip injects ballistic electrons (as minority carriers) into the surface region of a solid to depths of 100 to 1000 Å. Ballistic transport refers to those electrons that undergo no scattering or energy loss (Bell *et al.*, 1990). The transport of injected "hot" electrons and their subsequent recombination in the bulk is thought to give rise to the luminescence. There is a threshold voltage in the range of 1.5–100 eV for this process. For voltages greater than 5 eV, electron–hole production is thought to be associated with impact ionization. One could think of the vacuum STM as a very low-voltage SEM, e.g., it can be used for nanometer-scale luminescence profiling of quantum-well barriers (Renaud and Alvarado, 1991). Although one would expect carrier transport to determine the lateral spatial resolution, features as small as 5 nm have been resolved by STL in porous Si (Gu *et al.*, 1993) and GaAs quantum wires (Pfister *et al.*, 1994). This degree of luminescence resolution is achieved by geometrical confinement of the excitation.

3.2.7. Electron Energy Loss–CL Coincidence

Electron–photon coincidence, a technique derived from the field of nuclear physics, can be used to measure the lifetime of excited states of He atoms (Heron *et al.*, 1956). The method measures the time delay (referred to as coincidences) between excitation (by inelastically scattered electrons) and subsequent relaxation from an excited state (by photon emission). Imhof and Read (1971) have described a refinement where only electrons with specific energy losses are detected for coincidence. Further development that couples electron energy loss spectroscopy (EELS) and CL in a TEM has been described for the measurement of excited lifetimes in phosphors (Ahn and Krivanek, 1985) and semiconductors (Graham *et al.*, 1987). The correlated EELS–CL technique obviates the need for any high-speed beam-blanking. EELS–CL measurements offer interesting possibilities for fundamental work. For instance, since lifetime is inversely proportional to transition probability, it may be possible to correlate the efficiency of particular CL transitions with specific electron energy loss events.

3.3. Operational Considerations

Luminescence can be sensitive to very low concentrations of impurities in solids (10^{10}–10^{12} cm^{-3}). To achieve this degree of sensitivity requires consideration of the system detection efficiency, the nature of the sample, and the excitation conditions. The detection efficiency is normally a fixed and optimized quantity. High-purity and high-luminescence-efficiency samples facilitate detection and interpretation. However, excitation power densities in CL experiments vary from $mW\,cm^{-2}$ to $MW\,cm^{-2}$. This wide power range gives rise to additional physical effects and phenomena that influence the CL response of solids. The effects of excess carrier injection, thermal heating, bombardment damage, contamination, and detection artifacts will now be reviewed.

3.3.1. Excitation Level

The CL intensity response of a solid as a function of excitation power is typically nontrivial [Eqs. (13) and (14)]. Many factors can influence the CL process: the electronic and chemical nature of the surface, the impurity and defect content of the solid, and the temperature, to name but a few. It would be a formidable task to account for the behavior in all types of solids. The effects of excitation (injection level) will be illustrated by focusing on the relatively well-understood low-temperature luminescence response of high-purity semiconductors (GaAs). The injection level is viewed as *low* when the number of generated excess carriers is less than the equilibrium carrier concentration (doping level), *moderate* when the numbers are comparable, and *high* when the generated number is substantially greater. For a given excitation power, an equilibrium is established between excess carrier generation and recombination processes—the steady state.

Thermalization is the process by which excess electrons and holes come to initial equilibrium with the lattice. The rate at which energy can be dissipated and by what mechanism determines the sample attributes. "Hot" electrons can relax to the bottom of the conduction band through scattering mechanisms and phonon emission prior to recombination. The CL spectrum represents an optical snapshot of the excitation and relaxation kinetics of this quasi-equilibrium system, i.e., reflecting on the radiative component of the recombination traffic. The intensity of a spectral feature depends upon the transition efficiency relative to other relaxation modes. For example, transitions such as (e, A^0) or (D^0, h) must capture excited carriers from either the conduction or the valence band. Electron and hole capture rates are inherently different. Thus, the availability of carriers and, more significantly, the ability of an impurity state to capture the respective carriers determine the radiative efficiency. This efficiency is expressed in terms of the optical cross section for carrier capture. Electron and hole capture cross sections for shallow donor and acceptors are in the range 10^{-16}–$10^{-12}\,\mathrm{cm}^2$ (Bimberg *et al.*, 1985).

The presence of free excitons is often an indication of high-quality crystalline material. It is left to the reader to ponder whether a perfect crystal would actually exhibit FE luminescence. Exciton formation requires that both electrons and holes be present in large numbers. Excitons do not form readily when the electron–hole pair generation rate is less than the relaxation rate of one of the carriers. This imbalance leads to a shortage of one carrier type. The reason is usually associated with nonradiative recombination processes (Dean, 1968). Very low-level injection conditions thus favor single-particle radiative transitions. Figure 8a shows the CL spectrum from high-purity MOCVD *n*-type GaAs subject to very low excitation power. The spectrum is dominated by transitions involving background carbon residual impurities: band-to-acceptor (carbon) $(e, A^0)_C$ and donor–acceptor pair

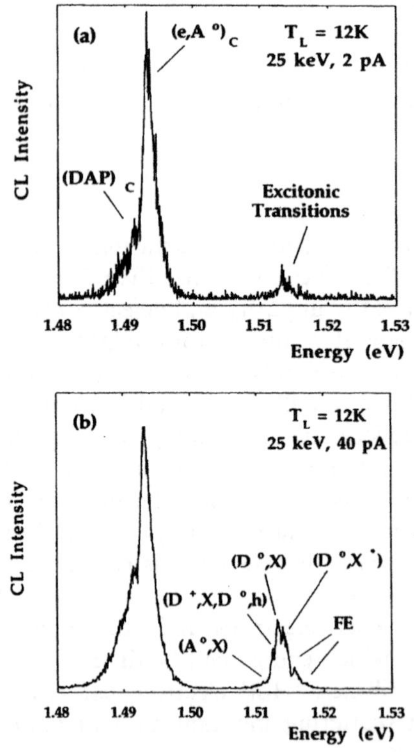

Figure 8. CL spectra from high-purity *n*-type MOCVD epitaxial GaAs (carrier concentration, $6 \times 10^{13}\,\mathrm{cm}^{-3}$) subject to 25-keV electrons. The effect of electron-beam power density on excitonic response at: (a) $1.65\,\mathrm{mW\,cm}^{-2}$ and (b) $33\,\mathrm{mW\,cm}^{-2}$.

$(DAP)_C$. Exciton-related emissions become favorable when the generation rate substantially exceeds the relaxation rate of both carriers. Electrons and holes can then exist in large numbers. This can be achieved by increasing the injection (power) level (see Fig. 8b): the spectrum exhibits FE and BE transitions (with neutral and ionized donor and acceptor impurities) and excited donor-bound exciton states (D^0, X^*). However, the injected carrier density is still relatively low since (1) the generation rate is low ($\approx 10^{12}\,s^{-1}$) [from Eq. (8)], and (2) the spectrum is dominated by $(e, A^0)_C$.

If one continues to increase the excitation power further, excitonic transitions will grow at a much faster rate than single-particle transitions. CL output as a function of input power will typically switch from a sublinear dependence to one that is slightly superlinear. Noticeable changes will also occur in the spectral lineshape of the band-to-acceptor and DAP transitions. Peak DAP emission will shift to higher energies as nearest neighbors saturate [Eq. (3)], and the high-energy tail of the band-to-acceptor (carbon) transition broadens. The latter is associated with carrier heating. In essence, the average kinetic energy of excess electrons in the conduction band increases because the rate of energy relaxation slows down. This "bottleneck" in the carrier relaxation process increases the effective temperature of the recombining electrons, T_e, over that of the lattice, T_L, nominally the sample temperature. This general situation is referred to as hot-carrier recombination (Munzel *et al.*, 1981) and also occurs with excitons (Shah, 1978). T_e characterizes the Maxwellian tail of the electron Fermi distribution, which modifies the spectral lineshape. The band-to-acceptor transition lineshape can provide a measure of carrier heating (Eagles, 1960). The (e, A^0) lineshape $I(h\nu)$ is interpreted in the form $I(h\nu) \propto E^{1/2} \exp(-E/kT_e)$, provided the excitation conditions are nondegenerate (low–medium injection level), and $E = h\nu - (E_g - E_A)$. The lattice (sample) temperature remains constant under these conditions. If one continues to pump in energy, the lattice will eventually heat up, i.e., the temperature will rise.

Figure 9 illustrates the effect of electron-probe focus on a semilogarithmic plot. Under focused conditions the spectrum is dominated by excitons and $T_e > T_L$. By defocusing the probe (i.e., flood illumination) one significantly reduces the injected excess carrier density. The spectrum is now dominated by DAP recombination (red

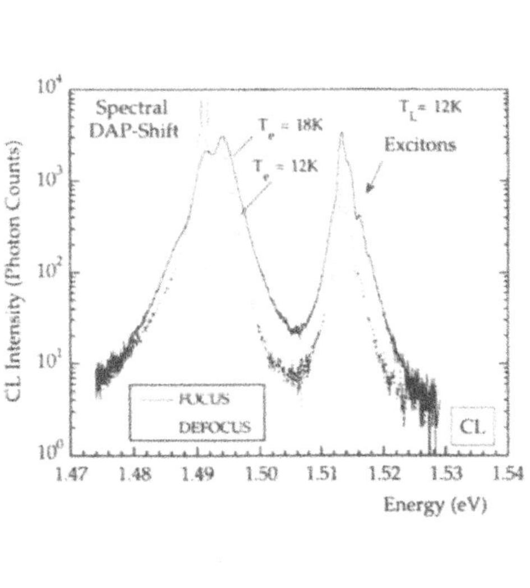

Figure 9. Hot-carrier recombination: The effect of electron probe focus on the CL emission spectrum of high-purity *p*-type GaAs (carrier concentration, $\approx 10^{14}\,cm^{-3}$) subject to 25-keV electrons (beam power = 7.5 μW, sample temperature = 12 K.

peak-shift) and $T_e = T_L$. The exciton temperature is unaffected in this experiment (as gauged from the slope of the high-energy FE tail). In general, the low-temperature CL response of high-purity material is very sensitive to beam voltage, beam current, and probe size. Any changes in injection conditions associated with changes in beam operating procedures are readily reflected in CL spectra.

3.3.2. Beam Heating

Electron irradiation can heat up the sample. The problem of calculating the temperature rise in a solid is similar to the determination of the excess carrier density. The calculation must include the function for depth-dependent energy loss as a volume source in the heat conduction equation (Carslaw and Jaeger, 1959). Analytical (Vine and Einstein, 1964) and Monte Carlo calculations (Merli, 1980) have been applied to this problem. The following modified form of the Vine–Einstein model includes the temperature dependence of thermal conductivity $K(T)$ in $\text{W cm}^{-1}\,\text{K}^{-1}$ (Touloukian, 1970):

$$(P/a)_{\text{eff}} = 4.27 \int_{T_S}^{T} K(T)dT \tag{16}$$

where P is the beam power in watts, a is the probe radius in cm, $(P/a)_{\text{eff}}$ represents the power corrected for beam penetration, and T_S is the initial temperature of the solid. High-current (μA) electron beams in SEMs are capable of raising the sample temperature by several hundred degrees. This heating can induce irreversible changes in the solid that are detrimental to CL. An imprint of the SEM scan-raster (grating-like contrast) onto the sample surface is indicative of such damage. However, beam heating can sometimes enhance the CL output of semiconductors by annealing of implantation damage in Si (Myhajlenko *et al.*, 1983*a*) and by redistributing traps in ZnS (Remond *et al.*, 1992). Other thermally induced effects include changes in CL contrast behavior at dislocations (Balk *et al.*, 1976) and generation of electrically active dislocations (Holt *et al.*, 1993).

Electron-beam heating of minerals is more complex because of possible charging, redistribution of trace activator/quencher impurities responsible for CL, and mixing of CL signals with thermoluminescence signatures. Thermoluminescence or thermally stimulated luminescence (TSL) results from the radiative relaxation of thermally liberated carriers. The temperature rise under high-power conditions is capable of generating TSL glow peaks (spectral emissions) from many common activators, e.g., Mn in calcite has glow peaks at 350 and 470 K. The time constants associated with the TSL process and electron-beam-induced trap filling are important here. However, the changes in CL normally observed in minerals with irradiation are more likely to be associated with electronic damage. For example, the enhancement in CL of calcite with conventional thermal heating is degraded by prolonged exposure to electrons, i.e., by defect creation (Mason, 1994). Sample charging in quartz can give rise to both enhancement and degradation of CL output with exposure time (Remond *et al.*, 1992). Charging-related effects can lead to a physical redistribution of activators and quenchers. The CL color changes of quartz are well documented in this respect (Ramseyer and Mullis, 1990).

3.3.3. Radiation Damage

A change in the coordination of an impurity in the crystal lattice can influence the CL. Atomic displacement damage, sometimes referred to as "knock-on" damage, is synonymous with high accelerating voltages. Knock-on damage results from the direct transfer of momentum between a fast electron and the nucleus. The displaced

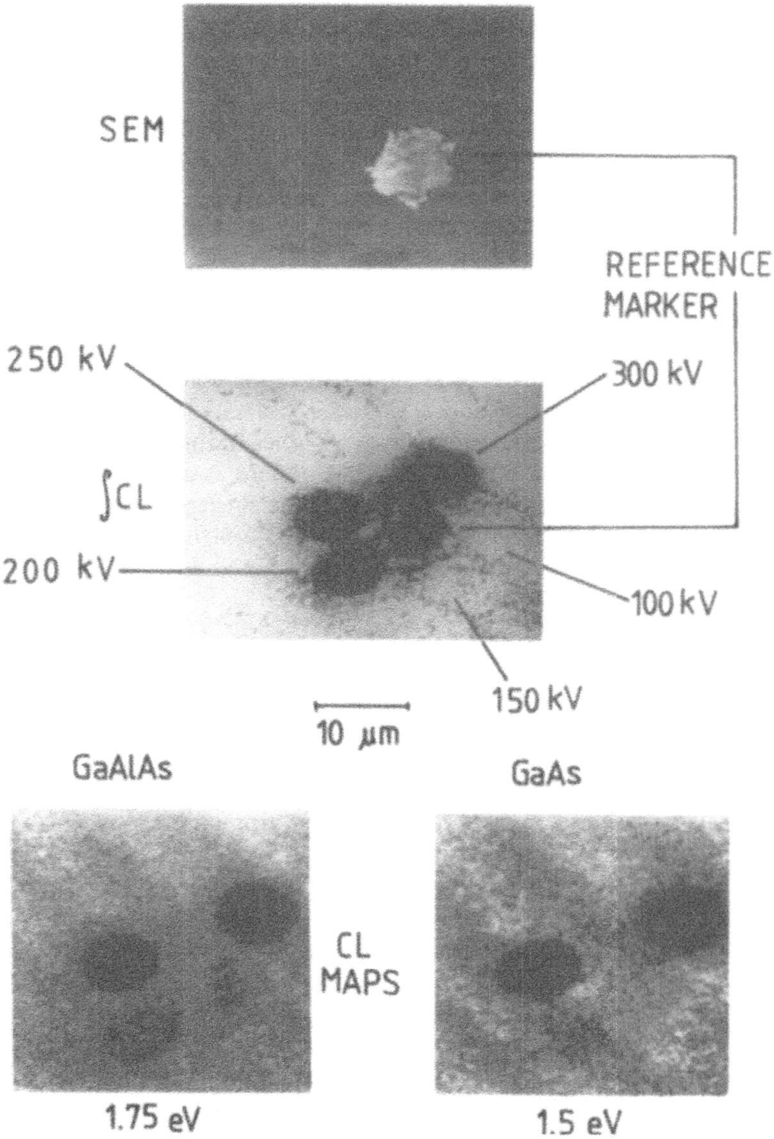

Figure 10. Electron irradiation damage (100–300 keV) in GaAs/GaAlAs laser heterostructure thinned to a thickness of $\approx 2\,\mu m$. Secondary electron (SEM), panchromatic, and spectrally resolved CL images at 1.75 eV (GaAlAs) and 1.5 eV (GaAs). Dark oval CL contrast is associated with radiation damage (Myhajlenko, unpublished).

atom moves into an interstitial lattice site leaving a vacancy behind—a Frenkel (vacancy–interstitial) pair. The resulting electronic defect state tends to be nonradiative in nature (Pons and Bourgoin, 1985). Threshold voltages for displacement damage are usually in excess of 150 keV. Figure 10 shows the effect of electron irradiation (up to 300 keV) on the CL properties of GaAs and GaAlAs. The quenched (dark) CL areas around the reference particle are most likely associated with the creation of nonradiative recombination centers. Damage threshold for GaAs and GaAlAs occurs between 150 and 200 keV. Similar damage has been observed in Si at voltages greater than 150 keV (Petroff, 1981) and in ZnSe at 300 keV (Batstone, 1985).

In noncentrosymmetric solids (*AB*-type semiconductors) there can be a crystal orientation dependence for displacement (Pons and Bourgoin, 1981). This is an anisotropy for defect generation because the atoms in one of the sublattices are easier to displace. Atomic displacement can also occur through indirect momentum transfer with electron excitations, e.g., between a fast electron and host atomic electrons (Hobbs, 1979). This mechanism is known as radiolysis and often occurs below the displacement threshold voltage. Radiolysis is more likely to occur with insulators (e.g., quartz, silicates, carbonates, and alkali halides) than with metals or semiconductors. Indirect displacement damage can sometimes lead to mass loss (electron-beam machining) in solids (Devenish *et al.*, 1989; Murooka and Yuan, 1993). This damage produces nanometer-scale holes in thin samples subject to highly focused electron beams in STEM or TEM. There is also the possibility of ion damage, i.e., residual gas ions that travel along the electron path and impinge on the sample. Ion damage can also occur in SEMs. In general, damage is not a well-defined term. For instance, displacement damage can occur at energies as low as 200 eV in GaAs (Nel and Auret, 1989)!

3.3.4. Contamination

Contamination of solids exposed to electron beams arises because vacuum systems typically contain residual water vapor and hydrocarbons. The polymerization (cross-linking) of organic molecules under electron irradiation produces carbonaceous films. These films tend to be insulating and can grow (thicken) under continuous electron bombardment by the electromigration of molecules from nearby unirradiated areas. Residual gases (water vapor, nitrogen and oxygen) can cause radiation-induced chemical etching of material. Contamination and etching effects are sensitive to temperature (Hren, 1986) and beam voltage (Hirsch *et al.*, 1994). Cryoshields and sample cooling can help reduce contamination. Adsorption and desorption processes affect luminescence because of modifications in such variables as work function, surface-state character and density, and surface charge (Wolkenstein, 1991). An adsorbed layer can either enhance or degrade the luminescence. Contamination typically degrades the CL of high-purity semiconductors through enhanced nonradiative recombination, i.e., produces characteristic dark areas in CL images (Warwick, 1987). Bright areas (CL enhancement) can occur when contamination results in the surface repulsion of carriers. The contamination rate is related to the depth dose function and is higher at lower beam voltages. In general, one is more likely to induce electronic damage than to build up a contamination film.

3.3.5. Artifacts

Where does the CL actually come from? The generated light travels in all directions in the crystal. Light very close to the front surface can escape with minimal spectral distortion owing to absorption (ray A in Fig. 11a). Ray B in the figure undergoes total reflection on the back surface prior to escape. These latter rays

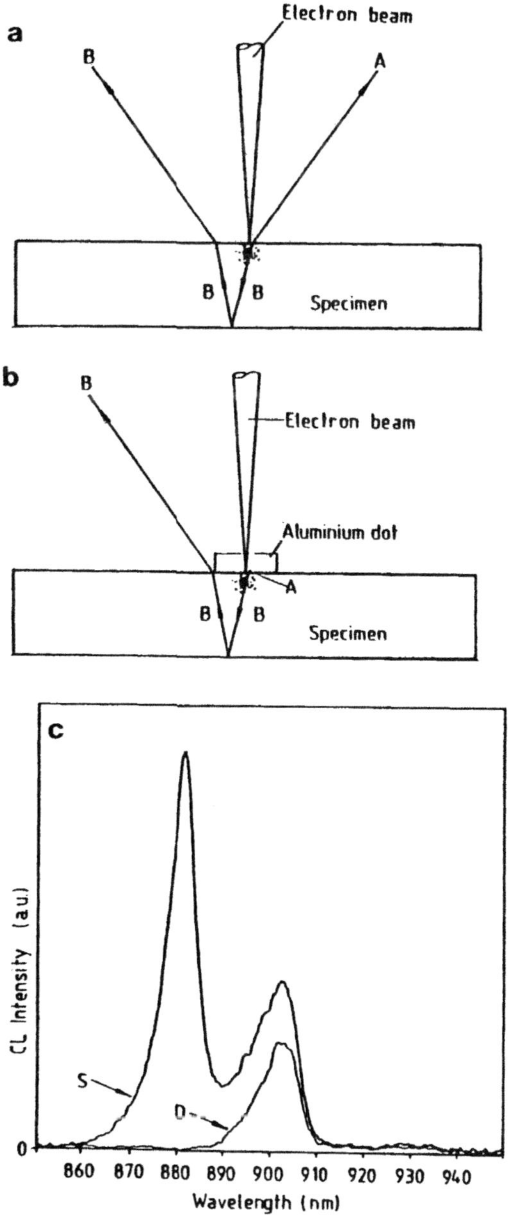

Figure 11. Example of ghost peak artifact in CL spectra from InP: (a) standard configuration showing rays *A* and *B* leaving the specimen; (b) modified situation with an aluminum dot to enhance the ghost peak; and (c) local spectra from a standard sample (*S*) and an aluminum-dot sample (*D*) (Warwick and Booker, 1983; with kind permission from IOP Publishing, Ltd., Bristol).

undergo selective absorption and lead to spectral distortions in peak shapes and peak ratios, a phenomenon known as "ghosting" (Carr and Biard, 1964). Ghosting can give rise to artifacts in emission spectra owing to anomalous optical collection. The degree to which ghosting occurs depends on the field of view of the collection optics (Warwick and Booker, 1983). This is demonstrated in Fig. 11c, where D is the spectral ghost peak associated with light rays B [the sample is configured with an Al-mask to prevent the escape of rays A (see Fig. 11b)], and S is the normal spectral output (configuration Fig. 11a with rays A and B emerging) from a sample of InP.

There is also evidence for remote CL, i.e., light emission from areas a considerable distance away from the point of excitation. Reabsorbed radiation (photon recycling) can excite material much deeper than the electron penetration depth. For example, higher-energy photons generated in the near-surface region of a GaAs/AlAs SQW by low-voltage electrons can be absorbed and give rise to CL emission from the GaAs substrate many micrometers deep (Holt *et al.*, 1991). Field-induced lateral drift of carriers caused by band-potential offsets in quantum structures can also give rise to remote CL (Norman and Ghisoni, 1993). Remote CL in bulk natural sphalerite (ZnS) is thought to be associated with (mirror) charge transport and cascade (carrier multiplication) phenomena (Vigouroux *et al.*, 1985).

Other interesting CL artifacts are associated with channeling patterns. Weak SEM electron channeling pattern (ECP) contrast can sometimes be superimposed on CL images of bulk samples (Schulson *et al.*, 1969). The effect is particularly noticeable in single-crystal optoelectronic material of poor luminescence quality (Edwards, 1991). This phenomenon is similar to the crystallographic CL contrast in TEM, where the forward-scattered electron intensity is subject to anomalous absorption (Pennycook and Howie, 1980). In bulk specimens, the CL signal is enhanced where the BSE signal is reduced by crystallographic channeling, i.e., directions in which more power is absorbed by the sample.

4. MATERIALS APPLICATIONS

4.1. Semiconducting Materials

Semiconducting materials as a group of solids have been extensively investigated by CL for over 30 years. Some of the early imaging (Wittry and Kyser, 1964) and spectroscopic (Cusano, 1964) studies were of bulk GaAs. There has been considerable development in materials growth and processing technologies since then. Within this framework, CL imaging is still informative (Yacobi and Holt, 1986, Holt and Joy, 1989). Quite often, CL imaging reveals inhomogeneities and defects in samples that have high-quality PL signatures. Therefore, it can be misleading to rely on PL alone for an evaluation of material quality. The two techniques clearly complement one another. The evolution in CL characterization can be ascertained from the prior proceedings of the *Oxford Meeting on the Microscopy of Semiconducting Materials* published by the Institute of Physics, Bristol, U.K. Not too long ago, quantum-structured materials and devices such as GaAs/GaAlAs and Si/SiGe were of topical interest. The usefulness of CL imaging is well-established; however, there are always new issues to be resolved. For example, can luminescence provide any objective information about interface roughness in mesoscopic structures (Bimberg *et al.*, 1987; Warwick *et al.*, 1990; Bimberg and Christen, 1993)? Samples that exhibit relatively sharp exciton lines (implication of atomic smooth interfaces) typically have roughness on the atomic scale when analyzed by lattice imaging in TEM. The situation is potentially more complex as compositional fluctuations on the atomic scale can also appear as structural roughening. The moral is to appreciate the spatial sensitivity of the selected characterization technique.

4.2. Minerals

CL observations of minerals and gems date back to the early work of Crookes (1879), who conducted spectral and polarization experiments. The book by Marshall (1988) covers CL history and applications, Waychunas (1988) describes basic luminescence processes and spectroscopies, and Remond *et al.* (1992) addresses recent developments in CL interpretation. The interpretation of CL from minerals is a major challenge at the moment. This is partly due to uncertainties concerning the origin of CL. More systematic studies similar to those already conducted with semiconductors will help toward an understanding of minerals. In these studies, the CL signatures from natural minerals (e.g., zircons) are compared with those from 'pure' artificial forms intentionally doped with rare-earth elements (Cesborn and Remond, 1996).

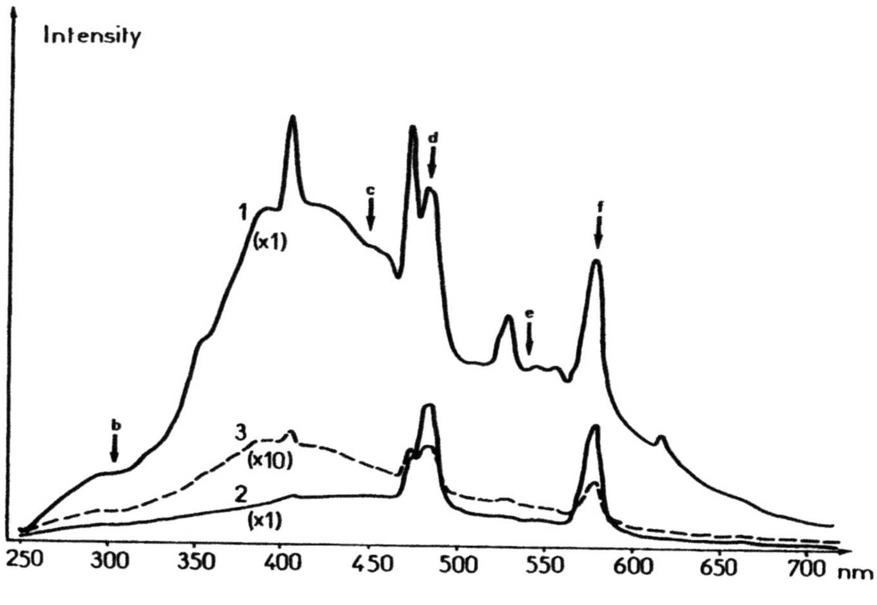

Figure 12. Room temperature CL emission spectra from natural Zr ($ZrSiO_4$) (see corresponding CL images in Fig. 13) (Redmond *et al.*, 1992; with kind permission from Scanning Microscopy International, Chicago).

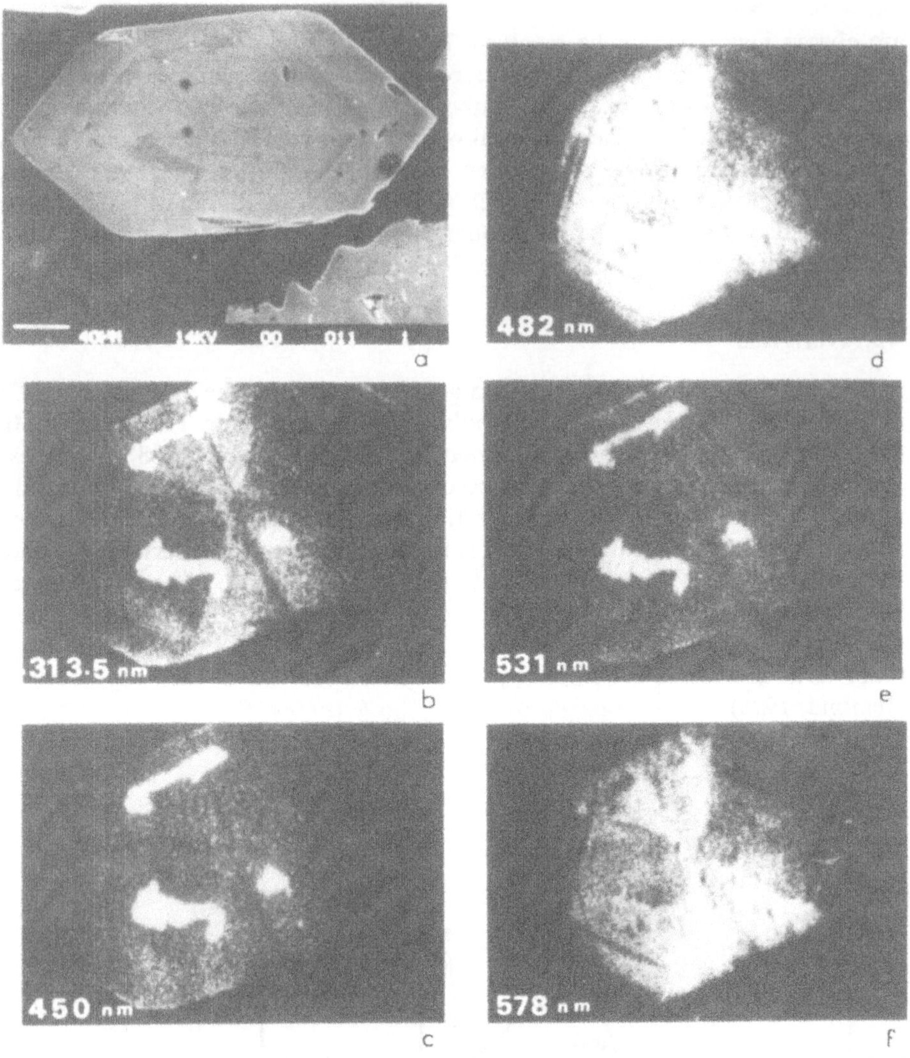

Figure 13. Backscattered electron image (a) and monochromatic CL images (b)–(f) from natural Zr (see Fig. 12 for corresponding spectral features). Images (d) and (f) correspond to emission from Dy^{3+} (Remond *et al.*, 1992; with kind permission from Scanning Microscopy International, Chicago).

Some recent CL applications can be summarized. Mason (1994) has reported on the potential use of CL as a geothermometer. The temperature of formation and subsequent thermal history of calcites can be estimated from CL signatures. Terrestrial, lunar, and meteoritic feldspars have distinctly different CL spectral attributes (Marshall, 1988). The origin of CL from quartz is unknown, but is thought to be defect-impurity related. Matters are made difficult by the fact that natural quartz is also susceptible to electron-beam damage (Ramseyer and Mullis, 1990). Nonetheless, CL imaging of quartz can reveal nonequilibrium phases produced by meteoritic impact shockwaves: shock-induced lamellae appear dark (Owen, 1989). CL spectroscopy can be used to identify rare-earth trace impurities. Figure 12 shows the CL emission spectrum from natural zircon. Figure 13b–f shows the corresponding monochromatic CL images. Of particular interest are the images at 482 nm (d) and 578 nm (f) that correspond to the presence of Dy^{3+} (Remond *et al.*, 1992). This correlation was made by comparing CL spectral signatures from Dy_2O_3 with that from natural zircon.

A knowledge of the CL behavior of minerals can also be applied to archaeology and fossils. The distribution and intensity of colors in CL micrographs can be used to fingerprint white (calcitic and dolomitic) marble. The variations in color contrast are due to different trace impurity distributions (Mn or Fe) that result from different geological histories. This information can help identify the geographical source of white marble used in ancient monuments and sculptures (Barbin *et al.*, 1992). In another interesting application, CL can be used to study the life cycle of animals. In particular, variations in Mn^{2+} activator concentrations give rise to intensity variations (zonation) in CL images of biogenic carbonates in mollusks (shells). These variations in concentration are related to environmental conditions and metabolic processes (physical growth) of these sea creatures (Barbin *et al.*, 1991).

4.3. Ceramics and Superconductors

There has been a considerable increase in the number of CL investigations of ceramics and related solids (Barrett and Page, 1992). This group of solids includes electronic oxides (e.g., ZnO, MgO, SiO_2, and Al_2O_3), ferroelectrics (e.g., $BaTiO_3$ and $LiNbO_3$), and high-temperature superconductors (e.g., Y-Ba-Cu-O, and Bi-Sr-Ca-Cu-O). Panchromatic CL imaging of the microstructures of these typically polycrystalline solids reveals grain boundaries, distribution of phases, precipitates, and mechanical damage (Holt, 1992). For example, Fig. 14 shows a BSE and CL image of a high-temperature W-Re-Hf superalloy. Any preferential precipitation of trace-impurity oxides (thoria) at grain boundaries at high temperatures is of interest. Panchromatic CL readily reveals the presence of thoria as bright particles against a nonluminescent metallic matrix. In another example, CL can delineate among different phases of zirconia (ZrO_2) in ceramics. The monoclinic phase of zirconia is more strongly luminescent than the cubic-plus-tetragonal phase. Monoclinic zirconia is thus readily visible in CL (Page, 1993).

The CL properties of ceramics are more perplexing than those of minerals. The origin of CL is usually attributed to fundamental emissions. However, recent results suggest that (point) defect–impurity complexes are primarily responsible for light emission. Specifically, oxygen and oxygen-related vacancies are thought to play a significant role in the luminescence properties of many ceramic materials: $BaTiO_3$ (Hübner *et al.*, 1993), MgO (Datta *et al.*, 1980), and ZnO (Lohnert and Kubalek 1983). However, a lack of any fundamental understanding about CL processes does not prevent useful observations (see e.g., Fig. 14). This is particularly true with respect to high-temperature superconductors. CL can distinguish between the superconducting (metallic) and semiconducting/insulating phases of Y-Ba-Cu-O (Miller *et al.*, 1990). The metallic phase has a lower CL efficiency relative to the other

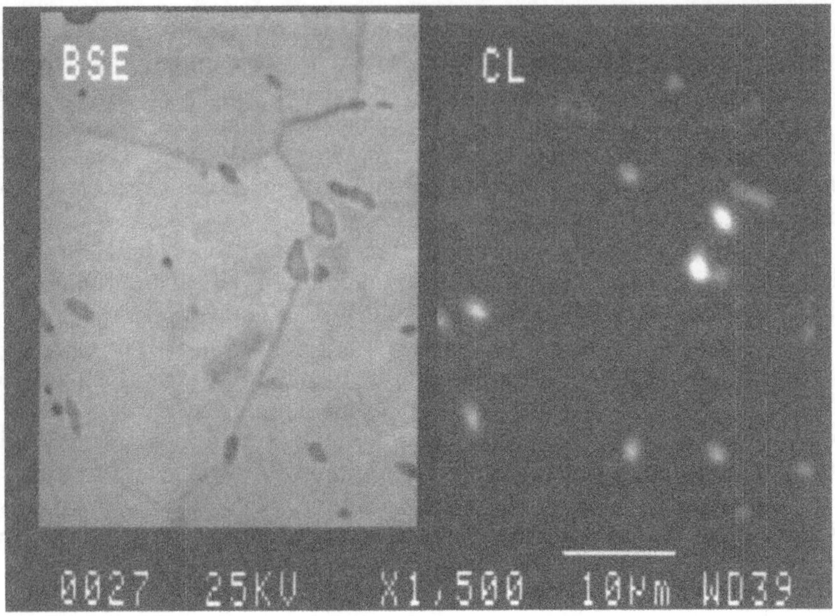

Figure 14. BSE and panchromatic CL images of thoria particles revealed at grain boundaries in W–Re–Hf high-temperature alloy (Myhajlenko, unpublished).

phases. The CL signal strength is thought to be determined by the oxygen content in the different phases. The CL properties of these materials are expected to be complex since they are formed by heating mixtures of rare-earth oxides and carbonates. There are possible contributions from thermoluminescence since the excitation powers used are relatively high. Barkay *et al.* (1992) have used CL to determine thickness variations in single-phase thin-film superconductors.

5. RECENT APPLICATIONS

A review of some recent developments in CL follows, including quantitation of stress tensors by polarized CL analysis; high-spatial-resolution imaging of defects and quantum structures by SEM, STEM, and STM; depth-resolved information; anomalous charging effects; and time-resolved measurements.

5.1. Stress Anisotropy in Thin Films

Uniaxial stress is known to modify both the band structure and energy of a semiconductor. In the case of GaAs, the fourfold degenerate valence band at $k = 0$ splits into two doubly degenerate components ($m_j = \pm\frac{3}{2}$ and $m_j = \pm\frac{1}{2}$). The valence-band splitting strongly influences the acceptor-impurity energy levels (Bhargava and Nathan, 1967). Stress (under ohmic contacts) can influence the threshold-voltage characteristics of GaAs field-effect transistors. The spectral broadening of CL linewidths attributed to tetragonal lattice distortion is found to correlate with the transistor threshold voltage (Warwick *et al.*, 1987). These variations in contact stress can lead to lack of uniformity in performance from one device to another.

More dramatic effects of stress on CL are observed with heteroepitaxial layers, e.g., GaAs grown on Si substrates (GaAs/Si). Interest in GaAs/Si is associated with monolithic integration of electronic and photonic technologies. However, this material system is hampered by significant mismatch in both lattice parameter ($\approx$4%) and thermal expansion coefficients. The lattice mismatch is relieved by the formation of threading dislocations. Thermal mismatch leads to severe tetragonal distortions on cooling to room temperature. The tensile stresses developed in these layers are high enough to generate microcracks that are visible in CL. The stress distribution can be qualitatively imaged by using excitons. Quantitation requires a knowledge of how stress affects excitons. The effect of uniaxial and biaxial stress on excitonic emission in GaAs/Si is similar to that for acceptor levels: the splitting at $k = 0$ results in light-hole (LH) ($m_j = \pm\frac{1}{2}$) and heavy-hole (HH) ($m_j = \pm\frac{3}{2}$) transition bands (Zemon *et al.*, 1986). In short, knowing the energy position and separation of the LH and HH exciton lines allows a quantitative description of the stress.

The CL mapping of stress variations around cracks and patterned regions reveals complex nonuniform behavior, including evidence for stress relief (Yacobi *et al.*, 1988). The nature of stress in GaAs/Si patterned in various rectangular and stripe geometries is observed to change from biaxial to uniaxial with reduced feature size (van der Ziel *et al.*, 1989). The stress distributions are made complex (relative to epilayers) by the free sidewalls ($\langle 011 \rangle$ facets) of the features. For example, with stripe patterns, the in-plane stress parallel to the stripe is found to remain constant while stress relief occurs in the perpendicular plane. A determination of the stress tensor requires deconvolution of the excitonic lineshape, i.e., for a precise assignment of the LH and HH energy splitting and peak positions. The biaxial stresses X_1 (longitudinal) and X_2 (transverse) along the orthogonal $\langle 001 \rangle$ directions of the (100) GaAs surface can be expressed as follows:

$$\begin{aligned} X_1\,(\text{kbar}) &= 0.138A + 0.5(0.0686S^2 - 0.0371A^2)^{1/2} \\ X_2\,(\text{kbar}) &= 0.267A - X_1 \end{aligned} \tag{17}$$

where $X_1 > X_2$, S (meV) is the excitonic peak splitting, and A (meV) is the average shift in excitonic peak energy from the unstressed position (van der Ziel *et al.*, 1989). Deconvolution of CL spectral data can provide high-spatial-resolution measurement of stress (Lingunis *et al.*, 1993).

Detection of CL under different polarization conditions provides an elegant alternative to visualizing strain-induced exciton splitting in epitaxial films (Rich *et al.*, 1991). Figure 15 shows monochromatic CL images of two microcracks (C_A and C_B) oriented perpendicular and parallel to the $\langle 110 \rangle$ crystal direction in GaAs/Si: (a) is recorded without a polarizer; (b) and (c) are taken with a linear polarizer set parallel and perpendicular to $\langle 110 \rangle$, $E_{\parallel}$ and $E_{\perp}$, respectively. The images reveal an anisotropy for CL emissions close to the cracks. This anisotropy is associated with

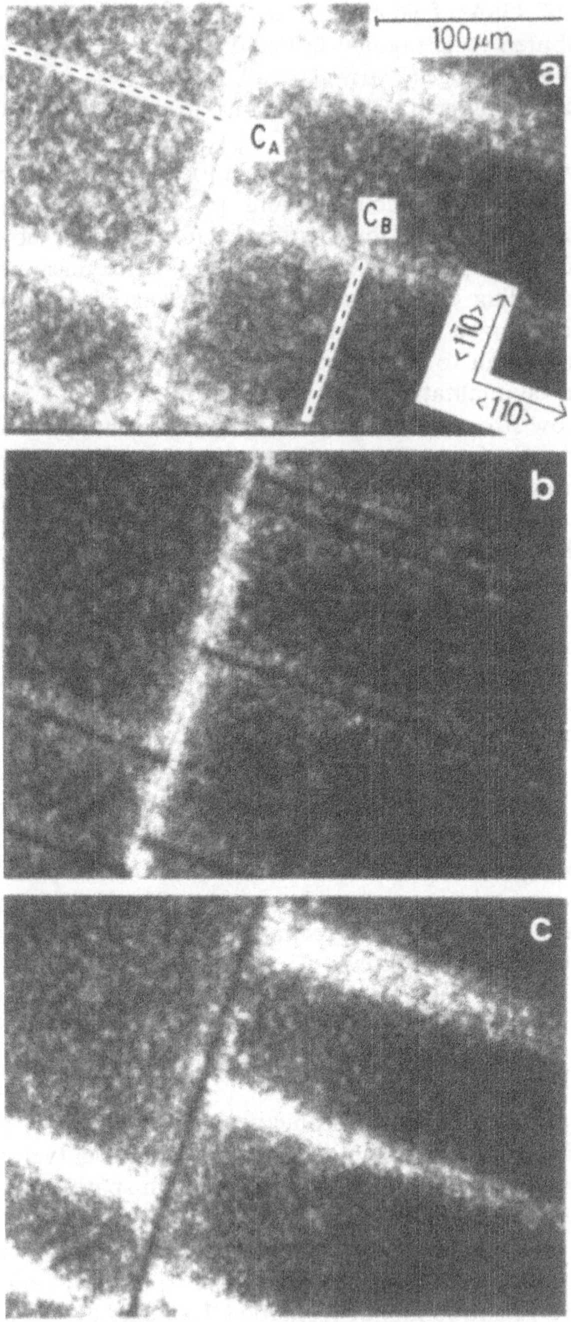

Figure 15. Monochromatic CL images of microcracks (C_A and C_B) in GaAs/Si revealing polarization anisotropy: (a) without a polarizer, and (b) and (c) with linear polarizers set parallel ($E_\parallel$) and perpendicular ($E_\perp$) to $\langle 110 \rangle$, respectively (Rich *et al.*, 1991; with kind permission from the American Physical Society).

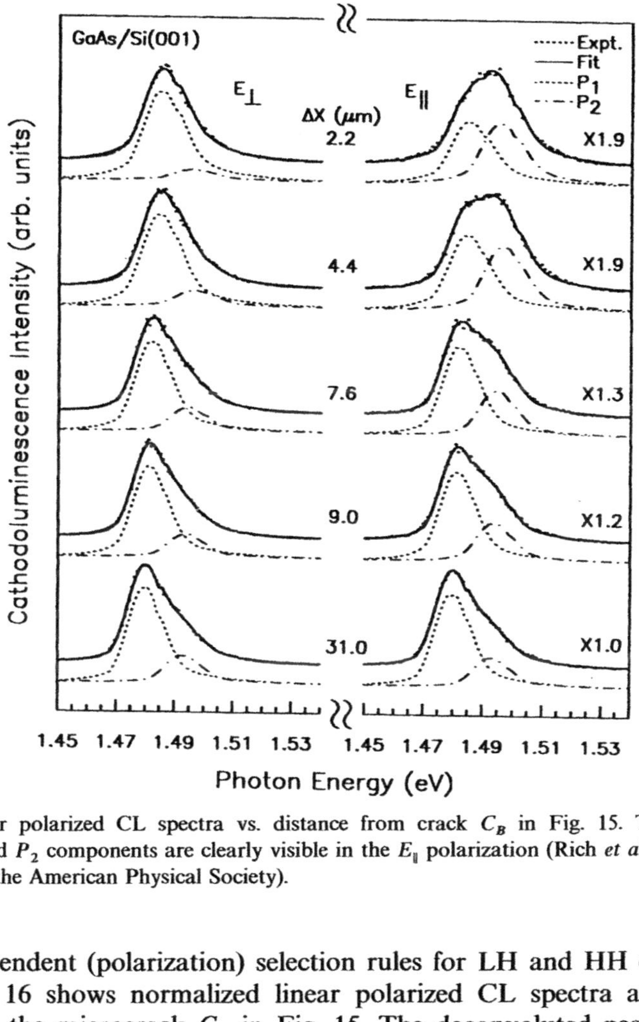

Figure 16. Linear polarized CL spectra vs. distance from crack C_B in Fig. 15. The deconvoluted strain-split P_1 and P_2 components are clearly visible in the $E_{\parallel}$ polarization (Rich *et al.*, 1991; with kind permission from the American Physical Society).

the strain-dependent (polarization) selection rules for LH and HH exciton transitions. Figure 16 shows normalized linear polarized CL spectra as functions of distance from the microcrack C_B in Fig. 15. The deconvoluted peaks P_1 and P_2 represent the strain-induced exciton LH and HH splitting. Polarization CL spectroscopy clearly facilitates stress calculations [(see, Eq. (17)] as two sets of data are analyzed simultaneously. Polarization spectroscopy can be applied to more realistic stress conditions that include peeling stresses and shear components (Tang *et al.*, 1994).

5.2. High-Spatial-Resolution Imaging of Defects and Epitaxial Films

Technology has recognized micromachines and mesoscopic structures. It is very important to be able to visualize and characterize ever smaller objects. To better understand how the physical properties of solids relate to their atomic and microstructural attributes requires one to have appropriate spatial resolution. Spatial resolution in CL is dependent on the kind of information one wishes to extract. For instance, nanometer-sized luminescent particles finely dispersed in an optically inactive matrix can be readily visualized by CL provided that the particle emissions are noninteracting. In bulk solids, the CL resolution ($\approx 1\ \mu m$) is usually limited by the generation volume or the carrier diffusion length (re: Sections 2.1 and 3.2.1). Confinement of either the excitation or the excess carriers can further improve the resolution to about 0.1 μm or better.

Three different approaches for improving spatial resolution are reviewed: (1) STEM/TEM, (2) LVSEM, and (3) STM. The improvement in resolution usually comes at the expense of reduced signal strength. In STEM/TEM, the use of electron-transparent films and high accelerating voltages reduces beam-spreading effects. The nonradiative recombination properties of two bound surfaces also restrict carrier diffusion. The same carrier confinement effect can be achieved on bulk samples with low accelerating voltages (≈ 1 keV) in a SEM. LVSEM injects carriers close to the surface with minimal lateral scattering. The detection of photons associated with prediffusion radiative recombination should reflect the spatial extent of the initial excitation volume. In GaAs, the surface can act as a picosecond "shutter," which effectively results in time-resolved CL imaging (Warwick, 1991). A similar effect occurs in Si, where most of the CL emanates from the generation volume (Cumberbatch *et al.*, 1981). To take the low-voltage approach one step further, STM can inject very low-energy electrons into solids. Nanometer-scale resolution has been reported in STL images of quantum-structured layers (Pfister *et al.*, 1994).

Simultaneous STEM/TEM CL offers high spatial resolution and potential correlation with crystallographic diffraction contrast. It is desirable to determine the optoelectronic properties of well-characterized individual dislocations (type, Burgers' vector, and degree of dissociation) from CL spectral signatures and in particular to distinguish between intrinsic and extrinsic dislocation properties. Intrinsic behavior is associated with core states, strain fields, or piezoelectricity (for noncentrosymmetric solids), whereas extrinsic characteristics are associated with point defects and impurities. In practice, these measurements are difficult to perform because CL signals are typically very weak, and the CL intensity and spectral response to thickness variations are nontrivial. Furthermore, samples with signal strengths suitable for production of high-quality CL spectra are generally too thick for weak-beam analysis at the same accelerating voltage (Myhajlenko *et al.*, 1984, 1985, 1986; Steeds, 1989*a*, *b*). As an example of the complex behavior, the CL intensity $I(t)$ as a function of film thickness t is predicted to be

$$I(t) = I_0\{t - 2L/[\coth(t/2L) + S]\} \tag{18}$$

where L is the carrier diffusion length, S is the reduced surface recombination velocity, and I_0 is a term related to the radiative recombination rate (Yuan *et al.*, 1989). The model predicts a cut-off thickness for CL emission that can be interpreted as a luminescence dead layer.

The small number of STEM/TEM CL studies conducted thus far indicates that there is no systematic relation between CL signature and dislocation character. Rather, there is considerable evidence to correlate specific luminescence bands with

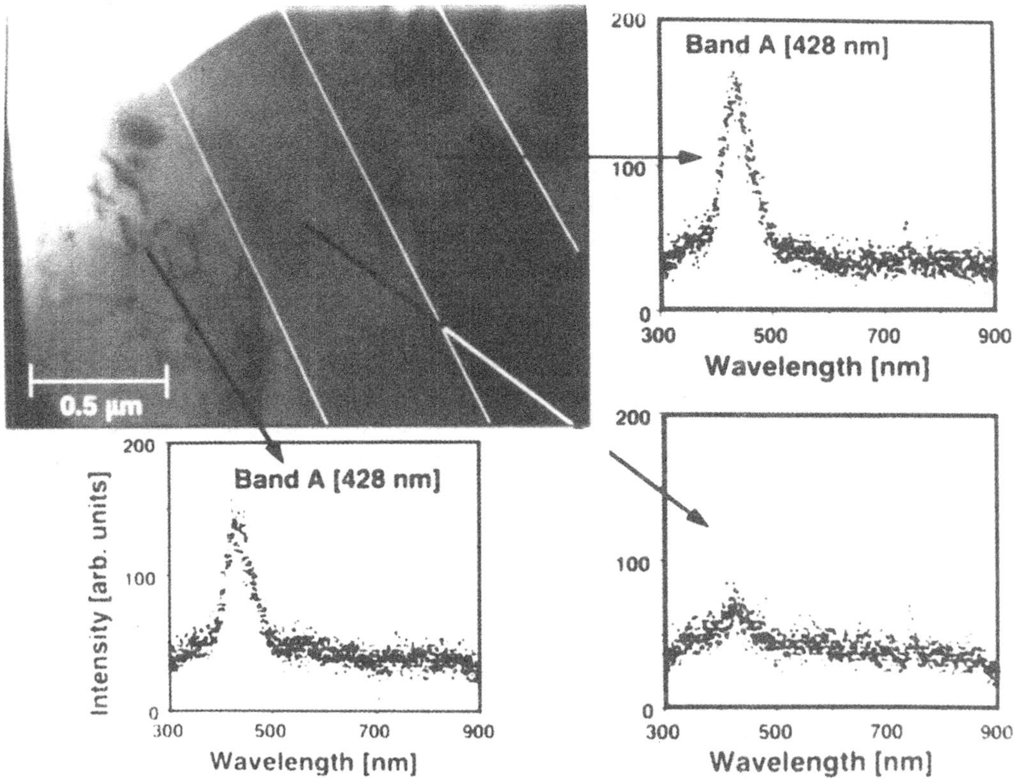

Figure 17. TEM image of diamond film containing defect-free grains and grains with dislocations. Local CL spectra reveal that band *A* (428 nm) is associated with dislocations (Graham *et al.*, 1991; with kind permission from the American Institute of Physics).

the presence of dislocations or dislocation clusters. In general, dislocation-related luminescence is associated with impurities. For instance, band *A* is observed in type IIb natural diamonds (Pennycook *et al.*, 1980; Yamamoto *et al.*, 1984), in diamond films grown by chemical vapor deposition (Graham *et al.*, 1991), and by the combustion-flame method (Graham and Ravi, 1992). Figure 17 shows a correlation of band *A* with the presence of individual dislocations and dislocation clusters. Band *A* is thought to be associated with DAP recombination at nitrogen-donor and boron-acceptor impurities, respectively (emission at 425–428 nm). In polarization CL measurements, the light emitted from dislocations in diamonds is found to be polarized parallel to the dislocation line and the degree of polarization is found to be independent of dislocation character (Yamamoto *et al.*, 1984).

Examples of complex luminescence behavior are the *Y* and *S* bands observed in ZnSe grown on GaAs. These bands are thought to be impurity related. At low temperatures (30–70 K), *Y* emission is observed at individual dislocations (independent of character) and complex dislocation tangles (Myhajlenko *et al.*, 1984; Batstone and Steeds, 1985, 1986). Although *S* emission is often observed together with *Y* emission, *S* is primarily associated with dislocation tangles. Similar complex behavior is observed with excitons. Generally, excitonic emissions are quenched at dislocations in semiconductors. There are exceptions: excitonic emissions are quenched at complex dislocations and at stacking faults in ZnSe, but not at individual dislocations that exhibit *Y* emission. Similarly, not all single dislocations quench excitonic emissions in InP (Myhajlenko *et al.*, 1984, 1985).

Room-temperature STEM/TEM CL observations of individual defects in semiconductors are rare because signal strengths are generally too weak. This is unfortunate since room-temperature behavior of defects is especially relevant to semiconductor manufacturers, as most fabricated devices operate at this temperature. In one example, exciton quenching at stacking faults in Al-doped ZnSe is associated with electrically active partial dislocations rather than the stacking fault ribbon (Batstone, 1985). Other interesting STEM/TEM CL observations include: strain-related dislocation luminescence bands in microhardness indented HgI_2 crystals (Petroff *et al.*, 1989); anticorrelation in exciton- and impurity-band emissions at intrinsic stacking faults in AlGaAs (Bailey *et al.*, 1987); a redshift in exciton emission energy at a group of dislocations in InP is interpreted in terms of dislocation-bound excitons (Myhajlenko *et al.*, 1986); a new luminescence band is observed at oval defects in AlGaAs MQW structures (Bailey and Steeds, 1989); and luminescence associated with the intrinsic electronic structure of dislocations in diamond films (Graham *et al.*, 1994).

The reported spatial resolution of STEM/TEM CL investigations of defects in solids is typically in the range 0.1–0.2 μm (see also Section 3.2.1). This relatively poor performance over what the technique can promise is due to the desire to record meaningful high-quality spectra. This signal threshold is determined by the sample thickness. It is often the behavior of excitons relative to other emissions that is of interest, as excitons provide a useful vehicle for probing the fundamental properties of solids. Unfortunately, exciton diameters can be relatively large ($\approx$15 nm in GaAs), which can limit resolution by defining a minimum thickness of sample that allows free excitons to exist. In general, single-particle transitions offer the best option for high-resolution imaging by STEM CL.

High-resolution imaging of bulk solids by low-voltage SEM is the next topic to be considered. Excess carriers can be confined by injecting electrons very close to the surface. The CL signal strength is (again) expected to be very weak. The best spatial resolution to date by field-emission-gun (FEG) LVSEM CL at about 1 keV is 60 nm from a well-characterized GaAs SQW (Warwick, 1991). Emission from this SQW sample is characterized by three well-resolved lines at 712, 720, and 726 nm. The energy positions correspond to thickness changes in the quantum well of about 0.8 monolayer (Kopf *et al.*, 1991). Figure 18 depicts the corresponding low-temperature (5 K) monochromatic SEM CL images of the quantum well in cross section. The variations in CL intensity along the length of the well at the different emission wavelengths suggests additional changes in well thickness (microroughness) on a fine atomic scale (see the region in the figure marked by the arrow) (Warwick and Kopf, 1992). On the other hand, what can be achieved with bulk samples in a conventional (W-filament) SEM at room temperature? The author has resolved CL contrast variations around dislocations in GaAs with a resolution of about 0.2 μm at 3 keV (electron range $\approx$0.15 μm). The spatial resolution is not significantly degraded by

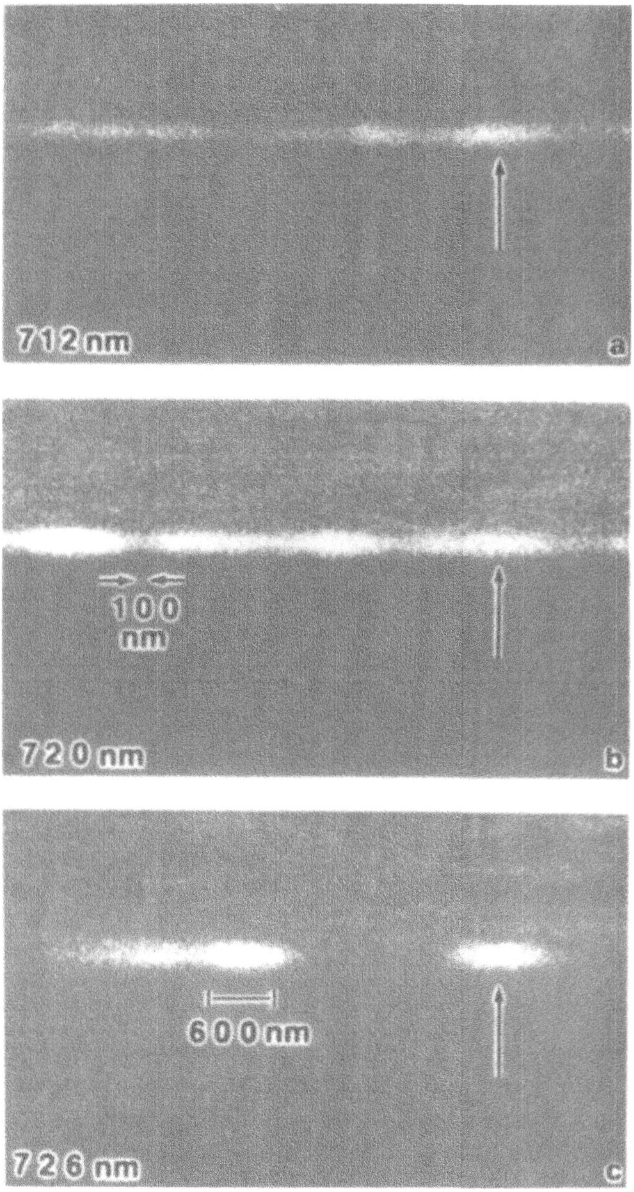

Figure 18. High-resolution SEM CL monochromatic images of a single quantum well recorded in cross section at: (a) 712 nm; (b) 720 nm; and (c) 726 nm. The contrast variations correspond to submonolayer thickness changes (see arrow). The beam voltage is $1 \approx 2$ keV (Warwick, 1991; with kind permission from IOP Publishing Ltd., Bristol).

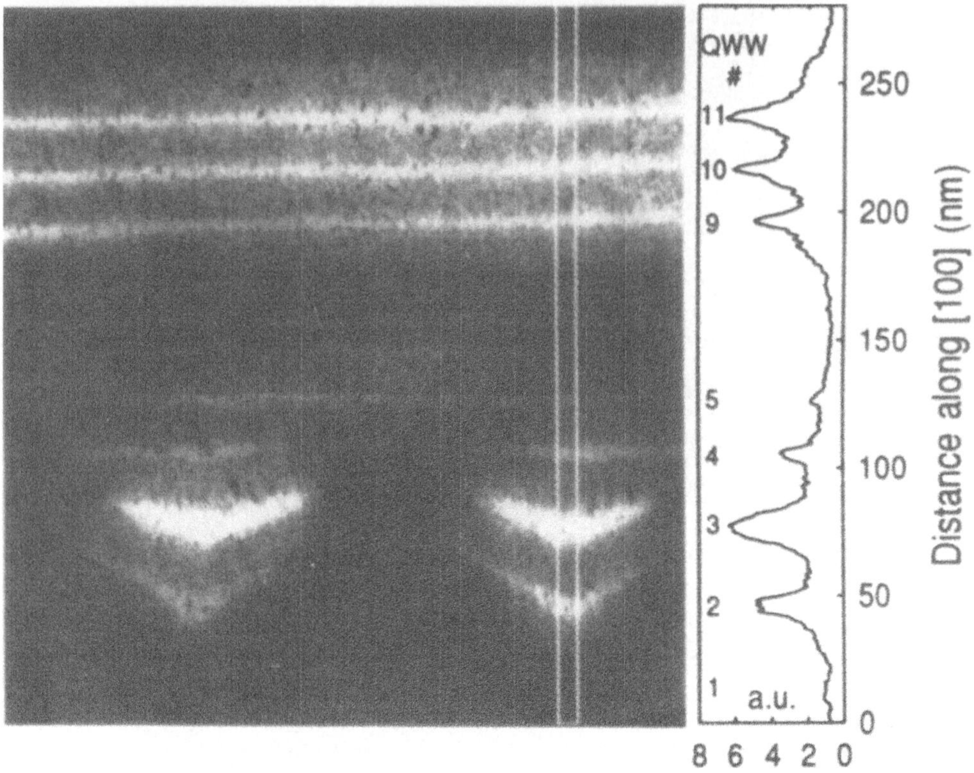

Figure 19. Scanning tunneling-induced luminescence (STL) cross-sectional image of two quantum-well wires. The STM bias voltage is 2.2 V and the tunneling current is 1.0 nA. The linescan profile on the right reveals luminescence intensity variations on a scale of ≈5 nm (Pfister *et al.*, 1994; with kind permission from the American Institute of Physics).

the diffusion of carriers. In fact, this resolution is comparable to typical STEM/TEM CL applications.

The imaging of quantum-well wire arrays by STM can yield nanometer-scale spatial resolution (≈5 nm) in luminescence. This permutation of the CL technique, STL, was described in Section 3.2.6. The injection of very low-energy (<50 eV) minority carriers into samples by STM can result in radiative recombination. The relative efficiency of individual quantum wires can be seen with STL in Fig. 19. The origin of the contrast is still a topic for debate. In addition to the competition between radiative and nonradiative processes, the induced luminescence generally depends on the injection efficiency into the sample. Nonetheless, this remarkable image reveals modulations in luminescence intensity on the nanometer scale that are probably related to doping and thickness variations in the quantum wells (Pfister *et al.*, 1994). The combination of STM and STL opens up an exciting world where correlation of structural inhomogeneities with electronic and optical properties may be feasible on the atomic scale.

5.3. Surface and Depth-Resolved Information

CL can differentiate between surface and bulk phenomena. Depth-sampling experiments can be performed on a wide variety of crystals by controlling the incident electron energy. CL studies have included characterization of ion implantation damage in GaAs (Norris *et al.*, 1973); the effect of surface recombination velocity, carrier diffusion length, and luminescence dead layers on depth-profile character (Wittry and Kyser, 1967; Puhlmann *et al.*, 1991); and observations of process-induced changes in the electronic structure of the surface (Brillson and Viturro, 1988). Surface sensitivity is achieved by the use of low accelerating voltages, which, as was pointed out earlier, can confine the excitation to a shallow region close to the surface. CL measurements performed in a UHV chamber allows the investigation of clean and chemically modified surfaces without the usual worry of contamination at low voltages. This makes possible the detection of electronic states localized at or near the surface including metal-induced deep states, defect levels, microroughness, and the band structure of localized compounds (Brillson and Viturro, 1988).

CL depth-profiling can be used in a variety of circumstances as the generation depth can be varied from nanometers to micrometers, e.g., selective excitation of multiple quantum-well layers (Trager-Cowan *et al.*, 1991), and visualization of defect propagation through epitaxial layers (Cleton *et al.*, 1993). From the quantitative perspective, one can estimate the surface recombination velocity or carrier diffusion length from CL depth profiles (Wittry and Kyser, 1967; Rao-Sahib and Wittry, 1969; Puhlmann *et al.*, 1991; Phang *et al.*, 1992). The author has evaluated chemically passivated GaAs surfaces by this method (Myhajlenko *et al.*, 1991). Passivation of GaAs surfaces also leads to changes in exciton–polariton behavior that may be related to surface microroughness (Myhajlenko *et al.*, 1994). Microscopic roughness refers to nonclassical scattering of light (Aspnes, 1985). The quantitative effect of microroughness on luminescence has yet to be fully explored.

The most difficult concept to accommodate in depth-profiling experiments is the so-called luminescence dead layer, which is a real phenomenon, as it represents a cut-off point in CL emission. The dead layer is usually attributed to residual surface electric fields, but can also be associated with surface damage or optical transition layers. The behavior of luminescence dead layers under special excitation conditions will now be described, with the assumption that the dead layer is related to the surface electric field. GaAs can exhibit unusual transient CL properties at liquid-helium temperatures under low-voltage excitation (0.2–2.0 keV). Specifically, prolonged electron-beam exposure (at least $0.05°\mathrm{C\,cm^{-2}}$) at the signal cut-off voltage can result in CL recovery (Myhajlenko, 1991). This dramatic change in CL behavior is shown in Fig. 20, where the transformation occurs at 1.5 keV (see arrow). The profile data can be modeled in terms of a reduction in the dead-layer thickness from about 500 to 50 Å. For all practical purposes the dead layer is effectively zero. A zero-thickness dead layer is the equivalent of the flat-band conditions often cited in PL. There is no band bending at the surface, i.e., no surface depletion region (Hollingsworth and Sites, 1982). The return of excitons at accelerating voltages as low as 200 eV and low beam power ($\approx 1\,\mu$W) supports this supposition, as excitons previously quenched at about 1.2 keV. These changes are related to surface charging

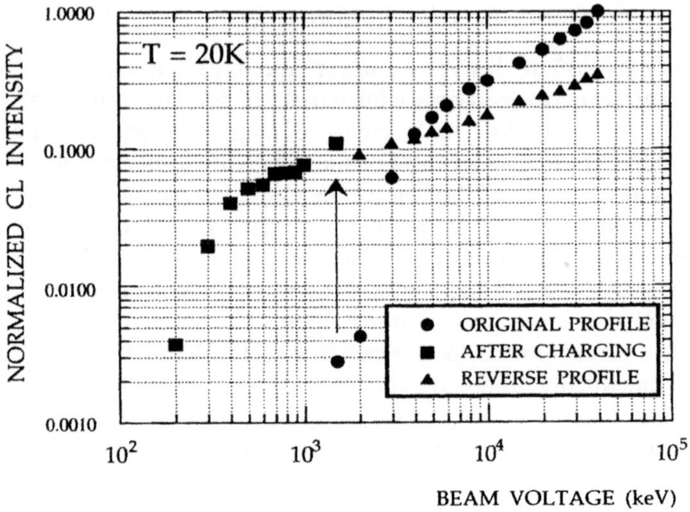

Figure 20. Normalized integrated band-edge CL depth profiles at 20 K of *n*-type GaAs. The original profile represents a typical result where signal cut-off occurs at ≈ 1.2 keV. Prolonged beam exposure at 1.5 keV results in anomalous charging and CL enhancement (marked by arrow). The subsequent after-charging profile tracks the CL intensity down to 200 eV. The reverse profile is a measurement of the CL intensity behavior back to 40 keV, following charging effects. Note for voltages higher than ≈ 3.5 keV the CL efficiency of the material is lower than before charging. The data are normalized to the original CL intensity at 40 keV (Myhajlenko, 1991; with kind permission from Scanning Microscopy International, Chicago).

(and possible crossover voltage phenomena) and/or saturation of deep-surface states since the enhancement degrades over time when the excitation is removed (Myhajlenko *et al.*, 1994). In all probability, the effect may also be linked to contamination and/or low-energy electron bombardment damage. The charging effect leads to permanent dark CL contrast features at room temperature consistent with contamination or damage artifacts. The features remain after sample removal from the vacuum. Anomalous charging effects are also observed in the CL of Ce:YAG crystals at energies less than 3 keV (Feng and Warde, 1993). In particular, luminescence relaxation occurs with charging and discharging at the surface owing to migration of charge from within the bulk of the crystal. The Ce-emission band and O vacancy-related defect luminescence band exhibit different behaviors with beam voltage and accumulated surface charge. The effects of electron yield and CL are difficult to separate because they are both related to energy loss processes of the "same" electrons.

5.4. Time-Resolved Measurements

To perform CL measurements with picosecond resolution requires instrumentation more sophisticated than that used for simple imaging or spectroscopy. Only a handful of laboratories have this capability, which is unfortunate as time-resolved luminescence can provide much useful information about the nature of impurities and defects in solids. For instance, the determination of minority carrier lifetimes from CL decays provides a quantitative basis for evaluation of material quality. The analysis of CL decays from semiconductors with surfaces, interfaces, and defects is nontrivial (Van Opdorp *et al.*, 1977). However, it is usual to assume that excess carriers (and hence the CL) decay exponentially, i.e., $CL(t) \propto \exp(-t/\tau)$. In general, the decay of luminescence in solids is rarely exponential in nature. This is evident from semilogarithmic plots of published data. A better description is given by a power law t^{-s} with exponents in the range $1 < s < 2$ (Jonscher and de Polignac, 1984). The physical basis for this model is beyond the scope of this work. Nonetheless, useful insight can still be gained from analysis of the exponential time constant.

Investigations of CL lifetimes around dislocations in a number of optoelectronic materials such as GaP, GaAs, and InP have revealed contradictory lifetime and intensity behaviors (Davidson and Dimitriadis, 1980; Steckenborn *et al.*, 1981). One can distinguish different capture processes from the transient behavior of spectral features and make a determination of impurity capture cross sections (Bimberg *et al.*, 1985). Time-resolved imaging and spectroscopy can directly visualize the lateral diffusion of carriers. One can measure exciton diffusion velocities in quantum islands (Christen and Bimberg, 1988) and capture times into quantum wires (Bimberg and Christen, 1993).

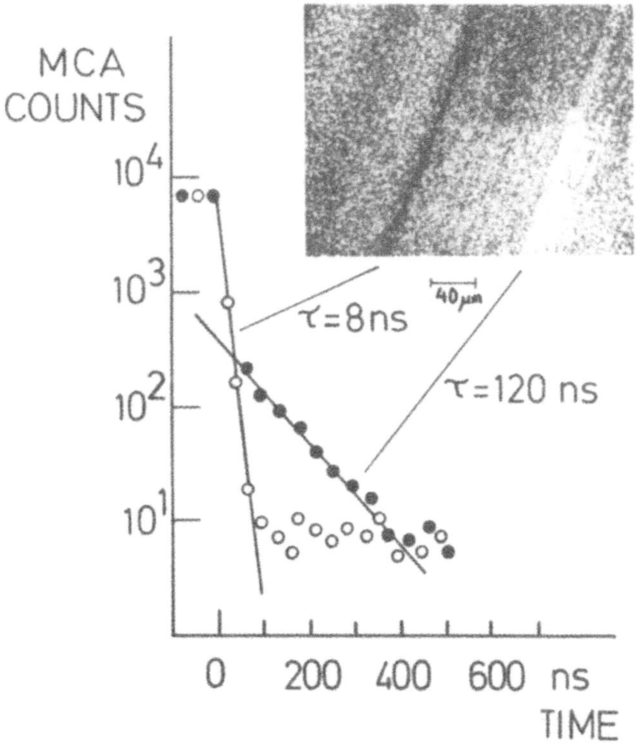

Figure 21. CL decay measurements at grain boundary features (bright and dark lines) in solar-grade ribbon silicon (Myhajlenko *et al.*, 1983; with kind permission from IOP Publishing Ltd., Bristol).

CL lifetime measurements performed in conjunction with imaging can assist in the interpretation of CL contrast. Figure 21 shows CL decays recorded by time-correlated photon counting at grain boundary features in solar-grade ribbon Si (Myhajlenko *et al.*, 1983*b*). The difference in CL lifetimes indicates that the contrast is associated with enhanced nonradiative recombination at the boundaries and not with differences in optical extraction efficiency. The surface morphology of ribbon Si is highly textured. The implication is that the dark-line feature is electrically active, which is deleterious to the current collection efficiency of solar cells.

6. FUTURE TRENDS

Areas of current interest and growth have already been covered: new materials, improvement in spatial resolution, complementary investigations, and quantitation. Growth in applications of CL over the entire broad range of solids, including organics, will continue for there are always new material systems to explore. Multidisciplinary and complementary investigations of artificial quantum-structured materials over the past decade have greatly augmented existing basic knowledge of real solids. This has led to a better appreciation of the interplay between atomic-scale morphologies and macroscopic properties of solids with their luminescence signatures. For instance, recent work by the author suggests that changes in surface microscopic roughness on a scale less than 10 Å [determined by atomic force microscopy (AFM)] may account for the anomalous behavior of exciton–polaritons in CL, PL, and photoreflectance (unpublished). It is expected that the use of CL with complementary techniques, such as EBIC and scanning deep-level transient spectroscopy (electrical), PL and ellipsometry (optical), and TEM, STM, and AFM (structural and chemical) will increase as the need for diverse information grows. Researchers are already combining STM/AFM instruments with SEM/TEM in order to extract a variety of complementary high-spatial-resolution information.

The use of FEG LVSEM or STM should continue to improve CL spatial resolution. It is not clear what ultimate resolution may be achieved with the right sample. Will one be able to detect CL from single dopant atoms with STM? The increased use of LVSEM will aid investigations of radiation- and charge-sensitive materials, such as polymers and biomedical specimens which have hitherto been excluded from routine CL. Interpretation of CL signatures will also improve as the benefits of increased activities in simulation and modeling are realized, and systematic studies are conducted on more "difficult" materials, such as minerals and ceramics. These advances, coupled with improved methods of numerical calculations, should push development in CL quantitation.

ACKNOWLEDGMENTS

The author appreciates many useful discussions and insights provided by John L. Edwards Jr., Laurence A. J. Garvie, John McHone, Rita-Ellyn Roberts, and Paul J. Wright during the writing of this manuscript. The author also gratefully acknowledges the following for use of research material and samples: Joanna L. Batstone, David B. Davito, Roger J. Graham, Henry J. Hutchinson, Mark Pfister, Guy Remond, Brian J. Skromme, John C. H. Spence, John W. Steeds, Bang-Hung Tsao, Colin A Warwick, and James A. Whiteaway.

REFERENCES

Ahn, C. C., and Krivanek, O. L. (1985), *Proc. 43rd EMSA*, p. 406.

Ashley, J. C., Tung, C. J., and Ritchie, R. H. (1979), *Surf. Sci.* **81**, 409.

Aspnes, D. E. (1985), *Handbook of Optical Constants of Solids* (Academic, New York), p. 89.

Bailey, S. J., and Steeds, J. W. (1989), *Inst. Phys. Conf. Ser.* **100**, 777.

Bailey, S. J., Preston, A. R., Steeds, J. W., and Morkoç, H. (1987), *Inst. Phys. Conf. Ser.* **90**, 273.

Bailey, S. J., Eades, J. A., and Olson, J. M. (1989), *Inst. Phys. Conf. Ser.* No **100**, p. 761.

Balk, L. J., Kubalek, E., and Menzel, E. (1976), *Scan. Elect. Microsc.* **I**, 257.

Barbin, V., Ramseyer, K., Debenay, J. P., Schein, E., Roux, M., and Decrouez, D. (1991), *Geol. Mag.* **128**, 19.

Barbin, V., Ramseyer, K., Decrouez, D., Burns, S. J., Chamay, J., and Maier, J. L. (1992), *Archaeometry* **34**, 175.

Barkay, Z., Deutscher, G., and Grunbaum, E. (1992), *Scan. Microsc. Supp.* **6**, 369.

Barrett, R., and Page, T. F. (1992), in: *Encyclopedia of Materials Science and Engineering* (R. W. Cahn, Ed.) (Pergamon, New York), p. 1553.

Bartram, R. H. and Stoneham, A. M. (1975), *Sol. St. Comm.* **17**, 1593.

Bastin, G. F., Heijlingers, H. J. M., and van Loo, F. J. J. (1986), *Scanning* **8**, 45.

Batstone, J. L. (1985), Cathodoluminescence and Transmission Electron Microscopy Characterisation of ZnSe, Ph.D. Thesis, University of Bristol.

Batstone, J. L., and Steeds, J. W. (1985), *Inst. Phys. Conf. Ser.* No. **76**, p. 383.

Batstone, J. L., and Steeds, J. W. (1986), *Proc. 46th EMSA*, p. 818.

Bell, L. D., Hecht, M. H., Kaiser, W. J., and Davis, L. C. (1990), *Phys. Rev. Lett.* **64**, 2679.

Benstock, E. J. (1991), Transmission Electron Microscope Cathodoluminescence of Meteoritic Forsterite, M.Sc. Thesis, Arizona State University.

Benstock, E. J., and Buseck, P. R. (1991), *Meteoritics* **26**, 317.

Berger, M. J., and Seltzer, S. M. (1964), *Nucl. Sci. Ser. Rep.* **39**, NAS-NRC Pub. No. 1133, 205.

Berger, S. D., McMullan, D., Yuan, J., and Brown, L. M. (1985), *Inst. Phys. Conf. Ser.* No. **78**, p. 137.

Berz, F., and Kuiken, H. K. (1976), *Sol. St. Electron.* **19**, 437.

Bethe, H. (1930), *Ann. Phys.* **5**, 325.

Bhargava, R. N., and Nathan, M. I. (1967), *Phys. Rev.* **161**, 695.

Bimberg, D. and Christen, J. (1993), *Inst. Phys. Conf. Ser.* No. 134, p. 629.

Bimberg, D., Munzel, H., Steckenborn, A., and Christen, J. (1985), *Phys. Rev. B.* **31**, 7788.

Bimberg, D., Christen, J., Fukunaga, T., Nakashima, H., Mars, D. E., and Miller, J. N. (1987), *J. Vac. Sci. Tech.*, **B5**, 1191.

Binnig, G., Rohrer, H., Gerber, C, and Weibel, E. (1982), *Appl. Phys. Lett.* **40**, 178.

Bishop, H. E. (1965), *Proc. Phys. Soc.* **85**, 855.

Blakemore, J. S. (1987), *Semiconductor Statistics* (Dover, New York).

Bongeler, R., Golla, U., Kassens, M., Reimer, L., Schindler, B., Senkel, R., and Spranck, M. (1993), *Scanning* **15**, 1.

Bonnell, D. A. (1993), *Scanning Tunneling Microscopy and Spectroscopy* (VCH, New York).

Boyes, E. D. (1989), *Inst. Phys. Conf. Ser.* No. **98**, p. 439.

Brillson, L. J., and Viturro, R. E. (1988), *Scan. Microsc.* **2**, 789.

Brunner, M., and Menzel, E. (1983), *J. Vac. Sci. Tech.* **B1**, 1344.

Bugajski, M., and Lewandowski, W. (1985), *J. Appl. Phys.* **57**, 521.

Carr, W. N., and Biard, J. R. (1964), *J. Appl. Phys.* **35**, 2776.

Carslaw, H. S., and Jaeger, J. C. (1959), *Conduction of Heat in Solids* (Clarendon, Oxford), p. 75.

Casey, H. C., and Jayson, J. S. (1971), *J. Appl. Phys.* **42**, 2774.

Cazaux, J. and Le Gressus, C. (1991), *Scan. Microsc.* **5**, 17.
Cesborn, F. and Remond, G. (1995), *Scan. Microsc. Suppl. Proc.* 13th Pfefferkorn Conference (Niagra-Falls, May 1994), *Scan. Microsc. Supp.* **9**, 57.
Chin, A. K., Temkin, H., and Roedel, R. J. (1979), *Appl. Phys. Lett.* **34**, 476.
Christen, J., and Bimberg, D. (1988), *JEOLNews* **26E**, 12.
Cleton, F., Sieber, B., Bensaada, A., Isnard, L., and Masut, R. A. (1993), *Inst. Phys. Conf. Ser.* **134**, 655.
Cocito, M., Franzosi, P., Salviati, G., and Taiariol, F. (1986), *Scan. Electron. Microsc.* **IV**, 1299.
Coombs, J. H., Gimzewski, J. K., Reihl, B., Sass, J. K., and Schittler, R. R. (1988), *J. Microsc.* **152**, 325.
Corrieg, X., Calderer, J., Blasco, E., and Alcubilla, R. (1990), *Sol. St. Electron.* **33**, 477.
Crookes, W. (1879), *Phil. Trans. Roy. Soc.* **170**, 641.
Cumberbatch, T. J., Davidson, S. M., and Myhajlenko, S. (1981), *Inst. Phys. Conf. Ser.* No. **60**, p. 197.
Cusano, D. A. (1964), *Sol. St. Electron.* **2**, 353.
Czyzewski, Z., and Joy, D. C. (1990), *Scanning* **12**, 5.
Datta, S., Boswarva, I. M., and Holt, D. B. (1980), *J. de Phys. Colloque* **C6**, 522.
Davidson, S. M. (1977), *J. Microsc.* **110**, 177.
Davidson, S. M. (1981), *Inst. Phys. Conf. Ser.* No. **61**, p. 39.
Davidson, S. M., and Dimitriadis, C. A. (1980), *J. Microsc.* **118**, 275.
Davidson, S. M., Cumberbatch, T. J., Huang, E., and Myhajlenko, S. (1981), *Inst. Phys. Conf. Ser.* **60**, 197.
Dean, P. J. (1968), *Trans. Metall. Soc. AIME* **242**, 384.
Dean, P. J., Herbert, D. C., Werhoven, C. J., Fitzpatrick, B. J., and Bhargava, R. N. (1981), *Phys. Rev. B* **23**, 4888.
Del Sole, R., D'Andrea, A., and Lapiccirella, A. (1988), *Excitons in Confined Systems* (Springer-Verlag, Berlin), Springer Proceedings in Physics Vol. 25.
Devenish, R. W., Bullough, T. J., Turner, P. S., and Humphrey, C. J. (1989), *Inst. Phys. Conf. Ser.* No. **98**, p. 215.
Donolato, C. (1978), *Optik* **52**, 19.
Eagles, D. M. (1960), *J. Phys. Chem. Sol.* **16**, 76.
Edwards, J. L. (1991), M.Sc. Thesis, p. 103 and p. 134, Arizona State University.
Feng, H. D., and Warde, C. (1993), *J. Appl. Phys.* **73**, 3926.
Frenkel, J. (1931), *Phys. Rev.* **37**, 17 and 1276.
Fujioka, H., and Ura, K. (1983), *Scanning* **5**, 3.
Garlick, G. F. J. (1949). *Luminescent Materials* (Clarendon, Oxford), p. 175.
Goldstein, J. L., Costly, J. L., Lorimer, G. W., and Reed, S. J. B. (1977), *SEM* **I**, 315.
Goulding, F. S. (1977), *Nucl. Instr. Meth.* **142**, 213.
Graham, R. J. and Ravi, K. V. (1992), *Appl. Phys. Lett.* **60**, 1310.
Graham, R. J., Myhajlenko, S., and Steeds, J. W. (1985), *J. Appl. Phys.* **57**, 1311.
Graham, R. J., Spence, J. C. H., and Alexander, H. (1987), *Mat. Res. Soc. Symp. Proc.* **82**, 235.
Graham, R. J., Moustakas, T. D., and Disko, M. M. (1991), *J. Appl. Phys.* **69**, 3212.
Graham, R. J., Shaapur, F., Kato, Y., and Stoner, B. R. (1994), *Appl. Phys. Lett.* **65**, 292.
Green, M. (1963), *Proc. Phys. Soc.* **82**, 204.
Gu, M., Syrykh, C., Halimaouli, A., Dumas, P., and Salvan, F. (1993), *J. Lumines.* **57**, 315.
Hackett, W. H. (1972), *J. Appl. Phys.* **43**, 1649.
Halbritter, J. (1983), *Scan. Electron. Microsc.* **III**, 511.
Hamilton, T. D. S., Munro, I. H. and Walker, G. (1978), in: *Luminescence Spectroscopy* (Lumb, M. D., Ed.) (Academic, London), p. 149.
Heron, S., McWhirter, R. W. P., and Rhoderick, E. H. (1956), *Proc. Roy. Soc.* **234A**, 565.
Hirsch, P. B., Howie, A., Nicolson, R. B., Pashley, D. W., and Whelan M. J. (1965), *Electron Microscopy of Thin Crystals* (Butterworths, London).
Hirsch, P., Kässens, M., Püttmans, M., and Reimer, L. (1994), *Scanning* **16**, 101.
Hobbs, L. W. (1979), in: *Introduction to Analytical Electron Microscopy* (J. J. Hren, J. I. Goldstein, and D. C., Joy, Eds.) (Plenum, New York), p. 437.
Hoenk, M. E., and Vahala, K. J. (1989), *Rev. Sci. Instr.* **60**, 226.
Hollingsworth, R. E. and Sites, J. R. (1982), *J. Appl. Phys.* **53**, 5357.
Holt, D. B. (1992), *Scan. Microsc.* **6**, 1.
Holt, D. B., and Napchan, E. (1994), *Scanning* **16**, 78.

Holt, D. B. and Joy, D. C. (1989), *SEM Microcharacterization of Semiconductors* (Academic, New York).
Holt, D. B., Norman, C. E., Salviati, G., Franchi, S., and Bosacchi, A. (1991), *Inst. Phys. Conf. Ser.* No. **117**, p. 689.
Holt, D. B., Napchan, E., Lazzarini, L., Salviati, G., and Urchulutegui, M. (1993), *Inst. Phys. Conf. Ser.* No. **134**, p. 661.
Hopfield, J. J. (1958), *Phys. Rev.* **112**, 1555.
Hopfield, J. J. (1966), *Jpn. Phys. Soc. J*, Suppl. **21**, 77.
Hren, J. J. (1986), Barriers to AEM: Contamination and Etching, in: *Principles of Analytical Electron Microscopy* (D. C. Joy, Ed.) (Plenum, New York), p. 353.
Huang, K., and Rhys, A. (1950), *Proc. Roy. Soc.* (London) **A204**, 406.
Hübner, T., Marx, U., and Hedrich, T. (1993), *Phys. Stat. Sol.* (*a*) **136**, K61.
Hwang, C. J. (1971), *J. Appl. Phys.* **42**, 4408.
ICRU (1984), International Commission on Radiation Units and Measurement, ICRU Report 37.
Imbusch, G. F. (1978), *Luminescence Spectroscopy* (Academic, London), p. 35.
Imhof, R. E., and Read, F. H. (1971), *J. Phys. B: Atom. Molec. Phys.* **4**, 450.
Jakubowicz, A. (1986), *J. Appl. Phys.* **59**, 2205.
Jonscher, A. K., and de Polignac, A. (1984), *J. Phys. C.* **17**, 6493.
Joy, D. C. (1987), *J. Microsc.* **147**, 51.
Joy, D. C. (1988), *Inst. Phys. Conf. Ser.* No. **93**, Vol. 1, p. 23.
Joy, D. C. (1995), *Scanning* **17**, 270.
Joy, D. C., and Luo, S. (1989), *Scanning* **11**, 176.
Kanaya, K., and Okayama, S. (1972), *J. Phys. D.* **5**, 43.
Klein, C. A. (1968), *J. Appl. Phys.* **39**, 2029.
Kopf, R. E., Schubert, E. F., Harris, T. D. and Becker, R. S. (1991), *Appl. Phys. Lett.* **58**, 631.
Koshek, G., and Kubalek, E. (1985), *Scanning* **7**, 199.
Le Gressus, C., Vigouroux, J. P., Duraud, J. P., Boiziau, C. and Geller, J. (1984), *Scan. Electron Microsc.* **I**, 41.
Lingunis, E. H., Haegel, N. M., and Karam, N. H. (1993), *J. Appl. Phys.* **74**, 2779.
Lohnert, K., and Kubalek, E. (1983), *Inst. Phys. Conf. Ser.* No. **67**, p. 303.
Lohnert, K., Hastenrath, M., Balk, L. J., and Kubalek, E. (1981), *Inst. Phys. Conf. Ser.* No. **60**, p. 179.
Lu, Z. H., Hanna, M. C., Szmyd, D. M., Oh, E. G., and Majerfeld, A. (1990), *Appl. Phys. Lett.* **56**, 177.
Lumb, M. D. (1978), *Luminescence Spectroscopy* (Academic, London).
Marek, J., Geiss, R., Glassman, L. M., and Scott, M. P. (1985), *J. Electrochem. Soc.* **132**, 1502.
Marshall, D. J. (1988), *Cathodoluminescence of Geological Materials* (Unwin Hyman, Boston).
Marshall, D. J. (1993), *Scan. Microsc.* **7**, 861.
Mason, R. A. (1994), *Chem. Geol.* **111**, 245.
Menzel, E., and Kubalek, E. (1979), *Scan. Electron Microsc.* **I**, 305.
Merli, P. G. (1980), *Optik* **56**, 205.
Miller, J. H., Hunn, J. D., Holder, S. L., DiBianca, A. N., and Bagnell, C. R. (1990), *Appl. Phys. Lett.* **56**, 89.
Montelius, L., Pistol, M. E., and Samuelson, L. (1992), *Ultramicroscopy* **42-44A**, 210.
Munzel, H., Steckenborn, A., and Bimberg, D. (1981), *J. Lumines.* **24/25**, 569.
Murooka, Y., and Yuan, J. (1993), *Inst. Phys. Conf. Ser.* No. **138**, p. 71.
Myhajlenko, S. (1991), *Scan. Microsc.* **5**, 603.
Myhajlenko, S., and Ke, W. K. (1984), *J. Phys. E.* **17**, 200.
Myhajlenko, S., Ke, W. K., and Hamilton, B. H. (1983*a*), *J. Appl. Phys.* **54**, 862.
Myhajlenko, S., Davidson, S. M., and Hamilton, B. (1983*b*), *Inst. Phys. Conf. Ser.* No. **67**, p. 327.
Myhajlenko, S., Batstone, J. L., Hutchinson, H. J., and Steeds, J. W. (1984), *J. Phys. C.* **17**, 6477.
Myhajlenko, S., Hutchinson, H. J., and Steeds, J. W. (1985), *Inst. Phys. Conf. Ser.* No. **76**, p. 377.
Myhajlenko, S., Hutchinson, J. L., and Steeds, J. W. (1986), *Proc. 46th EMSA*, p. 814.
Myhajlenko, S., Edwards, J. L., Rowley, K. L., and Roedel, R. J. (1990), *Scanning* **12**, 81.
Myhajlenko, S., Puechner, R. A., Edwards, J. L., and Davito, D. B. (1991), *Inst. Phys. Conf. Ser.* No. **117**, p. 707.
Myhajlenko, S., Wong, M., Edwards Jr., J. L., Maracas, G. N., and Roedel, R. J. (1992), *Scan. Microsc.* **6**, 955.

Myhajlenko, S., Davito, D. B., Puechner, R. A., and Edwards Jr., J. L. (1994), *Phil. Mag. B.* **69**, 553.
Napchan, E. (1989), *Rev. Phys. Appl., Colloq.* **6**, Suppl. No. 6, p. 15.
Napchan, E., O'Neill, D., and Zanoti-Fregonara, C. L. M. (1993), *Inst. Phys. Conf. Ser.* No. **134**, p. 693.
Nel, M., and Auret, F. D. (1989), *Jpn. J. Appl. Phys.* **28**, 2430.
Newbury, D. E. (1989), *SEM Microcharacterization of Semiconductors* (Academic, New York), p. 45.
Norman, C. E., and Ghisoni, M. (1993), *Scanning* **15**, 325.
Norris, C. B., Barnes, C. E., and Beezhold, W. (1973), *J. Appl. Phys.* **44**, 3209.
Ortega, J. M. and Poole Jr., W. G. (1981), *An Introduction to Numerical Methods for Differential Equations* (Pitman, Marshfield, MA).
Owen, M. R. (1989), Abstracts-with-Programs, Geological Society of America **21**, 278.
Page, T. F. (1993), *Inst. Phys. Conf. Ser.* No. **138**, p. 295.
Pankove, J. I. (1971), *Optical Processes in Semiconductor* (Dover, New York).
Pasemann, L., and Hergert, W. (1986), *Ultramicroscopy* **19**, 15.
Pennycook, S. J. (1981*a*), *Inst. Phys. Conf. Ser.* No. **61**, p. 55.
Pennycook, S. J. (1981*b*), *Ultramicroscopy* **7**, 99.
Pennycook, S. J., and Howie, A. (1980), *Phil. Mag. A.* **41**, 809.
Pennycook, S. J., Brown, L. M., and Craven, A. J. (1980), *Phil. Mag. B.* **41**, 589.
Petroff, P. M. (1981), *Inst. Phys. Conf. Ser.* No. **61**, p. 501.
Petroff, P. M., Lang, D. V., Strudel, J. L., and Logan, R. A. (1978), *Scan. Electron Microsc.* **I**, 325.
Petroff, P. M., Cibert, J., Gossard, A. C., Dolan, G. J., and Tu, C. W. (1987), *J. Vac. Sci. Tech.* **B5**, 1204.
Petroff, P. M., Hu, Y. P., and Milstein, F. (1989), *J. Appl. Phys.* **66**, 2525.
Petrov, V. I., and Gareeva, A. R. (1988), *Izves. Akad. Nauk SSSR: Ser. Fiz.* **52**, 1373.
Pey, K. L., Chan, D. S. H., and Phang, J. C. H. (1993*a*), *Inst. Phys. Conf. Ser.* No. **134**, p. 687.
Pey, K. L., Phang, J. C. H., and Chan, D. S. H. (1993*b*), *Scan. Microsc.* **7**, 1195.
Pey, K. L., Chan, D. S. H., and Phang, J. C. H. (1995*a*), *Scan. Microsc.* **9**, 355.
Pey, K. L., Phang, J. C. H., and Chan, D. S. H. (1995*b*), *Scan. Microsc.* **9**, 367.
Pfefferkorn, G., Brocker, W., and Hastenrath, M. (1980), *Scan. Electron Microsc.* **251**.
Pfister, M., Johnson, M. B., Alvarado, S. F., Salemink, H. W. M., Marti, U., Martin, D., Morier-Genoud, F., and Reinhart, F. K. (1994), *Appl. Phys. Lett.* **65**, 1168.
Phang, J. C. H., Pey, K. L., and Chan, D. S. H. (1992), *IEEE Trans. Electron Devices* **39**, 782.
Pines, D. (1964), *Elementary Excitation in Solids* (W. A. Benjamin, New York).
Pons, D., and Bourgoin, J. (1981), *Phys. Rev. Lett.* **47**, 1293.
Pons, D., and Bourgoin, J. (1985), *J. Phys. C.* **18**, 3839.
Priolo, F., Coffa, S., Franzo, G., Spinella, C., Carnera, A., and Bellini, V. (1993), *J. Appl. Phys.* **74**, 4936.
Puhlmann, N., Oelgart, G., Gottschalach, V., and Nemitz, R. (1991), *Semicond. Sci. Tech.* **6**, 181.
Raether, H. (1980), *Excitation of Plasmons and Interband Transitions by Electrons* Vol. 88 (Springer-Verlag, Berlin), p. 23.
Ramseyer, K., and Mullis, J. (1990), *Amer. Mineral.* **75**, 791.
Rao-Sahib, T. S., and Wittry, D. B. (1969), *J. Appl. Phys.* **40**, 3745.
Rasul, A., and Davidson, S. M. (1977), *Scan. Electron Microsc.* **I**, 233.
Reimer, L. (1985), *Scanning Electron Microscopy* (Springer-Verlag, Berlin).
Remond, G., Cesbron, F., Chapoulie, R., Ohnenstetter, D., Roques-Carmes, C., and Schvoener, M. (1992), *Scan. Microsc.* **6**, 23.
Renaud, P., and Alvarado, S. F. (1991), *Phys. Rev. B.* **44**, 6340.
Rich, D. H., Ksendzov, A., Terhune, R. W., Grunthaner, F. J., Wilson, B. A., Shen, H., Dutta, M., Vernon, S. M., and Dixon, T. M. (1991), *Phy. Rev. B* **43**, 6836.
Roberts, S. J. (1981), *Inst. Phys. Conf. Ser.* No. **61**, p. 51.
Roedel, R. J., and Keramidas, V. G. (1979), *J. Appl. Phys.* **50**, 6353.
Roedel, R. J., Myhajlenko, S., and Edwards, J. L. (1989), *J. Electrochem. Soc.* **136**, 1186.
Roedel, R. J., Rowley, K., Edwards, J. L., and Myhajlenko, S. (1991), *J. Electrochem. Soc.* **138**, 3120.
Rowley, K. L. (1989), Simultaneous Optical and Electrical Characterization in the Scanning Electron Microscope, M.Sc. Thesis, Arizona State University.
Saba, F. M. and Holt, D. B. (1985), *Inst. Phys. Conf. Ser.* No. **76**, p. 389.
Salviati, G. (1993), *Scanning* **15**, 350.
Schiller, C., and Boulou, M. (1975), *Philips Tech. Rev.* **35**, 239.

Schulson, E. M., van Essen, C. G., and Joy, D. C. (1969), *Scan. Electron Microsc.* (Chicago) p. 45.
Seiler, H. (1983), *J. Appl. Phys.* **54**, R1.
Shah, J. (1978), *Sol. St. Electron.* **21**, 43.
Sippel, R. F. (1965), *Rev. Sci. Instr.* **36**, 1556.
Smith, J. V., and Stenstrom, R. C. (1965), *J. Geol.* **73**, 627.
Spellward, P., Day, J. C. C., Ashenford, D. A., and Lunn, B. (1987), *Inst. Phys. Conf. Ser.* No. **90**, p. 277.
Steckenborn, A. (1980), *J. Microsc.* **118**, 297.
Steckenborn, A., Münzel, H., and Bimberg, D. (1981), *J. Lumines.* **24/25**, 351.
Steeds, J. W. (1989*a*), *Rev. Phys. Appl., Colloq.* **6**, Suppl. No. 6, p. 65.
Steeds, J. W. (1989*b*), *Inst. Phys. Conf. Ser.* No. **104**, p. 199.
Steiner, T., Thewalt, M. L. W., Koteles, E. S., and Salerno, J. P. (1986), *Phys. Rev. B* **34**, 1006.
Steyn, J. B., Giles, P., and Holt, D. B. (1976), *J. Microsc.* **107**, 107.
Stroscio, J. A., and Kaiser, W. J. (1993), *Scanning Tunneling Microscopy, Methods of Experimental Physics, Vol. 27* (Academic, New York).
Tajima, M. (1978), *Appl. Phys. Lett.* **32**, 19.
Tang, Y., Rich, D. H., Lingunis, E. H., and Haegel, N. M. (1994), *J. Appl. Phys.* **76**, 3032.
Titchmarsh, J. M., Booker, G. R., Harding, W., and Wight, D. R. (1977), *J. Mat. Sci.* **12**, 341.
Touloukian, Y. S. (Ed.) (1970), *Thermophysical Properties of Matter* (Plenum, New York).
Trager-Cowan, C., Parbrook, P. J., Clark, D., Henderson, B. and O'Donnell, K. P. (1991), *Inst. Phys. Conf. Ser.* No. **117**, 715.
Trigg, A. D. (1985), *Scan. Electron Microsc.*, **III**, 1011.
Tung, C. J., Ashley, J. C., and Ritchie, R. H. (1979), *Surf. Sci.* **81**, 427.
Vale, B. H. and Greer, R. T. (1977), *Scan. Electron Microsc.* **I**, 241.
van der Ziel, J. P., Chand, N., and Weiner, J. S. (1989), *J. Appl. Phys.* **66**, 1195.
Van Opdorp, C., Vink, A. T., and Werkhoven, C. (1977), *Inst. Phys. Conf. Ser.* No. **33b**, p. 317.
Van Roos, O. (1983), *J. Appl. Phys.* **54**, 1390.
Van Roosbroeck, W. (1953), *Phys. Rev.* **91**, 282.
Vigouroux, J. P., Duraud, J. P., Le Gressus, C., Petite, G., Agostini, P., and Bioziau, C. (1985), *Scan. Electron Microsc.* **I**, 179.
Vine, J., and Einstein, P. A. (1964), *Proc. IEE.* **111**, 921.
Vouk, M. A., and Lightowlers, E. C. (1977), *J. Phys. C.* **10**, 3689.
Wannier, G. H. (1937), *Phys. Rev.* **52**, 191.
Warwick, C. A. (1987), *Scan. Microsc.* **1**, 51.
Warwick, C. A. (1989), *Inst. Phys. Conf. Ser.* No. **98**, p. 447.
Warwick, C. A. (1991), *Inst. Phys. Conf. Ser.* No. **117**, 681.
Warwick, C. A., and Booker, G. R. (1983), *Inst. Phys. Conf. Ser.* No. **67**, p. 321.
Warwick, C. A., and Brown, G. T. (1985), *Appl. Phys. Lett.* **46**, 574.
Warwick, C. A., and Kopf, R. F. (1992), *Appl. Phys. Lett.* **60**, 386.
Warwick, C. A., Brown, G. T., Booker, G. R., and Cockayne, B. (1983), *J. Crys. Growth* **64**, 108.
Warwick, C. A., Kitching, S. A., Allenson, M., Gill, S. S., and Woodward, J. (1987), *Inst. Phys. Conf. Ser.* No. **87**, p. 757.
Warwick, C. A., Jan, W. Y., Ourmazd, A., and Harris, T. D. (1990), *Appl. Phys. Lett.* **56**, 2666.
Watanabe, M., Koishi, M., Fujiwara, M., Takeshita, T., and Cieslik, M. (1994), *J. Photochem. Photobiol. A: Chem.* **80**, 429.
Waychunas, G. A. (1988), *Rev. Mineral.* **18**, 639.
Wittry, D. B., and Kyser, D. F. (1964), *J. Appl. Phys.* **35**, 2439.
Wittry, D. B. and Kyser, D. F. (1967), *J. Appl. Phys.* **38**, 375.
Wolkenstein, T. (1991), *Electronic Processes on Semiconductor Surfaces during Chemisorption* (Consultants Bureau, New York), p. 367.
Wright, P. (1991), *Proc. 49th EMSA*, p. 838.
Yacobi, B. G., and Holt, D. B. (1986), *J. Appl. Phys.* **59**, R1.
Yacobi, B. G., and Holt, D. B. (1990), *Cathodoluminescence Microscopy of Inorganic Solids* (Plenum, New York).
Yacobi, B. G., Jagannath, C., Zemon, S., and Sheldon, P. (1988), *Appl. Phys. Lett.* **52**, 555.
Yamamoto, N., Spence, J. C. H., and Fathy, D. (1984), *Phil. Mag. B.* **49**, 609.

Yuan, J., Berger, S. D. and Brown, L. M. (1989), *J. Phys. Condens. Matter.* **1**, 3253.

Zarem, H. A., Sercel, P. C., Lebens, J. A., Eng, L. E., Yariv, A., and Vahala, K. J. (1989), *Appl. Phys. Lett.* **55**, 1647.

Zemon, S., Jagannath, C., Koteles, E. S., Shastry, S. K., Norris, P., Lambert, G., Choudhury, A. N. M., and Armiento, C. A. (1986), *Inst. Phys. Conf. Ser.* No. **83**, p. 141.

5

Ionoluminescence*

Debabrata Ghose and Rainer Hippler

1. INTRODUCTION

Ion bombardment of solid surfaces gives rise to a number of elastic and inelastic collision processes. The elastic energy losses experienced by the incident projectile are responsible for the displacement of target atoms and the formation of a collision cascade. The inelastic collisions also require the active participation of electrons and are the source of a wide variety of additional phenomena, e.g., emission of secondary electrons and ions as well as of electromagnetic radiation such as optical or light emissions and x-rays (Tolk *et al.*, 1977; Taglauer and Heiland, 1981). The optical emissions generated by the impact of heavy particles on surfaces are due to excited sputtered target atoms, ions, and molecules and to backscattered projectiles, which subsequently decay by light emission in the visible and ultraviolet range (White *et al.*, 1977). The collisional excitation of molecules on the surface of the solid and radiative recombination of electron–hole pairs inside the solid may also contribute to photon emission (White *et al.*, 1977). At moderate beam current densities ($\approx 10\,\mu A/cm^2$), a bright luminescence in front of the target surface is clearly visible with the naked eye. This halo usually extends to a distance of some millimeters from the target surface. The research in ionoluminescence or bombardment-induced light emission (BLE), as it is called, was initiated by Chaudhri and Khan (1961*a*, *b*) at Lahore and Kistemaker and Snoek (1962) at the FOM Institute in Amsterdam, simultaneously, in the early 1960s. Since then a number of review articles have appeared on the subject (Kerkdijk *et al.*, 1976a; White *et al.*, 1977; White, 1978; Van der Weg, 1978; Thomas, 1979; Thomas, 1980; MacDonald *et al.*, 1981; Tsong, 1981; MacDonald, 1982; Gruen *et al.*, 1983; Betz, 1987; Yu, 1991). The interest in BLE stems from intrinsic fundamental aspects of excitation processes as well as from its possible application in the characterization of material surfaces. Also, the study of sputtering spectra is useful for an understanding of the excitation mechanism in low-energy beam-foil-spectroscopy work (Braun *et al.*, 1974; Jensen and Veje, 1974)

*This chapter is dedicated to Prof. Hans O. Lutz on the occasion of his 60th birthday.

Debabrata Ghose • Saha Institute of Nuclear Physics, Bidhan Nagar, Calcutta 700064, India • *Rainer Hippler* • Fakultät für Physik, Universität Bielefeld, D-33501, Bielefeld, Germany. Present address: Institut für Physik, Universität Greifswald, Domstrasse 10a D-Germany.

Luminescence of Solids, edited by D. R. Vij. Plenum Press, New York, 1998.

because in both cases the interaction between the solid and the moving atom leads to excitation.

The sputtering of solids by energetic ion bombardment, where target atoms are ejected owing to momentum transfer by the projectile ions, is a universal phenomenon. It is also observed that a small fraction of the sputtered particles is emitted in an excited state rather than in the ground state. This chapter is mainly concerned with the emission of photons from those sputtered particles, but the emissions from other sources are also discussed briefly. The chapter is arranged as follows: Section 2 gives an account of the theoretical considerations of atomic excitation during sputtering. Section 3 describes some experimental observations related to the emission from excited sputtered particles, including Doppler-broadening, energy and angular dependences, and transient phenomena of the emitted radiation. Section 4 presents an application of BLE to surface analysis. As the optical spectra emitted from sputtered particles contain few sharp lines, identification of elements is a very rapid and easy process. Section 5 deals with other kinds of optical emissions, namely, continuum emission and emission from backscattered projectiles. Finally, there are some concluding remarks in Section 6.

2. MODELS FOR EXCITED ATOM FORMATION

Explanations for the mechanisms of atomic excitation in sputtering have been developed in parallel to those for secondary-ion formation. In contrast to secondary ions, the size of excited atoms is generally larger than the interatomic distance in the solid. Thus, it is plausible to assume that excited atoms do not exist inside a solid, and that excitation occurs at or near the target surface. Obviously, the physical and chemical states of the surface at the sputtering site will have considerable influence on the formation of the excited state. Models that are frequently discussed in the literature are: (1) electron-tunneling, (2) bond-breaking, and (3) local thermodynamic equilibrium.

2.1. Electron-Tunneling Model

The electron-tunneling or band-structure model dates back to the early work of Hagstrum (1954). It was first applied to the production of excited atoms following ion bombardment of solids by Van der Weg and Rol (1965), who noted that sputtered excited metal atoms have high kinetic energies and, hence, should originate from binary collisions between the projectile ion and lattice atoms at the surface rather than from the collision cascade. The excited atoms decay either by a radiationless deexcitation process such as resonance ionization or Auger deexcitation or by radiative photon emission. The nonradiative transition probabilities ($\sim 10^{15}\,s^{-1}$) in close proximity to the solid surface are usually several orders of magnitude higher than typical radiative transition probabilities ($\sim 10^{8}\,s^{-1}$). As a consequence, only a small fraction of the excited atoms can escape from a clean surface to decay by emission of line radiation.

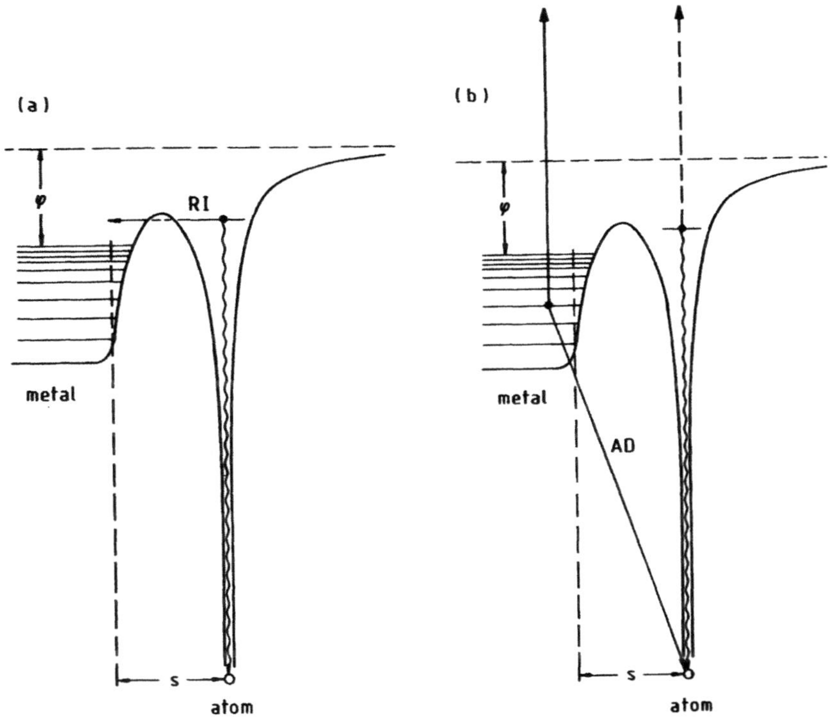

Figure 1. Schematic representation of the metal–excited atom energy-level system. Two possible non-radiative deexcitation processes, resonance ionization (RI) and Auger deexcitation (AD) are shown.

Figure 1 shows the potential energy diagram of a solid metal and an atom separated by a distance s. The metal surface is characterized by a square potential well of depth ε_0 filled with electrons up to the Fermi level ε_F. The Fermi level lies below the vacuum level by the work function φ. The nonradiative deexcitation processes that occur in close proximity to the surface (<50Å) can be described in a way similar to that developed by Hagstrum (1954) in ion neutralization spectroscopy measurements. In the resonance ionization (RI) process, an electron from an upper excited level of the atom tunnels through the surface potential barrier to an empty state with the same energy in the conduction band of the solid. With Hagstrum's nomenclature,

$$X^{0*} \xrightarrow{\text{RI}} X^+ + e_s^- \tag{1}$$

where X^{0*} is the sputtered excited atom, X^+ the resulting ion, and e_s^- the transfer electron. In the two-electron Auger deexcitation (AD) process, electrons from occupied states of the solid tunnel to energetically lower vacant states of the atom; the energy released in that transition may be used to eject the excited atomic electron into the continuum:

$$X^{0*} + e_s^- \xrightarrow{\text{AD}} X^0 + e_A^- \tag{2}$$

where e_A^- is the Auger electron.

The total probability P for an excited atom leaving the surface to decay by a radiative transition may be obtained from ab initio calculations (e.g., Wille, 1993; and references therein). For sufficiently large separations s from the solid surface, these calculations yield an exponential decrease of the transition probability for a nonradiative process k. The total nonradiative transition probability $A_{nr}(s)$ may thus be expressed as

$$A_{nr}(s) \approx \sum_k A_k \exp(-a_k s) \tag{3}$$

The survival probability $R(s)$ at a given distance $s = v_\perp t$ from the surface, where $v_\perp$ is the velocity component of the excited atom perpendicular to the surface (i.e., parallel with respect to the surface normal) depends on how fast the excited atom escapes from the surface. It can be calculated from the rate equation

$$\frac{dR(s)}{dt} = -R(s)A_{nr}(s) \tag{4}$$

with the boundary condition $R(0) = 1$. Its solution is given by

$$R(s) = \exp\left\{-\frac{1}{v_\perp}\sum_k \frac{A_k}{a_k}[1 - \exp(-a_k s)]\right\} \tag{5}$$

For long survival times ($t \to \infty$), Eq. (5) yields

$$R(\infty) = \exp\left(-\frac{1}{v_\perp}\sum_k \frac{A_k}{a_k}\right) \tag{6}$$

The quantity $A/a \equiv \Sigma_k A_k/a_k$ is called the survival parameter for k possible nonradiative deexcitation channels of a given excited state. Physically, A/a represents the escape velocity of the particle that is able to deexcite by photon emission. The value of A/a is found to be of the order of 10^5 and 10^4 m/s, respectively, for clean and oxygen-covered surfaces.

The electron exchange model explains the influence of surface oxidation and other chemical effects well. For example, it has been observed that line intensities are considerably enhanced when an oxide, nitride, or carbide surface is bombarded, as compared to the clean surface (Bhattacharya *et al.*, 1977). Also the admission of oxygen into the target camber during ion bombardment causes a considerable increase in light intensity from sputtered atoms. Figure 2 shows two photon spectra obtained from 300-keV Ar^+ bombardment of an Al target (Reinke *et al.*, 1991). The measurements were performed for a clean surface (Fig. 2a) and with a partial oxygen pressure of about 5×10^{-5} mbar inside the vacuum chamber (Fig. 2b); the estimated oxygen coverage (Kelly and Kerkdijk, 1974) was close to 100%. Most prominent in these spectra are the resonance transitions of the sputtered Al atoms around 3082,

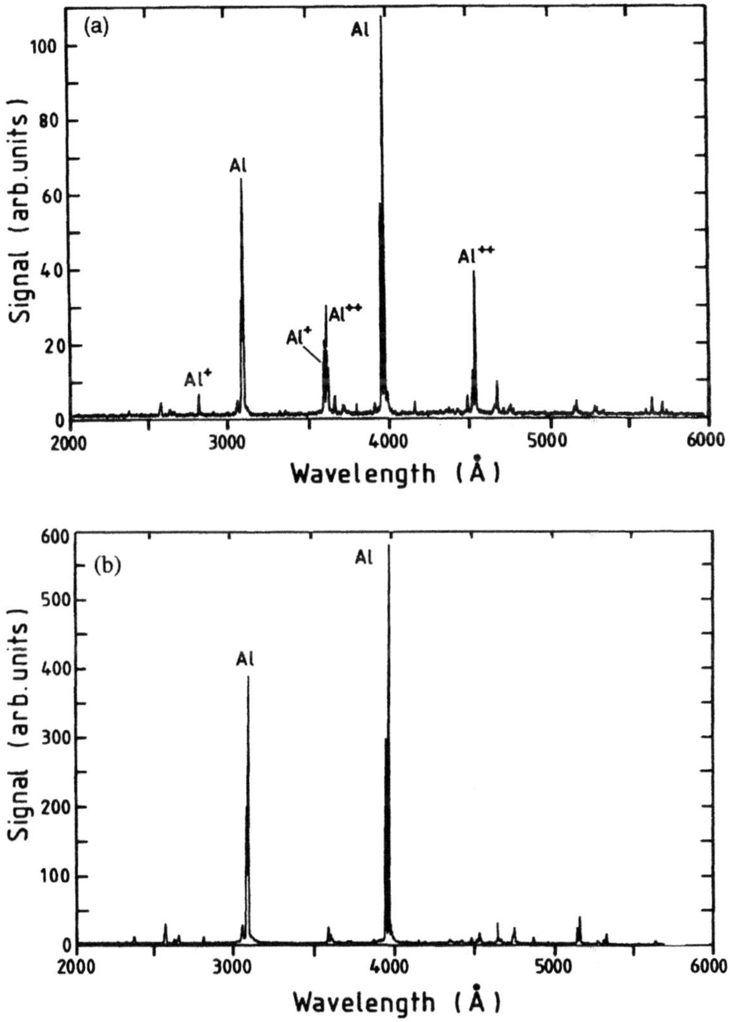

Figure 2. Light-emission spectrum following 300-keV Ar^+ bombardment of: (a) a clean and (b) an oxygen-covered Al surface (oxygen partial pressure 5×10^{-5} mbar; ion beam current density 2.2 $\mu A/cm^2$). Note the different intensity scales; the same arbitrary scale factor applies in both cases (Reinke *et al.*, 1991; with kind permission from Elsevier Science Ltd.)

3093 Å and 3944, 3962 Å. Also visible are lines from excited Al^+ (2816 and 3587 Å) and Al^{2+} ions (3602, 3612 Å and 4512, 4529 Å); the corresponding transitions are shown in Fig. 3. The observed increase in the photon intensity when the Al surface is exposed to oxygen of the most prominent transitions ($4s \rightarrow 3p$, $3d \rightarrow 3p$, $4d \rightarrow 3p$, and $5d \rightarrow 3p$) amounts to a factor of about 20. The increase is easily understood from the energy level diagram (Fig. 3), where it is clear that the bandgap of Al_2O_3 presumably formed by the oxygen exposure inhibits resonance tunneling, thereby resulting in an enhanced photon emission.

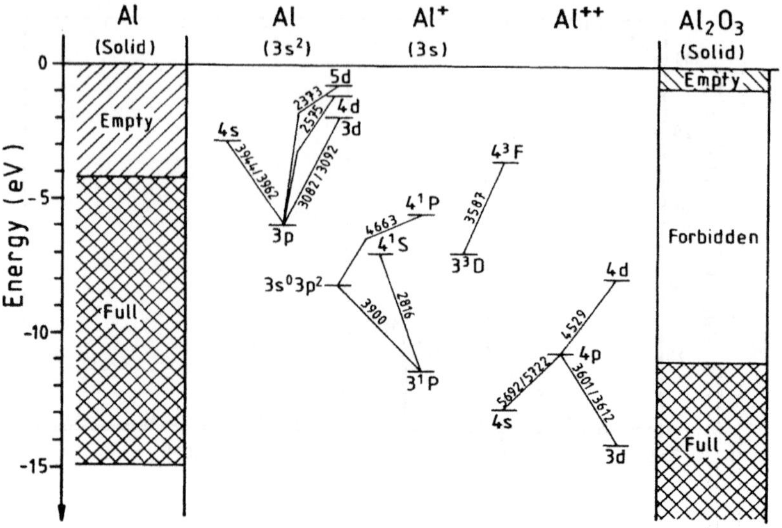

Figure 3. The atomic energies of several Al, Al^+, and Al^{2+} states in comparison to the band structure of metallic Al and of Al_2O_3. The wavelengths (in Å) of some transitions are indicated (Reinke *et al.*, 1991; with kind permission from Elsevier Science Ltd.)

The resonant-electron-transfer model has been used successfully to gain information about the Fermi levels and the width of the conduction bands of several solids (Gritsyna *et al.*, 1972; Kiyan *et al.*, 1976b, 1978). The model is also used to explain nonradiative deexcitation of slower particles with an upper energy level above the Fermi level and opposite to the empty conduction band, resulting in the higher kinetic energy of the light-emitting particles. Nevertheless, there are several doubts concerning the general validity of the model. For example, Thomas and de Kluizenaar (1976) deliberately lowered the work function of Cu and Al targets by adsorbing Cs on the surface. This apparently shifts the Fermi level upward with respect to the atomic levels by as much as the change in work function, and should enhance the photon emission by blocking off the RI process. However, this was not actually observed during the experiment. According to a different explanation in favor of the electron-exchange mechanism given by Yu (1979), in the vicinity of the surface, the image potential raises the excited level of the neutral atom, placing it above the Fermi level of the metal target, so that nonradiative transitions are always possible.

Inoue *et al.* (1989) investigated secondary-photon emission from excited Ga atoms sputtered from GaAs crystal as a function of Cs coverage. It was shown that a threshold coverage for the enhancement of photon yield exists and that the enhancement factor depends nonlinearly on the Cs coverage, in contrast to that for oxygen, where a linear dependence was observed (Inoue *et al.*, 1988). The authors attributed their results to the increasing importance of the AD process with the changing position of the Fermi level.

2.2. Molecular Dissociation Model

It is well-known that bombardment-induced photon emission and ion emission have several properties in common, in particular both are enhanced by the presence of oxygen in the target. One may, therefore, assume that the same mechanism is operative for excitation and ionization of atoms. With this in mind, Blaise (1976) put forward a so-called bond-breaking model, where metal–oxygen quasi-molecules leaving the surface by sputtering subsequently break up and populate excited states via a curve-crossing process in a transiently formed molecule. If it is assumed that a metal (M)–oxygen (O) bonding is responsible for the molecular curve-crossing mechanism, the potential energy curves $V(R)$ for various configurations of the M–O system behave as shown in Fig. 4 but tend toward the energy levels of the two separated M and O atoms at large internuclear distances R. Under this scheme, the crossings of the excited levels and the ionized level occur in the outward-going direction of the potential energy curves. One of the consequences of this model is that excited-state production is energetically more favorable than ion production because excitation energies are lower than ionization potentials. Also, this model defines an upper energy limit of the excitation energy that is of the order of the difference between the ionization potential of the metal atom and the electron affinity of oxygen. A similar mechanism was also discussed by Thomas (1977).

Tsong (1981) has modified this model on the basis of the following experimental observations: (1) secondary-ion emission is more efficient than photon emission, (2) a rather low fraction ($\sim 10^{-3}$–10^{-4}) is light-emitting sputtered atoms, and (3) the excited atoms and ions have higher kinetic energies than most other sputtered atoms and ions.

The highlights of Tsong's (1981) approach are that the crossing of the excited curve with the ionic curve occurs at the steep repulsive side of the potential energy

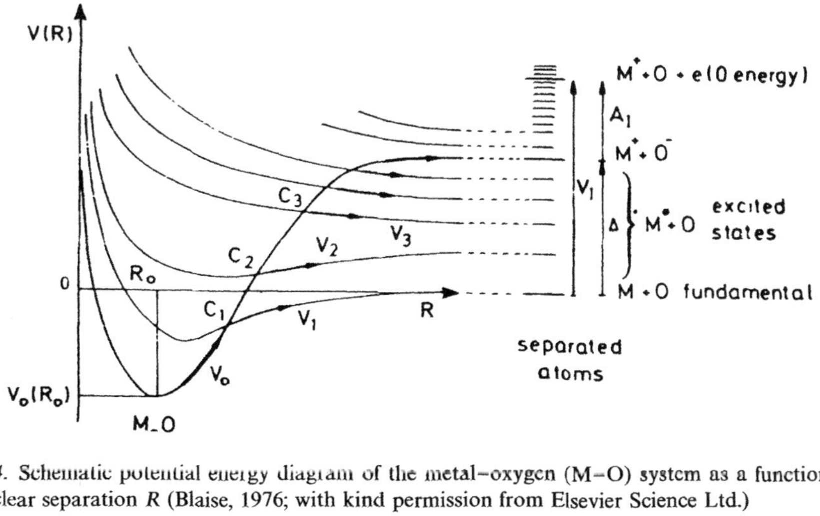

Figure 4. Schematic potential energy diagram of the metal–oxygen (M–O) system as a function of the internuclear separation R (Blaise, 1976; with kind permission from Elsevier Science Ltd.)

diagram and that the ionic curve lies below the excited-states potential curve by an amount determined by the Coulombic interaction between the solid surface and the departing quasi-molecule. According to this model, excited-state formation requires high-energy collisions in order to reach the crossing points.

Yu (1988) also proposed an elaboration of Blaise's concept for excited-state formation. He calculated the excitation probability using the Landau–Zener curve-crossing relation and showed that the excited-atom yield is much smaller than the secondary-ion yield.

2.3. Local Thermodynamic Equilibrium Model

In a number of papers excited-atom emission has been considered in the framework of the local thermodynamic equilibrium (LTE) model as introduced by Andersen and Hinthorne (1973) to describe secondary-ion emission phenomena. The LTE model is based on the assumption that during fast-ion bombardment of solids a local plasma layer is formed at the surface in which sputtered atoms, ions, molecules, and electrons are in local thermodynamic equilibrium. Consequently, the population n_i^* of atoms in the excited state i should conform to the Boltzmann distribution,

$$n_i^* \sim \exp(-E_i/kT) \tag{7}$$

provided that all the emitting particles are confined within the plasma volume so that excitation and deexcitation can proceed in equilibrium. The intensity I_{if} of a spectral line from such a source is given by (Tsong and Yusuf, 1979),

$$I_{if} \sim g_i A_{if} \tau_i \exp(-E_i/kT) \tag{8}$$

where g_i is the statistical weight of the upper level i, A_{if} the transition rate or the so-called Einstein coefficient for a transition between the upper level i and the final level f, E_i the excitation energy, τ_i the lifetime of the excited state i, k the Boltzmann constant, and T the temperature of the excitation source. If LTE exists, then a plot of $\log(I_{if}/g_i A_{if} \tau_i)$ vs. E_i should yield a straight line with slope $-(kT)^{-1}$. Some authors prefer to use another form of Eq. (8), which describes the distribution of the excited states in an arc discharge (Tsong and Yusuf, 1979).

The LTE model was tested for a number of elements (Martin and MacDonald, 1977; MacDonald and Martin, 1977a; MacDonald *et al.*, 1978; Tsong, 1977, 1978). The results show a considerable scatter of the data points, which makes it difficult to derive any conclusive temperature of the plasma. Tsong (1977, 1978) pointed out that the contribution from nonradiative deexcitation processes, especially RI, may cause a deviation from the straight-line fit. Also, the cascading processes from higher levels to the upper level of a considered transition can be of importance. Some authors have shown that for an alloy target the effective temperatures describing emission for each constituent are nearly the same (Arlinghaus and Bispinck, 1983; Tsong and Yusuf, 1979). Although such results go in favor of plasma-like conditions for excitation, the uncertainties involved in those measurements were quite large.

There are also serious objections relating to the physical justification for the existence of LTE plasma. Snowdon (1979) showed that under conventional ion-bombardment densities an optically thin plasma cannot exist inside the solid nor above the target surface because in that case the secondary-electron yield would be at least a factor of 10^4 higher than what is observed. Furthermore, the LTE model fails to explain the extreme sensitivity of the light emission to the surface chemical state. The validity of the LTE model becomes further questionable through the work of Snowdon *et al.* (1979). A comparision of the relative level populations of ZnI lines observed in Ar^+ impact on solid Zn, Zn^+ impact on Ar gas, and the arc spectrum of Zn shows no essential differences, although different physical mechanisms are responsible for excitation in each case. While the arc spectrum can be safely assumed to originate from LTE, the impact of Zn^+ on Ar is certainly an isolated biparticle collision.

Kelly *et al.* (1978) and Good-Zamin *et al.* (1978) put forward a somewhat different approach. They assumed that the sputtered particles are created by thermal spikes in which the deposited energy by the incident beam causes a high temperature to exist for a short time at the surface region. It is further assumed that the excited neutrals among other sputtered particles are initially in an equilibrium plasma-like environment, but then decay in free space. In such a description corrections for the transition probability or photon energy may be ignored. This idea was tested experimentally with Be, BeO, Al, Al_2O_3, Sc, GaAs, Y, In, and Tl, where quite acceptable LTE plots were obtained with temperatures in the range of 3600–5900 K. However, it has been pointed out that the derived temperatures are not dependent on the target mass, ion mass, ion energy, or target crystallinity contrary to what might be expected from a consideration of energy transfer and transport processes within the solid.

In conclusion, although there is no single piece of experimental evidence that definitely rules out the LTE concept, the model itself is not physically attractive, largely because the assumptions on which it is based are relatively weak.

3. EXPERIMENTAL OBSERVATIONS

3.1. Doppler-Broadening of Spectral Lines

The Doppler-broadening of spectral lines results from the nonzero velocity of the excited particles. Since the sputtered atoms have a distribution of velocities and directions with respect to the photon-detection system, the spectral lines are broadened in a manner consistent with Doppler-shifting. The presence of the solid surface causes an asymmetrical shape of the line profile owing to the lack of contributions to the photon yield from particles scattered into the target.

Following the first observation of Doppler-broadening of line spectra by Snoek *et al.* (1964), Van der Weg and Bierman (1969) investigated the profile of the atomic Cu 3247-Å spectral line, excited by the bombardment of a Cu crystal with 80-keV Ar^+ ions, in detail. Depending upon the geometry of the experiment, they found a peak shift of up to 0.6 Å. From the relation

$$\Delta\lambda = (v_{\parallel}/c)\lambda \tag{9}$$

where $\Delta\lambda$ is the Doppler shift, λ the unshifted wavelength, c the velocity of light, and

$v_\parallel$ the velocity component of the radiating atom along the observation direction of the photon emission, the velocity of the emitted Cu atoms was found to be about 5.5×10^4 m/s, corresponding to a kinetic energy of about 1 keV. They also obtained a reasonable agreement between the experimental line profile and the calculated profile based on binary collisions at the surface combined with radiationless deexcitation processes by electron tunneling. Subsequently, White *et al.* (1975) showed that the linewidth from a pure Si surface is considerably broader compared to that from an oxide surface, indicating that in the latter case the low-velocity Si* atoms are also able to radiate. This fact was also noted by Van der Weg and Bierman (1969).

Hippler *et al.* (1976) performed a systematic study of the lineshape measurements of the emission lines of neutral 3247.5-Å Cu, 4810.5- and 3075.9-Å Zn, and 3092.8-Å Al during 300-keV Ar^+ bombardment of Cu, Zn, and Al. The maximum Doppler shifts obtained for the investigated lines were: 1.4 Å for Cu, 1.3 and 2.1 Å for Zn, and 2.3 Å for Al, respectively. They also calculated the lineshape assuming a single recoil speed estimated from the maximum transferable energy between the incident ion and the target atom and the geometry of the collison experiment. Following the approach of Van der Weg and Bierman (1969), one approximates the transition rate $P(s)$ for radiationless deexcitation processes at a distance s from the surface by (Hagstrum, 1954)

$$P(s) = Ae^{-as} \tag{10}$$

from which the probability that the excited atom leaving the surface does not undergo a radiationless transition is obtained as

$$R(\infty) = \exp(-A/av_\perp) \tag{11}$$

where $v_\perp = v_\perp(\alpha, \phi)$ is the velocity component perpendicular to the surface. The quantity A/a is called the survival parameter and the two constants A and a are related to the height and width of the potential barrier between the escaping atom and the surface.

The production of excited atoms was assumed to arise from binary collisions between the projectile ion and a single target atom while higher contributions from the sputtering cascade were neglected. The probability P_{ex} for excitation to the level of interest of the sputtered target atom emitted at an angle α with respect to the ion beam was assumed to be of the form

$$P_{ex}(\alpha) \sim \cos\alpha \tag{12}$$

while the differential scattering cross section $\sigma_{ex}(\alpha)$ was taken as

$$\sigma_{ex}(\alpha) \sim (\cos\alpha)^{-2} \tag{13}$$

The lineshape was finally calculated by the integration (Van der Weg and Bierman, 1969)

$$N(\lambda + \Delta\lambda)d\lambda = \iint P_{ex}(\alpha)\sigma_{ex}(\alpha)R(\infty)\delta[(v_\parallel/c) - (\Delta\lambda/\lambda)]2\pi \sin\alpha \, d\alpha \, d\gamma \, d\lambda \tag{14}$$

where the angle ϕ represents the orientation of the target with respect to the ion beam, γ is the azimuthal angle, and $v_{\parallel}$ is the velocity component along the direction of observation. The experimental line profiles can be modeled by Eq. (14) for several values of the parameter A/a. For the Cu line the best fit was obtained for $A/a = 3 \times 10^4$ m/s, which deviates slightly from the value derived by Van der Weg and Bierman (1969). For the Zn line that has its excited level lying above the Fermi level of Zn, $A/a = 5 \times 10^4$ m/s, while an order of magnitude smaller A/a value was obtained for the line that has its excited level below the Fermi level. The main conclusion that emerged from this study is that the binary collision model yields only limited agreement with experimental lineshapes. One should also take into account the contributions from collision cascades and the influence of the surface binding energy as well.

In another approach, one can calculate the Doppler-broadened line profile from the energy distribution of sputtered atoms in the ground state, assuming that it is identical to that in the excited states. The energy distribution $\Phi(E, \theta_a)$ of sputtered atoms can be expressed by the well-known Thompson (1968) equation as

$$\Phi(E, \theta_a) \sim \frac{E}{(E + E_b)^k}\left(1 - \sqrt{\frac{E + E_b}{\gamma E_0}}\right)\cos\theta_a \tag{15}$$

where $k = 3$; E_0 is the projectile energy; E is the kinetic energy of the sputtered atom; E_b is its binding energy; $\gamma = 4m_1m_2/(m_1 + m_2)^2$, with m_1 and m_2 the projectile mass and target mass, respectively; and θ_a is the emission angle with respect to the surface normal. Betz (1987) analyzed Doppler-broadened line profiles of sputtered excited Ca, Al, and Cr atoms within the framework of Thompson's distribution results, taking the binding energy E_b as a free parameter. The best fit for the Ca 4227-Å line was obtained for $E_b = 12$ eV for both clean and oxidized Ca surfaces. The corresponding values for the Al 3962-Å line from a clean and oxidized Al surface were $E_b = 4$ and 20 eV, respectively. For Cr, on the other hand, the profile for the oxidized surface was narrower than for the clean surface, indicating a higher E_b value for the latter surface. A comparative study of normal ion incidence and grazing ion incidence shows that the excited atoms of high characteristic emission energies correspond to direct-recoil sputtering, while those of low emission energies are of cascade origin (Betz, 1987). More recently, Bissessur and Tsong (1990) measured Doppler-broadening as a function of perpendicular distance from the target surface for the K-doublet at 7665 and 7699 Å emitted by excited K atoms sputtered from KCl and KI targets following Ar^+ and Kr^+ ion bombardment. The experimental line profiles, although qualitatively similar to those predicted by Thompson's distribution, were found to be significantly broader. Also, the full widths at half-maxima (FWHM) of the emission profiles increase consistently with distance from the surface and so too the value of E_b. For example, in the case of 7665-Å K-line from the KCl target, the best fit for the E_b value is 14.8 eV when the lineshape is measured at the surface and 64.2 eV when it is measured at a distance of 4 mm from the surface. These values are much higher than the physical surface binding energy, which is 7.2 eV for KCl. However, the kinetic energies of excited K^* atoms obtained from linewidth measurements were rather low ($\sim$10–40 eV).

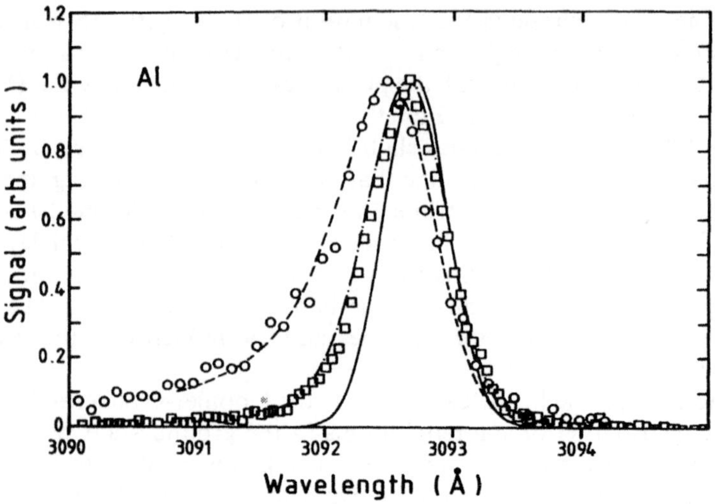

Figure 5. Lineshape of the $3s^23d \rightarrow 3s^23p$ resonance transition ($\lambda = 3092.7$ Å) of Al following 300-keV Ar^+ bombardment of a clean (o) and an oxygen-covered (□) surface. The solid line represents a reference line measured with a hollow-cathode lamp. Dashed and dash-dotted lines are calculations based on the resonance tunnel model for a clean and oxygen-covered surfaces, respectively (Reinke *et al.*, 1991; with kind permission from Elsevier Science Ltd.)

Reinke *et al.* (1991) found that a better fit to Thompson's (1968) distribution, Eq. (15), is obtained if the exponent of the denominator, k, is chosen to be 2.6 rather than 3. Combining Eqs. (11) and (15), they also determined the value of the parameter A/a. Figure 5 shows the Doppler-broadened Al 3093-Å transition following 300-keV Ar^+ bombardment of a clean and of an oxygen-covered Al surface. Also shown is a reference line obtained from a hollow-cathode lamp. The line resulting from the clean surface is considerably broader than that of an oxygen-covered surface. For the clean surface, the ratio $A/a = 4 \times 10^4$ m/s was adjusted to yield the best fit with experiment while $A/a = 0$ was best for the oxygen-covered surface. This choice not only reproduces the shape of the measured lines but also the measured intensity increase of about 20 for an oxygen-covered surface. A similar procedure for the 3247.6-Å Cu resonance line profile study showed that the values of A/a are, respectively, 9×10^4 and 6×10^4 m/s for the clean and oxygen-covered Cu surfaces. The relatively small reduction in the A/a value in the case of an oxygen-covered surface has been interpreted as being due to the relatively small change in the Cu work function upon adsorption of oxygen.

The measured lineshape of the ionic Al^{2+} 4529-Å transition shows comparatively small changes when the Al surface is exposed to oxygen. This is not unexpected since owing to the energetic position of the involved Al^{2+} state with respect to the band energies of the Al bulk (see Fig. 3), the nonradiative deexcitation processes cannot take place. The Al^{2+} 4529-Å line is considerably broadened, however; if the spectral resolution of the monochromator (0.91 Å) is taken into account, the corresponding intrinsic linewidth $\Delta\lambda^{(\mathrm{FWHM})}$ is about 1.6 and 1.4 Å for the clean and oxygen-covered surface, respectively. When the line profiles are fitted to the Thompson distribution, the parameter E_b yields a high value of 500 eV (Reinke, 1991; Reinke *et al.*, 1991). All these observations imply that an appreciable fraction of the excited Al^{2+} ions has kinetic energies of more than 1 keV. This is considerably higher than the kinetic energies with which the excited neutral Al atoms leave the surface and may be taken as evidence of different production mechanisms for excited Al atoms and Al^{2+} ions. Presumably, the Al^{2+} ions are produced in close and, hence, more violent encounters than the neutral atoms, which are believed to result from the sputtering cascade. In passing, it should be noted that line profile studies under grazing-incidence conditions during 5-keV $Kr^+ \rightarrow Mg$ showed that excited Mg^{+*} ions are ejected with one order of magnitude higher velocity than excited neutral Mg^* atoms (Fournier *et al.*, 1992).

3.2. Projectile Energy Dependence of the Photon Yield

Since a fraction of the sputtered particles leaves in some excited state and participates in the photon emission, it may be thought that a variation in the projectile energy and/or of the angle of ion incidence would affect the photon yield in the same way that it affects the sputtering yield. White and Tolk (1971) measured the intensity of the optical line produced in 60- to 3000-eV Ar^+ impact on Cu. The results showed a slow increase of intensity with increasing energy but the threshold energy was poorly defined. Assuming that the deexcitation probability is of the type given by Eq. (11), they determined the excited-state formation efficiency as a function of energy. A good agreement with the experimental data was obtained for $A/a = 2 \times 10^4$ m/s.

Bhattacharya *et al.* (1979) measured the photon intensities of neutral Si and ionic Si^+ and Si^{2+} lines as a function of ion energy for Ar^+ incident on Si in the energy range 50–500 keV. They studied both visible and ultraviolet photon emission. The photon yields did not correlate with the theoretical sputtering yield (Sigmund, 1969) for Si. Similar results were also obtained for Si emission owing to 75- to 300-keV SF_5^+ impact on Si(100) (Ghose *et al.*, 1997). Bhattacharya *et al.* (1979) interpreted the origin of light-emitting sputtered atoms as being generated from a few violent collisions near the surface. Thus, the intensity of each line was normalized to the direct knock-off sputtering yield as calculated by Winters and Sigmund (1974) and plotted as a function of the reciprocal of the maximum ejection velocity component normal to the surface. The results show that all the lines follow approximately the same exponential dependence as the survival probability from various radiationless deexcitation processes given by Eq. (11). The slope of the curves gives values of A/a in the range of 1.7 to 2.6×10^5 m/s. Heiland *et al.* (1977) employed a similar procedure to analyze the photon yield data from Mo, Zr, and Mg bombarded by 0.4- to 8.0-keV He^+, Ne^+, and Ar^+ ions. Here, both the direct knock-off collisional sputtering and the reflected-ion contribution to the sputtering yield were considered to be the source of excited-state atom formation (Winters and Sigmund, 1974).

Wright and Gruen (1980*a*, *b*) studied photon emission from neutral Be and ionic Be^+ transitions as a function of the incident energy of 200- to 3000-eV Kr^+, Ar^+,

Ne^+, He^+, O_2^+, and N_2^+ ions bombarding a Be metal foil. The results have been analyzed in terms of a model that is based on a velocity-dependent excitation process and the linear collision cascade theory of sputtering. The secondary-photon emission yield Y_{if}^* of a given emission line $i \rightarrow f$, can be expressed as

$$Y_{if}^* \propto JY \exp(-A_i'/v_m) \tag{16}$$

where J is the ion current density, Y the sputtering yield, and v_m the velocity corresponding to the maximum energy transferred between the incident ion and the target atom. A_i' is the nonradiative deexcitation parameter, which depends on the form of the excitation function $R_i^{ex}(v, \theta_a)$, where v is the velocity and θ_a the angle of emission of the escaping particle.

If R_i^{ex} is a constant or power of the velocity then $A_i' = \Sigma_k A_k/a_k$, where k represents all possible nonradiative deexcitation channels. If $R_i^{ex} = \exp[-B_i/(v \cos\theta_a)]$, then $A_i' = \Sigma_k A_k/a_k + B_i$. Fitting Eq. (16) to the experimental data yielded values of $A_i' = 1$ to 3×10^5 m/s for Be and Be^+ transitions. It decreased to 5 to 7×10^4 m/s upon O_2^+ bombardment for the neutral Be states, but remained unchanged for the Be^+ state. More recently, Ghose (1995) studied Ga spectral lines originating from the sputtering of GaAs under 6- to 10-keV Ar^+ ion bombardment. The values of A_i' obtained from Eq. (16) were found to be approximately 1.2 to 2×10^5 m/s. Wright and Gruen (1980*a*, *b*) noted that the photon yield is not proportional to the physical sputtering yield as would be expected from the assumption that the energy distributions of sputtered atoms in the excited state and in the ground state are identical. However, their results show that the electron-exchange mechanism between the particle and surface is the dominant effect in the production of excited atoms or ions. They also found that the total probability for the nonradiative deexcitation processes is the product of the probabilities, i.e., the sum of the A_k/a_k values, for all of the allowed deexcitation channels for a given excited state.

Reinke and Hippler (1992) investigated the energy dependence of neutral Al 3962-Å and ionic Al^{2+} 4529-Å lines from polycrystalline Al and the atomic Cu 3248-Å resonance transition from polycrystalline Cu for 50- to 350-keV Ne^+, Ar^+, Kr^+, and Xe^+ ions. For Al, the largest photon yield is observed with Xe^+ projectiles, whereas it is smallest for Ne^+. In the energy range studied, the measured photon yield increases with increasing energy of the Xe^+ projectiles, but decreases with increasing Kr^+, Ar^+, and Ne^+ energy. Since it is believed that the excited atoms and ions are produced by a sputtering process, the measured projectile and energy dependences are compared to theoretical predictions for the sputtering yield. According to Sigmund (1969), the sputtering yield Y may be expressed as

$$Y = 0.042\alpha_0 4\pi \frac{Z_1 Z_2 e^2 a_{12}}{E_b} \frac{m_1}{m_1 + m_2} s_n(\varepsilon) \tag{17}$$

where Z_1 and Z_2 are the atomic numbers of incident ion and target atom, respectively, α_0 is a dimensionless quantity depending on the mass ratio m_2/m_1 and the angle of ion incidence, a_{12} is the Thomas–Fermi screening radius, and $s_n(\varepsilon)$ is the reduced nuclear stopping cross section at the reduced energy ε:

$$\varepsilon = \frac{m_2}{m_1 + m_2} \frac{a_{12}}{Z_1 Z_2 e^2} E_0 \tag{18}$$

The theoretical values of $s_n(\varepsilon)$ as tabulated by Lindhard *et al.* (1968), together with the experimental results (Reinke and Hippler, 1992), which were converted according to Eq. (17) and arbitrarily normalized to the theoretical curve, show that in the projectile and energy range of interest, the predicted sputtering yield displays a maximum at $\varepsilon_{max} \approx 0.3$ after which it decreases with incident energy. This behavior qualitatively agrees with the experimental results for the light emission. On a more quantitative basis, the production of excited Al^{2+} ions is in fair agreement with the theoretical prediction for the sputter yield, except at reduced energies ε, where the observed photon yield Y^* for Xe^+ ions is smaller than anticipated. In contrast, excited neutral Al atoms generally display a steeper energy dependence than predicted theoretically. This behavior appears to be projectile-dependent and is most pronounced for the lightest projectile Ne^+. For the Cu resonance transition, the energy dependence of the lighter projectiles Ne^+ and Ar^+ is found to agree well with the calculated sputtering yield. Discrepancies are observed for Kr^+ projectiles at reduced energies for which the observed photon yield increases more steeply than predicted theoretically. Although the origin of the differences is not clear, one possible reason might be the neglect of inelastic collisions in the sputtering process.

Andersen *et al.* (1982) studied relative level excitations for a number of levels in Mg, Cu, Zn, Ga, and Au sputtered by Xe^+ ions. For Cu, Zn, and Mg targets, the relative level excitation is independent of the projectile energy. The same also holds for several neutral Au levels. Furthermore, the excitation of individual Au lines follows the total sputtering yield in the energy range 20–110 keV. In the case of Ga, however, the population of the Ga 6*p* and the Ga^+ 5*p* states decreases as the projectile energy increases from 20 to 40 keV, while for all other levels the population increases.

In an another experiment, Veje (1988) measured the yields of sputtered excited neutrals and ions from 10- to 160-keV Ar^+ bombarded Mg, MgO, Al, and Si targets. Here, in no case, did the experimental excitation yield curves follow the corresponding total sputtering yield curves. This presumably indicates that all the excited atoms are not the end products of the linear collision cascade but that some of them originate from direct-recoil sputtering. Also, unlike with sputtered neutrals, the energy and angular distributions of sputtered excited atoms may vary with projectile energy, atomic number, and angle of ion incidence, as was pointed out by Veje (1988). The experimental data, however, show that the relative excitation of different levels of the same charge state and element is independent of the projectile energy, similar to previous results (Andersen *et al.*, 1982). At the same time, the curves for excited ions are found to be steeper than those for excited neutral atoms. The author attributed the results to different excitation mechanisms in jellium-like metals Mg and Al and semiconductor Si. Veje (1990) showed that the atomic levels having binding energies smaller than the work function of the solid have a lower probability of being excited by a resonant process because the electron transfer from the valence band of the solid to the atomic level in question involves a change in the binding energy. In those cases, the relative photon yield is not found to be proportional to the sputtering yield.

Assad *et al.* (1991, 1994) measured light yields from Ag and Na atoms, respectively, as a function of projectile energy when Ag and NaCl targets were bombarded with Ar ions of incident energy 8–99 keV. Various charge states of the projectile ion were used in the investigated energy range. Over that range, the light yield of Ag showed an increase by a factor of ten while the Na light yield after an initial rise of 35% slowly decreased In order to understand the possible mechanisms contributing to the light-emission process, the authors performed sputtering yield simulations using the MARLOWE computer code (Robinson, 1990). These calculations indicate a strong correlation between the total sputtering yield and the light yield only in the case of Na, although for both targets the energy distributions of the sputtered particles were found to be largely independent of the projectile energy.

3.3. Target Angular Dependence of the Photon Yield

The angular dependence of the photon yield curve is found to be similar to that of the neutral sputtering yield, i.e., the yield first rises monotonically with ϕ, the angle of ion incidence with respect to the surface normal, passes through a maximum at $\hat{\phi}$ typically 60–80° and then decreases sharply as ϕ approaches 90°. Such a variation was obtained for neutral Zn triplet lines (4680 Å, 4723 Å, and 4810 Å) during 8-keV He^+ bombardment of a Zn target (Gabla *et al.*, 1977), the neutral 3962-Å Al line, the 3415-Å Ni line, and the 3248-Å Cu line for 40-keV Ar^+ bombardment of Al, Ni, and Cu targets (Jimenez-Rodriguez *et al.*, 1984), 3247-Å and 3338-Å Cu and 3687-Å Cu^+ lines for 27-keV Ar^+ bombardment of Cu (Borisov, 1986), and 3962-Å neutral Al and 4529-Å Al^{2+} lines for 300-keV Ar^+ bombardment of an Al target (Reinke and Hippler, 1992).

When the target surface is tilted, the enhancement of the photon yield is mainly the result of two effects: the increase of (1) the total sputtering yield, and (2) the number of high-velocity sputtered atoms. For not-too-oblique incidence, Sigmund (1969) predicted a $(\cos\phi)^{-f}$ dependence, where $1 < f < 2$, rather than the normally expected $(\cos\phi)^{-1}$ trend.

Using computer simulation calculations of sputtering yield and the normal velocity $v_\perp$ distributions and incorporating a velocity-dependent survival probability, Jimenez-Rodriguez *et al.* (1984) calculated values of the parameter A/a for which the experimental results best conform to the theoretical predictions. For Cu and Ni, A/a values are, respectively, 3.8×10^4 and 5.4×10^4 m/s. Gabla *et al.* (1977) obtained $A/a = 8 \times 10^2$ m/s for the Zn line, which is quite low compared to that reported by Hippler *et al.* (1976). Presumably, the target used by Gabla *et al.* (1977) was oxidized under poor vacuum conditions, which considerably diminished nonradiative deexcitation processes.

Larsen and Veje (1983) studied the angular dependence of the atomic excitation during sputtering of Be, B, Mg, and Au by 80-keV Ar^+ ion bombardment. In general, the excitation intensities increase with increasing angle of incidence as expected, but a few levels show much stronger angular variations. It was proposed that these levels are selectively excited through molecular-orbital electron-promotion processes during binary collisions among energetic recoil target atoms at the surface.

3.4. Transients

It is well-known that the intensity of spectral lines changes abruptly with time after the incident beam is interrupted or there is sudden change of beam current or oxygen pressure. The transient, i.e., the variation of line intensity as a function of time is, generally, explained as being due to a change in oxygen coverage on the surface. Transients are frequently observed right after the beginning of the ion bombardment. As an example, Fig. 6 shows a typical time dependence of the neutral Si 2516-Å line intensity, when a 300-keV Kr^+ beam falls for the first time on a Si(100) target (Ghose *et al.*, 1995). The target was initially clean and lying at base pressure ($\sim 10^{-9}$ mbar) for several hours before bombardment. The light intensity first rises sharply and then declines exponentially to an equilibrium level corresponding to the intensity for the clean surface. Such an effect is not only caused by oxygen and oxygen-containing molecules; N_2 and NH_3 adsorptions, e.g., are also known to affect the photon yield in a similar way (Betz, 1987; Bhattacharya *et al.*, 1977). It has been shown that the existence of transients is related to the development of an oxide-like or nitride-like band structure at the surface. A detailed discussion on transients can be found in an article by Blaise (1976).

A number of authors have investigated the influence of oxygen on the equilibrium photon intensity (Thomas and de Kluizenaar, 1974; Bhattacharya *et al.*, 1977; Braun, 1979; Loxton *et al.*, 1980; Mériaux *et al.*, 1975; Tsong and Tsuji, 1980; Hippler and Reinke, 1992; Ghose *et al.*, 1995; Kelly and Kerkdijk, 1974). Most of the results show an initial increase in intensity proportional to increasing oxygen pressure, leveling off at high oxygen pressure. However, for some target elements, a fall in photon yield at high oxygen exposure was observed (Ghose *et al.*, 1995; Loxton *et al.*, 1980; Tsong and Tsuji, 1980). The observed decrease is thought to be related to the decrease in sputtering yield of the target. In passing, it should be noted that the presence of nitrogen has a comparatively small effect on the photon intensity at moderate pressures (Mériaux *et al.*, 1975; Bhattacharya *et al.*, 1977; Ghose *et al.*, 1995; Kelly and Kerkdijk, 1974). For example, during 300-keV Kr^+ bombardment of Si(100), Ghose *et al.* (1995) observed that the introduction of nitrogen up to about

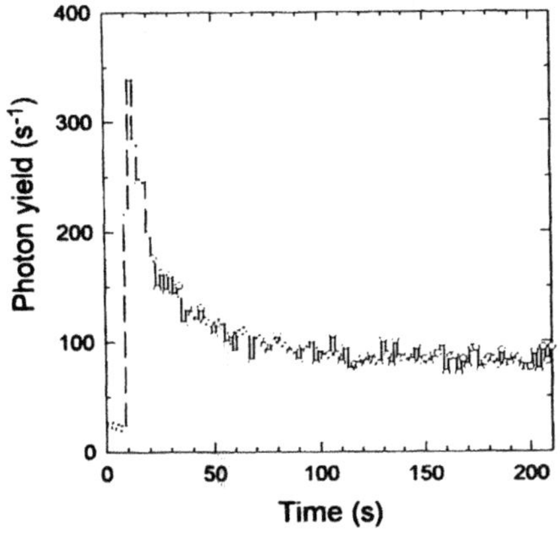

Figure 6. Light intensity variation of the neutral Si 2516-Å line vs. time for 300-keV Kr^+ bombardment of a clean Si surface lying several hours at base pressure ($\sim 10^{-9}$ mbar) (Ghose *et al.*, 1995; with kind permission from Elsevier Science Ltd.)

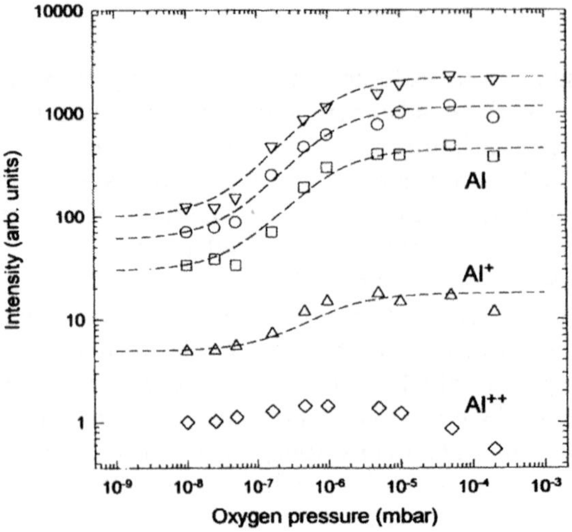

Figure 7. Light intensity vs. oxygen pressure for 300-keV Ar^+ bombardment of an aluminum target and for various Al (∇, ○, □), Al^+ (Δ) and Al^{2+} (◇) transitions. The dashed lines are corresponding calculations according to Eqs. (20) and (21) (Hippler and Reinke, 1992; with kind permission from Elsevier Science-NL.)

10^{-5} mbar has little influence on excited Si* atom emission except at about 10^{-4} mbar, where there was a sharp rise in the photon yield. The insensitivity of the photon yield to increasing nitrogen pressure (up to $\sim 10^{-5}$ mbar) may be explained by the fact that nitrogen molecules in the ground state do not chemisorb on Si (Winters and Kay, 1972; Mogab and Lugujjo, 1976), whereas nitrogen atoms and ions can form a chemical bond with Si atoms at the solid surface. Thus at very high pressure ($\sim 10^{-4}$ mbar) there is a possibility that a large fraction of the surface area will be covered by nitrogen atoms from the beam-induced dissociation of nitrogen molecules, thereby causing an increase in the photon yield.

The relative intensities of several atomic and ionic Al lines as a function of oxygen pressure for 300-keV Ar^+ bombardment of an Al target are shown in Fig. 7 (Hippler and Reinke, 1992). A pronounced oxygen dependence is observed for lines from neutral Al atoms, whereas ionic lines are less affected by the oxygenated surface. This indicates that either suppression of the nonradiative transitions is less complete for deeper energy levels (ionic lines) or that the production mechanisms of excited ions and atoms are different. The increase in photon yield for atomic lines has been interpreted as being due to increasing oxygen coverage at the surface. The oxygen adsorption on the target surface is governed by the rate of arrival of the oxygen atoms Φ_O multiplied by the sticking probability c_s and the rate of removal of oxygen atoms owing to sputtering by the incident beam. The time dependence of the oxygen coverage θ is given by (Van der Weg and Lugujjo, 1975; Rausch and Thomas, 1978; Kelly and Kerkdijk, 1974)

$$\frac{d\theta}{dt} = -\frac{1}{n_s}(\Phi_i Y_O + \Phi_O c_s) \cdot \theta + \frac{\Phi_O c_s}{n_s} \tag{19}$$

where Y_O, is the sputtering yield of oxygen, n_s the surface atom density, and Φ_i the flux density of the incident ion beam. For long durations $d\theta/dt \to 0$ and the equilibrium oxygen coverage Θ is given as

$$\Theta = \frac{\Phi_O}{\Phi_O + (Y_O/c_s)\Phi_i} \tag{20}$$

Thus Θ ranges from zero (a clean surface) to unity (complete oxygen coverage). The photon signal $I(\Theta)$ for a given equilibrium coverage Θ may be expressed as the sum of two contributions I_C and I_O from the clean and from the oxygen-covered part, respectively, of the surface:

$$I(\Theta) = I_O\Theta + I_C(1 - \Theta) \tag{21}$$

Combining Eqs. (20) and (21), one can predict the pressure dependence of photon intensity provided that the oxygen sputtering yield Y_O and the sticking coefficient c_s are known. Vice versa, a fit to the pressure dependence of the experimental data may yield the ratio Y_O/c_s. From Fig. 7, for 300-keV Ar^+ bombardment of a polycrystalline Al target at an ion incidence angle of 45°, this ratio is $Y_O/c_s = 30$. If it is assumed that $c_s = 0.1$ (Kelly and Kerkdijk, 1974), the sputtering coefficient of adsorbed oxygen atoms on an Al surface comes out as 3 atoms/ion for an angle of ion incidence of 45° corresponding to about 2.1 atoms/ion at normal ion incidence. Similar measurements for 300-keV Kr^+ bombardment on Si(100) yield $Y_O \approx 1$ atom/ion (Ghose *et al.*, 1995).

One can also estimate the sputtering coefficient of oxygen from the temporal decay of the emission intensity when the ion beam is directed onto a surface carrying a single monolayer of adsorbed oxygen. Such a time dependence is shown in Fig. 8 for the Si 2882-Å line following 300-keV Kr^+ bombardment of a Si(100) target. If one assumes that the decrease in the photon intensity is related only to the reduction in surface coverage owing to sputtering, it follows from Eq. (19) that, under appropriate boundary conditions (Ghose *et al.*, 1995),

$$\theta = \exp\{-[(\Phi_i Y_O t)/n_s]\} \tag{22}$$

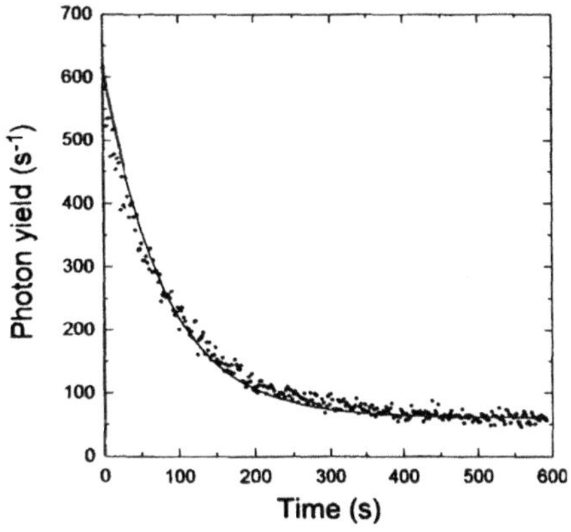

Figure 8. Light intensity variation of the Si 2882-Å line vs. time for 300-keV Kr^+ bombardment of a Si surface carrying a layer of adsorbed oxygen. The experimental data points (·) are compared with a theoretical prediction (solid line) according to Eq. (22) (Ghose *et al.*, 1995; with kind permission from Elsevier Science-NL.)

The decay constant (half-time) is given by

$$t_{1/2} = 0.693n_s/\Phi_i Y_O \tag{23}$$

Equation (23) shows that $t_{1/2}$ varies inversely with the sputtering coefficient of oxygen as well as with the beam current density. The higher the beam current density is, the narrower the half-width of the decay curve, which is indeed observed in the experiment.

Several authors reported a sudden increase in light intensity when the ion beam was interrupted (Kelly and Kerkdijk, 1974; Van der Weg and Lugujjo, 1975). In all these experiments, the base pressure was high enough (typically 10^{-6} mbar) to develop an oxygen coverage on the surface from the residual gases, which is believed to be responsible for the change in line intensity. A transient is characterized by two parameters: namely, the height, which is the ratio between the initial value and the equilibrium value, and the half-time for decay. For a fixed ion current density, the transient height first increases with increasing interruption time and then levels off. The falling transient behavior after the sharp rise is well decribed by Eq. (22). The beam-off transients were observed as long as the oxygen coverage at the surface was less than 1.

Figure 9 illustrates a comparison of beam-off transients at low oxygen coverage ($\Theta \ll 1$) and at maximum oxygen coverage ($\Theta = 1$), respectively, for the Si 2882-Å line during 300-keV Kr^+ bombardment of Si(100). The condition $\Theta < 1$ means that the surface is partially covered with oxygen. During the interruption of the bombardment, the oxygen coverage will increase. As a result, the photon intensity will suddenly rise to a higher level when the bombardment is resumed. As the bombardment proceeds, the excess oxygen will be removed by sputtering. As a consequence, the photon intensity decreases with time and finally reaches its steady-state value (Fig. 9a). On the other hand, for the $\Theta = 1$ condition, the oxygen coverage will remain unaltered during the interruption of the bombardment, since all the adsorption sites are occupied. Therefore, when the bombardment is resumed,

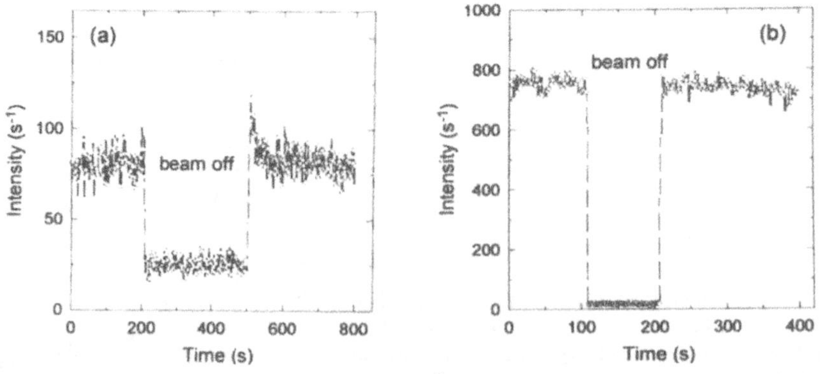

Figure 9. Transient in the photon yield of the Si 2882-Å line from a Si target owing to the interruption of the 300-keV Kr^+ ion beam at: (a) low partial oxygen pressure (1.2×10^{-7} mbar) and (b) high partial oxygen pressure (3.2×10^{-5} mbar) (Ghose *et al.*, Unpublished.)

Table 1. Correlation between Observed Transients in the Yield of Excited Atoms after Sudden O_2 or NH_3 Pressure Increase and the Bandgap of the Phase Formed under O_2 or N_2 Implantation. The Signal Increase at 1×10^{-5} mbar O_2 or NH_3 as Compared to a Clean Surface Is Given (Betz, 1987; with kind permission from Elsevier Science, B.V.)

	Phase formed at implantation of				*Signal*
Target	*Oxygen*	*Nitrogen*	*Bandgap*	*Transients observed*	*increase* ×
Al	Al_2O_3		yes	yes	35
Al		AlN	yes	yes	45
Si	SiO_2		yes	yes	100
Si		Si_3N_4	yes	yes	45
Mo	MoO_2		no	no	4
Mg	MgO		yes	yes	3
Cr	Cr_2O_3		yes	yes	5
Cr		CrN	no	no	4
Pb	PbO		yes	yes	8
Be	BeO		yes	yes	50
Ti	TiO		no	no	3
Ti		Ti_2N	no	no	3
V	VO		no	no	3
In	In_2O_3		yes	yes	3

the intensity will return to the previous equilibrium value, without any significant transient regime (Fig. 9*b*).

A variation in the ion beam current also produces transients, even when the oxygen pressure is maintained constant. It has been noted that transients are ill-defined or absent at current densities that are low enough to sputter oxygen atoms at a slower rate than that for adsorption (Kelly and Kerkdijk, 1974).

Finally, transients owing to pressure variations at constant ion current density will be discussed. Oxygen-pressure-induced transients have been the most frequently studied. Introduction or withdrawal of oxygen leads to an instantaneous rise or fall, respectively, in the photon signal. The duration of the transient regime, however, may vary to a large extent depending upon the nature of the reactions that take place on the surface, e.g., random adsorption with or without the accompaniment of recoil implantation. Betz (1987) studied pressure-induced transients in a number of elemental targets under oxygen or ammonia exposure, which are listed in Table 1. He found that transients occur only when the bandgap is formed by recoil-implanted oxygen or nitrogen atoms. Van der Weg and Lugujjo (1975) noted that elements whose *d*-shells are filled or nearly filled are susceptible to strong transient effects caused by oxygen. The transient measurement owing to incremental changes in oxygen pressure is quite interesting. In the experiment for the Si 2882-Å line following 300-keV SF_5^+ ion bombardment of Si (Ghose *et al.*, 1997), a small amount of high-purity oxygen gas was introduced into the target chamber through a pulsed valve and a leak valve in series. The pulsed valve operates for 150 μs for single-pulse generation. The pressure first rises to about 5×10^{-5} mbar almost instantaneously and then gradually decreases to the base pressure ($\sim 10^{-9}$ mbar). The line intensity, on the other hand, shows an initial increase, passes through a maximum, and then decreases exponentially to an equilibrium level. It is interesting to note that there was a sharp rise in photon intensity at times when the oxygen pressure was falling rapidly. This is not surprising if one keeps in mind that the degree of oxygen coverage is the consequence of two competing effects, namely, the adsorption of oxygen atoms from the ambient oxygen atmosphere and the sputter removal of oxygen atoms by the incident ion beam. Thus, in the rising part of the photon yield, the rate of adsorption exceeds the rate of erosion. The yield proceeds to a maximum corresponding to an instantaneous equilibrium oxygen coverage, and finally falls owing to both sputtering of oxygen atoms and an excessive decline in the adsorption rate.

Tsong and Tsuji (1980) studied the transient behavior of line intensities arising from 18- to 20-keV Ar^+ ion bombardment of Si, Ni, Zr, Ti, and Cr targets by interrupting the oxygen flow into the target chamber. They observed transients of two different categories depending on the initial oxygen pressure at the moment the oxygen supply was terminated. If the initial oxygen pressure is high ($\sim 10^{-4}$ mbar), the photon intensity rises rapidly to a maximum peak and then decays to an equilibrium value. On the other hand, when the initial oxygen pressure is low ($\lesssim 10^{-5}$ mbar), the intensity drops immediately to the equilibrium yield. Exceptions are atomic Ti lines, where only transients of the first category are observed irrespective of the value of the initial oxygen pressure. Such behavior is in conformity with the nature of the variation in photon intensity as a function of the oxygen partial pressure for the respective elements (Tsong and Tsuji, 1980).

The relative importance of adsorbed or recoil-implanted oxygen atoms in the transient can be distinguished from the measurement of the half-life of the decay following the initial rise. For Si and Zr, the $t_{1/2}$ values are consistent with the one obtained from Eq. (23). This indicates that for these elements the intensity enhancement is caused primarily by an adsorbed monolayer of oxygen. On the other hand, the rather large $t_{1/2}$ values for Ti and Cr reflect an additional contribution of recoil-implanted oxygen in the enhancement of photon emission. In the latter case, longer times are required to sputter away oxygen atoms beneath the surface. Tsong and Tsuji (1980) discussed the role of recoil-implanted oxygen atoms further in the emission process. These atoms serve as the sources of electron-emissive sites. The metal ions, which are supposed to form during the metal–oxygen bond-breaking process, pick up electrons through interaction with these polarized surface sites, thereby creating more excited neutrals and giving rise to enhanced emission.

To summarize, transients can arise whenever there is a change in equilibrium surface oxygen coverage as a result of a change in the experimental parameters, e.g., gas pressure. Even in oxides, e.g., transition metal oxides, where oxygen adsorption is plausible, transients could be observed (Blaise, 1976). Kelly and Kerkdijk (1974) reported that the transient in Al_2O_3 is of opposite character to that in Al, i.e., a sudden oxygen exposure exhibits a momentarily low yield before rising slowly to an equilibrium level. This is consistent with the phenomenon of a lower sputtering yield of Al_2O_3 when it is bombarded in an oxygen atmosphere. Finally, if the oxygen adsorption modifies only the first atomic layer, the transient will be of short duration. Transients of longer half-lives are observed if there is a phase change in the near-surface region by recoil implantation or oxidation.

4. APPLICATION TO SURFACE ANALYSIS

The phenomenon of light emission from sputtered atoms serves as the basis for surface chemical analysis, where the species present at the surface can be identified by their characteristic line emissions. White *et al.* (1972, 1974) proposed the acronym SCANIIR (surface composition by analysis of neutral and ion impact radiation) for this surface analytical technique and discussed at length its potential for material characterization in the wavelength region from 2000 to 8500 Å. Though the yield of photons is generally lower than the yield of ions, SCANIIR has the advantage of a very high resolution (closely separated optical lines can be distinguished) in comparison with its counterpart the SIMS (secondary-ion mass spectrometry) technique.

The secondary-photon yield Y_{if}^* for a transition $i \rightarrow f$ at a wavelength λ can be expressed as

$$Y_{if}^*(\lambda) = \varepsilon Y \dot{n}_{\text{ion}} b_{if} R(\infty) \eta \, \Delta\Omega \tag{24}$$

where Y is the sputtering yield, $\dot{n}_{\text{ion}}$ the rate of incident ions, ε the excitation probability, b_{if} the branching ratio for the investigated photon transition, $R(\infty)$ the survival probability, $\Delta\Omega$ the solid angle, and η the quantum efficiency of the detection system. Tsong and Yusuf (1978) estimated the detection limit for 34 elements present in a NBS glass sample. The chosen emission lines have short lifetimes ($\lesssim 10^{-8}$ s) so that the decay takes place close to the surface. The detection limit, however, varies widely from 1 to 7500 ppm by weight for the elements studied. SCANIIR can also be used for depth-profiling. Since layer by layer of the target surface is successively sputtered away, the distribution of elements as one goes deeper can be found from the variations of the analytical signal with time. Of particular interest is the capability of performing hydrogen depth profiles in solids using this technique. Tsong and Liebert (1978) claim a detection limit of about 1500 ppm in the analysis of hydrogen in amorphous Si films, topaz minerals, and NBS Ti standards. In passing, it should be noted that depth-profiling based on sputtering is associated with various processes, such as recoil implantation, cascade mixing, and the development of surface topography, which may distort the original (true) distribution (Emmoth *et al.*, 1976; Braun *et al.*, 1976: Kuduk *et al.*, 1989, 1991).

Quantitative elemental analysis by SCANIIR is rather difficult to perform because the intensity of emission is matrix-dependent; there is also no proper theoretical formulation of excited-state formation and the related emission intensity. Nevertheless, attempts have been made by several authors to obtain quantitative results by using a LTE-like model (MacDonald and Martin, 1977*a*, *b*; MacDonald and Garrett, 1978; MacDonald *et al.*, 1978; Martin and MacDonald, 1977; Okutani and Shimizu, 1979); however, their validity is questionable, as discussed earlier.

An interesting application of SCANIIR is that it can provide information about the lateral distribution of elements across the sample surface. Mériaux *et al.* (1972)

pioneered in the construction of a microanalyzer based on sputtering-induced light emission. They developed two types of instruments for imaging SCANIIR microanalysis. In the optical microanalyzer, an image of the distribution of a particular element is obtained by bombarding all points of a surface area with a broad primary ion beam and by collecting the luminescence on a photographic film through an optical filtering device such as a Fabry-Perot etalon. In the scanning ion probe microanalyzer, a finely focused primary ion beam is raster-scanned over the surface. The emitted optical radiation is subsequently collected by a spherical mirror, analyzed with an interference filter, and detected by a cooled photomultiplier. In the former type of microanalyzer, all points of the bombarded surface contribute to the final image simultaneously, whereas in the latter, the picture elements are recorded in sequence. The relative advantages and disadvantages of these two methods were discussed by Mériaux *et al.* (1972).

Inoue *et al.* (1992) developed a scanning SCANIIR microscope employing a photon-counting technique and a ratemeter for lateral and in-depth composition analysis of solids (Fig. 10). In this system, the primary beam current is kept low enough to minimize the damage of the specimen surface. Figure 11 illustrates brightness-modulated line-scan images of a 200-mesh Cu grid with secondary electrons (Fig. 11a) and Cu 3248-Å photons (Fig. 11b), respectively. A spatial resolution of 30 μm was achieved. In the study of depth profiles of the SiO_2/Si interface measured employing the Si 2882-Å line with this instrument, they obtained a depth resolution of 40 nm. A special feature of this type of microscope is that it can analyze insulators with much less difficulty than with conventional scanning SIMS because the kinetic energy of light-emitting excited sputtered neutrals is not affected by the charging-up of the sample.

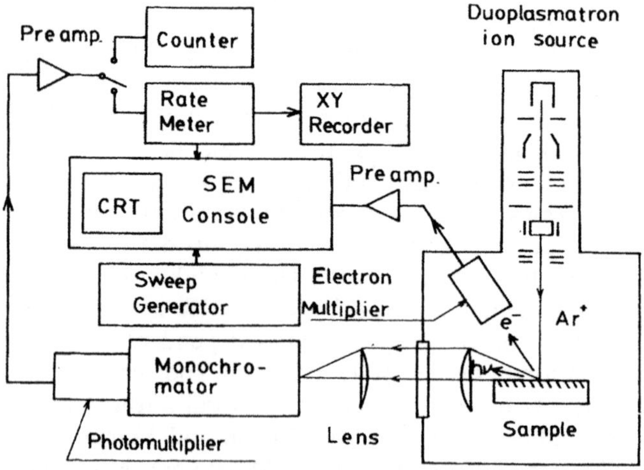

Figure 10. A schematic diagram of the experimental setup of a scanning SCANIIR microscope (Inoue *et al.*, 1992; with kind permission from the *Japanese Journal of Applied Physics.)*

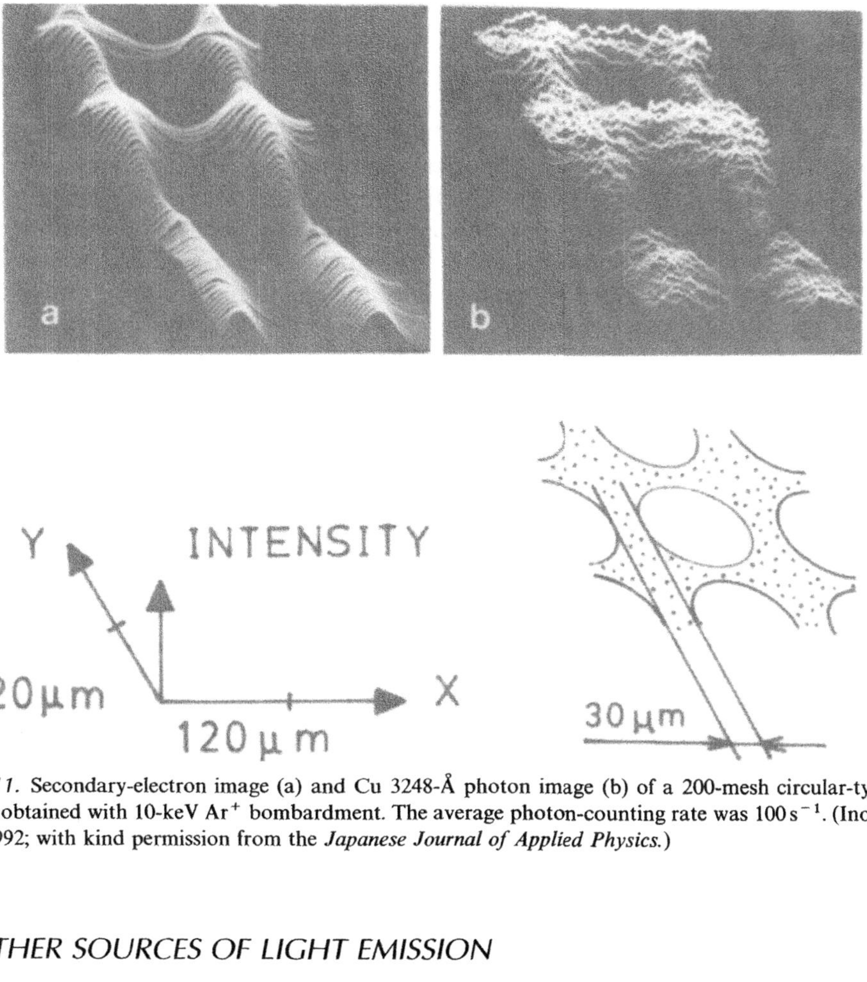

Figure 11. Secondary-electron image (a) and Cu 3248-Å photon image (b) of a 200-mesh circular-type Cu grid obtained with 10-keV Ar^+ bombardment. The average photon-counting rate was $100\,s^{-1}$. (Inoue *et al.*, 1992; with kind permission from the *Japanese Journal of Applied Physics.*)

5. OTHER SOURCES OF LIGHT EMISSION

5.1. Continuum Emission

Ion bombardment of solids often exhibits broadband unresolved or continuum radiation with some discrete lines superimposed. Three types of continua have been identified.

5.1.1. Emission from Impurity Molecules on Complexes

The first of this kind of continuum is generally observed for transition metals with unfilled *nd*-shells (Mo, Ta, W, Ti, Nb, and Cr). Although there are a number of speculations as to the origin of this continuum (Tolk *et al.*, 1973; White *et al.*, 1976; Kiyan *et al.*, 1976*a*; Kerkdijk *et al.*, 1976*b*; Bayly *et al.*, 1976; Rausch *et al.*, 1976; Martin and Loxton, 1980; Braun and Emmoth, 1980; Emmoth *et al.*, 1976), it is believed that it is due to the radiative decay of excited metal oxide molecules or clusters ejected as a result of collision processes occurring in the near-surface region. Similar to line emission, this type of continuum emission is also enhanced by the presence of oxygen or the bombardment of oxide targets. Fournier *et al.* (1991) noted a weak continuum extending over the visible and near-ultraviolet regions during sputtering of an $YBa_2Cu_3O_7$ superconductor, but a strong continuum having a decreasing intensity with bombarding time during sputtering of an Y_2O_3 target. These continua have been attributed to the band emission from oxide molecules ejected from the surface. Bazhin *et al.* (1990) performed a computer simulation study taking into account the molecular excitation in a repulsive state owing to binary collisions to explain the projectile energy dependence and oxygen dependence of emission intensity.

Kelly *et al.* (1981) studied the continuum radiation from ion-bombarded B and B_2O_3 in the presence of reactive gases in detail. For B_2O_3, the continuum extending from 3300 to 5300 Å did not depend on the type of the gas present. This wavelength region corresponds to the $A^2\Pi \rightarrow X^2\Sigma^+$ system of BO molecules. However, in the case of elemental B strong molecular emission in the region 4100–4500 Å was observed only in the presence of hydrogen or methane, which is attributed to the $A^1\Pi \rightarrow X^1\Sigma^+$ system of rotationally excited BH molecules. Kelly *et al.* (1981) noted that for such systems, the excited states of the molecules correlate adiabatically with ground-state atoms or with low-lying metastable states. In a study with Si having an acetone overlayer by Thomas and de Koning (1978), the observed broadband emission was identified as the CH 4300-Å $A^2\Delta \rightarrow X^2\Pi$ molecular emission. Similarly, Braun and Emmoth (1980) observed a strong BeH band spectrum around 5000 Å during ion bombardment of Be under impure conditions. Fournier *et al.* (1993) found that the continuum radiation in the wavelength region 4800–5010 Å observed during 5-keV Kr^+ bombardment of an oxygenated Mg surface appears well resolved under high resolution, and was identified as being due to the $B^1\Sigma \rightarrow X^1\Sigma$ ($\Delta v = 0$) transition in the rovibrationally excited MgO molecule. Vasiljeva (1991) discussed the mechanisms of the formation of oxide molecules during ion bombardment of rare-earth elements, which also emit continuous emission bands. It is assumed that the escaping metal atoms, which are excited in the triplet state during the course of electron exchange between the atom and the surface, react with adsorbed oxygen atoms to form excited diatomic oxide molecules.

5.1.2. Emission from Implanted Species

The second type of continuum arises during bombardment of energetic inert gas ions to a solid. The projectile ions are implanted and trapped in the target material and subsequently released as excited molecules. From the results of a number of experiments (Braun and Emmoth, 1980; Gritsyna *et al.*, 1969, 1974; Knudson *et al.*, 1978; Bhattacharya *et al.*, 1978; Hill *et al.*, 1979), the general properties of these continua can be summarized as follows:

1. The continua emitted from excited He_2, Ne_2, and Ar_2 are found to be in the vacuum ultraviolet region.

2. The continuum originates inside the target or on the surface.
3. The intensity increases with increasing ion dose until the saturation point is reached. Also, the higher the beam energy is the longer the saturation time, which is consistent with the implantation theories.
4. The implanted ions are believed to form gas bubbles inside the target. The pressure inside these bubbles can be remarkably high as evident from the formation of solid Ar, Kr, or Xe bubbles in some fcc metals (vomFelde *et al.*, 1984; Templier *et al.*, 1984; Evans and Mazey, 1985). Once the bubbles develop, the contained gas can also be excited by other projectile impacts.

The interpretation has been offered that these continua occur owing to transitions from a transient bound excited state to the repulsive ground state of the rare-gas dimer. Similar continuous emission from molecular inert gases has also been detected in other kinds of experiments, e.g., in high-pressure gas discharges (Tanaka and Yoshino, 1963). Recently, broad luminescence bands in the vacuum ultraviolet centered at 1110 and 1265 Å were observed when solid Ar films were electronically excited by MeV He ions (Reimann *et al.*, 1992). The 1265-Å band spectrum has striking similarities to the one associated with implanted Ar ions in a solid target (Braun and Emmoth, 1980). This band is identified from the decay of vibrationally relaxed $Ar_2^*({}^{1,3}\Sigma_u^+)$ molecules trapped in the bulk of the Ar film (Reimann *et al.*, 1992).

5.1.3. Luminescence from the Solid

The third kind of radiative continuum occurs when insulating targets such as alkali halides are bombarded with energetic ions (Bazhin *et al.*, 1976; Tolk *et al.*, 1973). In this case, the origin of ionoluminescence is within the bulk of the solid and is ascribed to radiative electron–hole recombinations at defects. Such a continuum has also been observed in metals, e.g., Al during bombardment by low-Z projectiles (Zivitz and Thomas, 1976*a*, *b*). A discussion on the experimental results for this type of luminescence can be found in an article by White *et al.* (1977). More recently, Suchańska *et al.* (1994) studied ion-beam-induced luminescence in a number of pure and impurity-doped alkali halide crystals, namely, NaCl, KCl, KI, KBr, LiF, and NaF. The results show that apart from activated luminescence bands, there are intrinsic luminescence bands, which are more prominent and characteristic of pure crystals.

The band positions and half-widths of the intrinsic luminescence spectra remain invariant with the change of ion beam parameters and also with target temperature. However, a blueshift in the peak intensity occurs for crystals with wider bandgap energies. It has been proposed that the ionoluminescence is caused by the radiative recombination of electrons from the conduction band with holes at the V_2 center. A large number of such centers are formed with ion doses greater than $10^{20}\,m^{-2}$, and these are found to be stable at room temperature. At higher temperatures ($> 300\,K$) or doses ($> 10^{21}\,m^{-2}$), the V_2 centers tend to agglomerate and transform into V_3 centers. These changes are reflected in a decrease in the luminescence intensity.

5.2. Light Emission from Excited Projectiles

A fraction of the projectiles backscattered from the surface may also be in an excited state, either as scattered ions or as neutrals, and may decay radiatively. The integrated photon intensity in a spectral line is related to the backscattered flux of excited atoms, whereas the line profile yields information on the distribution in regard to the speed and direction of the particles. Light particles such as hydrogen or helium because of their higher reflection coefficients are mostly used for projectile excitation studies. The excited reflected particles are quite fast, which is manifested in the Doppler-broadening of the spectral lines. As with sputtered excited particles radiationless processes such as AD and RI can significantly alter the proportion of the light-emitting species when the departing particles are in the immediate vicinity of the surface. Here, also, one can estimate the value of the survival parameter A/a either from the lineshape analysis or line-intensity measurements as a function of the projectile energy. However, the theoretical modeling to fit the experimental data is quite sensitive to the choice of the value of A/a. The results of various published works are summarized by White *et al.* (1977) and Thomas (1980), where the mean values of A/a quoted for different projectiles lie in the range $1\text{–}3 \times 10^6$ m/s. The emission intensity from backscattered projectiles is also influenced by the presence of adsorbed oxygen, giving rise to transient effects as well as to an increase in the line intensity (Rausch and Thomas, 1978), which are qualitatively very similar to those for emission from sputtered particles.

Erickson and Smith (1975) observed the oscillatory yield of backscattered He^+ ions for low-energy He^+ impact on Pb. Such an oscillation is attributed to the quasi-resonant charge transfer between the empty $He^+(1s)$ state and the energetically close Pb(5*d*) state. A similar search for the oscillatory structure of the 3888-Å ($3p\,^3P^0 \rightarrow 2s\,^3S$) radiation of backscattered He from a Pb target in the energy range 0.7–2 keV was attempted by Taglauer *et al.* (1979). In this case, also, a quasi-resonant electron exchange between the excited *p* state of He and the conduction band of the solid is possible. No such oscillations were recorded, however; instead, the yield of photons increased monotonically with energy. Later, Veje (1984) extended the measurements to higher energies and obtained the same results. He also studied the energy dependence of the $3d\,^3D$ and $5d\,^3D$ excited levels in addition to the $3p\,^3P$ level of neutral helium backscattered from Au, Tl, and Pb targets. In all cases, the yield first rises to about 40 keV and then reaches a plateau. It was concluded that the excited particles are predominantly fast particles and the excitation results from the direct capture of electrons from the valence band when the projectile leaves the surface. The excited levels are found to be populated according to an n^{-4} dependence, where n is the principal quantum number.

There are several interesting investigations of projectile excitation in grazing ion–surface collisions. Here the scattered particle trajectory is governed by the motion of the projectile ions parallel to the surface with an energy nearly equal to the incident ion energy and also by the motion of ions normal to the surface, the energy being orders of magnitude less. Consequently, no penetration of projectiles into the bulk of the solid is expected to occur. However, at the minimum distance of approach to the surface, there is a probability of continuous creation and destruction of possible projectile states (Winter and Zimny, 1988). For projectile velocities lower than the Bohr velocity, resonant electron transfer and Auger-type mechanisms play a dominant role in the population of excited terms so that the binding energy of an atomic term in comparison to the work function of the target is important.

The most striking feature of grazing ion–surface collisions is a pronounced orientation in the spatial distribution of orbital angular momenta of excited terms, which results in a large degree of circular polarization of the emitted light (Andrä, 1975). Martin *et al.* (1980) showed that the degree of polarization is strongly surface-sensitive. During glancing angle collisions of 50-keV H_2^+ with a Nb surface, they found that the normalized Stokes parameter S/I measured for the Balmer α-hydrogen line drops from 48 to 17% under high oxygen exposure. It has been suggested that at the later stages of the oxygen adsorption process, where the H_α intensity is found to deviate from its uniform variation with the oxygen coverage, the observed polarization is associated with H^+–O rather than H^+–Nb collision.

Finally, it should be noted that it will be of great interest to study photon emission from highly charged ions having vacancies in the outer shells and interacting with a target surface. However, it has been pointed out that for higher-Z projectiles, the probability of photon emission owing to inner-shell vacancies is more likely (e.g., Varga, 1987), resulting in the emission of x-ray photons, as was observed during, e.g., slow collisions of Ar^{17+} with Be (Donets, 1983). Photon and particle emission by highly-charged Xe^{q+} ion impact on SiO_2 was investigated by Schiwietz *et al.* (1993). The specific energy was 2.2 keV/q and the incident charge q was varied between 30 and 50. No spectral analysis was performed for the emitted photons, which were detected with an open channel-plate detector sensitive to the vacuum ultraviolet part of the electromagnetic spectrum. A pronounced increase in the photon yield with increasing charge state was observed while at the same time the total sputtering yield displayed a significantly weaker q-dependence. The absolute photon yield was from 1 to 10 photons per incident ion. The physical origin of the emitted photons did not become clear from this investigation, although it was expected that the major fraction could result from backscattered projectiles rather than from sputtered excited atoms.

6. CONCLUSIONS

The present chapter has covered some aspects of bombardment-induced light-emission phenomena. In general, light emission is found to be highly sensitive to the chemical state of the surface. Therefore, to understand the basic mechanisms of excitation processes satisfactorily, experiments have to be performed under well-controlled bombardment conditions. The various theoretical models proposed so far are incomplete in the sense that they do not provide information that can be subjected to a quantitative test. It is unfortunate that researchers have paid relatively little attention to light-emission phenomena in comparison to the number of studies devoted to SIMS, the main reason for this disparity being the lower yield of photons, although, the simple experimental setup and less expensive equipment make the former field much more attractive than SIMS studies. It should be noted, however, that both secondary-photon-emission and ion-emission studies are equally important for detailed understanding of inelastic ion–solid collision processes.

REFERENCES

Andersen, N., Andersen, B., and Veje, E. (1982), *Rad. Effects* **60**, 119.
Andersen, C. A., and Hinthorne, J. R. (1973), *Anal. Chem.* **45**, 1421.
Andrä, H. J. (1975), *Phys. Lett.* **54A**, 315.
Arlinghaus, H., and Bispinck, H. (1983), *Surf. Sci.* **134**, 567.
Assad, C., and Tribble, R. E. (1994), *Nucl. Instr. Meth.* **B93**, 427.
Assad, C., Liu, W., and Tribble, R. E. (1991), *Nucl. Instr. Meth.* **B62**, 201.
Bayly, A. R., Martin, P. J., and MacDonald, R. J. (1976), *Nucl. Instr. Meth.* **132**, 459.
Bazhin, A. I., Rausch, E. O., and Thomas, E. W. (1976), *Phys. Rev. B* **14**, 2583.
Bazhin, A. I., Suchanska, M., and Teplov, S. V. (1990), *Nucl. Instr. Meth.* **B48**, 639.
Betz, G. (1987), *Nucl. Instr. Meth.* **B27**, 104.
Bhattacharya, R. S., van der Veen, J. F., Kerkdijk, C. B. W., and Saris, F. W. (1977), *Rad. Effects* **32**, 25.
Bhattacharya, R. S., Hasselkamp, D., and Schartner, K. H. (1979), *J. Phys. D* **12**, L55.
Bhattacharya, R. S., Lang, K. G., Scharmann, A., and Schartner, K. H. (1978), *J. Phys. D* **11**, 1935.
Bissessur, V., and Tsong, I. S. T. (1990), *Nucl. Instr. Meth.* **B52**, 129.
Blaise, G. (1976), *Surf. Sci.* **60**, 65.
Borisov, A. M. (1986), *Rad. Effects* **97**, 85.
Braun, M. (1979), *Phys. Scripta* **19**, 33.
Braun, M., and Emmoth, B. (1980), *Nucl. Instr. Meth.* **170**, 585.
Braun, M., Emmoth, B., and Martinson, I. (1974), *Phys. Scripta* **10**, 133.
Braun, M., Emmoth, B., and Buchta, R. (1976), *Rad. Effects* **28**, 77.
Chaudhri, R. M., and Khan, M. Y. (1961*a*), *Nature* **189**, 996.
Chaudhri, R. M., and Khan, M. Y. (1961*b*), Nature, **192**, 646.
Donets, E. D. (1983), *Phys. Scripta* **T3**, 11.
Emmoth, B., Braun, M., and Palenius, H. P. (1976), *J. Nucl. Mater.* **63**, 482.
Erickson, R. L., and Smith, D. P. (1975), *Phys. Rev. Lett.* **34**, 297.
Evans, J. H., and Mazey, D. J. (1985), *J. Phys. F* **15**, L1.
Fournier, J., Fournier, P. G., Kaddouri, A., and Dunet, H. (1991), *J. Appl. Phys.* **69**, 2382.
Fournier, P. G., Fournier, J., and Bellaoui, B. (1992), *Nucl. Instr. Meth.* **B67**, 604.
Fournier, P. G., Fournier, J., Bellaoui, B., Benoist d'Azy, O., and Taieb, G. (1993), *Nucl. Instr. Meth.* **B78**, 144.
Gabla, L., Pedrys, R., and Szymoński, M. (1977), *Acta Phys. Pol. A* **51**, 311.
Ghose, D. (1995), *Vacuum* **46**, 13.
Ghose, D., Brinkmann, U., and Hippler, R. (1995), *Surf. Sci.* **327**, 53.
Ghose, D., Brinkmann, U., and Hippler, R. (1997), *Phys. Rev. B* **55**, 13989.
Good-Zamin, C. J., Shehata, M. T., Squires, D. B., and Kelly, R. (1978), *Rad. Effects* **35**, 139.
Gritsyna, V. V., Kiyan, T. S., Koval', A. G., and Fogel', Ya. M. (1969), *Sov. Phys. JETP Lett.* **9**, 124.
Gritsyna, V. V., Kiyan, T. S., Koval', A. G., and Fogel', Ya. M. (1972), *Rad. Effects* **14**, 77.
Gritsyna, V. V., Kiyan, T. S., Koval', A. G., and Fogel', Ya. M. (1974), *Opt. Comm.* **10**, 320.
Gruen, D. M., Pellin, M. J., Young, C. E., Mendelsohn, M. H., and DeWald, A. B. (1983), *Phys. Scripta* **T6**, 42.
Hagstrum, H. D. (1954), *Phys. Rev.* **96**, 336.
Heiland, W., Kraus, J., Leung, S., and Tolk, N. H. (1977), *Surf. Sci.* **67**, 437.
Hill, K. W., Comas, J., Nagel, D. J., and Knudson, A. R. (1979), *Phys. Scripta* **20**, 652.
Hippler, R., and Reinke, S. (1992), *Nucl. Instr. Meth.* **B68**, 413.
Hippler, R., Krüger, W., Scharmann, A., and Schartner, K. H. (1976), *Nucl. Instr. Meth.* **132**, 439.
Inoue, M., Sugiyama, Y., Nishigaki, S., and Noda, T. (1988), *Nucl. Instr. Meth.* **B33**, 519.
Inoue, M., Sugiyama, Y., Nishigaki, S., and Noda, T. (1989), *Surf. Sci.* **209**, L157.

Inoue, M., Kunitomo, S., Nishigaki, S., and Noda, T. (1992), *Jpn. J. Appl. Phys.* **31**, L200.
Jensen, K., and Veje, E. (1974), *Z. Phys.* **269**, 293.
Jimenez-Rodriguez, J. J., Karpuzov, D. S., and Armour, D. G. (1984), *Surf. Sci.* **136**, 155.
Kelly, R. and Kerkdijk, C. B. (1974), *Surf. Sci.* **46**, 537.
Kelly, R., Good-Zamin, C. J., Shehata, M. T., and Squires, D. B. (1978), *Nucl. Instr. Meth.* **149**, 563.
Kelly, R., Dzioba, S., Tolk, N. H., and Tully, J. C. (1981), *Surf. Sci.* **102**, 486.
Kerkdijk, C. B., Kistemaker, J., and Saris, F. W. (1976*a*), in: *Physics of Ionized Gases* (B. Navinsek, Ed.) (J. Stefan Institute, University of Ljubljana, Yugoslavia), p. 357.
Kerkdijk, C. B., Schartner, K. H., Kelly, R., and Saris, F. W. (1976*b*), *Nucl. Instr. Meth.* **132**, 427.
Kistemaker, J., and Snoek, C. (1962), *Le Bombardement Ionique* (C.N.R.S., Paris), p. 51.
Kiyan, T. S., Gritsyna, V. V., and Fogel', Ya. M. (1976*a*), *Nucl. Instr. Meth.* **132**, 415.
Kiyan, T. S., Gritsyna, V. V., and Fogel', Ya. M. (1976*b*), *Nucl. Instr. Meth.* **132**, 435.
Kiyan, T. S., Gritsyna, V. V., and Fogel', Ya. M. (1978), *Z. Phys.* **A285**, 257.
Knudson, A. R., Nagel, D. J., Comas, J., and Hill, K. W. (1978), *Nucl. Instr. Meth.* **149**, 507.
Kuduk, R., Kulik, M., Maczka, D., Pyszniak, K., and Zuk, J. (1989), *Phys. Stat. Sol. (a)* **112**, 799.
Kuduk, R., Kulik, M., Maczka, D., and Kobzev, A. P. (1990), in: *Proc. Int. Conf. Ion Implantation and Ion Beam Equipment*, Elenite, Bulgaria, 1990 (D. S. Karpuzov, I. V. Katardjiev, and S. S. Todorov, Eds.) (World Scientific, Singapore), p. 527.
Larsen, P., and Veje, E. (1983), *Phys. Rev. B* **28**, 5011.
Lindhard, J., Nielsen, V., and Scharff, M. (1968), *Kgl. Danske Videnskab. Selskab, Mat.-Fys. Medd.* **36**, No. 10.
Loxton, C. M., MacDonald, R. J., and Martin, P. J. (1980), *Surf. Sci.* **93**, 84.
MacDonald, R. J. (1982), *Appl. Surf. Sci.* **13**, 63.
MacDonald, R. J., and Garrett, R. F. (1978), *Surf. Sci.* **78**, 371.
MacDonald, R. J., and Martin, P. J. (1977*a*), *Surf. Sci.* **66**, 423.
MacDonald, R. J., and Martin, P. J. (1977*b*), *Surf. Sci.* **67**, 237.
MacDonald, R. J., Garrett, R. F., and Martin, P. J. (1978), *Surf. Sci.* **75**, L155.
MacDonald, R. J., Loxton, C. M., and Martin, P. J. (1981), in: *Inelastic Particle–Surface Collisions* (E. Taglauer and W. Heiland, Eds.) (Springer-Verlag, Berlin), p. 224.
Martin, P. J., and Loxton, C. M. (1980), *Rad. Effects Lett.* **50**, 161.
Martin, P. J., and MacDonald, R. J. (1977), *Surf. Sci.* **62**, 551.
Martin, P. J., Berzins, L., and MacDonald, R. J. (1980), *Surf. Sci.* **95**, L277.
Mériaux, J. P., Goutte, R., and Guillaud, C. (1972), *J. Radioanal. Chem.* **12**, 53.
Mériaux, J. P., Goutte, R., and Guillaud, C. (1975), *Appl. Phys.* **7**, 313.
Mogab, C. J., and Lugujjo, E. (1976), *J. Appl. Phys.* **47**, 1302.
Okutani, T., and Shimizu, R. (1979), *Surf. Sci.* **88**, L51.
Rausch, E. O., and Thomas, E. W. (1978), *Nucl. Instr. Meth.* **149**, 511.
Rausch, E. O., and Bazhin, A. I., and Thomas, E. W. (1976), *J. Chem. Phys.* **65**, 4447.
Reimann, C. T., Brown, W. L., Grosjean, D. E., and Nowakowski, M. J. (1992), *Phys. Rev. B* **45**, 43.
Reinke, S. (1991), Diplomarbeit, Universität Bielefeld.
Reinke, S., and Hippler, R. (1992), *Nucl. Instr. Meth.* **B67**, 620.
Reinke, S., Rahmann, D., and Hippler, R. (1991), *Vacuum* **42**, 807.
Robinson, M. T. (1990), *Nucl. Instr. Meth.* **B48**, 408.
Schiwietz, G., Schneider, D., Clark, M., Skogvall, B., DeWitt, D., and McDonald, J. (1993), *Rad. Effects Def. Sol.* **127**, 11.
Sigmund, P. (1969), *Phys. Rev.* **184**, 393.
Snoek, C., Van der Weg, W. F., and Rol, P. K. (1964), *Physica* **30**, 341.
Snowdon, K. J. (1979), *Rad. Effects* **40**, 9.
Snowdon, K. J., Carter, G., Armour, D. G., Andresen, B., and Veje, E., (1979), *Surf. Sci.* **90**, 429.
Suchańska, M., Bazhin, A. I., and Konopelko, E. I. (1994), *Phys. Stat. Sol. (b)* **182**, 231.
Taglauer, E., and Heiland, W. (Eds.) (1981), *Inelastic Particle–Surface Collisions* (Springer-Verlag, Berlin).
Taglauer, E., Heiland, W., MacDonald, R. J., and Tolk, N. H. (1979), *J. Phys.* **B12**, L533.
Tanaka, Y., and Yoshino, K. (1963), *J. Chem. Phys.* **39**, 3081.
Templier, C., Jaouen, C., Rivière, J. P., Delafond, J., and Grilhé, J. (1984), *Comp. Rend., Acad. Sci. Paris* **299**, 613.

Thomas, E. W. (1980), *Prog. Surf. Sci.* **10**, 383.
Thomas, G. E. (1977), *Rad. Effects* **31**, 185.
Thomas, G. E. (1979), *Surf. Sci.* **90**, 381.
Thomas, G. E., and de Koning, B. R. (1978), *Chem. Phys. Lett.* **55**, 418.
Thomas, G. E., and de Kluizenaar, E. E. (1974), *Int. J. Mass Spectrom. Ion Phys.* **15**, 165.
Thomas, G. E., and de Kluizenaar, E. E. (1976), *Nucl. Instr. Meth.* **132**, 449.
Thompson, M. W. (1968), *Phil. Mag.* **18**, 377.
Tolk, N. H., Simms, D. L., Foley, E. B., and White, C. W. (1973), *Rad. Effects* **18**, 221.
Tolk, N., Tully, J. C., Heiland, W., and White, C. W. (Eds.) (1977), *Inelastic Ion-Surface Collisions* (Academic, New York).
Tsong, I. S. T. (1977), *Surf. Sci.* **69**, 609.
Tsong, I. S. T. (1978), *Surf. Sci.* **75**, 159.
Tsong, I. S. T. (1981), in: *Inelastic Particle–Surface Collisions* (E. Taglauer and W. Heiland, Eds.) (Springer-Verlag, Berlin), p. 258.
Tsong, I. S. T., and Liebert, R. B. (1978), *Nucl. Instr. Meth.* **149**, 523.
Tsong, I. S. T., and Tsuji, S. (1980), *Surf. Sci.* **94**, 269.
Tsong, I. S. T., and Yusuf, N. A. (1978), *Appl. Phys. Lett.* **33**, 999.
Tsong, I. S. T., and Yusuf, N. A. (1979), *Surf. Sci.* **90**, 417.
Van der Weg, W. F. (1978), in: *Material Characterization Using Ion Beams* (J. P. Thomas and A. Cachard, Eds.) (Plenum, London), p. 81.
Van der Weg, W. F., and Bierman, D. J. (1969), *Physica* **44**, 206.
Van der Weg, W. F., and Lugujjo, E. (1975), in: *Atomic Collisions in Solids,* Vol. 2 (S. Datz, B. R. Appleton, and C. D. Moak, Eds.) (Plenum Press, New York), p. 511.
Van der Weg, W. F., and Rol, P. K. (1965), *Nucl. Instr. Meth.* **38**, 274.
Varga, P. (1987), *Appl. Phys.* **A44**, 31.
Vasiljeva, E. K. (1991), *Nucl. Instr. Meth.* **B54**, 441.
Veje, E. (1984), *Nucl. Instr. Meth.* **B2**, 218.
Veje, E. (1988), *Nucl. Instr. Meth.* **B33**, 543.
Veje, E. (1990), *Nucl. Instr. Meth.* **B48**, 581.
vomFelde, A., Fink, J., Müller-Heinzerling, Th., Pflüger, J., Scheerer, B., Linker, G., and Kaletta, D. (1984), *Phys. Rev. Lett.* **53**, 922.
White, C. W. (1978), *Nucl. Instr. Meth.* **149**, 497.
White, C. W., and Tolk, N. H. (1971), *Phys. Rev. Lett.* **26**, 486.
White, C. W., Simms, D. L., and Tolk, N. H. (1972), *Science* **177**, 481.
White, C. W., Simms, D. L., and Tolk, N. H. (1974), in: *Characterization of Solid Surfaces* (P. F. Kane and G. R. Larrabee, Eds.) (Plenum, New York), p. 641.
White, C. W., Simms, D. L., Tolk, N. H., and McCaughan, D. V. (1975), *Surf. Sci.* **49**, 657.
White, C. W., Tolk, N. H., Kraus, J., and Van der Weg, W. F. (1976), *Nucl. Instr. Meth.* **132**, 419.
White, C. W., Thomas, E. W., Van der Weg, W. F., and Tolk, N. H. (1977), in: *Inelastic Ion–Surface Collisions* (N. H. Tolk, J. C. Tully, W. Heiland, and C. W. White, Eds.) (Academic, New York), p. 201.
Wille, U. (1993), *Nucl. Instr. Meth.* **B79**, 132.
Winter, H., and Zimny, R. (1988), in: *Coherence in Atomic Collision Physics* (H. J. Beyer, K. Blum, and R. Hippler, Eds.) (Plenum, New York), p. 283.
Winters, H. F., and Kay, E. (1972), *J. Appl. Phys.* **43**, 794.
Winters, H. F., and Sigmund, P. (1974), *J. Appl. Phys.* **45**, 4760.
Wright, R. B., and Gruen, D. M. (1980*a*), *Nucl. Instr. Meth.* **170**, 577.
Wright, R. B., and Gruen, D. M. (1980*b*), *J. Chem. Phys.* **73**, 664.
Yu, M. L. (1979), *Surf. Sci.* **90**, 442.
Yu, M. L. (1988), in: *Secondary Ion Mass Spectrometry (SIMS)*, Vol. VI (A. Benninghoven, A. M. Huber, and H. W. Werner, Eds.) (Wiley, New York), p. 41.
Yu, M. L., (1991), in: *Sputtering by Particle Bombardment*, Vol. III (R. Behrisch and K. Wittmaack, Eds.) (Springer-Verlag, Berlin), p. 91.
Zivitz, M., and Thomas, E. W. (1976*a*), *Phys. Rev. B* **13**, 2747.
Zivitz, M., and Thomas, E. W. (1976*b*), *Nucl. Instr. Meth.* **132**, 411.

6

Electroluminescence

H.-E. Gumlich, A. Zeinert, and R. Mauch

1. INTRODUCTION

In contrast to earlier definitions it is understood nowadays that electroluminescence (EL) is the direct, nonthermal generation of light resulting from the application of an electric field to a material. The word "nonthermal" implies that incandescence is excluded. Also the luminescence owing to gas discharges, which was considered part of EL at the beginning of this century in the relevant chapters in some books, is no longer included within this definition. Today, the defining feature of EL is that the source of energy is the electric field and the conversion of electric energy into light is unlike that in cathodoluminescence (CL), where there is a multistage process with acceleration of electrons in a vacuum and the generation of secondary electrons.

EL occurs basically in two forms, known as injection EL and high-field EL. In injection EL, light is emitted upon recombination of minority and majority carriers across the bandgap of crystals. High-field EL consists of excitation of luminescence centers by majority charge carriers accelerated under the action of strong electric fields of the order of 1 to 2 MeV/cm. The field may be either ac or dc. The active layer can consist of a doped semiconductor and in addition either a powder (embedded in a matrix) or an organic or inorganic thin film.

The history of EL starts as early as the beginning of this century. It is not easy to judge who was really the first to observe the emission of light owing to the direct action of electric fields on semiconductors. The first reports on the subject were all purely phenomenological. One of these earliest descriptions was given by Round (1907). The temporary emission of light under the action of an electric field (Gudden-Pohl-Effect) was described in 1920 (Gudden and Pohl, 1920). Nowadays this effect is explained by the emptying of electron traps, that are filled probably by visible or ultraviolet irradiation, by electric fields.

Coustal (1934) reported on a permanent diminution of the luminescence intensity as long as an electric field was applied during excitation. The opposite effect

---

H. E. Gumlich • Institut für Festkörperphysik, Technische Universität, D-10623, Berlin, Germany. *A. Zeinert* • Université de Picardie Jules Verne, F80035 Amiens, France. *R. Mauch* • Heinrich Hertz Institut, 10117 Berlin, Germany.

Luminescence of Solids, edited by D. R. Vij. Plenum Press, New York, 1998.

was described by Destriau and Destriau (1954), who observed an enhancement of the luminescence of x-ray excited II–VI compounds containing Mn during the action of an ac field. The real beginning of EL is certainly connected with the name of Georges Destriau, who discovered the permanent excitation of luminescence of II–VI powders by ac fields (Destriau, 1936). In the 1950s there was some confusion in terminology concerning the combined action of irradiation and electric fields. Nowadays, the expression *photoelectroluminescence* is used whenever the excitation mechanism of the electric field is influenced by additional irradiation, and *electrophotoluminescence* is the term employed when the excitation of the luminescence by irradiation is controlled by electric fields, leading either to an enhancement or to a quenching effect of the luminescence emission (Gumlich, 1970). A comprehensive description of the state of the art including a complete bibliography of the first period of EL research can be found in the books of Ivey (1966) and Henisch (1962).

A turning point in the studies on EL came with the work of Inogouchi, who was able to establish thin-film EL (TFEL) cells based on ZnS:Mn, which had both high efficiency and a reasonable lifetime for the flat panels (Inoguchi *et al.*, 1974). The key point of the Inoguchi structure is that the insulators completely encapsulate the active layers and serve as buffers, shielding the device from breakdown when local current densities are high (Fig. 1). The highly reliable operation ensured by this kind of structure has two important consequences:

1. From the point of view of applications these structures are promising candidates to serve as flat videoscreens. Yellow-emitting flat panels with ZnS:Mn as the active layer have been commercialized for more than 15 years. More recently a polychrome display (green, yellow, red) based on a filter system was put on the market by PLANAR (Barrow *et al.*, 1994).
2. From the aspect of fundamental research, the reliability of the structure and the fair reproducibility of the device characteristics that are due to improved thin-film deposition techniques help one to investigate the physical aspects of high-field EL more precisely.

EL research is not restricted to ZnS as the host matrix for the active layer. A group of wide bandgap sulfides and selenides that show good EL qualities are now under consideration. As it has turned out that a full-color EL display could not be achieved using only ZnS or ZnSe, alkaline-earth chalcogenides (AECs), such as SrS, SrSe, and CaS were also investigated. However, despite considerable progress in recent years, the problem of finding an efficient blue luminophor in order to get

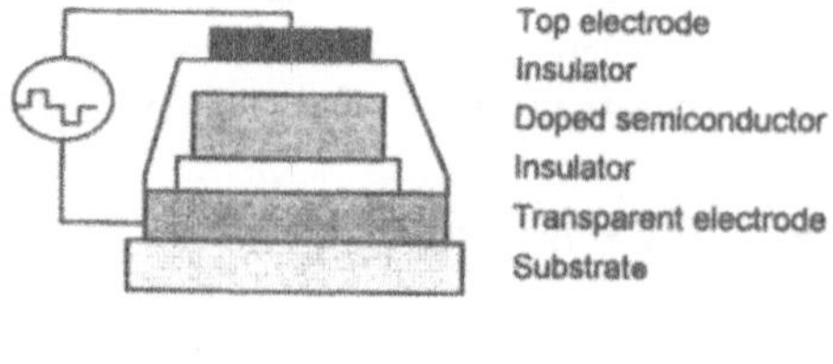

Figure 1. A cross-sectional view of an ac thin-film EL device.

full-color displays is still not totally solved, as the AECs are not sufficiently efficient in the blue range.

Although a slightly different EL behavior for these materials as compared to that with ZnS has been reported in some cases, the underlying mechanisms of high-field EL remain the same: Charge carriers are (1) injected into the active layer, (2) accelerated toward optical energies, and (3) excite luminescence centers.

Therefore, much of the physics of ZnS-type devices can be applied to devices based on AECs. However, it should be noted that even though the ZnS matrix is by far the most studied and best understood EL material, there are still many open questions, in particular, questions concerning the electronic behavior under high-field operation.

For several reasons the primary concern of the present chapter should be high-field EL of ZnS:Mn: First, this substance still shows the most efficient and most stable luminescence by excitation in ac fields. Second, most studies on the physics of EL have been done on ZnS:Mn, and even though some basic problems are not yet definitely solved, it nonetheless provides a model suitable for use with other substances, e.g., ZnSe and SrS. Third, the literature on the EL of light-emitting diodes based on III–V compounds is so vast that there is not enough space available for comprehensive coverage of the subject. From the field of injection luminescence only organic EL is considered briefly, because at present this topic is normally included in the workshops and conferences on EL in order to compare its development with regard to the future of flat panels. The reader interested in light-emitting diodes of III–V compounds is referred to special review articles, but the present chapter focuses on the mechanisms of high-field EL.

2. ZnS-TYPE ELECTROLUMINESCENCE

In this section, basic aspects of high-field EL in ac-driven thin-film ZnS-type devices will be reviewed. After an overview of the main physical processes (Section 2.1) there will be a closer look at some topics that are presently under discussion. The following steps will be considered: injection of carriers from interface states, acceleration and multiplication of carriers, excitation of luminescence centers, and finally the emission of light (Fig. 2).

2.1. Phenomenology and Basic Physical Processes

2.1.1. Electric Processes

Above a certain threshold voltage of the order of 100 V (the exact value depends on the specific sample characteristics and the chosen drive waveform) one observes a charge transfer across the semiconductor layer. Just above the threshold region, the charge transfer level increases by many orders of magnitude with increasing voltage. At high transferred charge one can observe an approximately linear increase with V. This region is sometimes called the "saturation region" in the

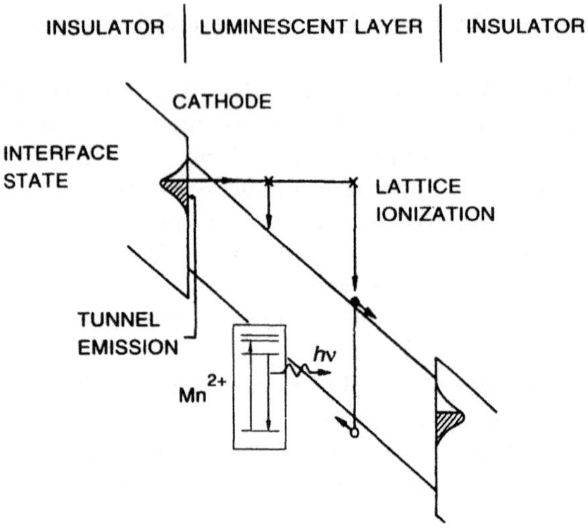

Figure 2. Band diagram illustrating the main steps in inorganic EL.

literature, because of the shape of the semilogarithmic representation. An increase in V beyond this region leads to destructive breakdown in the sample.

The current density measured in the external circuit is (Neyts and De Visschere, 1992; Bringuier and Geoffroy, 1992):

$$j = \frac{C_i C_s}{C_i + C_s}\frac{dV}{dt} + \frac{C_i}{(C_i + C_s)d_s}\int_0^{d_s} j_s(x, t)dx \tag{1}$$

where d_s designates the semiconductor thickness and C_i and C_s are, respectively, the total insulator and phosphor capacitances per unit area. The first term in Eq. (1) corresponds to the displacement current. This component can be removed by means of an electric bridge (Chen and Krupka, 1972). By connecting a reference capacitor with the capacitance $C = C_i C_s/(C_i + C_s)$ in series with the sample, one substracts the current density j_r flowing in the reference capacitor from the total current j measured in the external circuit. This permits one to get the second term of Eq. (1), which is due to the conduction current density j_s in the semiconductor layer.

During the current flow, the charge carriers cross the ZnS layer under the action of the electric field and are captured either in the bulk or at the opposite insulator–semiconductor (IS) interface: anodic in the case of electrons and cathodic in the case of holes. In contrast to dc-driven EL devices the carriers cannot leave the semiconductor because of the IS interface potential barrier. At the end of the voltage pulse, there will be an electric polarization in the semiconductor, which stems from stored charge at the IS interfaces and in the bulk. This polarization will add to the field across the ZnS layer on the next opposite polarity pulse. The polarization is related to an internal dipole moment per unit area (Bringuier and Geoffroy, 1992):

$$D = \int_0^{d_s} x\rho(x)dx \tag{2}$$

where $\rho(x)$ designates the charge density per unit volume.

In most cases, D will not give the number of the implied charge carriers during the charge transfer. If it is assumed that two electrons start from the "middle" of the semiconductor layer at $d_s/2$ (leaving two positive charges in the bulk) and are finally stored at the anodic IS interface, one would measure the same value for D if only electrons have been injected from the cathodic IS interface and are stored at the anodic counterpart.

The dipole moment D can be measured by a sensing capacitor C_r in series with the EL sample (Chen and Krupka, 1972; Smith, 1981) (inset of Fig. 3). This kind of circuit was used for the first time by Sawyer and Tower (1930) to study the behavior of ferroelectrics. The relation between the stored charge Q collected at the sensing capacitor electrodes in the external circuit and the charge distribution D (per unit area) in the semiconductor is (Neyts and De Visschere, 1992; Bringuier and Geoffroy, 1992):

$$\frac{Q}{S} \equiv Q_r = \frac{C_i}{C_i + C_s}\frac{D}{d_s} \tag{3}$$

where S is the surface of the EL sample.

In the steady state, there is a residual polarization charge P_1 still acting at a given pulse because of the earlier charge transfers. During the charge transfer under the electric pulse, the sign of the polarization changes (Fig. 3). If D_1 is the dipole moment just before a given pulse, and D_2 the one just after, then the transferred charge Q_t can be defined as the difference between P_1 and the polarization charge after a given pulse P_2:

$$Q_t \equiv P_2 - P_1 = \frac{(Q_{r_2} - Q_{r_1})(C_i + C_s)}{C_i} = \frac{D_2 - D_1}{d_s} \tag{4}$$

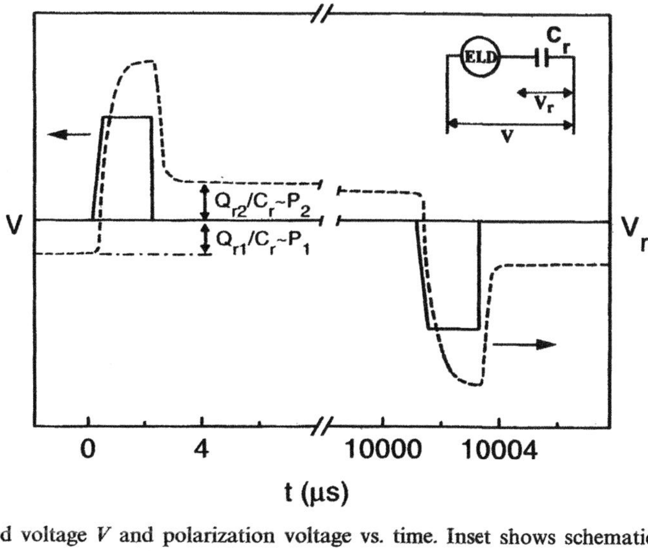

Figure 3. Applied voltage V and polarization voltage vs. time. Inset shows schematically the Sawyer–Tower circuit. From V_r one can deduce the residual polarization charge in the semiconductor before (P_1) and after the electric pulse (P_2).

The charge transfer is related to another aspect concerning the electric processes in ac-driven EL samples: the determination of the electric field inside the semiconductor. When no charge is stored in the sample, the relation between the applied voltage V and the field F_s in the semiconductor layer is given simply by the distribution of the voltage among the different capacitors:

$$F_s = \frac{C_i}{(C_i + C_s)d_s} V \tag{5}$$

The relation between F_s and V is rather simple even in the presence of charge transfer, provided that carrier supply and trapping is restricted to the IS interfaces. In this case, F_s is homogeneous along the ZnS layer depth and is counterbalanced by the homogeneous polarization field of the transferred charge:

$$F_s = \frac{V_s}{d_s} = F_s(t) = \frac{1}{(C_i + C_s)d_s}\left[C_i V(t) - \int_0^t j_s(t)dt - P_1\right] \tag{6}$$

with

$$Q_t = \int_0^{t=T} j_s(t)dt \tag{7}$$

where T is the pulse duration.

In general, the bulk is not electrically neutral owing to stored charge (ionized donors, filled traps) and the electric field is no longer uniform. In this case, Eq. (6) provides only a mean value of F. In Section 2.4 some consequences of the field nonuniformity will be described.

2.1.2. Light-Emission Mechanisms

The light emission that is observed parallel to the charge transfer in ZnS doped with luminescent impurities is characteristic for the dopant (Fig. 4). For the most frequently investigated luminescence centers of ZnS-based ACTFELD (ac thin-film EL devices), the Mn ion (Mn^{2+}) and the Tb ion (Tb^{3+}), the emissions correspond to transitions of internal orbitals. The two ions, Mn^{2+} and Tb^{3+}, substitute for the Zn ion in the semiconductor. In order to compensate for the surplus charge of Tb^{3+} one often introduces Tb as a complex center such as TbF or TbF_3 in the ZnS matrix (Krupka and Mahoney, 1972).

One important aspect of high-field EL concerns the nature of the excitation process of the luminescence centers. In fact, there are two possible ways for the charge carriers to excite the dopants: (1) direct excitation per (inelastic) impact, or (2) indirect excitation (the incident charge carrier excites a center of the acceptor–donor type, which subsequently recombines). The recombination energy is then transferred to a close luminescent center. The latter process is important with ZnS:Mn,Cu powder EL (PEL) samples (Gumlich, 1981) but for ACTFELD direct-impact excitation seems to be the dominant mechanism. In the case of ZnS:Mn, Tanaka *et al.* (1976) compared the photoluminescence (PL) power efficiency η_{PL} (Mn luminescence stimulated by band-to-band excitation 3.68 eV/illumination input power) with the EL power efficiency η_{EL} (luminescence/power). They concluded that

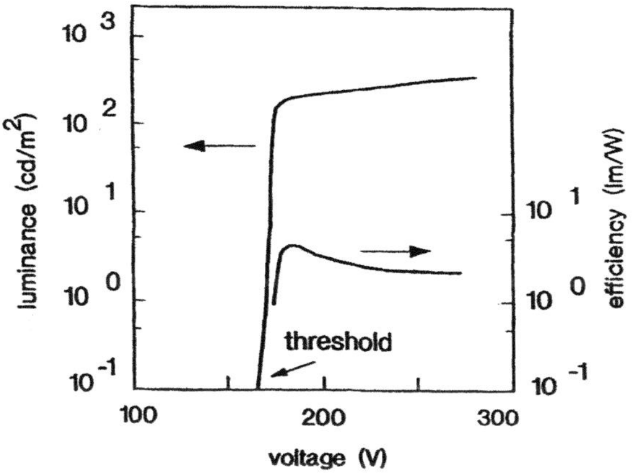

Figure 4. Luminance and efficiency vs. applied voltage.

the values of η_{PL} and η_{EL} are incompatible with an indirect-excitation mechanism for EL. In the case of Tb, the evaluation is complicated by the nature of the center complex (TbF, TbF_3). Benalloul (1989) showed by means of time-resolved spectroscopy that at least in ZnS:TbF_3 samples, impact excitation prevails.

Apart from the luminescence owing to intentionally introduced impurities such as Mn or Tb one can observe a broadband light emission extending from around 330 nm to the red spectral region (Smith, 1981). This emission is also observed in undoped samples and has a very short decay time (< 50 ns), which implies that it is only observable during the charge transfer across the ZnS. Owing to its emission maximum in the blue region, it is often called blue emission, but one has to stress that the spectrum is not identical to the well-known PL spectrum of the blue self-activated emission in ZnS (Prener and Weil, 1959). EL emission is referred to in the following as ZnS emission. The ZnS-emission intensity is very weak compared to the luminescence observed for centers such as Mn or Tb. Its absolute intensity decreases with increasing doping concentration. It is not probable that activator–donator centers are involved in this luminescence emission because of the constancy of the spectrum in time-resolved spectroscopy. It has been suggested that the emission is due to interband transitions of high-energy electrons (Skolnick, 1981) but its nature is not completely understood.

2.1.3. *Efficiencies*

In order to discuss luminescence efficiency (Fig. 4) the main processes of high-field EL in ACTFELD can be summarized as follows: (1) electron injection into the conduction band of the semiconductor, (2) subsequent acceleration of the electrons in the high electric field to optical energies, (3) impact excitation of the luminescence center, (4) (non-)radiative deexcitation of the centers, (5) recapture of the electrons, and finally (6) the generated photons must "find the way out" of the sample and only a fraction of the generated photons will be outcoupled through the surface. These processes are reflected by three partial efficiencies, which have been defined in the case of the simplified model, where charge injection and storage are assumed to occur only at the IS interfaces (Mach and Müller, 1984):

THE EXCITATION EFFICIENCY

$$\eta_{\text{exc}} = \frac{\text{Number of excited luminescence centers (cm}^{-3})}{Q_t/q \ (\text{cm}^{-2})}, \tag{8}$$

where Q_t is the transferred charge defined in Eq. (4), and q is the elementary charge.

THE RADIATIVE EFFICIENCY

$$\eta_{\text{rad}} = \frac{\text{Number of centers decaying radiatively (cm}^{-3})}{\text{Number of excited centers (cm}^{-3})} \tag{9}$$

THE OUTCOUPLING EFFICIENCY

$$\eta_{\text{out}} = \frac{\text{Number of photons emitted through the surface (cm}^{-2})}{\text{Generated photons (cm}^{-3})} \tag{10}$$

The product of these partial efficiencies is the total efficiency:

$$\eta_{\text{exc}}\eta_{\text{rad}}\eta_{\text{out}} = \eta_{\text{tot}}$$
$$\equiv \frac{\text{Number of photons emitted through the surface (cm}^{-2})}{Q_t/q \ (\text{cm}^{-2})} \tag{11}$$

When electron supply and trapping occurs only at the IS interfaces, Q_t/q is equal to the number of transferred electrons. In the general case, where the bulk contributes to electron supply and trapping, Q_t/q is no longer the number of transferred electrons but rather the sum of the distances x_i, covered by the individual electrons during their transfer across the semiconductor layer.

From different definitions it follows that η_{exc} and η_{rad} relate to physical processes in the semiconductor layer while the outcoupling η_{out} can be considered as an external partial efficiency. The value of η_{out} depends on the optical index of the semiconductor and to some extent on the surface roughness of the active layer (grain size): a rougher surface (coarse grains) favors light outcoupling as evidenced by surface–plasmon reflection measurements (Mach *et al.* 1990). In most cases η_{out} lies between 0.1 and 0.4, but it is difficult to determine the exact value.

2.1.4. Open Issues

Among the different physical processes noted above only the transitions that are responsible for the observed emissions of the Mn^{2+} and Tb^{3+} ions are well understood (Gumlich, 1981; Bénalloul, 1989). The phenomenology of the non-radiative processes has been established in ZnS:Mn even if the nature of the nonradiative recombination centers remains speculative (Benoit *et al.*, 1990). All of these aspects concern the radiative efficiency defined in Section 2.1.3. More recent research has been focused on questions related to the excitation efficiency: (1) the origin of the injected charge carriers, (2) the energy distribution of the carriers, (3) the impact cross section of the luminescence centers, and (4) the distribution of the electric field in the semiconductor. These aspects are treated in the following sections.

2.2. Free Charge Carrier Generation and Trapping

2.2.1. Preliminary Considerations

The presence of free charge carriers is a necessary condition for high-field EL. In capacitive-coupled devices such as ACTFELD, charge carriers cannot be injected from the external electric circuit. Only the semiconductor and the IS interfaces remain as potential carrier sources. The same argument holds for charge carrier trapping after the transfer in the semiconductor layer. What kind of mechanisms are conceivable in the two cases?

From the charge–voltage curves of ACTFELD it is clear that one must look for a generation mechanism that is related to the presence of a high-threshold field. As far as the interfaces are concerned, tunneling injection or field-assisted thermal ionization (Schottky emission) from deep states are the main possible processes. These two injection mechanisms are also possible in the bulk as Poole–Frenkel emission, which is the bulk analogue of Schottky emission (Shockley and Last, 1957). Furthermore, as far as the bulk is concerned, charge carriers can be also furnished by impact ionization of deep centers and electron–hole pair generation. Among all these processes, field-assisted thermal ionization seems to be less probable in ACTFELDs because the set-in of the charge transfer curves is rather insensitive to temperature (Smith, 1981) compared to what one would expect for Schottky or Poole–Frenkel emission. It is mainly for this reason that tunneling has been assumed to be the predominant carrier injection mechanism.

2.2.2. Interface States

For most authors, liberation of charge carriers trapped at the IS interfaces is thought to be the main carrier generation mechanism in ACTFELD for the following reasons: The main argument is based on the influence of the nature of the insulator. It is well-established that the threshold field for charge transfer can be modified by the nature of the insulators and its deposition methods (Smith, 1981). Moreover one observes a polarity-dependent efficiency when two different insulators are used (Yang and Owen, 1983). However, these arguments hold only to a certain point. In fact, by choosing another insulator or deposition method, the growth conditions of ZnS on the first insulator can be modified and influence the bulk quality, and so the bulk flaw density. This effect may be restricted to a region near the bottom IS interface, resulting in an asymmetric behavior. There are several reasons for expecting a significant density of states at the IS interface: (1) the periodic lattice structure is interrupted and there are dangling chemical bonds, which cause electronic states within the forbidden gap (Tamm states); and (2) there is a mismatch between the insulator material (amorphous or crystalline) and the semiconductor lattice.

Beyond these rather general considerations the specific nature of the interface states is virtually unknown. Assuming that the injection mechanism is Fowler–Nordheim tunneling (temperature-insensitive), one can estimate the energy-level depth corresponding to typical threshold fields between 1 and 2 MV/cm. It turns out that the energy levels should be located between 0.6 and 1 eV below the bottom of the conduction band (Smith, 1981). Thermally stimulated current measurements provided trap-level depths between 0.8 and 1.8 eV for the Y_2O_3–ZnS interface (Sano and Nunomura, 1989). For the SiON–ZnS interface trap levels deeper than 0.7 eV below the conduction-band minimum have been reported (Kobayashi *et al.*, 1992). Sahn *et al.* (1991) calculated the energy levels caused by anion vacancy states in the insulator–ZnS interface, and found thermal activation energies to be between 0.6 and 1.7 eV, depending on the insulator material.

2.2.3. Bulk States

Similar to the case of interface tunneling, charge carriers can be generated in the semiconductor layer by tunneling from bulk states in the forbidden gap into the conduction band. As in the former, the relevant states must be deep ($E_c - E \gg kT$) because of the high-threshold fields (~ 1 MV/cm) for charge transfer in ACTFELD.

Energy levels can be generated in the bulk by un-intentionally introduced impurities [e.g., O, F, Al, and Cl (Landoldt-Börnstein, 1988)] or by intrinsic flaws in the ZnS lattice. The term "flaw" is employed for any permament perturbation of the crystal lattice that results in a localized state. Grain boundaries may add to the generation of deep levels. Intrinsic flaws stem from defects involving Zn^{2+} and S^{2-} ions. Displaced Zn^{2+} ions in interstitial sites act as donors while Zn^{2+} vacancies behave as acceptors (Swang *et al.*, 1969; Kukimoto *et al.*, 1976). A sulfur vacancy acts as a divalent donor with activation energies of $E(V_S^{2-} \rightarrow V_S^-) = 0.14\,\text{eV}$ and $E(V_S^- \rightarrow V_S) = 0.6\,\text{eV}$ (Landoldt-Börnstein, 1988). It should be noted that virtually none of the experiments on flaws in ACTFELD can distinguish between interface and bulk as the polarization measured in the Sawyer–Tower circuit [Eq. (3)] is only a mean value of the whole insulator–ZnS system. The same is true for traps detected by different techniques (thermal-released, photon-released charges). These traps are currently ascribed to IS interface states although none of the experiments has ruled out the possibility that at least some of the levels are localized in the bulk.

2.2.4. Impact Ionization

Two conditions have to be fullfilled for ionization processes by hot carriers: The kinetic energy has to be high enough and the total momentum of the impact partners has to be conserved. In contrast to tunneling injection, ionization by hot carriers is a multiple-feature process. The impact ionization is by definition a three-body process and conservation of momentum implies that the threshold energy is in general higher than E_{gap}. The accurate calculation of the ionization probability depends on the band structure but, in some cases, a simplified band-structure approximation provides reliable estimates.

The phenomenological description of impact ionization by charge carriers can be provided by the ionization coefficient α_i, designating the number of generated electron–hole pairs per unit distance. One distinguishes α_i^n for electrons from α_i^p for holes. Measurements on ZnS Schottky diodes suggest that band-to-band ionization occurs at electric fields around 2 MVcm^{-1}. The value of α_i^n varies between 7×10^3 and 5×10^4 cm^{-1} in a field range F between 1.6 and 2 MV/cm (Thompson and Allen, 1987). Recent transient measurements on partly doped ACTFELD indicate that α_i^n does not exceed this order of magnitude in ZnS-based samples (Zeinert *et al.*, 1997).

2.3. High-Field Charge Transport in ZnS

For a given dopant concentration and free charge carrier density, the impact excitation probability depends on the energy-dependent cross section $\sigma(E)$ of the particular luminescence center and the energy distribution of the charge carriers. Apart from the specific problem of determining $\sigma(E)$, it is evident that charge carriers that are able to excite luminescent centers in the visible range must have energies of the order of several electron volts. As these energies are much higher than the mean energy at thermal equilibrium $\frac{3}{2}kT(\cong 0.04\,\text{eV}$ for $T = 300\,\text{K}$, k being the Boltzmann constant) the name "hot" charge carriers is justified (Reggiani, 1985). Central problems in EL are to explain how charge carriers can heat up to such high energies in a material and to determine the fraction of hot charge carriers for a given electric field.

It should be emphasized here that a transport theory alone cannot automatically predict the energy distribution of the free charge carriers in ACTFELD because of the uncertainty of the electric field. Practically, it is difficult to separate the two aspects: transport mechanism and inhomogeneous field. For clarity's sake, however, it was preferable in this chapter to treat them separately.

The description here will be limited to a single-carrier picture in the sense that only the electron transport will be treated, but this should be adequate to explain the observations in an n-type material such as ZnS. In n-type materials, hole injection is predominantly realized by the generation of electron–hole pairs in lattice ionization. Generally, one assumes that these holes are rapidly captured by hole traps in the bulk so that they cannot drift to the cathode. However, one should note that more recent measurements under ultraviolet irradiation suggest that there are hints of a contribution of holes to the charge transfer in ZnS-based ACTFELD (Corlatan *et al.*, 1994).

2.3.1. Framework of Transport Theories

By far, most theories about charge carrier transport in semiconductors are based on the band model, where electronic transport is controlled by diverse scattering mechanisms, some of which are sketched out in the following paragraphs.

(a) Electron–Electron Interaction. The electron–electron scattering rate depends on the density n of conduction electrons. It has been estimated that a density of at least 10^{14} cm^{-3} is necessary for competing scattering times compared to other scattering mechanisms (Fröhlich, 1947). Typical values of current density J in

ACTFELD does not exceed 4 A/cm^2. By taking a rather low drift velocity v of $0.5 \times 10^7\,\text{cm}^{-1}$ one obtains $n = J/ev = 5 \times 10^{12}\,\text{cm}^{-3}$ (e is the electron charge). This is only a rough estimate but it can be assumed that electron–electron interaction does not play a significant role in high-field EL.

(b) Electron–Phonon Interaction. The electron–phonon (e–ph) interaction energy depends on the lattice strain. In materials with more than one atom per primitive unit cell one distinguishes acoustic and optical strains. The strains influence electrons in two distinct ways: Short-range deviations from the periodic potential cause practically instantaneous changes in energy, and one can evaluate this type of lattice scattering by means of a deformation potential. In the case of polar semiconductors the vibrations of oppositely charged ions give rise to long-range macroscopic electric fields in addition to deformation potentials. The interaction of an electron with these fields produces additional scattering. In the case of polar optical scattering only longitudinal optical (LO) modes are important because the field associated with transverse optical modes is small. Polar LO scattering is referred to as Fröhlich scattering (Fröhlich, 1954). This type of interaction favors scattering with phonons of low momentum so that only small changes in the electron momentum owing to the collisions are likely. Consequently the electrons will be scattered along the direction of the electric field.

(c) Impurity and Defect Scattering. In addition to e–ph scattering, electrons can scatter with defects. Scattering with neutral impurities is negligible at 300 K as the corresponding scattering rates are very low compared to e–ph scattering rates (Bhattacharyya *et al.*, 1993). Ionized-impurity scattering plays an important role for mobility in the low-field region (< 0.1 MV/cm) but the scattering rate decreases with electron energy and does not affect the transport behavior for fields $F \geqslant 1$ MV/cm, even for rather high concentrations ($10^{18}\,\text{cm}^{-3}$). (Bhattacharyya *et al.*, 1993).

(d) Semiclassical Approach. For a given band structure the semiclassical approach to charge carrier transport consists of solving the Boltzmann equation. One assumes that the deviation from a perfect periodic potential (scattering sources) can be described by a perturbation potential.

While the Boltzmann formalism is successful for the description of transport processes in the low-field range, there are some specific problems in the hot-electron regime. The Boltzmann approach assumes that collision events are localized in space and time. In fact, the scattering electron is actually moving under the combined force of the electric field and the scattering potential. Scattering occurs within and between accelerating states. In a high electric field the energy absorption by a charge carrier during this finite extent of the collision is $eF\tau v_D$ where τ is the time of the event, and v_D the drift velocity, and may become comparable to the uncertainty broadening $\Delta E = \hbar\tau^{-1}$. In this case, Fermi's Golden rule does not provide the correct scattering probability, and one has to include higher-order corrections (Barker, 1981; Arora, 1984). For a detailed description of the approximations and the limitations of the Boltzmann framework, the reader is referred to Dresden's (1961) exhaustive review.

In order to determine the transport behavior one often uses the Monte Carlo method, a procedure that consists of a simulation of the dynamics of charge carriers inside the semiconductor under given scattering probabilities. The times between two collisions and the scattering process involved in the simulation are selected according to a given probability describing the microscopic process (e.g., optical e–ph scattering) (Reggiani, 1985).

2.3.2. Models for High-Field Charge Transport in ZnS-Based ACTFELD

ACTFELD operate at electric fields above 1 MV/cm. In this range, e–ph interaction is the predominant scattering mechanism. All the models that rely on the band structure (see Sections 2.3.2a–c) converge at the same kind of scenario: at low electron energies, polar e–ph scattering is predominant; as the scattering rate decreases slightly with electron energy, the average electron energy increases with increasing field until electrons can occupy satellite valleys. Scattering to those valleys cannot be assured by polar scattering because of the required large change $\Delta\mathbf{k}$ of the crystal momentum for intervalley scattering. However, nonpolar deformation potential scattering, which allows large $\Delta\mathbf{k}$ is possible and this kind of interaction stabilizes the energy distribution. A very different approach to high-field transport in ZnS-like materials was proposed more recently in which the band structure is neglected and only polar e–ph scattering is present. This model predicts quasi-ballistic electron acceleration resulting in average electron energies greater than E_{gap} (Section 2.3.2d).

(a) Monte Carlo Calculations: Full-Band Structure. Brennan (1988) presented Monte-Carlo simulations for high-field electron transport in ZnS and ZnSe. The calculations include the details of the first two conduction bands derived from earlier pseudopotential calculations in the case of ZnS (Cohen and Bergstrasser, 1966). Furthermore, full-order treatment of the e–ph scattering mechanisms is taken into account, i.e., collision-broadening is included but the calculation provides "only" a 30% correction from the adiabatic approximation. This is a remarkable result because the scattering rates attain as much as $7 \times 10^{14}\,\text{s}^{-1}$, i.e., an electron mean free time around 3 fs (!), which indicates that the range of applicability of the Golden rule is wider than expected (Peierls, 1979; Bringuier, 1994*a*). Brennan (1988) found a steady-state drift velocity around 0.7×10^7 cm/s for fields $F > 100$ kV/cm. For $F = 1$ MV/cm, about 1% of the electrons have an energy $E \geqslant 2$ eV. With this very low fraction of hot electrons it is difficult to explain the high luminescence efficiency of ZnS:Mn and ZnS:Tb devices. The contradiction with the empirical data could be due to the specific band-structure calculation used by Brennan. In fact, this calculation underestimates the energy of the X-valleys and as a consequence intervalley scattering starts at lower electron energies as it should do. This yields a cooler energy distribution.

(b) Monte Carlo Calculations: Multivalley Approximation. Bhattacharyya *et al.* (1993) presented Monte Carlo calculations in ZnS, approximating the conduction band by nonparabolic multivalleys (Γ, L, X). In addition to e–ph scattering (polar optical, acoustic, intervalley), the model includes impurity scattering. As in the case of Brennan's calculations, a stable drift velocity is found around 0.7×10^7 cm/s. The impurity scattering does not affect the drift velocity v_D at high fields ($F > 10^5$ V/cm). At low fields, it results in lower values of v_D and is in better

agreement with experimental mobilities than Brennan's data. As to energy distribution, the authors came to the conclusion that at 2 MV/cm a significant fraction of electrons have energies that exceed 2 eV: some 26% at 1 MV/cm and 65% for 2 MV/cm. However, essentially no electrons with energies above 4 eV are found owing to the rather high density of states in the multivalley, which gives rise to an enhanced scattering rate. This would rule out band-to-band impact ionization but one has to keep in mind that the band structure is only approximated.

(c) Lucky Drift. An analytical approach to the high-field transport problem came from Bringuier (1991, 1994*a*), who used the lucky-drift theory developed by Ridley (1983). This theory distinguishes two different modes: a ballistic mode within the momentum-relaxation time τ_m and a drift mode during the energy-relaxation time τ_e. The model can only be applied if $\tau_e \gg \tau_m$, i.e., if each collision results in an appreciable loss of the momentum component in the transport direction but only a small loss of energy. That means, in this context, that one has to assume that the dominant cooling mechanism at high fields is not polar optical e–ph scattering, which is known to favor forward scattering, especially for high fields (Section 2.3.1) but deformation potential scattering. This is in accordance with the scenarios described by the Monte Carlo calculations described above; however, except for the effective mass, no details of the band structure are included. The mean free pathlength $\lambda = v_g \tau_m$ is derived from the group velocity v_g of a parabolic band and numerical calculations of the e–ph scattering rate $1/\tau$ (Brennan, 1988), which is approximately the momentum relaxation rate $1/\tau_m$ because each collision is assumed to result in an appreciable loss of momentum. The energy relaxation length λ_E depends on λ, the field F, and the phonon distribution. The probability for an electron to attain a certain energy depends in the lucky-drift formalism only on F, λ, and λ_E. In a generalized version of the lucky-drift model, where the nonparabolicity of the conduction band is taken into account, λ depends on the energy range (Bringuier, 1994*b*). For a mean field of $F = 2$ MV/cm about 50% of the electrons reach at least 2.7 eV. For recent developments of the lucky drift model of high-field transport the reader is referred to Bringier (1995, 1996, 1997).

(d) Ballististic Acceleration. According to some authors the conduction-band structure disappears under thin-film EL (TFEL) operating conditions (Mach *et al.*, 1990) and the polar optical e–ph interaction is the only energy loss mechanism unless the electrons have enough energy to generate electron–hole pairs. As the scattering rate of the polar optical e–ph interaction remains constant for increasing electron energy, above a critical field electrons gain more energy from the field than they lose to optical phonons, and the carriers "run away" to very high energies. The Monte Carlo calculations that are based solely on the polar optical e–ph interaction parabolic conduction-band structure show for ZnS, CaS and SrS "a ballistic spike transport" of the electrons above a critical field that depends on the material (Fitting and von Czarnowski, 1986; Fitting *et al.*, 1990). In the case of ZnS, nonstationary heating sets in at 0.2 MV/cm, and at 1 MV/cm the electrons are virtually ballistic (Fitting and von Czarnowski, 1986). After a drift distance of 100 nm the mean energy is about 9 eV.

The main objections expressed by the critics of the ballistic model is that the band structure and the nonpolar e–ph interaction are completely ignored. From the viewpoint of fundamental physics, one would like to understand why under TFEL operating conditions the conduction-band structure should disappear in certain II–VI materials. It should be kept in mind that the applied fields around $F \approx 1$ MV/cm = 0.01 V/Å are negligible compared to atomic fields and as indicated in Section 2.3.2a, the adiabatic approximation seems to hold even for high scattering rates.

(e) Conclusions. To sum up this overview of the transport models, almost all current theories of high-field transport in ZnS predict that at fields of the order of 2 MV/cm, a large fraction of the electrons can attain kinetic energies of 2 eV or more. For 1 MV/cm the values given by Brennan's (1988) Monte Carlo simulation are the lowest: only 1% of the electrons would have the impact energy for Mn. The Monte Carlo simulation of Bhattacharyya *et al.* (1993) as well as the lucky-drift approach provide a significant fraction of hot electrons even at 1 MV/cm. Band-to-band ionization is negligible in both Monte Carlo calculations and in the lucky-drift model (for $F = 2$ MV/cm).

In the models relying on the band structure, the nonpolar e–ph interaction described by the deformation potential is responsible for the stabilization of the electron energy distribution at high fields around 2 MV/cm. Despite this important result one should emphasize that the presented models can only be considered as reasonable approximations since the $\Gamma - X$ and $\Gamma - L$ deformation-potential constants are unknown in ZnS. Γ, X, and L are prominent points in the band-structure model (Kittel, 1971). Monte Carlo calculations generally use GaAs and InP values. However, these parameters are critical for the assessment of the electron energy distribution as a function of the applied field. These known constants for GaAs and InP typically vary by a factor of two. As a consequence, in the worst case, the electric field necessary for a given mean energy would be uncertain by a factor of four (!) (Bringuier, 1994*b* and references therein). Although this worst-case scenario is probably too pessimistic, it is a clear indication that any refinement of the present transport models seems questionable unless crucial parameters of ZnS are better known.

The model of Mach and Müller (Mach *et al.*, 1990) is the only one that predicts "all electrons hot" for fields around 1 MV/cm and a strong band-to-band ionization rate. There is as yet no experimental evidence to support the assumption that the influence of band structure is obsolete in ACTFELD or that the predicted band-to-band ionization rate is incompatible with present experimental data.

2.4. Nonuniform Field

2.4.1. General Remarks

In the previous section, it was more or less tacitly assumed that the electric field in the semiconductor is uniform. This picture holds if: (1) only IS interface states are involved in free charge carrier supply and trapping, and (2) the field distortion provoked by the charge flow is low. This is a direct consequence of the Poisson equation, which relates the electric field F_s and the charge density ρ:

$$\frac{\partial}{\partial x} F_s(x, t) = -\frac{\partial^2}{\partial x^2} U(x, t) = \frac{\rho(x, t)}{\varepsilon} \tag{12}$$

Taking into account the fact that in ACTFELD a free charge carrier density of $n = 10^{13}\ \text{cm}^{-3}$ is never exceeded, one calculates a field gradient of about 2 MV/cm^2. This means that for a 1-μm-thick layer the difference ΔF_s between the cathodic and anodic field is about 200 V/cm, i.e., 0.01% of the fields encountered in ACTFELD. Hence, it follows that field distortion by the current flow is negligible. However, as was outlined in Section 2.2, when the bulk contributes to the current, ionized centers remain in the semiconductor, generating a "fixed" space charge and as will be seen below, this charge can be large enough to create a significant field distortion. In the following "space charge" will be referred to only in this sense, with the caveat that it is associated with bound centers and not with free charge carriers.

From Eq. (12) one can immediately draw two conclusions about the field shape and the minimum charge density necessary for significant field gradients: (1) A positive space charge enhances the cathodic field region and reduces the anodic field region compared to the mean field; of course, the opposite is true for a negative space charge. Examples of positive space charges are trapped holes or ionized impurities, and negative space charges can be generated, for instance, by trapped electrons. (2) If one assumes a homogeneous density n of ionized centers, one needs $n = 10^{16}\ \text{cm}^{-3}$ for $\Delta F_s = 0.1 F_s$ (for $F_s = 2$ MV/cm and $d_s = 1\ \mu$m). This means about 10^{12} centers/cm^2 are necessary to introduce a 10% deviation from a typical mean field $F_s = 2$ MV/cm in a 1-μm-thick semiconductor layer.

2.4.2. Some Experimental Evidence

There is experimental evidence for nonuniformity of the electric field inside the ZnS layer under certain ACTFEL drive conditions. This phenomenon was first reported for ZnS:Mn devices (Marello and Onton, 1979; Marello *et al.*, 1981). The authors investigated probe layer samples. This kind of particular sample structure was proposed first by a French group (Benoit *et al.*, 1979). Instead of the whole ZnS layer only one thin layer is doped with the luminescent ions, a structure that permits one to investigate the efficiency as a function of the position in the semiconductor layer. Marello *et al.* (1981) found that the efficiency decreases across the ZnS layer downstream from the cathodic to the anodic region. The authors concluded that the electric field in the ZnS layer is higher near the cathode interface as the electrons impact-excite the luminescence centers more efficiently.

Benoit *et al.* (1993) went a step further, investigating the excitation efficiencies of ZnS:Mn and ZnS:TbF_3 samples as a function of the transferred charge Q_t in layers with good ZnS crystallinity grown by electron-beam evaporation. As for the internal efficiency, Marello *et al.* (1981) found the excitation efficiency η_{exc} to be lower in the anodic field region. Moreover, the difference increases with transferred charge and generally attains its maximum value before the breakdown point of the sample. The same type of results are observed in atomic layer epitaxy (ALE)-grown devices

(Zeinert *et al.*, 1997). Hence probe layer devices have provided direct experimental proof of the presence of a positive space charge in the semiconductor, which apparently varies with Q_t.

Space charge has been made responsible for the hysteretic behavior that can be observed under certain conditions in highly doped (>1 mol%) ZnS:Mn ACTFELD (Howard *et al.*, 1982). These samples exhibit a hysteresis slope of the charge–voltage and luminescence–voltage curves, which has been ascribed to the action of a positive space charge related to deep hole traps. This model has been reviewed by Jarem and Singh (1988), who introduced an ideal step function for the ionization coefficient.

The presence of a positive space charge is undoped ZnS is suggested by current characteristics over broad temperature ranges that exhibit behavior that is incompatible with simple charge transfer between IS interfaces (Bringuier and Geoffroy, 1992).

Some authors have claimed to have determined the nature of the positive space charge in ALE-deposited ZnS:Mn devices. They identify the space charge as arising from impact ionization of Zn vacancies (Douglas *et al.*, 1993). Their indirect argumentation is based on the assumptions that the observed blue-light emission in their devices can be ascribed to a donor–acceptor (Cl_s–V_{Zn}) recombination and that the Zn vacancy density is higher near the bottom IS interface.

As space charge is related to the ZnS layer, one expects that it can also be observed in dc-driven devices although the carrier injection mechanism is different and the space charge may not have the same consequences for device behavior as in ACTFELD. Beale (1990) showed indeed that dc-driven ZnS:Mn metal–insulator–metal (MIM) structures exhibit conduction by field ionization of deep traps in dynamic equilibrium with their repopulation by Coulomb-assisted capture.

3. HOST MATERIALS AND ACTIVATORS

3.1. Zinc Sulfide

3.1.1. ZnS:Mn

As ZnS:Mn is by far the most stable and efficient material in both EL and PL, its properties should be discussed in more detail. ZnS can crystallize in both Blende and Wurtzite structures, depending on the preparation conditions and on the amount of Mn. The forbidden gap, E_g of the two structures differ only slightly. For the Blende structure $E_g \simeq 3.7\,\text{eV}$, and for the hexagonal Blende structure some authors assume $E_g \simeq 3.6\,\text{eV}$, both at $T = 4.2\,\text{K}$ (data collected in Landoldt-Börnstein, 1982). The luminescence emission owing to Mn is assigned to the internal transition ${}^4T_1(G) \rightarrow {}^6A_1(S)$ of Mn^{2+} in tetrahedral coordination of the neighboring sulfur ions.

With an increasing amount x of Mn in $Zn_{1-x}Mn_xS$ at $x > 0.001$, five absorption and excitation bands are observed, which are also assigned to internal transitions of Mn (Gumlich *et al.*, 1966) and are reasonably explained by ligand field theory (Pohl and Gumlich, 1989, Fig. 5). There is only a small shift in the energy of the zero-phonon lines going from the Blende structure to the Wurtzite (Fig. 6). This

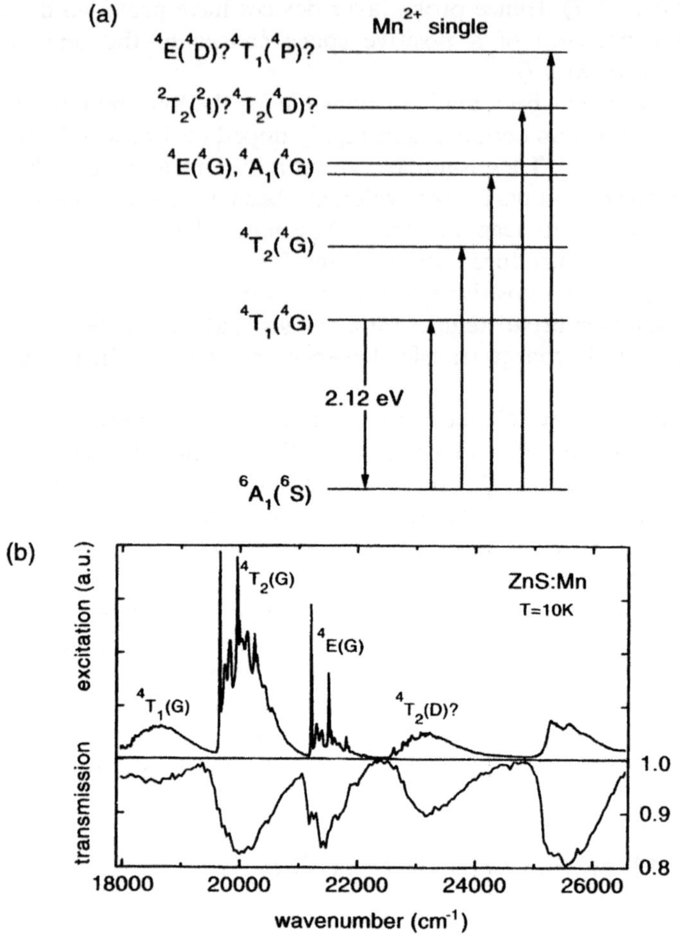

Figure 5. (a) Crystal-field split ground and excited states of Mn^{2+} in ZnS (Pohl, 1988). (b) Excitation and transmission spectra of ZnS:Mn (Pohl, 1988).

can be explained by the fact that the next neighbors (nn) and the next nearest neighbors (nnn) are the same in both structures, and only the third nearest neighbors of the Mn are different (Pohl *et al.*, 1984).

As a matter of fact, normally neither structure appears in pure form, but each in most cases is accompanied by stacking faults giving rise to characteristic zero-phonon lines (Busse *et al.*, 1979).

Emission occurs at a somewhat lower energy than the first excited level ($^4T_1$) above the ground state (2.24 eV) owing to a Stokes shift. Since the transition is spin- and parity-forbidden, the lifetime τ_0 of the isolated ion is rather long: 1.15 to 1.77 ms, depending on the ZnS crystalline structure (hexagonal, cubic) for PL (Gumlich, 1981). The observed decay times of TFEL correspond to these data. One finds 1.3 ms for thin films that have been grown by ALE (Benoit *et al.*, 1990). ALE films seem to

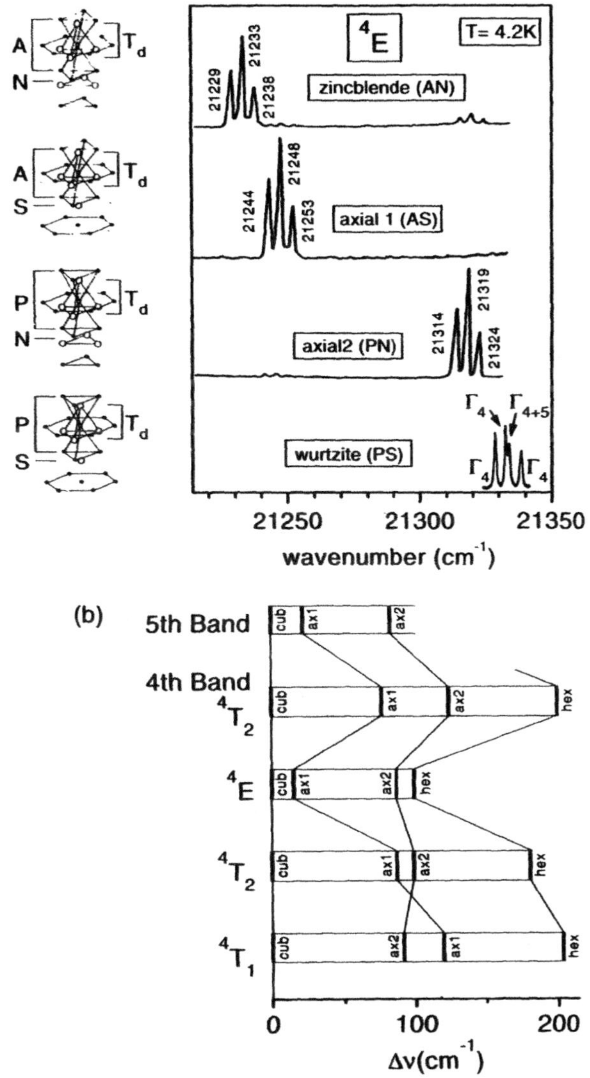

Figure 6. (a) Zero-phonon lines of excitation of the transition $^6A_1(S) \rightarrow {}^4E(G)$ of Mn^{2+} in polymorphic ZnS [emission due to the transition $^4T_1 \rightarrow {}^6A_1(S)$] (Gumlich, 1989). (b) Energy shift of the Mn^{2+} levels in polymorphic ZnS with respect to Mn^{2+} at cubic sites (Pohl *et al.*, 1984).

have a predominantly hexagonal structure (Busse *et al.*, 1983). Films grown by electron-beam evaporation, which are expected to have cubic structure, lead to 1.7 ms (Benoit *et al.*, 1988). When the Mn concentration exceeds about 0.1 mol%, the decay curve becomes nonexponential with a decay time $t < \tau_0$, owing to energy diffusion between Mn centers. This process favors nonradiative deexcitation, and as a consequence the radiative efficiency η_{rad} (ratio of the emitting centers to the excited ones) decreases compared to η_{rad} for the isolated ion. When a critical density of excited Mn centers is reached (at a high drive level), interaction between excited centers becomes relevant, resulting in a new nonradiative deexcitation channel and a drop in the radiative efficiency (Benoit *et al.*, 1990).

With increasing amounts of Mn ($x > 0.001$) additional zero-phonon lines are observed, which show a much faster decay ($t = 0.2$ ms). These lines are attributed to Mn–Mn pairs (Pohl and Gumlich, 1989).

When the amount x of Mn exceeds 0.01, broad additional emission bands are observed in $Zn_{1-x}Mn_xS$ (Fig. 7), which tentatively are attributed to local domains of MnS (Gumlich *et al.*, 1967; Goede and Heimbrodt, 1988). Similar low-energy emission bands are observed in all big bandgap II–VI compounds containing large amounts of Mn (Benecke and Gumlich, 1991).

The electronic structure of ZnMnS is obviously decisive for its high efficiency in EL. First of all, it is important to note that Mn stays in the 2^+ state during the entire EL cycle. This is in contrast to Cu activators which lose and capture charges within the EL cycle. Second, the ground state of Mn^{2+} lies below the top of the valence band. By now it is clear by photoelectron spectroscopy that the peak of the contribution of the Mn ^6S-state to the valence band is to be found 3.5 eV below the top of this band (Weidemann *et al.*, 1992). This implies that Mn does not act as an acceptor in ZnS.

With resonant photoelectron spectroscopy for II–VI compounds containing Mn, based on Fano resonances (Bressler, 1994), it turned out that the Mn has not only one contribution to the valence band peaking at 3.5 eV, but two more, which peak at 1.5 and 7.5 eV below the top of the valence band. Here the concept of configurational interaction (CI), which was developed by Fujimori and co-workers (see Fujimori and Mizokawa, 1993) offers a much better explanation of "hybridized but localized" character of Mn in ZnS. The work was elaborated mainly by the cluster calculations of MnTe by Ley *et al.* (1987) and the experiments of Bressler (1994). Following these authors, one describes the ground state of the system as a mixture of d^5 and d^6L states, where d^6L means a combination of six electrons within the $3d$-shell and a hole within the p states of the ligand.

Very recently the Mn transition $3d^5 \rightarrow 3d^4\,4p^1$ has also been detected by using Fano resonances. Mertins (1995) excited the Mn $^4T_1(G) \rightarrow {}^6A_1(S)$ emission by

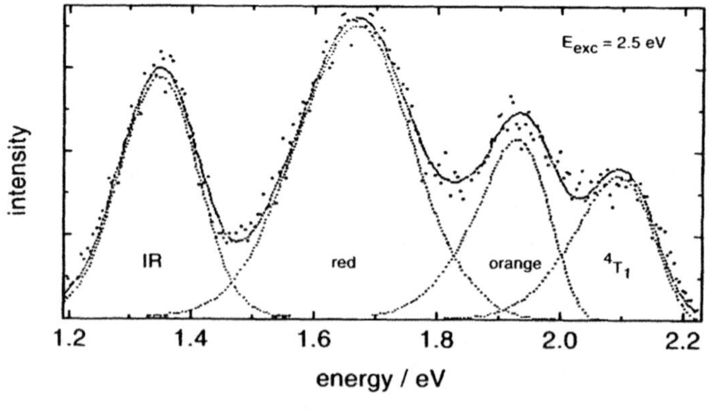

Figure 7. Emission spectrum of ZnMnS at 100 K. The fitted band shapes are represented by dashed lines (Benecke and Gumlich, 1991).

synchrotron radiation. Obviously the empty Mn $4p$ states are degenerate with the continuum of the ZnS conduction band, which allows the excitation in Fano resonances. The energies of the transitions $3d^5 \rightarrow 3d^4\ 4p^1$ is tentatively determined to lie within the range 5–10 eV.

In view of the discussion on the energy distribution of hot electrons in EL it is interesting to note that some steps could be observed in the excitation spectrum of ZnS:Mn within the range 3–25 eV. These steps are to be expected at energies at which the following conditions are fulfilled: from a theoretical point of view carrier multiplication is favored at all points of the Brillouin zone with a high density of states, provided that final states are accessible and that the relaxation energy of hot electrons is equal to the impact ionization energy. From the experiments of Mertins (1995) it can be concluded that this is found to be true at $E = 13.6$, 14.8, and 16.4 near the X-point and at 18.1 eV near the Λ line, if one uses the results of band structure calculations of Walther *et al.* (1970).

3.1.2. ZnS:Tb

The emission lines of the Tb ion in ZnS are due to the internal transitions $^5D_3 \rightarrow {}^7F_j$ and $^5D_4 \rightarrow {}^7F_j$ ($j = 3,4,5,6$). The $^5D_4 \rightarrow {}^7F_5$ line dominates the spectrum with a maximum at 2.28 eV (545 nm) (Krupka and Mahoney, 1972). The intensities of the $^5D_3$ emission peaks are very low (some orders of magnitude) compared to the intensities of the $^5D_4$ emission peaks. One reason for this difference is the fact that the $^5D_4$ level can be populated via cross-relaxation between Tb neighbor ions from the $^5D_3$ level: $(^5D_3 \rightarrow {}^5D_4)_{\mathrm{Tb}_1} \Rightarrow (^7F_6 \Rightarrow {}^7F_{0\ \mathrm{or}\ 1})_{\mathrm{Tb}_2}$. The probability of this event increases with decreasing mean distance among the Tb ions. The critical distance is about 13 Å (Robbins *et al.*, 1976), which corresponds to a concentration of 1.4 mol%, if one assumes a random-like distribution (Benalloul, 1986).

3.2. Alkaline-Earth Thiogallates

Some years ago, a new group of blue-emitting TFEL luminophors was introduced. The host matrices of these luminophors consist of alkaline-earth thiogallates (AETs): $CaGa_2S_4$, $SrGa_2S_4$, or $BaGa_2S_4$ (Benalloul *et al.*, 1993*a*,*b*; Barrow *et al.*, 1993; Sun *et al.*, 1994), which belong to the large group of $M^{II}AM_2^{III}(S, Se)_4$ compounds, where M^{IIA} = Ca, Sr, or Ba and M^{III} = Al, Ga, In, or Y (Donohue and Hanlon, 1974). The luminescence properties of these compounds doped with different dopants, especially Eu^{2+} and Ce^{3+}, have been investigated for powders or single crystals (Peters and Baglio, 1972; Peters, 1972, 1975; Donohue and Hanlon, 1974; Garcia *et al.*, 1982; Davolos *et al.*, 1989; Le Thi *et al.*, 1992). The energy emission of the luminescence centers is determined by the choice of the M^{II} and M^{III} elements. For example, the energy of the lower Eu^{2+} level in these compounds decreases with increasing size of M^{III} and with decreasing size of the substitutional M^{II} cation. Thioaluminates and thiogallates show high luminescence efficiency at 300 K, owing to the increase in the energy bandgap in the following order in trivalent elements: In < Ga < Al (Donohue and Hanlon, 1974; Davolos *et al.*, 1989; Garcia, 1984). Only a weak PL has been reported for homologous Se compounds of thiogallates (Garcia *et al.*, 1982). The Eu^{2+} emission in MY_2S_4 phases lies in the red-orange region and is markedly affected by thermal quenching at 300 K (Garcia and Fouassier, 1994).

In conclusion, from their known luminescent properties, the thiogallates and thioaluminates are the most promising candidates for EL applications among the $M^{IIA}M_2^{III}(S, Se)_4$ compounds. The thioaluminates, however, are rather sensitive to the action of moisture in air as are the alkaline-earth sulfides (AESs) so that research focuses at present on the thiogallates. The thiogallates in fact show a series of interesting optical properties:

1. They are known as efficient CL luminophors when doped with rare-earths (Peters, 1972, 1975; Peters and Baglio, 1972; Davolos *et al.*, 1989).
2. They provide a high solubility of rare-earth centers, which make them attractive as EL luminophor host matrices.
3. Concentration quenching is much less pronounced compared to the doped AESs (Peters, 1972, 1975).
4. All three host matrices present deep-blue emissions when doped with Ce, with significantly better color coordinates than can be attained with the binary sulfides mentioned above.

The Ca and Sr thiogallates have an orthorhombic crystal lattice while $BaGa_2S_4$ has a cubic structure. The melting point is about 1200°C for $SrGa_2S_4$ and $BaGa_2S_4$, and 1150°C in the case of $CaGa_2S_4$ (Eisenmann *et al.*, 1982; Donohue and Hanlon, 1974). Some of the thiogallates were studied in dc-driven EL powder samples with poor efficiency (Vecht, 1982).

AET energy bandgaps vary between 4.1 eV ($BaGa_2S_4$) and 4.4 eV ($SrGa_2S_4$) (Garcia, 1984). No data on the mobility and the band structure have yet been published. The relative dielectric constants of thin films are about 15 (Barrow *et al.*, 1993).

The first deposition method reported in the literature was rf-sputtering. The target was either a stoichiometric mixture of various sulfides, e.g., SrS, Ga_2S_3, and EuS (99.9%) for $SrGa_2S_4{:}Eu^{2+}$ (Benalloul *et al.*, 1993*a,b*) or a single-phase doped thiogallate phosphor (Barrow *et al.*, 1993). Recently, some alternative deposition methods have been reported: molecular beam epitaxy (Inoue *et al.*, 1994), multi-source vapor deposition (Oberacker *et al.*, 1994), and electron-beam evaporation (Miura *et al.*, 1994).

When the thiogallate films are prepared by sputtering, the substrate temperature must be below 350°C. The layers are amorphous and nonluminescent as they are deposited. The films must be annealed after deposition above 550°C, in order to get them crystallized, as revealed by CL and x-ray diffraction measurements (Benalloul *et al.*, 1993*a,b*; Barrow *et al.*, 1993). Full crystallization is only attained with postdeposition annealing above 650°C (Barrow *et al.*, 1993).

The electric properties of the thiogallates under high fields were investigated by putting the doped thiogallated layer in the Inoguchi EL device structure: the thiogallate layer is sandwiched between two insulator layers. Since the dielectric constant of the thiogallates is high (about 15, compared to 8.4 for ZnS), high-dielectric insulators are preferable in order to increase the field across the luminophor layer for a given voltage.

Above a threshold field around 1.5 MV/cm, one can observe a charge transfer in the devices accompanied by luminescence with the particular characteristics of the dopant. Figure 8 shows the EL spectra at 300 K for the $SrGa_2S_4$ host matrix doped with different luminescence centers. In $SrGa_2S_4$:Eu, one observes a broad band centered at 532 nm that is due to the $4f^65d$-$4f^7$ transition of Eu^{2+}, in accordance with data in the literature (Davolos *et al.*, 1989). The Eu emission in $Sr_2Ga_2S_5$ has its maximum at 550 nm (Davolos *et al.*, 1989), and this indicates that no other phase of the Sr thiogallate is present in the layer. Despite the existence of three crystalline sites for Sr, the emission spectrum consists of one band.

$SrGa_2S_4$:Ce^{3+} exhibits two bands corresponding to the 5d-$4f(^2F_{5/2}, {}^2F_{7/2})$ transitions. Ce doping generally requires charge compensation because the Ce^{3+} ion substitutes for a M^{2+} cation. There is an analogous problem in the case of SrS:Ce. Different co-dopings have been investigated: Cl, Na, F. It seems that the spectrum is independent of the co-doping (Benalloul *et al.*, 1993*b*; Sun *et al.*, 1994) but that F co-doping leads to lower efficiency. As to the Pb-doped sample, one observes a broadband emission, which covers the whole spectrum and which is in accordance with CL spectra published on Pb-doped $SrGa_2S_4$ powders (Peters, 1975).

The electric behavior shows many of the characteristics known from ZnS-based TFEL devices. By means of the bridge method one can observe the time-resolved charge transfer in the thiogallate layer.

The optical response follows the transferred charge owing to the short decay time of the optical transition. However, in the second part of the light response, the luminescence decreases faster than the conduction current, indicating that the electrons are not as efficient at this time as at the beginning of the charge transfer. This phenomenon has to be investigated.

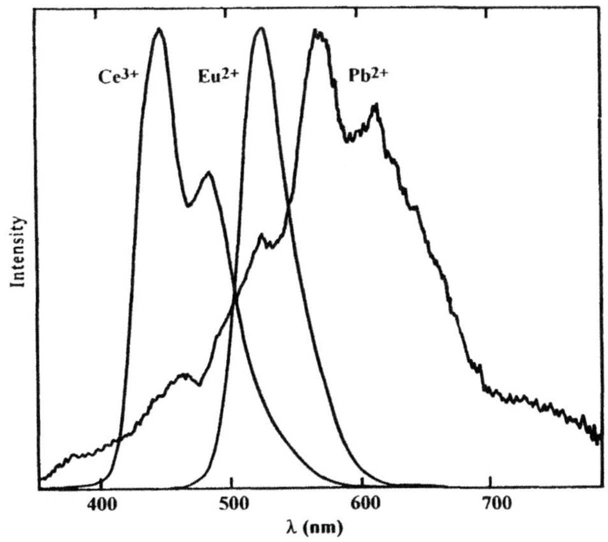

Figure 8. EL spectra for differently doped $SrGa_2S_4$ thin-film devices; $T = 300$ K (Benalloul *et al.*, 1993*b*).

At the trailing edge of the pulse, when the applied field to the sample decreases, no second current peak is observed, even with a large transferred charge. This result indicates that the investigated luminescence centers in $SrGa_2S_4$ are not ionized under the influence of the electric field and not excited via a recombination process of charge carriers as for Eu^{2+} in the AESs (Ando and Ono, 1990). The luminescence centers may be direct electron impact or indirect via energy transfer from a recombination center that has been excited by hot electrons.

The last years have shown that the thiogallates allow the acceleration of charge carriers to optical energies and provide EL when doped with Eu, Ce, or Pb. The panel performances attained with these materials are very good. The blue and green color coordinates are the best available at present for a trichrome panel: $SrGa_2S_4$:Ce—$x = 0.14$, $y = 0.10$ (Benalloul *et al.*, 1993*a,b*; Barrow *et al.*, 1993); $CaGa_2S_4$:Ce—$x = 0.15$, $y = 0.19$ (Barrow *et al.*, 1993); $SrGa_2S_4$:Eu—$x = 0.26$, $y = 0.69$ (Benalloul *et al.*, 1993*a,b*).

3.3. SrS:Ce and Other Rare-Earth-Doped Alkaline-Earth Sulfides

As ZnS:Mn is among the chalcogenides, SrS:Ce is by far the best performing AES-based EL material in terms of luminance and efficiency. Hence, AES-type EL basic materials properties as well as what is understood about them to date will be discussed using SrS:Ce as an example.

3.3.1. Basic Materials Properties

A general difference compared to ZnS-based EL is the fact that only sparse bulk data are available since only a few attempts to grow single crystals have been reported (Kaneko and Koda, 1988; Helbing and Feigelson, 1994). Therefore, band-structure data are mainly theoretically determined and detailed experimental confirmation is still missing. Hasegawa and Yanase (1980) published a band structure for SrS that is generally used as a reference.

SrS is a highly ionic material and grows in a rocksalt structure. The band structure is an indirect one with the bottom of the conduction band at the X-point and the top of the valence band at the Γ-point. The indirect bandgap energy is 4.3 eV, while the direct transition is 4.8 eV.

Since the cation radii fit perfectly to those of the rare-earth ions, a high solubility of dopants is obtained for such EL phosphors. The luminescent transition of Ce^{3+} from $^2T_{2g}(5d)$ to the spin–orbit split ground state $^2F_{7/2}, {}^2F_{5/2}(4f)$ is dipole-allowed, resulting in a very fast radiative decay time of 27 ns, which is at least four orders of magnitude shorter than that of Mn in ZnS (Hüttl *et al.*, 1993).

Only rare data on single crystalline AESs are available; therefore, powders that have been developed as efficient CL phosphors have to be used to judge the material quality of the thin films. The powders that were used for comparisons were from state-of-the-art, and yield CL efficiencies exceeding 30 lm/W.

The radiative τ_{rad} as well as the luminescent decay time τ_{lum} can be calculated by fitting the time-resolved PL decay after the Ce ion has been directly excited by a 500-ps-wide laser pulse (Hüttl *et al.*, 1994). It turns out that in excess of 0.1 mol%

Ce, the decay starts to become nonexponential and is associated with a pronounced phosphorescence. This is often called concentration-quenching. At this concentration level an energy transfer from excited Ce^{3+} to traps becomes measurable. The energy is stored in the traps for a while and retransferred to Ce^{3+}, resulting in the phosphorescence. While at doping concentrations lower than 0.1 mol% τ_{rad} is almost maintained ($\tau_{lum} \approx \tau_{rad}$), higher doping causes a reduction of τ_{lum}. Hence, 0.1 mol% can be seen as the concentration at which quenching starts becoming severe.

If one looks for possible reasons for concentration-quenching, one can see that EPR measurements performed on powder samples indicate that at doping levels below 0.1 mol% an undisturbed incorporation of Ce^{3+} takes place on Sr sites, as shown in Fig. 9 (Hüttl *et al.*, 1994). Higher doping results in a significant portion of Ce being incorporated on the wrong sites or in a nonoptimal vicinity. Two reasons can be stressed: either the limit of solubility is reached or the incorporation of Ce in a Ce^{4+} valency state is increased. The latter explanation is supported by thin-film data obtained by x-ray photoelectron spectroscopy (XPS). Higher doping concentrations result in an enhanced incorporation of Ce^{4+} (Mauch *et al.*, 1992).

3.3.2. AES Peculiarities

As has been reported by several authors, a distinct interaction between the luminescent centers and the host material is observed in AESs (Yoshiyama *et al.*, 1989; Ando and Ono, 1989; Ando, 1992; Troppenz *et al.*, 1994*a*) since the luminescent centers are energetically located in the forbidden gap of the host material. Such an interaction has been observed first by checking the time-resolved emission of the device being employed during pulsed-driving. Unlike ZnS:Mn, in addition to the emission at the leading edge of the voltage pulse there is a second emission pulse when the voltage is switched off. Most likely, a loss mechanism induced by the high electric field, so-called field-quenching, is responsible for the trailing edge emission. In order to check such a mechanism, photoexcitation during normal EL operation has been carried out by Troppenz *et al.* (1994*b*) on SrS:Ce, Cl TFEL devices.

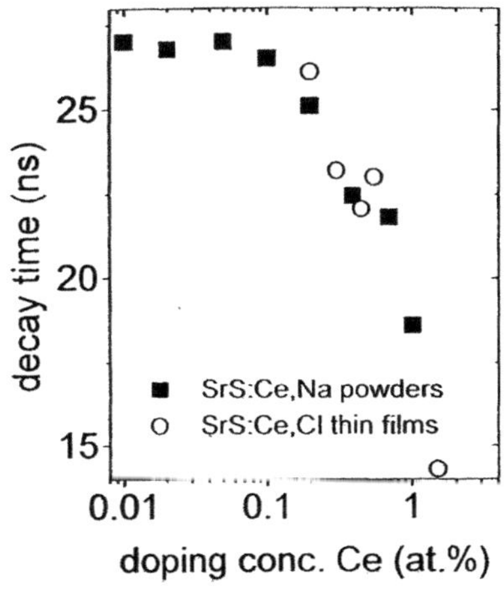

Figure 9. Luminescence decay time τ_{lum} vs. Ce concentration for powders and thin films.

The laser pulse wavelength was adjusted to the absorption band of Ce^{3+}. As also observed in CaS:Eu by Ando and Ono (1989), a photoinduced transferred charge (*PQ*) is generated by photoexcitation in the presence of the high electric field, confirming the field-quenching theory. The transferred charge is generated by the excitation of Ce^{3+}. However, it is not obvious whether the transfer of the electron into the conduction band is due directly to an interaction between Ce^{3+} and the conduction band or rather than an additional energy transfer via traps is involved. In addition to the observed photoinduced transferred charge, a pronounced trailing edge emission far below the actual EL threshold is found, indicating the retransfer of energy to the luminescent centers if the electric field turns in the opposite direction. Most likely, the Coulomb attraction of the remaining ionized traps leads to the capture of these electrons followed by photon emission.

If one compares these findings with the reported strong phosphorescence in SrS:Ce powders, one finds several similarities. The basic mechanism leading to phosphorescence is an energy transfer from Ce^{3+} to traps and vice versa, in this case thermally stimulated. Hence, adapting this understanding to the thin-film case, the trailing edge can be described in the frame of a field-induced phosphorescence. Furthermore, the nature of the trap involved is probably the same for both thin films and powders. These mechanisms are responsible for a distinct doping-dependent behavior of EL characteristics as compared to ZnS:Mn.

3.3.3. Doping-Dependent EL Characteristics

The active contribution of Ce to the charge generation by *PQ* measurements having been proved, the influence of Ce concentration on the transferred charge, depending on the applied voltage ($Q-V$ characteristics) under EL driving, should be discussed. Chubachi and Aoyama (1992) demonstrated the influence of Ce on the steepness of the $L-V$ curve (Fig. 10). A comparison of the $Q-V$ characteristics of undoped and Ce-doped SrS indicates that in the undoped case only a low transferred

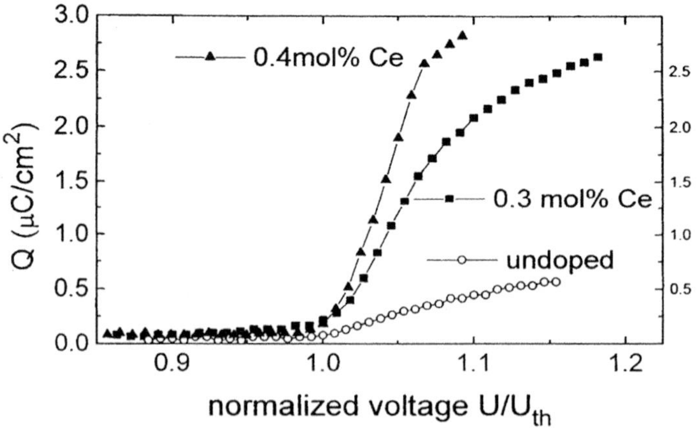

Figure 10. Transferred charge vs. voltage characteristics for undoped and differently doped SrS:Ce,Cl EL devices.

charge level can be obtained, probably owing to the fact that there are only a few host defects (Mauch, 1994), something not observed in ZnS:Mn, where transferred charge is almost independent of the doping level.

On the other hand, if Ce-doping is introduced, the well-known $Q-V$ characteristics are observed. Up to the optimum level of 0.4 mol%, it turns out that the higher the doping level the steeper the $Q-V$ characteristics. Doping levels exceeding the optimum margin result again in a softening of the $Q-V$ curve. In the optimum doping range transfer charge levels as high as 4 $\mu C/cm^2$ are possible.

The dependence of emission spectra on the Ce concentration has not yet been discussed. As noted earlier, the increase in doping concentration results in a spectral shift toward green (Mauch *et al.*, 1992) with the consequence of a reduced blue portion. Possible explanations are either a different crystal-field splitting (Vercaemst *et al.*, 1993) induced by the enhanced lattice distortion or simply reabsorption. The latter is explained by referring again to the emission of Ce^{3+} that originates from the dipole-allowed spin–orbit split transition $^2T_{2g}(5d) \rightarrow {}^2F_{7/2}, {}^2F_{5/2}(4f)$. If the Ce^{3+} concentration is increased, the overlap of the emission and the absorption bands is more pronounced owing to spectral-broadening. Hence, $^2F_{5/2}$ photons are reabsorbed, leading to the observed greenshift. As a matter of fact, these findings have to be taken into account if luminance and efficiency data of different groups are compared, since a greenshift of the spectrum may improve the performance only by the greater sensitivity of the naked eye to the particular emission.

4. MATERIALS REQUIREMENTS AND DEPOSITION TECHNIQUES

If one looks at the structure of a MISIM device or display, the respective thin-film layers have to fulfill different requirements. For all layers, perfect adherence to either the substrate or to its predecessing film is a general prerequisite. In addition, depending on the function there are more demanding property requirements, which will be discussed in this section for different materials. A short introduction to the various thin-film technologies currently being used in EL production and research is also included.

4.1. Electrode Material

4.1.1. Transparent Conductor

In all display applications where thin films are used, the transparent electrode is one of the most critical layers. As no metal is suitable for this purpose owing to the need for high transmittivity for the generated light, transparent conductive oxides have to be used.

It turns out that the resistivity of the transparent conductor determines the size of the display because each individual light-emitting point (pixel) needs a certain time for excitation. The simplest equivalent electrical circuit of a pixel is given by a series resistor and a series capacitor. Hence, the time constant for charging a pixel is determined by the product of the resistance R of the electrode and the capacitance

C of the pixel, and while the latter is fixed, the former can be minimized by using the most conductive material. The lower *R* is the lower the *RC* constant of the line and, therefore, the higher the number of pixels that can be excited in a certain time frame, and, finally, the larger the possible display size.

Further requirements for the transparent conducting material are a high light transmission in the visible range, high temperature stability without loss of conductivity and/or transmittivity, and well-fitting expansion coefficients to the substrate and the subsequent layers. Moreover, there must also be the possibility for simple patterning by photolithography and etching.

Several materials are currently under investigation. The best-performing material so far is indium–tin–oxide (ITO) in typical compositions of 90% In_2O_3 and 10% SnO_2, resulting in conductivities of $10^{-4}\Omega^{-1}\,cm^{-1}$ and optical transmissions exceeding 90%. Other materials such as ZnO:Al, SnO_2:F, or $CdSnO_3$ known from their uses in thin-film photovoltaics are not well suited to TFEL because their conductivity is not high enough. Despite the fact that ITO can be satisfactorily employed, even further improved display properties would be possible if higher-performing transparent conductors could be found.

4.1.2. Metal Electode

The counterelectrode to the transparent conductor is generally a metal. However, the choice of the metal depends on whether a standard or an inverted device structure is used. In the case of the standard structure highly reflective Al is usually chosen, whereas in the inverted case Mo is the one used most often, as the high-temperature processing of the phosphor layer does not allow an Al-covered substrate and its adherence to the subsequent layers is not satisfactory.

4.2. Insulator

In terms of the reliability of EL devices, the insulating layers are of predominant importance. Because of the very high-breakdown field strengths in EL devices during operation, the insulating layers, which must stabilize the dielectric breakdown of the phosphor layer, need to be capable of withstanding several megavolts per centimeter. In addition, the charge that is transferred through the device, and hence through the insulating layer, has to be as high as possible (typically 3–4 $\mu C/cm^2$) in order to achieve high enough luminance levels.

Therefore, apart from good adhesion and a low density of pinholes, the insulating layers have to be characterized by a high-breakdown field E_b and a high relative dielectric constant ε_r. The product of both multiplied by the dielectric constant ε_0 is the so-called figure of merit *F*, which determines the maximum achievable transferred charge and can be used to judge the applicability of the respective insulating materials. A selection of materials that are currently under investigation is summarized in Table 1.

Furthermore, suitable insulating materials have a small loss angle ($\tan\delta$). Consequently, only very dense amorphous layers can be applied since in polycrystalline films the grain boundaries will cause high conduction and a high loss angle,

Table 1. Several Insulating Materials That Are Currently Used in Thin-Film EL Research and Products

Dielectric material	ε_r	E_b(MV/cm)	F(μC/cm^2)
SiO_2	4	6	2
SiON	6	8	5
Si_3N_4	8	4	4
Al_2O_3	8	5	4
AlON	10	5	5
Al_2O_3-TiO_2	10	5	5
SiAlON	8	4	4
Y_2O_3	12	3–4	4
$BaTiO_3$	14	3	4
$BaTa_2O_6$	22	3–4	7
Ta_2O_5	24	2	3

resulting in power dissipation in the insulating film. The materials listed in Table 1 show the noted properties only if they are in an amorphous state. Some of them, such as $BaTiO_3$, tend to crystallize at deposition temperatures higher than 300°C, and so are not suitable for use in high-field EL.

4.3. Phosphor Materials

The phosphor materials in general consist of a host material and a dopant, a situation that is comparable to CL and ultraviolet-lamp phosphors. Therefore, the requirements for the phosphor material have to be discussed for the host and the dopant separately.

In the case of the host, the material should have a large enough bandgap to allow the generation of emission in the visible range, in particular, in the blue region, with the lowest wavelengths at around 400 nm. Therefore, the minimum bandgap of the host material should be about 3.1 eV. The second requirement is high field conductivity at electric field strengths greater than 1 MV/cm. Finally, the incorporation of appropriate luminescent centers with sufficiently high doping levels up to 1 at.% should be possible without major lattice distortions.

It turns out that II–VI host materials, either II^b–VI^b compounds (chalcogenides), such as ZnS and ZnSe, or II^a–VI^b compounds (AESs, such as SrS and CaS, satisfy these requirements perfectly. More recently, AETs, such as $CaGa_2S_4$ and $SrGa_2S_4$, have also been found to be suitable EL host materials.

The luminescent centers that are incorporated into the host material as dopants should have a well-fitted cation or ionic radius in order to substitute either the II^a or the II^b cation with high solubility. Furthermore, these centers have to emit visible light after excitation. Since the most likely mechanism in TFEL is direct-impact excitation by high-energy electrons, the luminescent centers have to have a large cross section for this event. Finally, the luminescent transition should be stable in high electric fields of several megavolts per centimeter; otherwise severe quenching occurs. If all of these requirements are taken into account, transition metals such as Mn as well as rare-earth ions like Eu or Ce are the most suitable for use.

4.4. Deposition Methods

As almost all the currently available thin-film technologies have been applied to TFEL, only a few important points will be touched upon here.

1. *Electrodes*: In the case of electrode materials, transparent conductors as well as metals, standard techniques such as sputtering and electron beam evaporation are used. The most important factors in a successful ITO fabrication, apart from control of the oxygen partial pressure during film growth, are the treatment conditions after the deposition of the ITO layer. It turns out that a subsequent temperature step (in vacuum) of more than 400°C for 1 h results in optimum performance, independent of the fabrication technique.
2. *Insulators*: As already stressed in Section 4.3, the insulating layers have to be very dense and amorphous in order to achieve optimum performance. Hence, electron beam evaporation results mostly in low-quality insulators compared to those produced by sputtering. Therefore, rf-magnetron sputtering is a widely used technique for this purpose, but the best results up to now have been obtained with ALE. This chemical vapor deposition (CVD) technique allows the fabrication of dense multilayer insulators consisting of stacked Al_2O_3–TiO_2 (Antson, 1982) with superior properties.

The phosphor fabrication techniques currently being used in research, as well as in production, are based on either physical vapor deposition (PVD) or CVD methods, not only for the classical ZnS:Mn but also for SrS:Ce. For the case of ZnS:Mn only references to literature will be included here because almost all the techniques in use lead to successful thin-film growth, but for the AESs the situation is more complicated owing to the ionic nature of their bonds, so they will be dealt with in more detail.

4.4.1. Deposition Techniques for ZnS-Type EL

The most often used PVD methods are electron-beam evaporation (Hurd and King, 1979; Sasakura *et al.*, 1981), multisource deposition (Nire *et al.*, 1987), and sputtering (Abdalla *et al.*, 1984; Ohnishi *et al.*, 1985; Matsuoka *et al.*, 1988). If CVD methods are concerned, among others, the ALE has to be singled out (Suntola *et al.*, 1980; Hyvärinen *et al.*, 1988; Leskelä, 1989) because it is the only CVD technique in EL that has been successfully adapted for production purposes. Low-pressure hydride-transport CVD (Mikami *et al.*, 1989) and metal–organic CVD (Cattel *et al.*, 1983; Hirabayashi and Kozawaguchi, 1985) have also been used.

As already stressed, PVD as well as CVD methods result in good-quality ZnS:Mn and ZnS:RE thin films as has been proved by the fact that both are used nowadays to fabricate monochrome TFEL products. However, when the crystallographic structure was investigated, it was found that PVD methods result in ZnS that crystallizes in a cubic zincblende structure whereas CVD methods usually lead to hexagonal ZnS. This difference occurs owing to the fact that PVD depositions are done at substrate temperatures below 300°C, whereas in the CVD case the substrate temperatures exceed even 400°C. The luminance and efficiency levels of ZnS:Mn TFEL emission has been reported to be larger in the latter case. This can be attributed to the improved outcoupling of such devices via scattering induced by the coarse grain structure of the ZnS:Mn films. The coarse grains are a direct consequence of the high-temperature thin-film growth.

4.4.2. Deposition Techniques for AES Type EL

If one looks at the techniques that have been used to fabricate AES-based phosphor layers, in particular, for the purpose of SrS:Ce growth, it turns out that as with ZnS a huge range was tested. However, unlike with ZnS, the quality of the SrS:Ce thin films and consequently the performance of the EL devices vary dramatically depending on the deposition technique. This is mainly due to the more complicated ionic nature of AES bonds. As a matter of fact, much higher substrate temperatures are needed in order to promote a proper nucleation of the thin film and to control the stoichiometry of the growing surface.

Tanaka *et al.* (1985) used electron-beam evaporation with sulfur coevaporation, and Ar–S posttreatment (Tanaka *et al.*, 1991). Others introduced $CuBr_2$ co-doping during sputtering (Ohnishi and Okuda, 1990; Ohnishi, 1994), whereas Tanda *et al.* (1989), Mauch *et al.* (1992), Chubachi and Aoyama (1992) and Nakanishi *et al.* (1993) used multisource deposition without any further treatment. An advanced ALE technique was used by Leskelä *et al.* (1991) and Soininen *et al.* (1993). Even rf-ion plating was tried (Isida *et al.*, 1992).

More recent improvements in luminance (Velthaus *et al.*, 1994*a,b*) can be attributed directly to an improved stoichiometry and crystallinity of the SrS:Ce layers achieved by matching the S/Sr evaporation rates during the reactive evaporation. The best results so far have been obtained with this technique, reaching a luminous efficiency that exceeds 1.63 lm/W (Mauch *et al.*, 1995). However, many process improvements as well as further basic studies on the growth mechanisms are needed in order to optimize this type of EL.

5. COLOR ELECTROLUMINESCENCE EMISSION OF THIN FILMS

This section will summarize the status of color TFEL with respect to its applicability in multi- and full-color products. Not only the respective materials and material combinations are stressed but also the color concepts that are currently under investigation.

5.1. ZnS:Mn-Based Color Electroluminescence

Despite the fact that the intrinsic yellowish emission of ZnS:Mn has only limited applicability in color EL, ways to make use of this efficient phosphor in color products have been discussed (Okibayashi *et al.*, 1991; Haaranen *et al.*, 1992; Laakso *et al.*, 1991) and used in various demonstration modules and even products. The basic idea of all these approaches is to use red and green color filters on top of the EL device in order to generate colors subtractively. The principle is illustrated in Fig. 11.

Based on this structure, several multicolor products up to 640 × 480 VGA panel (EL640.480-AA1 from Planar International) are currently available. The highest luminance levels reported thus far are summarized in Table 2.

5.2. ZnS:RE-Based Color Electroluminescence

As ZnS has proved its superior properties as an EL host material in the case of Mn-doping, it is obviously worthwhile to investigate the possibilities of making use of this material in color EL as well.

The transitions are from the 4*f* inner-shell type and are mainly in the visible range. The rare-earth ions are usually in 3^+ valency state. Hence, charge compensation is required once a rare-earth ion replaces Zn^{2+}. Therefore, in the late 1960s, Kahng (1968) introduced the so-called LUMOCEN type of EL, in which the rare-earth ions are incorporated as halide-containing molecules such as TbF_3 (see also Chase *et al.*, 1969).

Although there were some early reports, extensive studies were carried out mainly in the 1980s when thin-film deposition methods had reached a certain performance level and the incorporation of such complicated molecules could be done in a more controlled way. It turns out that up to now only ZnS:Tb can be developed to meet the requirements for use. This might be due to the cation and covalent radii of the rare earths, which do not fit perfectly to ZnS and, hence, may cause severe lattice distortions if they are incorporated.

ZnS:Tb, F is a green-emitting luminescent layer (Kobayashi *et al.*, 1984; Hirabayashi *et al.*, 1988). It has been found, particularly if sputtering is employed,

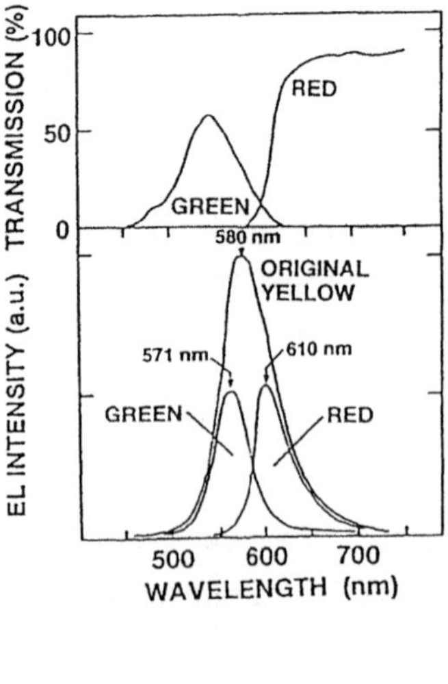

Figure 11. Filtered red and green emission of a ZnS:Mn thin-film EL device and respective filter characteristics (Okibayashi *et al.*, 1991).

Table 2. Performance of ZnS:Mn-Based EL (60 Hz)

Material	Emission color	CIE coordinates x	CIE coordinates y	Luminance (cd/m^2)	Luminous efficiency (lm/W)
ZnS:Mn	Yellowish-red	0.51	0.50	400	4
ZnS:Mn/filter	Green	0.45	0.55	90	1
ZnS:Mn/filter	Red	0.65	0.35	75	0.8

that the material can be further optimized when in addition to Tb and F, O is also incorporated (Ohnishi and Okamoto, 1989; Okamoto *et al.*, 1989). In the case of ALE, Härkönen *et al.* (1990) demonstrated that the performance can be improved by fabricating a multilayer structure of ZnS- and Tb-containing layers.

To summarize the data that have been obtained so far for ZnS:RE-based EL, in particular for red, green, and blue, the emission spectra of the best candidates are shown in Fig. 12. Further, the luminance levels of these phosphor layers, obtained under display driving conditions, are given in Table 3.

5.3. Alkaline-Earth-Sulfide-Based Color Electroluminescence

As already discussed, AESs are well-suited host materials for rare-earth-doping owing to the optimum fitting cation radii. However, they behave differently as is known from ZnS-type EL. Recent advances in phosphor depositions have led to

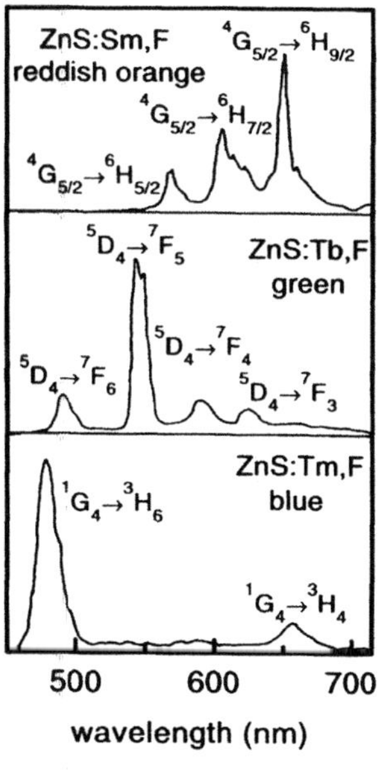

Figure 12. Emission spectra of various rare-earth-doped ZnS EL devices.

Table 3. Performance of ZnS:RE-Based EL (60 Hz)

Material	Emission color	CIE coordinates		Luminance	Luminous
		x	y	(cd/m^2)	efficiency (lm/W)
ZnS:Sm, Cl	Red	0.60	0.35	12	0.08
ZnS:Tb, F	Green	0.32	0.60	125	0.8
ZnS:Tm, F	Blue	0.11	0.09	<1	<0.01

significantly improved device performance, which has allowed them their introduction into products. Among them all, SrS:Ce has to be noted in this context as it generates comparatively high luminance in the blue after filtering. In comparison to ZnS:Tm, F, by assuming the same color coordinates, it is roughly 50 times brighter. But Eu-doped CaS might also replace ZnS:Mn as a red emitter in the future. In Fig. 13, the emission spectra of CaS:Eu, CaS:Ce, and SrS:Ce are plotted (Ono, 1993), demonstrating the broad emission spectra, which are due to the dipole-allowed transitions.

In parallel with previous sections, the highest luminance levels obtained so far for various phosphor materials are given in Table 4.

5.4. Alkaline-Earth-Thiogallate-Based Color Electroluminescence

As discussed in Section 3.2, the rare-earth-doped AETs were developed by Barrow *et al.* (1993). In particular, Ce-doping of (Ca, Sr)Ga_2S_4 leads to a deep blue emission without filtering.

If Ca is replaced by Sr, the emission is shifted to the blue, though at the expense of luminance. The performance obtained up to now is summarized in Table 5.

The luminance data of $SrGa_2S_4$:Ce has been measured at a driving frequency of

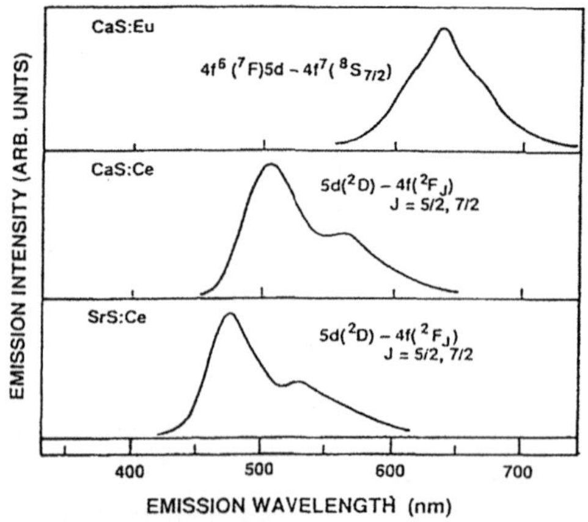

Figure 13. EL emission spectra of various rare-earth-doped AESs (Ono, 1993).

Table 4. Performance of AES:RE-Based EL (60 Hz)

Material	Emission color	CIE coordinates x	CIE coordinates y	Luminance (cd/m^2)	Luminous efficiency (lm/W)
CaS:Eu	Red	0.68	0.31	12	0.05
CaS:Ce	Green	0.27	0.52	10	0.1
SrS:Ce, Cl	Greenish-blue	0.28	0.49	200	1.63
SrS:Ce	Bluish-green	0.19	0.38	65	0.44

Table 5. Performance of AET:RE-Based EL (60 Hz)

Material	Emission color	CIE coordinates x	CIE coordinates y	Luminance (cd/m^2)	Luminous efficiency (lm/W)
$CaGa_2S_4$:Ce	Blue	0.15	0.19	10	—
$SrGa_2S_4$:Ce	Blue	0.10	0.12	32 (1 kHz)	—

1 kHz (Oberacker *et al.*, 1994). An estimate of the luminance at 60 Hz results in roughly 2 cd/m^2, which has to be compared to the 10 cd/m^2 obtained with $CaGa_2S_4$:Ce.

5.5. Color Concepts

Two different approaches are currently under investigation for full-color EL: the structured-phosphor concept and the color-by-white concept. Both are shown schematically in Fig. 14.

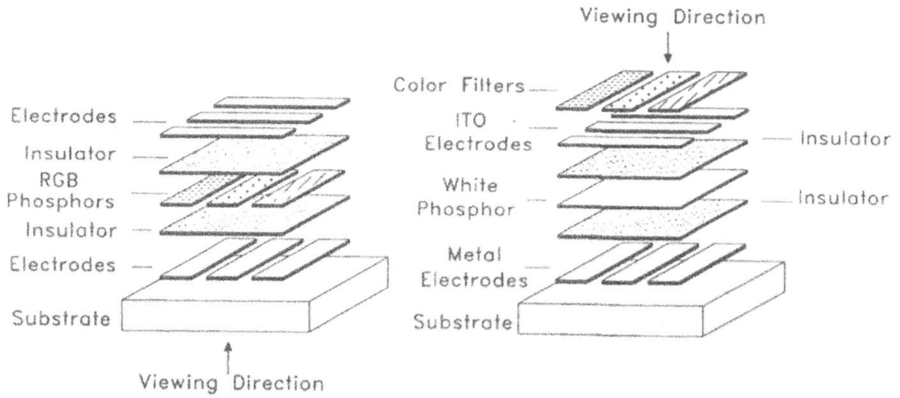

Figure 14. Structured-phosphor and color-by-white display approaches.

Several differences characterize these concepts. On the one hand, the viewing direction in the case of the structured-phosphor approach is through the glass substrate, whereas in the color-by-white approach it is from the opposite side (inverted structure). The inverted structure in the latter case is needed to avoid parallax artifacts. The second major difference is found in the active layer itself.

In the structured-phosphor approach the pixel color is generated by a dedicated red-green-blue (*RGB*) light-emitting layer, which requires subsequent deposition and photolithographic structuring of the active layers. In the color-by-white display only one broadband light-emitting layer is needed. The respective pixel color is subtractively determined by a RGB color filter matrix as in liquid-crystal device (LCD) displays. It is obvious that both concepts have their benefits and trade-offs. In terms of luminance and efficiency the structured-phosphor concept performs better than the color-by-white solution owing to the latter's loss in the color filters. Hence, the same aerial luminance for both display types requires in the color-by-white case a pixel luminance before filtering that is several times higher than that needed for the structured-phosphor display. However, if manufacturing issues are considered, the color-by-white approach offers a lot of advantages since monochrome processing can be used. This leads to the expectation that the high yields that are currently obtained in monochrome production are also possible for full color. On the other hand, in the case of the structured-phosphor displays the additional photolithographic steps have to be considered as primary yield- and cost-determining factors.

Since no decisive parameter for either of the two concepts has been reported so far, research is being pursued on both, demonstrations for both cases are currently available. In the case of the patterned phosphor structure commercialization has been announced starting with a $\frac{1}{4}$ VGA panel as introduced by Barrow *et al.* (1994). For the color-by-white approach, significant improvements have been obtained by optimizing the SrS:Ce, which is responsible for the bluish-green part of the spectrum. If such a SrS:Ce is combined with ZnS:Mn to a yellowish-white emitting structure, efficiencies of 2 lm/W can be achieved which result in the luminance levels given in Table 6.

Table 6. Luminance Levels Obtained for the Structured-Phosphor and the Color-by-White Approaches (60 Hz)

Material	Emission color	CIE coordinates *x*	CIE coordinates *y*	Luminance (cd/m^2)	Luminous efficiency (lm/W)
ZnS:Mn/filtered	Red	0.65	0.35	75	0.8
ZnS:Tb	Green	0.32	0.60	125	0.8
$CaGa_2S_4$:Ce	Blue	0.15	0.19	10	—
ZnSMn/SrS:Ce	Unfiltered	0.39	0.48	340	2
	Red	0.65	0.34	39	—
	Green	0.20	0.61	90	—
	Blue	0.10	0.26	18	—

6. STABILITY OF ACTFEL DISPLAYS AND DEVICES

Since ZnS:Mn-based displays have already been used commercially they can be discussed on the basis of practical experience. In the case of new phosphor materials, such as SrS:Ce and (Ca, Sr)Ga_2S_4:Ce, only data acquired during accelerated aging are available.

In general, if reliability is an issue several factors are important. On the one hand, the device has to reliably withstand high voltages in order to achieve electric field strengths in the phosphor layer exceeding 1 MV/cm. The most important point in this context is the quality of the insulating layers, since they have to stabilize the electric breakdown of the light-emitting layer. On the other hand, the long-term reliability is determined primarily by the stability of the phosphor layer, e.g., stability of the luminescence yield and the proper sealing of the display against moisture and humidity. A well-matched combination of phosphor and dielectric layers in terms of sticking and stress is needed in order to avoid peeling. Moreover, if, for instance, a gray-scale display is considered, stable $L-V$ characteristic, which are not achievable in a straightforward manner, are a further prerequisite.

6.1. Reliability of ZnS:Mn-Based Electroluminescence

One of the best examples for long-term reliable EL displays are the large information boards at Helsinki Airport (Duncker, 1983), which consist of 3000 EL submodules. These displays have been operating 24 h a day, since 1983. After 50,000 h of continuous operation, the mean time between failure for one submodule was calculated at more than 2,000,000 h. This result clearly demonstrates that TFEL provides one of the most reliable displays available. Usually, TFEL displays are guaranteed more conservatively for a lifetime of 40,000 h.

6.1.1. Long-Term Stability

In order to achieve a high enough safety margin for electrical operation, a lot of work was done to optimize the dielectric layers. In the case of sputtered insulating films, Al_2O_3 and SiO_xN_y are widely used since they offer physically and chemically stable insulating properties and the process technology is well known. On the other hand, if ALE is employed, a multilayer stack of Al_2O_3 and TiO_2 is used, which offers a high figure of merit (see Section 4). Nowadays, the dc breakdown voltage of EL panels is higher (300 V), which results in a safety margin of around two since the threshold voltages for the operation of EL panels is between 140 and 180 V, depending on the display structure.

Electrical reliability is a prerequisite for operation. However, long-term stability is determined by other factors; one of these is the defect density. Defects cause tiny pinholes by a partial electrical breakdown the first time the electric field is applied on the display (burn-in). These pinholes are usually only a few micrometers in width and do not disturb the visual appearance of the pixel, but if proper encapsulation of the display is not maintained, moisture can penetrate the thin-film stack, resulting in peeling and finally in severe damage. Hence, in order to improve long-term stability, both a reduction of the defect density has to be achieved by improving the thin-film processes and method for careful sealing of the thin-film package has to be developed. As already noted, in the case of ZnS:Mn these problems have been solved.

6.1.2. Stability of L–V Characteristics

The drive method of the display is very important for its reliability, in particular, for the stability of the *L–V* characteristics of the pixels. The first field-refresh method developed (Takeda *et al.*, 1980), where the display is asymmetrically driven, results in the generation of latent images caused by a deterioration in the *L–V* characteristics (luminance and threshold voltage) of several pixels (Flegal and King, 1986; Müller *et al.*, 1988). However, if symmetric drive methods are employed this problem vanishes since all the pixels are symmetrically driven and there is no charge accumulation or damaging of one interface (Shoji *et al.*, 1989).

A general problem for the reliability of EL displays is found in the shift of the *L–V* characteristics, especially, if gray scales are concerned. Herein, the issue is a homogeneous decrease or increase in the threshold voltage and/or a softening of the *L–V* characteristics over the whole display area. These problems have been discussed in detail in the literature (Inoguchi and Mito, 1977; Törnqvist and Koppela, 1982) and it turns out that postdeposition heat treatments and the control of crystallinity of the phosphor layers as well as a proper choice of insulating films help to overcome them (Mikami *et al.*, 1992).

6.2. Reliability of Color Electroluminescence

While in the case of ZnS-based EL reliability has been proved in the products, alkaline-earth-based phosphor layers have not yet been used. Hence, only laboratory samples can be analyzed, often under accelerated-aging conditions. It is obvious that the problems discussed for ZnS:Mn are also relevant for the new phosphors. Furthermore, unknown degradation phenomena arise. For instance, the loss of luminescent centers during operation causing a reduction of the luminescent yield accompanied by a general increase of nonradiative relaxation was found to be a further aging problem (Müller, 1993).

In the case of green-emitting ZnS:Tb, only a few reports on stability are available and no detailed analyses have yet been carried out. Rather, promising data have been reported for sputtered ZnS:Tb (Tuenge, 1989). A luminance decrease of only 8% after 3000 h continuous operation at 1 kHz, which corresponds to about 50,000 h under display driving conditions could be realized by matching the deposition conditions of the ZnS:Tb layer.

More severe degradation has been observed in alkaline-earth-based EL, but, encouraging data have resulted from optimizing the phosphor layers. In the case of sputtered and subsequently rapid-thermal-annealed $SrGa_2S_4$:Ce, only a 10% decrease in luminance at 40 V above threshold was found after operation of 500 h at 1 kHz accelerated aging (Barrow *et al.*, 1993).

A promising stability result was also obtained in the case of SrS:Ce by employing a reactive evaporation process for Sr, S, $CeCl_3$ and co-doping with ZnS or Mn (Velthaus *et al.*, 1994*a,b*; Mauch *et al.*, 1995) (Fig. 15).

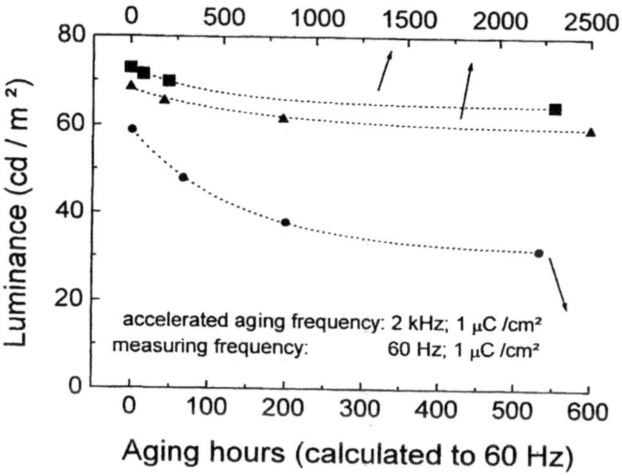

Figure 15. Aging behavior of reactively evaporated SrS:Ce EL devices prepared by conventional means and a novel process using ZnS or Mn as co-doping materials (Mauch *et al.*, 1995).

7. POWDER ELECTROLUMINESCENCE

7.1. Introduction

As noted earlier, the luminescence excitation of powder phosphors that was discovered by Destrian is, apart from some more or less precise descriptions of other authors, the oldest electroluminescence phenomenon (Destrian, 1936). The first report on dc-driven PEL (DCPEL) was given by Zalm *et al.* (1954), but an important part in the development of this kind of EL is due to Vecht and his co-workers (1968). Besides the surveys of the early history of PEL compiled by Ivey (1966) and Henisch (1962), Chada (1993) reported on the present state of the art of PEL and its history.

7.2. Materials

7.2.1. Phosphors

There is no doubt that the II–VI compounds also play a central role in the powder branch of EL, and among these materials ZnS again dominates.

As for the particle size of the powders, there is a significant difference between the requirements of the application in ac-driven PEL (ACPEL) and DCPEL. While in ACPEL there is no real restriction on the size of the particle as long as its crystallinity is not destroyed by milling, the requirements in DCPEL are more rigid. In DCPEL the particles must be both small and of uniform size, and this requires special precipitation methods when the phosphors are prepared (Chada, 1993).

7.2.2. Activators

Stoichiometric II–VI compounds without impurities show only the PL of excitons, but no EL. Only certain impurities, traditionally called *activators* give rise to EL emission. The most prominent activators in PEL are Cu and Mn (Gumlich, 1981). Some others, especially rare-earth ions are used to obtain different emission spectra, but they are less efficient than Cu and Mn (Chada, 1993). It should be noted that other ions with incomplete $3d$-shells, such as Fe, Ni, and Co, in the II–VI compounds reduce the EL intensity and are called *killers* (Gumlich and Moser, 1965).

The two prominent activators, Cu and Mn, each have a different role and should be considered separately in both ACPEL and DCPEL.

(a) Cu in ACPEL. Cu in ACPEL has two functions. It plays a key role in the mechanism of excitation and behaves as a classic activator in ZnS, giving rise to blue and green luminescence emission.

Most authors assume that the excitation mechanism in PEL has three steps: (1) the liberation of carriers from traps or from special phases, (2) the acceleration of carriers by the electric field leading to carrier multiplication, and (3) the impact excitation or ionization followed by energy transfer processes and the emission of light.

Cu plays a key role in the first step. There are good reasons to assume that at the surface of ZnS grains either metal–semiconductor contacts or pn-junctions like p-Cu_2S–n-ZnS are established. These Cu-rich layers supply electrons or holes by field emission (Zener effect) and/or by thermal emission (Alfrey and Taylor, 1955; Piper and Williams, 1955). The electrons freed by one of these mechanism are accelerated by the field (step 2) and excite activators (step 3).

A variant on this model has been proposed by Fischer (1971). He also based his model on the existence of conductive precipitates of $Cu_{2-x}S$ on the grain boundaries or on interfaces, estimating the bandgap of these precipitates to be between 1.25 and 1.7 eV. The carriers are liberated by field emission into the ZnS phase. He assumed injection of both electrons and holes from adjacent ranges, which recombine during a field cycle giving rise to emission. This suggestion, which is backed up by microscopic observations (comets), is not too far from the proposal of Goldberg and Nickerson (1963), who assumed an injection of holes from $Cu_{2-x}S$ into ZnS. These holes are supposed to be trapped in the activator centers. It seems very likely that one injection mechanism will not explain the whole variety of experimental findings, but that several effects may come into the play, depending on the specific experimental conditions.

In regard to the emission mechanism owing to Cu centers, it has to be noted that the impact of electrons and Cu leads to ionization of the centers while the free electrons are swept away by the field. This behavior is in contrast to Mn, which remains in the same valence state during the excitation process.

As also observed in PL, the color of ZnS:Cu can be varied by changing the amount of coactivators (e.g., Cl, Al). In ACPEL the ratio of blue to green emission can also be changed by changing the field frequency. An explanation based on the kinetics of holes was given by Broser *et al.* (1965).

(b) Mn in ACPEL. As noted earlier, Mn will not be ionized during the ac cycle, and the spectral distribution of the emission is not influenced by any coactivator. The yellow emission due to the internal transition ${}^4T_1(G) \rightarrow {}^6A_1(S)$ may be either

excited directly by impact or indirectly by first exciting some other activator or coactivator (such as Cu or Cl) followed by an energy transfer to the Mn ion.

When the concentration of Mn exceeds about 1%, additional emission bands are observed peaking at 657, 606, and 940 nm, which are interconnected by temperature-dependent energy transfer. These low-energy emission bands are tentatively attributed to local phase transitions in ZnMnS leading to Mn^{2+} in octrahedral coordination and to Mn in connection with some impurities (Benecke and Gumlich, 1991).

The Mn in ZnS and ZnCdS shows another unique effect. The yellow PL emission that is due to the internal transition $^4T_1 \rightarrow {}^6A_1$ of Mn can be enhanced or quenched by the action of ac fields far below the voltage threshold of ACPEL. The ratio of enhancement and quenching by electric fields depends mainly upon the kind of excitation (x-rays, ultraviolet, or visible), on field frequency, and on temperature. A summary of the field enhancement effects, which were first reported by Destriau and Destriau (1954), is given by Gumlich (1970).

7.2.3. DCPEL

Also in DCPEL Cu plays a decisive role, acting as a luminescence activator, giving rise to green emission, and forming *p*-type $Cu_{2-x}S$ phases at surfaces and interfaces of ZnS. Possibly it acts as a sensitizer to feed the characteristic yellow Mn emission via energy transfer. In contrast to ACPEL the blue emission is normally not observed in DCPEL. The need for a conducting phase on the surface of the particles and the physical contact of particles in DCPEL was demonstrated as early as the 1950s (Homer *et al.*, 1953; Zalm *et al.*, 1954). However, it is not clear whether this Cu-rich surface layer is a metallic Cu film or, as it is generally assumed in the case of ACPEL, a *p*-conducting $Cu_{2-x}S$ semiconductor.

The Cu-rich coating plays an important role in the forming process. In DCPEL the powders require a special treatment before they emit light. During the preparation, all particles are coated in an identical way with a conductive surface and so are electrically identical. This implies that they give rise to a uniform field distribution. In order to allow impact excitation at reasonable field strengths, an inhomogeneous field distribution is needed, which can be established by the application of an electric field during formation, with the transparent electrode positively charged. By this treatment a highly resistive phosphor layer is formed, which is able to emit light even at low voltages ($U < 10$ V). A comprehensive description of the forming process has been given by Chada (1993).

While some details of the forming process are still controversial, there seems to be an agreement that the change of the Cu-rich surface is the main point of this procedure, either by Cu-ion migration or by chemical change of the surface coating.

8. ELECTROLUMINESCENCE OF ORGANIC MATERIALS

Organic EL is a low-field injection-type EL with a structure similar to that utilized in light-emitting diodes. The advantage of organic EL as compared to other types of solid state light emitters is found in the possibility of realizing low-voltage devices on large areas deposited at room temperature. Even low-cost flexible and pliable plastic-type substrates can be used. Furthermore, it has been demonstrated that structuring in the submicron range is possible. Therefore, at present organic EL seems to be a real competitor for inorganic EL.

EL of organic materials was reported as early as several decades ago (Bernanose, 1955). At first it was more or less a curiosity in some laboratories, but organic materials were always favored for their high quantum efficiency in the visible part of the spectrum, the blue range included. Pope *et al.* (1963) reported on EL of single crystals of pure anthracene and of anthracene doped with tetracene produced by applying a dc voltage of about 400 V. The authors assumed that an acceleration–collision mechanism excited the characteristic emission.

Many early attempts to use organic materials in EL were restricted by the need for high voltages. A short survey of these attempts and the literature of this period was given by Tang and Van Slyke (1987). The prospects for devices were interesting as was determined by systematic investigations on organic semiconductors (Naarmann, 1969).

Tang and co-workers also succeeded in achieving better operating conditions by using double layers (Tang and Van Slyke, 1987). In contrast to most organic EL cells, which use a single layer of organic material sandwiched between two injecting electrodes, their EL diode consisted of a double layer or organic thin films, with one layer of only monopolar transport. It is also common to use a band-structure model in organic materials, where the π-electrons are distributed all over the polymer molecules, and in that way establish a conduction band, whereas the σ-electrons remain strictly localized, forming a valence band. A low-work-function alloy was used as the cathode for efficient electron injection. The scheme of the cell is shown in Fig. 16.

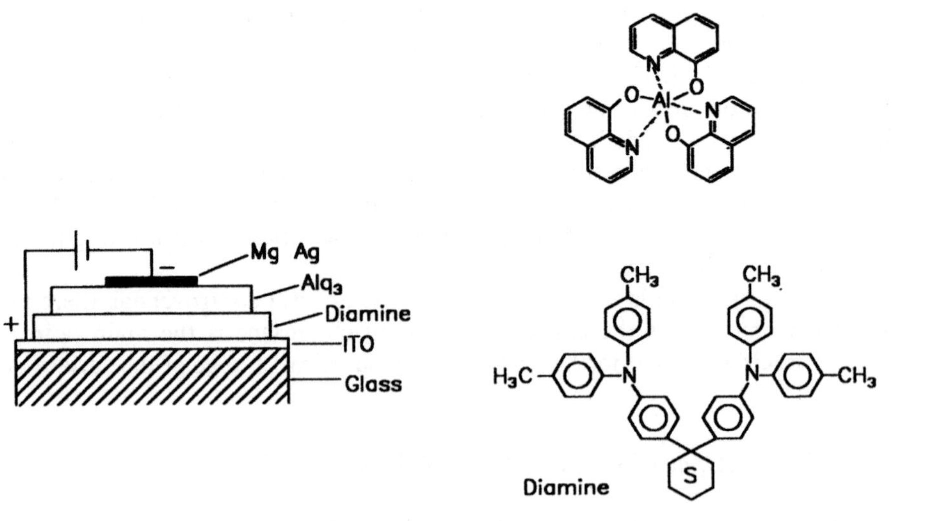

Figure 16. Configuration of an EL cell and molecular structure (Tang and Van Slyke, 1987; with kind permission from the American Institute of Physics).

The novel EL device consisted of a light-emission layer of 8-quinolinol Al complex (Alq) and a hole-transporting layer. Higher luminance, higher efficiency, and lower voltage were obtained with this structure compared to earlier devices. In this way the authors were able to produce high brightness even with a dc voltage of 10 V. The stability of the cell was restricted by degradation, leading to a loss of brightness amounting to 30% within 10 h.

Another breakthrough came in 1990 through publications of Burroughes *et al.* (1990) on light-emitting diodes based on conjugated polymers. In conjugated polymers the semiconduction behavior is associated with π-molecular orbitals delocalized along the polymer chain. The advantage over nonpolymeric organic semiconductors is the possibility of forming useful and robust structures. It should be remembered that the response of this system to electronic excitation is a nonlinear one. Double injection of electrons and holes can lead to a self-localized state. Burroughes and co-workers demonstrated for the first time that poly-(*p*-phenylene-vinylene) (PPV) can be used as the active element in a large-area light-emitting diode with a spectral distribution in the green-yellow range.

Meantime, Scherf and Müller (1994) succeeded in developing new conjugated polymers with an emission that peaked within the blue range (quoted after Schwörer, 1994). Friend and his group reported on systems with a quantum efficiency of more than 1% (Burroughes *et al.*, 1990; Friend, 1993). Schwörer and co-workers were able to prepare organic EL diodes based on (PPV), leading to light-emitting surfaces of 50 cm^2 (Schwörer, 1994). The voltage threshold was reported to be as low as 3 V (Gmeiner *et al.*, 1993).

One of the most important features of the organic EL materials, however, is their ability to form microstructures by photolithography (Schmid *et al.*, 1993) with nearly rectangular edges (< 100 nm). It should be noted that the spectral distributions of PL and EL of organics are always equal. This means that the luminescence mechanism including the formation of singlet excitons is the same in both cases.

Adachi *et al.* (1988) used multistructures to prepare a cell with three active layers: a hole-transport layer, a luminescent layer, and an electron-transport layer. In this way they were able to select luminescent layer materials freely in order to obtain different emission colors. The spectral distribution of three possible luminescent layers is shown in Fig. 17. In this example one finds the full color range of blue, green, and red.

Finally two more recent developments should be noted. The first is the organic light-emission device (LED), emitting white light, which is based on a superposition of three thin films: a triphenyldiamine film showing blue emission, an Al complex (Alq) emitting green light, and a doped Alq film that emits red light (Kido *et al.*, 1995). The special features of these diodes are films in between of the emitting layers, which restrict the mobility of charge carriers in order to control the light intensity of the emitting components. Light intensity of about 2000 cd/m^2 is reported, which is nearly half the value of a commercial fluorescent tube. However, in this case the stability and, thus, the lifetime are still problems.

The second development concerns the organic small-scale LEDs. Granström *et al.* (1995) report on diodes with diameters within the micro- and even nanometer ranges. They used the pores of microfilters to build hole-injecting films that emit red

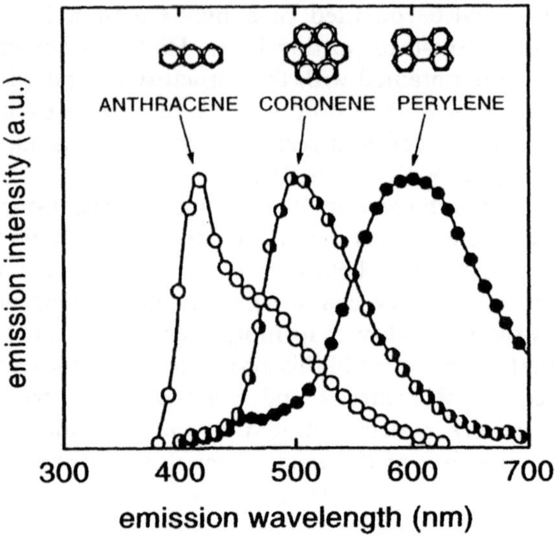

Figure 17. Spectral distribution of three different layered organic EL diodes (Adachi *et al.*, 1988; with kind permission from *Japanese Journal of Applied Physics*).

luminescent light. The authors discuss possible application of these micro-LEDs in near-field microscopy. However, the emission is very weak compared to other light sources used in near-field microscopy.

In conclusion, the organic semiconductors are promising materials for EL. Their great advantage seems to be the possibility of emitting all colors, even in the blue range, where there are still big problems with the inorganics. Another prospect is the creation of flexible and robust panels.

On the other hand, the lifetime of the organic panels is still a decisively limiting factor. In this respect, the organics cannot yet compete with the inorganics. In addition, panel driving schemes also have to be investigated, as the low-voltage device requires current switching, which might end up leading to the necessity of active matrix driving.

Generally speaking, the limits of quantum efficiency, lifetime, maximum surface, and flexibility of organics are not yet definitely known. In view of the significant progress in research during the last years the competition between organics and inorganics is still wide open.

ACKNOWLEDGMENT

The authors are very much obliged to Wolfgang Busse for critical reading of the manuscript and for expert technical assistance.

REFERENCES

Abdalla, M. I., Plumb, J. L., and Hope, L. L. (1984), *Tech. Digest Soc. Information Display*, p. 245.

Adachi, C., Tokito, S., and Saito, S. (1988), *Jpn. J. Appl. Phys.* **27**, L269.

Alfrey, G. F., and Taylor, J. B. (1955), *Proc. Phys. Soc.* **B 68**, 775.

Ando, M. (1992), *Proc. 6th Intern. Workshop on EL* (V. P. Singh and J. C. McClure, Eds.) (Cinco Puntos, El Paso), p. 85.

Ando, M., and Ono, Y. A. (1989), *J. Appl. Phys.* **65**, 3290.

Ando, M., and Ono, J. A. (1990), *J. Appl. Phys.* **68**, 3578.

Antson, J. (1982), *Tech. Digest Soc. Information Display*, p. 124.

Arora, V. K. (1984), *Jpn. J. Appl. Phys.* **24**, 537.

Barker, J. R. (1981), *J. Lumines.* **23**, 101.

Barrow, W. A., Coovert, R. C., Dickey, E., King, C. N., Laakso, C., Sun, S. S., Tuenge, R. T., Wentrass, R., and Kane, J. (1993), *SID 93, Digest*, p. 761.

Barrow, W. A., Coovert, R., Dickey, E., Flegal, T., Fullmann, M., King, C., and Laakso, C. (1994), *Proc. 14th Intern. Display Res. Conf.*, Monterey, p. 448.

Beale, M. (1990), *Acta Polytech. Scand. Appl. Phys. Ser.* **170**, p. 123.

Bénalloul, P. (1986), Ph.D. Thesis, Université Pierre et Marie Curie (Paris VI).

Bénalloul, P. (1989), in: *Electroluminescence* (S. Shionoya and H. Kobayashi, Eds.) (Springer-Verlag, Berlin), p. 36 and references therein.

Bénalloul, P., Barthou, C., Benoit, J., Eichenauer, L., and Zeinert, A. (1993*a*), *Appl. Phys. Lett.* **63**, 1954.

Bénalloul, P., Barthou, C., Benoit, J., Eichenauer, L., and Zeinert, A. (1993*b*), *Proc. 13th Intern. Display Res. Conf.*, Strasbourg, p. 609.

Benecke, C., and Gumlich, H.-E. (1991), The Luminescence of Wide Band Gap II-Mn-VI-Semiconductors, in: *Diluted Magnetic Semiconductors* (M. Jain, Ed.) (World Scientific, Singapore), p. 85.

Benoit, J., Bénalloul, P., Parrot, R., and Mattler, J. (1979), *J. Lumines.* **18/19**, 739.

Benoit, J., Bénalloul, P., Geoffroy, A., and Barthou, C. (1988), *Phys. Stat. Sol.* (*a*) **105**, 637.

Benoit, J., Bénalloul, P., Barthou, C., Casette, C. S., and Soret, J. C. (1990), *Phys. Stat. Sol.* (*a*) **122**, 427.

Benoit, J., Barthou, J., and Bénalloul, P. (1993), *Appl. Phys.* **73**, 1435.

Bernanose, A. (1955), *J. Chim. Phys.* **52**, 396.

Bhattacharyya, K., Goodnick, S., and Wagner, J. F. (1993), *J. Appl. Phys.* **73**, 3390.

Brennan, K. (1988), *J. Appl. Phys.* **64**, 4024.

Bressler, P. R. (1994), *Electronic Structure of Epitaxially Grown Magnetic Semiconductors on II-VI-Compounds: Manganese Chalcogenides* (Verlag Shaker, Aachen), p. 1.

Bringuier, E. (1991), *J. Appl. Phys.* **70**, 4505.

Bringuier, E. (1994*a*), *Phys. Rev. B.* **49**, 7974.

Bringuier, E. (1994*b*), *J. Appl. Phys.* **75B**, 4291.

Bringuier, E. (1995), *Phys. Rev. B.* **52**, 8092.

Bringuier, E. (1996), *Phys. Rev. B.* **54**, 1799.

Bringuier, E. (1997), *Phys. Rev. B.* **56**, 5328.

Bringuier, E., and Geoffroy, A. (1992), *Appl. Phys. Lett.* **60**, 1256.

Broser, I., Gumlich, H.-E., and Moser, R. (1965), *Z. Natur.* **20a**, 1648.

Burroughes, J. H., Bradley, D. D. C., Brown, A. R. Marks, R. N., Makey, K., Friend, R. H., Burnes, P. L., and Holmes, A. B. (1990), *Nature* **347**, 539.

Busse, W., Gumlich, H.-E., Geoffroy, A., and Parrot, R. (1979), *Phys. Stat. Sol.* (*b*) **93**, 591.

Busse, W., Gumlich, H.-E., Törnqvist, R., and Tanninen, V.-P. (1983), *Phys. Stat. Sol.* (*a*) **76**, 553.

Cattel, A. F., Cockayne, B., Dexter, K., Kirton, J., and Wright, P. J. (1983), *IEEE Trans. Electr. Dev.* **ED-30**, 471.

Chada, S. S. (1993), Powder Electroluminescence, in: *Solid State Luminescence* (A. H. Kitai, Ed.) (Chapman and Hall, London), p. 159.

Chase, E. W., Heppelwhite, R. T., Krupka, D. C., and Kahng, D. (1969), *J. Appl. Phys.* **40**, 2512.

Chen, Y. S., and Krupka, D. C. (1972), *J. Appl. Phys.* **43**, 4089.

Chubachi, Y., and Aoyama, K. (1992), *Jpn. J. Appl. Phys.* **31**, 60.

Cohen, M. L., and Bergstrasser, T. K. (1966), *Phys. Rev.* **141**, 789.

Corlatan, D., Neyts, K., De Visschere, P., and Van den Bossche, J. (1994), *Proc. 7th EL Workshop*, Beijing (Science Press, Beijing), p. 211.

Coustal, R. (1934), *C. R. Acad. Sci. Paris* **198**, 1403 and 1597.

Davolos, M. R., Garcia, A., Fouassier, C., and Haggenmuller, P. (1989), *J. Sol. St. Chem.* **83**, 316.

Destriau, G. (1936), *J. Chim. Phys.* **33**, 589.

Destriau, G., and Destriau, M. (1954), *C. R. Acad. Sci. Paris* **238**, 2298.

Donohue, P. C., and Hanlon, J. E. (1974), *J. Electrochem. Soc.* **121**, 137.

Douglas, A. A., Wager, J. F., Morton, D. C., Koh, J. B., and Hogh, C. P. (1993), *J. Appl. Phys.* **73**, 296.

Dresden, M. (1961), *Rev. Mod. Phys.* **33**, 265.

Duncker, J. (1983), *Tech. Digest Soc. Information Display*, p. 42.

Eisenmann, B., Jakowski, M., and Schaefer, H. (1982), *Mat. Res. Bull.* **17**, 1169.

Fischer, A. G. (1971), *J. Electrochem. Soc.* **118**, 139.

Fitting, H.-J., and von Czarnowski, A. (1986), *Phys. Stat. Sol.* (*a*) **93**, 385.

Fitting, H.-J., Müller, G. O., Mach, R., Reinsperger, G. U., Hingst, Th., and Schreiber, E. (1990), *Phys. Stat. Sol.* (*a*) **121**, 305.

Flegal, R. T., and King, C. N. (1986), *Tech. Digest Soc. Information Display*, p. 177.

Friend, R. (1993), Mini-Symposium LED, quoted after Schwöerer, M. (1994), *Phys. Bl.* **49**, 52.

Fröhlich, H. (1947), *Proc. Roy. Soc.* **A188**, 521.

Fröhlich, H. (1954), *Adv. Phys.* **3**, 325.

Fujimori, A., and Mizokawa, T. (1993), Electronic Structure of Transition Metal Impurities in Semiconductors, in: *II-VI-Semiconductor Compounds* (M. Jain, Ed.) (World Scientific, Singapore), p. 103.

Garcia, A. (1984), Ph.D. Thesis, Bordeaux.

Garcia, A., and Fouassier, C. (1994), Unpublished results.

Garcia, A., Fouassier, C., and Dougier, P. (1982), *J. Electrochem. Soc.* **129**, 2063.

Gmeiner, J., Karg, S., Meier, M., Riess, W., Strohriegel, P., and Schwoerer, M. (1993), *Acta Poly.* **44**, 201.

Goede, O., and Heimbrodt, W. (1988), *Phys. Stat. Sol.* (*b*) **146**, 11.

Goldberg, P., and Nickerson, J. W. (1963), *J. Appl. Phys.* **34**, 1601.

Granström, M., Berggren, M., and Inganäs, O. (1995), *Science* **267**, 1479.

Gudden, B., and Pohl, R. W. (1920), *Z. Phys.* **2**, 192.

Gumlich, H.-E. (1970), *Der Energietransport in der Elektrolumineszenz und Elektrophotolumineszenz von II-VI-Verbindungen*, Vol. 133 (Sammlung Vieweg, Braunschweig).

Gumlich, H.-E. (1981), *J. Lumines.* **23**, 73.

Gumlich, H.-E., and Moser, R. (1965), *Z. Natur.* **20a**, 1490.

Gumlich, H.-E., Pfrogner, R. L., Shaffer, J. C., and Williams, F. E. (1966), *J. Chem. Phys.* **44**, 3929.

Gumlich, H.-E., Moser, R., and Neumann, E. (1967), *Phys. Stat. Sol.* **24**, K13.

Gumlich, H.-E. (1989), Some aspects of impurities in wide band gap II–VI-compounds, in: *Growth and Optical Properties of Wide-Gap II–VI Low-dimensional Semiconductors* (T. C. McGill, C. M. Sotomayor Torres, and W. Gebhardt, Eds.) (Plenum, New York), p. 200.

Haaranen, J., Törnqvist, R., Koponen, J., Pitkänen, T., Surma-aho, M., Barrow, W., and Laakso, C. (1992), *Tech. Digest Soc. Information Display*, p. 348.

Härkönen, G., Härkönen, K., and Törnqvist, R. (1990), *Tech. Digest Soc. Information Display*, p. 232.

Hasegawa, A., and Yanase, A. (1980), *J. Phys. C: Sol. St. Phys.* **13**, 1995.

Helbing, R., and Feigelson, R. S. (1994), *J. Cryst. Growth* **138**, 1075.

Henisch, H. K. (1962), *Electroluminescence* (Pergamon, Oxford).

Hirabayashi, K., and Kozawaguchi, H. (1985), *Jpn. J. Appl. Phys.* **24**, 1484.

Hirabayashi, K., Kozawaguchi, H., and Tsujiyama, P. (1988), *Jpn. J. Appl. Phys.* **27**, 587.

Homer, H. M., Ruolon, R. M., and Butler, K. H. (1953), *J. Electrochem. Soc.* **100**, 566.

Howard, W. E., Sahni, O., and Alt, P. M. (1982), *J. Appl. Phys.* **53**, 639.

Hurd, J. M., and King, C. N. (1979), *J. Electr. Mat.* **8**, 879.

Hüttl, B., Müller, G. O., Mach, R., Fouassier, C., and Benalloul, B. (1993), *Adv. Mat. Opt. Electr.* **3**, 131.

Hüttl, B., Troppenz, U., Venghaus, H., Mauch, R. H., Kreissl, J., Garcia, A., Fouassier, C., Benalloul, P., Barthou, C., Benoit, J., Gendron, F., and Ronda, C. (1994), *European Workshop on II–VI Compounds*, Linz, *Mat. Sci. Forum*, Vol. 182–4, p. 263.

Hyvärinen, J., Sonninen, M., and Törnqvist, R. (1988), *J. Cryst. Growth* **86**, 695.
Inoguchi, T., and Mito, S. (1977) *Electroluminescence: Topics in Applied Physics Vol. 17* (J. I. Pankove, Ed.) (Springer-Verlag, Berlin), p. 196.
Inoguchi, T., Takeda, M., Kakihara, Y., Nakata, Y., and Yoshida, M. (1974), *SID '74 Digest*, p. 86.
Inoue, Y., Tanaka, K., Okamoto, S., and Kobayashi, K. (1994), *Proc. 14th Intern. Display Res. Conf.*, Monterey, Cal., p. 169.
Isida, K., Uehara, T., Kikuchi, N., Shiano, Y., Takabashi, S., and Kwita, S. (1992), *Proc. 6th Intern. EL Workshop* (V. P. Singh and J. C. McClure, Eds.) (Cinco Puntos, El Paso), p. 292.
Ivey, H. F. (1966), Electroluminescence and Related Effects, in: *Advances in Electronics and Electron Physics, Suppl. 1* (Academic, New York), p. 1.
Jarem, J. M., and Singh, V. P. (1988), *IEEE Trans. Electron Dev.* **35**, 1834.
Kahng, D. (1968), *Appl. Phys. Lett.* **13**, 210.
Kaneko, Y., and Koda, T. J. (1988), *J. Cryst. Growth* **86**, 72.
Kido, J., Kimura, M., and Katsutoshi, N. (1995), *Science* **267**, 1332.
Kittel, Ch. (1971), *Introduction to Solid State Physics* (Wiley, New York), p. 442.
Kobayashi, H., Tanaka, S., Kunou, T., Shiiki, M., and Sasakura, H. (1984), *Tech. Digest Soc. Information Display*, p. 187.
Kobayashi, S., Wager, J. F., and Abu-Dayah, A. (1992), *Proc. 6th Intern. EL Workshop* (V. P. Singh and J. C. McClure, Eds.) (Cinco Punto, El Paso), p. 234.
Krupka, D. C., and Mahoney, D. M. (1972), *J. Appl. Phys.* **43**, 2314.
Kukimoto, H., Oda, S., and Katayama, H. (1976), *J. Lumines.* **12**, 923.
Laakso, C., Khormaei, R., King, C. N., Härkönen, G. Pakkala, A., Pitkänen, T., Surma-aho, M., and Törnqvist, R. (1991), *Proc. 11th Intern. Display Res. Conf.*, p. 43.
Landoldt-Börnstein (1982), *New Series* **17**, 61.
Landoldt-Börnstein (1988), *New Series Group III*, Vol. 22a (Springer-Verlag, Berlin), p. 168.
Le Thi, K. T., Garcia, A., Guillen, F., and Fouassier, C. (1992), *Mat. Sci. Eng.* **B14**, 393.
Leskelä, M. (1989), *Electroluminescence* (S. Shionoya and H. Kobayashi, Eds.) (Springer-Verlag, Berlin), p. 204.
Leskelä, M., Niinistö, L., Nykänen, E., Soininen, P., and Tiitta, M. (1991), *Mat. Res. Soc. Symp. Proc.* **222**, 315.
Ley, L., Taniguchi, M., Gijsen, J., and Jonson, R. L. (1987), *Phys. Rev. B.* **35**, 2839.
Mach, R., and Müller, G. O. (1984), *Phys. Stat. Sol. (a)* **81**, 609.
Mach, R., Müller, G. O., Schrottke, L., Benalloul, P., and Benoit, J. (1990), *Acta Polytech. Scand. Appl. Phys. Ser.* **173**, 291.
Marello, V., and Onton, A. (1979), *Appl. Phys. Lett.* **34**, 525.
Marello, V., Samuelson, L., Onton, A., and Reuter, W. (1981), *J. Appl. Phys.* **52**, 3590.
Matsuoka, T., Kuwata, J., Nishikawa, M., Fujita, Y., Tohda, T., and Abe, A. (1988), *Jpn. J. Appl. Phys.* **27**, 592.
Mauch, R. H., Velthaus, K. O., Bilger, G., and Schock, H. W. (1992), *J. Cryst. Growth* **117**, 964.
Mauch, R. H. (1994), *Proc. 7th Intern. EL Workshop*, Beijing (Science Press, Beijing), p. 86.
Mauch, R. H., Velthaus, K. O., Hüttl, B., Herrmann, R., and Troppenz, U. (1995), *Tech. Digest Soc. Information Display* (to be published).
Mertins, H. Ch. (1995), Doctoral Thesis, Technische Universität, Berlin, D83.
Mikami, A., Terada, K., Tanaka, K., Taniguchi, K., Yoshida, M., and Nakajima, S. (1989), *Tech. Digest Soc. Information Display*, p. 309.
Mikami, A., Terada, K., Okibayashi, K., Tanaka, K., Yoshida,, M., and Nakajima, S. (1992), *J. Appl. Phys.* **72**, 773.
Miura, N., Ogura, A., Ohashi, K., Namiki, T., Matsumoto, H., and Nakano, R. (1994), *Proc. 7th Intern. EL Workshop*, Beijing (Science Press, Beijing), p. 420.
Müller, G. O. (1993), *Tech. Digest Soc. Information Display*, p. 773.
Müller, G. O., Mach, R., Reetz, R., and Reinsperger, G. U. (1988), *Tech. Digest Soc. Information Display*, p. 23.
Naarmann, H. (1969), *Naturwiss.* **56**, 308.
Nakanishi, Y., Ito, T., Hatanaka, Y., and Shimadra, G. (1993), *Appl. Surf. Sci.* **65/66**, 515.
Neyts, K., and De Visschere, P. (1992), *Sol. St. Electr.* **35**, 933.

Nire, T., Watanabe, T., Tsurumaki, N., Miyakoshi, A., and Tanda, S. (1987), *Tech. Digest Soc. Information Display*, p. 242.
Oberacker, T. A., Velthaus, K. O., Mauch, R. H., Schock, H. W., and Tuenge, R. T. (1994), *Proc. 7th Intern. EL Workshop*, Beijing (Science Press, Beijing).
Ohnishi, H. (1994), *Tech. Digest Soc. Information Display*, p. 129.
Ohnishi, H., and Okamoto, K. (1989), *Proc. 9th Intern. Display Res. Conf.*, p. 32.
Ohnishi, H., and Okuda; T. (1990), *Proc. SID* **31**, p. 31.
Ohnishi, H., Yamamoto, K., Katayama, Y. (1985), *Proc. 5th Intern. Display Res. Conf. Society for Information Display* (SID), p. 159.
Okamoto, K., Yoshimi, T., Nakamura, K., Kobayashi, T., Sato, S., and Miura, S. (1989), *Jpn. J. Appl. Phys.* **28**, 1378.
Okibayashi, K., Ogura, T., Terada, K., Taniguchi, T., Yamashita, T., Yoshida, M., and Nakajima, S. (1991), *Tech. Digest Soc. Information Display*, p. 275.
Ono, Y. A. (1993), SID Lecture notes.
Peierls, R. (1979), *Surprises in Theoretical Physics* (Princeton University Press, Princeton, NJ), p. 121.
Peters, T. E. (1972), *J. Electrochem. Soc.* **119**, 1720.
Peters, T. E. (1975), *J. Electrochem. Soc.* **122**, 98.
Peters, T. E., and Baglio, J. A. (1972), *J. Electrochem. Soc.* **119**, 230.
Piper, W. W., and Williams, F. E. (1955), *Brit. J. Appl. Phys.* (*Suppl.*) **6**, 39.
Pohl, U. W. (1988), Doctoral thesis Technical University D83, Berlin, Germany.
Pohl, U. W., and Gumlich, H.-E. (1989), *Phys. Rev. B* **40**, 1194.
Pohl, U. W., Gumlich, H.-E., and Busse, W. (1984), *Phys. Stat. Sol.* (*b*) **125**, 773.
Pope, M., Kallmann, H. P., and Magnante, P. (1963), *J. Chem. Phys.* **38**, 2042.
Prener, J. S., and Weil, D. J. (1959), *J. Electrochem. Soc.* **106**, 409.
Reggiani, L. (1985), *General Theory*, in: *Hot-Electron Transport in Semiconductors* (L. Reggiani, Ed.) (Springer-Verlag, Berlin), p. 1.
Ridley, B. K. (1983), *J. Phys. C: Sol. St. Phys.* **16**, 3373.
Robbins, D. I., Cockayne, B., Lent, B., and Glasper, J. L. (1976), *Sol. Stat. Com.* **20**, 673.
Round, H. J. (1907), *Electric World* **XIX**, 309.
Sahn, S. H., Hyun, D. G., Noma, M., Hosomi, S., and Hamakawa, Y. (1991), *Jpn. J. Appl. Phys.* **30**, 2526.
Sano, Y., and Nunomura, K. (1989), in: *Electroluminescence* (S. Shionoya and H. Kobayashi, Eds.) (Springer-Verlag, Berlin), p. 77.
Sasakura, H., Kobayashi, H., Tanaka, S., Mita, J., and Nakayama, H. (1981), *J. Appl. Phys.* **52**, 6901.
Sawyer, C. B., and Tower, C. H. (1930), *Phys. Rev.* **35**, 269.
Scherf, U., and Müller, K. (1994), German Patent, quoted after Schwörer, M. (1994), *Phys. Bl.* **49**, 52.
Schmid, W., Dankesreiter, R., Gmeiner, J., Vogtmann, Th., and Schwörer, M. (1993), *Acta Polymer.* **44**, 208.
Schwörer, M. (1994), *Phys. Bl.* **50**, 53.
Shockley, J., and Last, J. T. (1957), *Phys. Rev.* **107**, 392.
Shoji, K., Ohba, T., Kanatani, Y., and Uede, H. (1989), in: *Electroluminescence* (S. Shionoya and M. Kobayashi, Eds.) (Springer-Verlag, Berlin), p. 324.
Skolnick, M. S. (1981), *J. Phys. D., Appl. Phys.* **14**, 301.
Smith, D. H. (1981), *J. Lumines.* **23**, 209.
Soininen, E., Leppänen, M., and Pakkala, A. (1993), *Proc. 13th Eurodisplay, Society for Information Display (SID)*, Strasbourg, p. 233.
Sun, S.-S., Tuenge, R. T., Kane, J., and Ling, M. (1994), *J. Electrochem. Soc.* **141**, 2877.
Suntola, T., Antson, J., Pakkala, A., and Lindfors, S. (1980), *Tech. Digest Soc. Information Display*, p. 108.
Swang, R. K., Aven, M., and Devine, J. Z. (1969), *J. Appl. Phys.* **40**, 89.
Takeda, M., Kanatani, Y., Kishishita, H., Inoguchi, T., and Okuno, K. (1980), *Tech. Digest Soc. Information Display*, p. 66.
Tanaka, S., Kobayashi, H., Sasakura, H., and Hamakawa, Y. (1976), *J. Appl. Phys.* **47**, 5391.
Tanaka, S., Shanker, V., Shiiki, M., Deguchi, H., and Kobayashi, H. (1985), *Tech. Digest Soc. Information Display*, p. 218.
Tanaka, S., Morita, H., Yamada, K., and Kobayashi, H. (1991), *Proc. 11th Intern. Display Res. Conf.*, p. 137.

Tanda, S., Miyakoshi, A., and Nire, T. (1989), in: *Electroluminescence* (S. Shionoya and H. Kobayashi, Eds) (Springer-Verlag, Berlin), 180.

Tang, C. W., and Van Slyke, S. A. (1987), *Appl. Phys. Lett.* **51**, 913.

Thompson, T. D., and Allen, J. W. (1987), *J. Phys. C: Sol. St. Phys.* **20**, L499.

Törnqvist, R., and Koppela, S. J. (1982), *J. Cryst. Growth* **59**, 395.

Troppenz, U., Hüttl, B., Velthaus, K. O., and Mauch, R. H. (1994*a*), *J. Cryst. Growth* **138**, 1017.

Troppenz, U., Hüttl, B., Velthaus, K. O., and Mauch, R. H. (1994*b*), *Proc. 7th Intern. EL Workshop*, Beijing (Science Press, Beijing), p. 162.

Tuenge, R. T. (1989) *Electroluminescence* (S. Shionoya and H. Kobayashi, Eds.) (Springer-Verlag, Berlin), p. 132.

Vecht, A. (1982), Department of Materials Science Report, Thames Polytechnic, London.

Vecht, A., Werring, N. J., and Smith, P. J. F. (1968), *Brit. J. Appl. Phys. Ser. 2* **1**, 134.

Velthaus, K. O., Hüttl, B., Troppenz, U., Herrmann, R., and Mauch, R. H. (1994*a*), *Proc. 7th Intern. EL Workshop*, Beijing (Science Press, Beijing), p. 451.

Vercaemst, R., Poelman, D., Van Meirhaeghe, R. L., Laflère, W., and Cardon, F. (1993), *Proc. 13th Eurodisplay*, Strasbourg, p. 479.

Walther, J. P., Cohen, M. L., Petroff, Y., and Balkanski, M. (1970), *Phys. Rev. B* **1**, 2661.

Weidemann, R., Gumlich, H.-E., Kupsch, M., Middelmann, H.-U., and Becker, U. (1992), *Phys. Rev. B* **45**, 1172.

Yang, K.-W. C., and Owen, S. J. T. (1983), *IEEE Trans. El. Dev.* **ED-30**, 452.

Yoshiyama, H., Sohn, S. H., Tanaka, S., and Kobayashi, H. (1989), in: *Electroluminescence* (S. Shionoya and H. Kobayashi, Eds.) (Springer-Verlag, Berlin), p. 48.

Zalm, P., Diemer, G., and Klasens, H. A. (1954), *Philips Res. Rep.* **9**, 81.

Zeinert, A., Benalloul, P., Barthou, C., and Benoit, J. (1997), *Semicond. Sci. Technol.* **12**, 1479.

7

Thermoluminescence

D. R. Vij

1. INTRODUCTION

Thermoluminescence (TL) in solids is the light emission (mainly visible) that takes place during heating of a solid following an earlier absorption of energy from radiation. It is in fact the release, in the form of light, of previously absorbed energy and is quite different from incandescence light emission from a substance that is heated at high temperatures. The essential condition for TL to occur in an insulator or a semiconductor is that the material must have been previously exposed to radiation. Once TL emission has been observed, the material will not show it again after simply cooling the specimen and reheating it, but has to be reexposed to radiation to obtain TL again. TL, although based upon the same fundamental principles as other luminescence processes, is conventionally a misnomer since the heat radiation is only a stimulant and not an exciting agent.

Urbach (1930) is usually credited with suggesting TL as a potentially useful research tool for trap-level analysis. The basic mathematical treatment of TL was first given by Randall and Wilkins (1945) and Garlick and Gibson (1948), after which it started gaining favor for applications in archaeology, geology, health physics, medical sciences, radiation dosimetry such as clinical dosimetry, and personnel and environmental monitoring, as well as for analyzing defect structures in solids (Townsend, 1994). Daniels *et al.* (1953) were the first to use TL as a technique for radiation dose measurements in radiation dosimetry. Their proposal also led to its use in determining the age of archaeologic and geologic specimens.

A large number of dielectric materials exhibit TL emission, including minerals, rocks, inorganic (amorphous, single-crystal and polycrystalline) semiconductors and insulators, glasses and ceramics, organic compounds, biological materials, and biochemicals. Materials such as LiF, $CaSO_4$, CaF_2, BeO, and Al_2O_3, $Li_2B_4O_7$ are the most extensively studied TL materials because of their applications in dosimetry.

This chapter presents a brief description of the physical and mathematical treatment of TL emission and the methods of analysis of TL data. TL characteristics

---

D. R. Vij • Department of Physics, Kurukshetra University, Kurukshetra, India.

Luminescence of Solids, edited by D. R. Vij. Plenum Press, New York, 1998.

of various important useful materials, along with their methods of preparation, phenomenological models, and mechanisms are discussed. There is also a treatment of instrumentation for measuring TL emission. The chapter concludes with a discussion of applications of TL.

2. THERMOLUMINESCENCE MODELS

Depending on the duration, one can distinguish two classes of luminescence—fluorescence and phosphorescence—the line of demarcation being about 10^{-8} s, which is the lifetime of the excited atoms in the gaseous state. Phosphorescence and thermoluminescence are due to one and the same process, the only difference being the fixed and the rising temperature, respectively, of the emitting material during the time the emission is observed.

Generally speaking, the presence of traces of an impurity (activator) thermally inducted into the host lattice and the structural defects create discrete energy states within the forbidden gap of the material through which radiative recombination may take place, giving rise to luminescence. These defects are called "luminescent centers." The presence of vacant lattice sites or other impurities, lattice defects, and/or irregularities in the host lattice providing unoccupied states (traps) delay the luminescence (phosphorescence) by detaining (trapping) the charge carriers (electrons/holes) before their radiative recombination with the luminescent centers.

2.1. Jablonski Model

According to the Jablonski (1935) model, the luminescent centers (emitting system) of the solid are raised to their excited state e by the absorbed energy of the radiation (which may be ultraviolet, visible, x-rays, γ-rays, heavy ions, cosmic rays, etc.). The return to the ground state g occurs with the emission of light called fluorescence (see Fig. 1a). The system in the excited state can also make a transition to a metastable level (also called a trap), where it may remain until it is returned again to e by having received enough energy E (either thermal or optical), and finally goes back to the ground state g, giving rise to phosphorescence (delayed emission). Fluorescence is temperature-independent, whereas phosphorescence is not. The phosphorescence emission is delayed in time by an amount τ equal to the time spent in the metastable/trapping state. In the case where the energy E needed to raise the system from the state m to the state e is thermal, the following relation generally holds true:

$$\frac{1}{\tau} = s \exp\left(\frac{-E}{kT}\right) \tag{1}$$

where s is a constant, k is Boltzmann's constant, and T is the absolute temperature. In this case $\tau \geqslant t$, where t is the time spent by the system in the excited state and is equal to about 10^{-8} s.

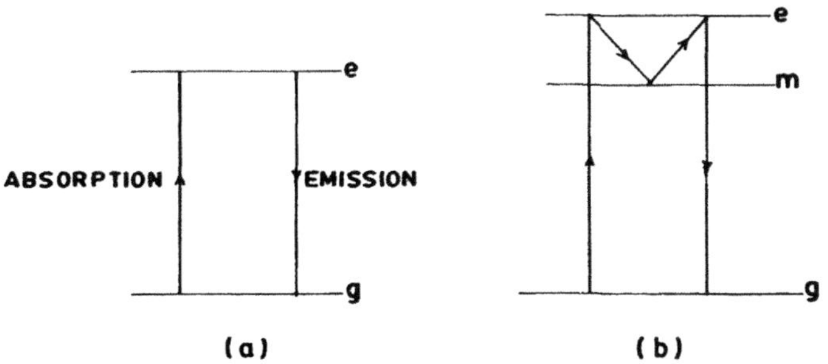

Figure 1. Jablonski's model for (a) fluorescence, and (b) phosphorescence.

2.2. Configurational–Coordinate Model

The configurational-coordinate (CC) model was first suggested by von Hippel (1936) and applied by Seitz (1939) for the description of luminescence. The curves in Fig. 2 represent the potential energy of the luminescent center as a function of the distance between an impurity atom and the first nearest neighbor. For example, in KCl:Tl phosphors, where the Tl^+ ion is substituted for a K^+ ion, Tl^+ is surrounded by six neighboring Cl^- ions; this system state creaked by the Cl^- ions interacting with the Tl^+ ion can be characterized by a distance called the configurational coordinate. The configuration diagram defines the potential energy $U(x)$ of the system Tl^+ plus lattice as a function of the coordinate x. The two curves (Fig. 2) are the ground state U_g ($^1S_0$) and the excited state U_e ($^3P_1$) of Tl^+ ions (Curie, 1960). For organic molecules, however, the vibrational levels are introduced in the curves to represent their vibrational states in the ground and excited states of the emission center.

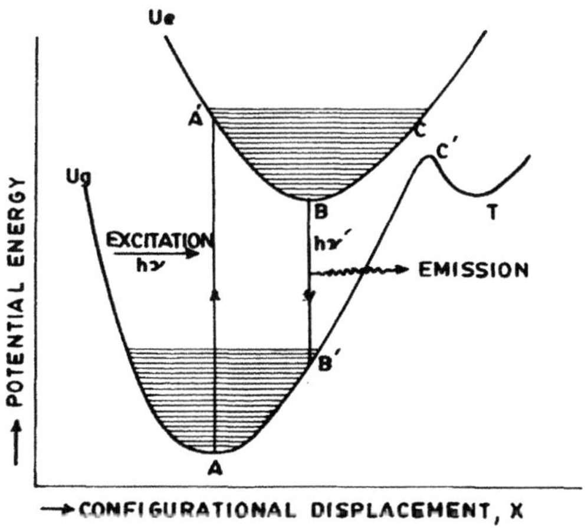

Figure 2. Configurational coordinate diagram for an impurity atom in a semiconductor/insulator.

During irradiation/excitation (absorption of energy), the center will be raised to its excited state U_e (transition AA'), where the system will be in a higher vibrational state and so relaxes to the stable configuration B, dissipating the excess energy to the lattice in the form of heat. The center will emit light through transition BB' and will again be in its ground state U_g. The center still being in the higher vibrational level in the ground state will come to its minimum energy position A with a further nonradiative loss of energy. The case governed by Stokes law (i.e. emission occurring toward the higher-wavelength side of absorption) can also be explained with this model. The excited center may also get trapped at T through CC', wherein after remaining there for some time τ it might escape to the excited state via $C'C$ emitting light via transition BB'.

Rejumping of the center from T via $C'C$ to the excited state depends upon the probability of escape p per unit time denoted by

$$p = 1/\tau = s\exp(-E/kT) \tag{2}$$

where s is the frequency factor having the dimension of reciprocal seconds and E is the energy required for the release of the center from the trap to reach its excited state. The emission is now delayed (phosphorescence) by an amount of time that depends upon the time τ that the carrier spends in the trap. In the case of deep traps, the heat energy at room temperature may not be sufficient to raise the center to its excited state. Here heating of the luminescent material (phosphor) after the irradiation has been removed will raise the trapped center via route $C'C$, giving rise to emission via transition BB' (Fig. 2). It is this emission during heating that is called TL, and a plot of luminescence intensity versus temperature is called glow curve (see Fig. 3).

Quenching or nonradiative emission of absorbed energy can also be interpreted with this model (Dexter *et al.*, 1955). The center may return to the ground state via $CC'B'A$ giving a radiationless transition, transferring the absorbed energy as vibrational energy to the crystal lattice. Williams (1955) successfully determined CC curves for luminescent centers by measuring absorption and emission spectra, decay constants, and luminescence efficiency as well as their dependence over a wide range of temperatures.

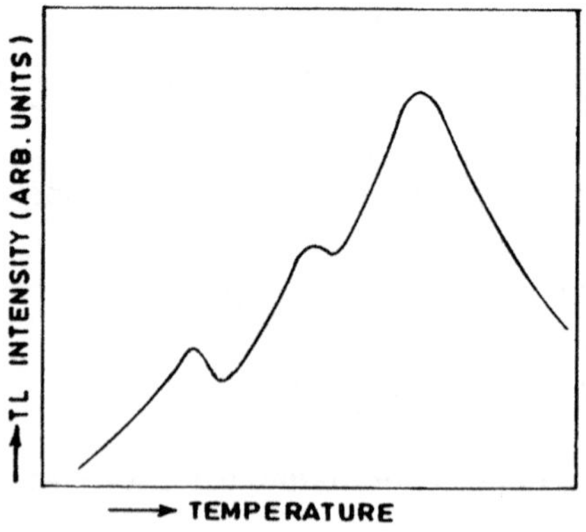

Figure 3. A schematic representation of a glow curve.

2.3. Energy-Band Model

According to the energy-band model, the trace of foreign impurities and defects responsible for luminescence in solids, introduced into the lattice of the host material by heat treatment, can be imagined to form discrete energy levels (donor/acceptor levels) within the forbidden energy gap. Other impurities and the presence of vacant lattice sites or other lattice structural defects provide unoccupied energy levels (called traps) that have the capability of detaining the charge carriers before their recombination with the luminescent centers, thereby delaying the luminescence (i.e., phosphorescence). These traps are denoted as hole traps or electron traps (depending on whether the trapped carriers are holes or elections).

A schematic representation of a simplified energy-band model is shown in Fig. 4. During irradiation, if the luminescent center C is raised to its excited state C' without going to the conduction band, its return to the ground state will emit fluorescence (transitions 1 and 2 are for absorption and emission, respectively). On the other hand, if the electron is raised to the conduction band, either from the valence band (transition 3) or luminescent center C via transition 4, the electron will be free to move with the much more likely possibility of its being trapped in the trapping state T (transition 5). The electron may spend some time τ in the trap before being released to the C-band (transition 6) for its radiative recombination with the luminescent center C exhibiting phosphorescence. Holes play the same role as electrons when they are trapped (transition 7) and released (transition 8) to recombine radiatively (transition 9) at the recombination centers (Schön, 1942; Prener and Williams, 1956; Lambe and Klick, 1955; Klasens, 1953). Direct combination of free electrons and holes across the bandgap is a less likely process than indirect combination, especially in wide-gap semiconductors or insulators.

In the case of shallow traps, detrapping and recombination may occur at a substantial rate around room temperature (the temperature at which the material is

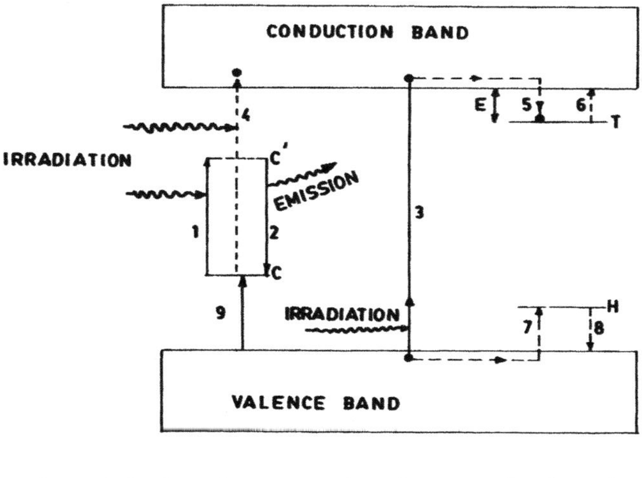

Figure 4. A schematic representation of an energy-band model.

usually irradiated) giving rise to phosphorescence, which is nothing but TL at room temperature. If the traps are deep enough, the charge carriers (electrons/holes) will be retained for a long period of time even after the irradiation is removed. Now if the luminescent material (phosphor) is supplied with continuously increasing heat energy, the charge carriers will be freed from their respective traps and may recombine with the luminescent centers, giving rise to TL. The intensity of TL emission increases as the temperature is raised (as the charge carriers start escaping from their traps), reaches a maximum, starts decreasing as the traps are depleted, and reduces to zero when all the traps are empty. The temperature at which the glow curve exhibits a maximum (see Fig. 3) is related to the trap depth E below/above the edge of the C-band/V-band, respectively. The area under the glow curve is proportional to the number of filled traps and hence to the amount of radiation that the material is subjected to initially.

A single maximum in the glow curve predicts trapping levels located at a fixed energy E below the edge of the C-band (electron traps) or above the V-band (hole traps). More than one maxima in a TL curve reveals that the traps are distributed in separate groups at different depths, each maximum representing a particular set of trapping levels. The trap depth can be calculated from a knowledge of the temperature at which the maximum TL occurs. As TL is the result of excitation using ionizing radiation and one can also detect photons with high efficiency, the TL phenomenon constitutes a sensitive technique for monitoring radiation doses.

Several theoretical and experimental exercises have been carried out to modify the simple energy-band model in order to examine the phenomenon of conductivity occurring simultaneously with TL (Hagebeuk and Kivits, 1976; McKeever, 1985; Lewandowski *et al.*, 1994; Castiglioni *et al.*, 1994; Bindi *et al.*, 1994; Böhm *et al.*, 1994; Mandowski and Swiatek, 1994), and for the specimen showing TL but no conductivity (Chen and Kirsh, 1981; Bräunlich, 1979; Halperin and Braner, 1960; Hagekyriakou and Fleming, 1982). Gasiot and Fillard (1977) have also described models for the TL process that occurs as a result of the thermal release of electrons and their recombination with trapped holes as well as the thermal release of trapped holes and their recombination with trapped electrons. Other models, such as the associated-pair-model (Cooke and Close, 1982). The competing-trap model (Mayhugh *et al.*, 1968; Fain *et al.*, 1994), the track-interaction model (Claffey *et al.*, 1968), the dipole model (Grant and Cameron, 1966), Mayhugh's (1970) model, and the z-center model (Nink and Kos, 1980), as applicable to TL and related phenomena, such as phototransferred TL (Pradhan *et al.*, 1984; Sunta, 1979) have been proposed for different materials. For more details of models as they apply to TL in various materials the reader is referred to Vij (1993).

3. THERMOLUMINESCENCE MECHANISMS

The release of charge carriers from the trapping levels by a thermal process depends on the probability of their escape [Eq. (2)]. The charge carriers (electrons/holes) released from the traps on heating recombine radiatively via the C-band or the V-band, giving rise to TL whose intensity I_t at any time t during heating is proportional to the rate of recombination of holes and electrons at the recombination center level C (see Fig. 4), and is given by

$$I_t = I_0 \exp(-pt) \tag{3}$$

where I_0 is initial intensity at $t = t_0$, and $p = s\exp(-E/kT)$. The frequency factor s varies slowly with temperature and can be constant for a particular trap.

In practice, an electron/hole when released from the trap has a finite probability of getting retrapped. If the electrons/holes released on heating are not retrapped, and all of them recombine radiatively, the TL intensity I_{TL} will be proportional to the rate of release of the trapped carriers, and is given by

$$I_{\mathrm{TL}} = N_{t_0} s\exp(-E/kT) \exp\left[-(s/\beta)\int_{T_0}^{T} \exp(-E/kT)dT\right] \tag{4}$$

where N_{t_0} is the number of electrons trapped at initial temperature T_0 and β the rate of heating. Equation (4) is the Randall and Wilkins (1945) expression for the first-order (monomolecular) kinetics under conditions of no retrapping.

It can be easily observed that I_{TL} builds up as the temperature increases as the second exponential term in Eq. (4) approximates unity; reaches a maximum at particular intermediate temperatures as the second term starts decreasing; and then falls off for any further increase of temperature as the second exponential decreases rapidly to reduce I_{TL} to zero.

By setting $dI_{\mathrm{TL}}/dT = 0$ at $T = T_m$, one obtains

$$\beta E/kT_m^2 = s\exp(-E/kT_m)$$

or

$$E = kT_m \log_e(skT_m^2/\beta E) \tag{5}$$

At low temperatures, however, one gets

$$I_{\mathrm{TL}} = N_{t_0} s\exp(-E/kT) \tag{6}$$

Equation (6) can be used to obtain the value of the trap depth E from the initial portion of the TL curve (the so-called initial rise method) without determining s.

A graph of $\ln(I_{\mathrm{TL}})$ against $1/T$ will give a straight line whose slope is equal to E/k, from which the value of E can be obtained.

From Eq. (5) it is obvious that: (i) with the increase of β (heating rate), the T_m shifts to higher temperatures for a given trap (Bräunlich, 1967); and (ii) for a constant value of β, T_m shifts toward higher temperatures as E increases or s decreases.

In the case of second-order (bimolecular) kinetics (Garlick and Gibson, 1948), the released free carrier is considered to have an equal probability for going to recombination centers or returning to the same trap (retrapping). I_{TL} can now be expressed as

$$I_{\mathrm{TL}} = \frac{N_{t_0}^2}{N} s\exp\left(\frac{-E}{kT}\right)\left[\frac{N_{t_0}}{N}\left(\frac{s}{\beta}\right)\int_{T_0}^{T}\exp\left(\frac{-E}{kT}\right)dT + 1\right]^{-2} \tag{7}$$

where N is the total number of traps of which N_t are occupied at a time t. The second-order curve shows more TL during the second half of the first-order curve (Chen and Kirsh, 1981). The second-order kinetics will be exhibited by those phosphors in which the concentrations of traps and recombination centers are almost equal. For larger concentrations of recombination centers (say, ~100 times) the glow curve falls off rapidly beyond the TL maximum conforming to the first-order kinetics (see Fig. 5). The TL curve for second-order kinetics behaves in a similar fashion as that of first-order except that the curve is of greater half-width.

The initial rise of the TL curve can be used for obtaining E values (when s is not known) since Eq. (7) reduces to $I_{TL} = N_{t_0}^2 s \exp(-E/kT)$ at low temperatures.

In a case of general-order kinetics (where a process follows neither first-order nor second-order kinetics) the TL intensity can be represented as

$$I_{TL} = N_{t_0} s \exp\left(\frac{-E}{kT}\right)\left[\frac{(r-1)s}{\beta}\int_{T_0}^{T}\exp\left(\frac{-E}{kT}\right)dT + 1\right]^{-(r/r-1)} \tag{8}$$

where r is the order of the kinetics, and may not be an integer value. Equation (8) reduces to first-order [Eq. (4)] for $r \to 1$, and to second-order [Eq. (7)] for $r = 2$ (Partridge and May, 1965).

4. METHODS OF ANALYSIS

Trapping parameters such as order of kinetics r, trap depth energy E, and the frequency factor s have appreciable influence on the TL properties of a phosphor. Hence, knowledge of them is of paramount importance for understanding the TL phenomenon in the phosphor, and there have been many approaches to determining these parameters experimentally (Aramu *et al.*, 1975; Bräunlich, 1968; Chen, 1969, 1976; Chen and Kirsh, 1981; Christodoulides, 1985; Kivits and Hagebeuk, 1977; Luthra, 1989; Nicholas and Woods, 1964; Sathyamoorthy *et al.*, 1976; Bindi *et*

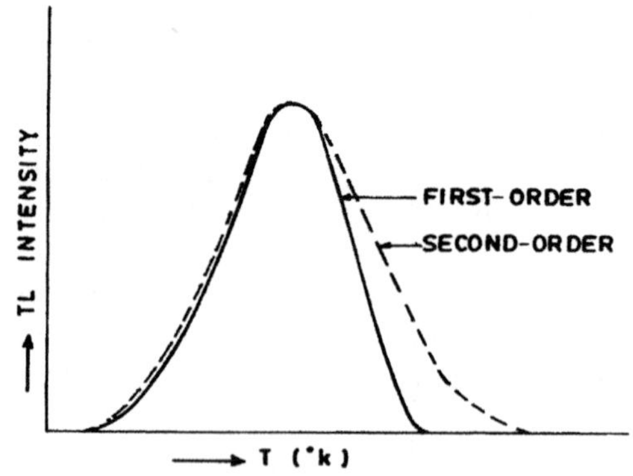

Figure 5. Computed slow curves corresponding to first- and second-order kinetics.

al., 1994). In spite of the fact that these parameters can be experimentally determined using various techniques, there is hardly any method that is completely independent of the order of the kinetics except the initial-rise method, for which prior knowledge of the recombination kinetics is required. Though exact TL kinetics equations can be written and solutions obtained, the TL analysis can be meaningful only if the defect structure of the solid is known and some associated temperature-dependent parameters, such as mobility, luminescence efficiency, and band structure, are reasonably well understood.

Generally, in practice, a material does not exhibit a single TL peak but a multipeak curve. Before starting any analysis, the peak under study has to be isolated from the neighboring overlapping peaks. This method of isolation, the so-called *cleaning technique*, has been proposed by various researchers (Hoogenstraaten, 1958; Gartia and Ratnam, 1975). With this technique, a proper thermal treatment is chosen to remove all temperature peaks lower than the peak under study (Hoogenstraaten, 1958). The TL peak can also be isolated by high-temperature irradiation, room-temperature optical bleaching, or photostimulation with an appropriate frequency. Various methods of analysis developed for obtaining TL parameters, based on such parameters as heating rate, area measurement under the curve, glow-curve shape, curve-fitting technique, and initial position of the glow curve, are discussed briefly below.

4.1. Method Using Different Heating Rates

As described in Section 3, the heating rate β has a pronounced effect on the position of the TL curve maximum T_m. Using two different linear rates β_1 and β_2 and finding the corresponding peak temperatures T_{m_1} and T_{m_2}, respectively, and using Eq. (5),

$$\frac{\beta E}{kT_m^2} = s\exp\left(\frac{-E}{kT_m}\right)$$

one can calculate the trap depth (Booth, 1954) as

$$E = \frac{kT_{m_1}T_{m_2}}{T_{m_1} - T_{m_2}} \ln\left[\frac{\beta_1}{\beta_2}\left(\frac{T_{m_2}}{T_{m_1}}\right)^2\right] \tag{9}$$

By using several heating rates, and plotting $\ln(T_m^2/\beta)$ vs. $1/T_m$, a straight line is obtained with the slope E/k and intercept $\ln(E/sk)$, from which E and s can be calculated.

Apart from being simple, this method of evaluating E is insensitive to retrapping effects, and the calculation of E is not affected by problems of thermal quenching as with the initial-rise method. As the value of T_m is not very sensitive to changes in β, this method is only approximate and results in a possible error of 20–30%. E can be obtained within 5% if the temperature measurement can be achieved within 1 K accuracy. Reliable results, however, are not obtained for overlapping peaks, since T_m is appreciably affected while peaks are isolated by the cleaning process.

Chen and Winer (1970) have considered general-order kinetics to obtain an equation of the form

$$\frac{\beta E}{kT_m^2} = s\left[1 + (r-1)\frac{2kT_m}{E}\right]\exp\left(\frac{-E}{kT_m}\right) \tag{10}$$

where r is the order of the kinetics.

A plot of $\ln(T_m^2/\beta)$ vs. $(1/T_m)$ still yields a good value of E. The only disadvantage in this case is that one needs to know r beforehand. Recently, Kitis *et al.* (1994) observed that values of trapping parameters calculated by different methods decrease as the heating rate is increased.

4.2. Half-Width Method

The shape of the TL glow curve is strongly influenced by the order of the kinetics. In half-width methods, the temperatures T_m, T_1, and T_2, which are, respectively, the peak temperature and temperatures on the lower (T_1) and upper (T_2) sides corresponding to half the peak intensity (see Fig. 6), and dependent on the shape of the glow curve, are utilized to form equations to relate E to all or some of these temperatures (Balarin, 1979; Chen, 1969; Christodoulides, 1985; Grossweiner, 1953; Halperin and Braner, 1960; Lushchik, 1956). Using only the ascending part of a glow peak, one finds the value of E for the first-order kinetics (Grossweiner, 1953)

$$E = \frac{1.51kT_mT_1}{T_m - T_1} \tag{11}$$

whereas when the descending part of the glow peak is used (Lushchik, 1956) the value of E is expressed as

$$E = kT_m^2/(T_2 - T_m) \tag{12}$$

Chen (1969, 1974) has shown empirically from analysis of the experimental data that in case of first-order kinetics

$$E = 1.52(kT_m^2)/\delta_1 - 1.58(2kT_m) \tag{13}$$

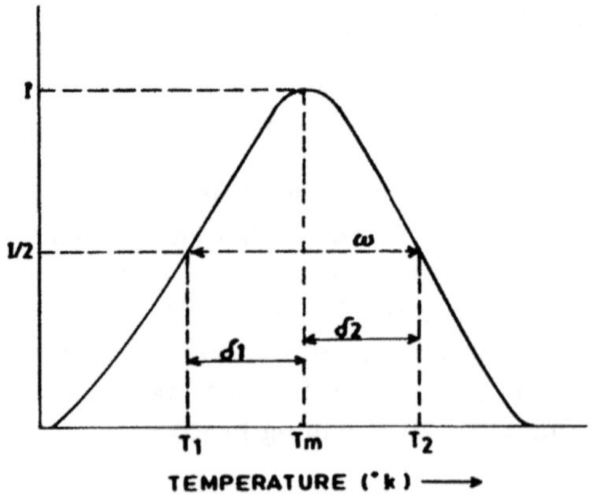

Figure 6. A representative diagram of different quantities to be utilized from a glow curve to determine trap depth energy E from the half-width method.

and for the second-order kinetics

$$E = 1.81(kT_m^2)/\delta_1 - 2kT_m \tag{14}$$

Equations (13) and (14) have been considered where the frequency factor s is independent of temperature. In the case where s depends on the temperature as T^α, then the correct result will be equal to $E - \alpha kT_m$, where E is obtained from Eqs. (13) and (14) (Chen, 1969). An accurate value of E for both the kinetics and the order of the kinetics can also be obtained by using both T_1 and T_2. Half-width methods, though easier, give less accurate values of E and s.

4.3. Initial-Rise Method

The initial-rise method can be used when the sample has a single glow curve or when there is no overlapping of glow peaks belonging to different trapping states. The method is based on the fact that as the glow curve initially begins to rise, the density of unoccupied recombination centers and the density of trapped electrons remain approximately constant, and hence the TL intensity is strictly proportional to $\exp(-E/kT)$.

In the case where the factor s is temperature-dependent, by the factor T^α, the actual trap depth will be $E - \alpha kT$.

4.4. Other Methods

4.4.1. Isothermal-Decay Method

Isothermal decay is a general technique for determining E and s and does not employ any particular heating cycle. The preexcited TL material is quickly heated to a particular constant temperature and the light emission (i.e., phosphorescence), which decays exponentially as a function of time, is monitored. The slope m of the linear plot between $\ln(I_t/I_0)$ and t will be equal to $s\exp(-E/kT)$. The decay is observed at several temperatures and their respective slopes m are calculated. The slope of the linear plot of $\ln(m)$ vs. I/T will be E/k and the intercept is $\ln(s)$, from where E and s are calculated (Moharil, 1981; Manam, 1994).

4.4.2. Curve-Fitting Technique

Curve-fitting can be used to evaluate the parameters E, s, and r by numerical computations based on a specific model and assuming certain initial approximate values for them (Kirsh *et al.*, 1979). A theoretical curve is computed by the estimated values of E, s, and r for a TL peak of the glow curve, which have been calculated using one of the analytical methods already described for first-order ($r = 1$) and general-order ($r \neq 1$) kinetics. The computed curve and the actual experimental curve are compared and a root-mean-square (rms) deviation between the two is calculated. The procedure is sequentially continued by changing E, s, and r values until a minimum value of the rms deviation is obtained. These computed values of E, s, and r, for minimum rms deviation, are the required values of TL parameters for the specimen.

4.4.3. Area-Measurement Technique

Area measurement is actually an extension of the initial-rise method (Aramu and Maxia, 1975), and is based on the fact that the density of the filled traps at any temperature T is proportional to the remaining area A_T of the glow peak. The conventional Arrhenius diagram is modified with each value of I_{TL} normalized to the total area A_0 of the saturated glow peak. The linear plot of $\ln[I_{TL}/A_0]$ vs. I/T for larger values of $1/T$ (i.e., in the initial-rise portion of the curve) is governed by

$$\ln(I_{TL}/A_0) = \frac{-E}{kT} + b \tag{15}$$

where b is a constant. The value of s is given by qe^b.

5. THERMOLUMINESCENCE-RELATED PHENOMENA

5.1. Thermally Stimulated Conductivity

Thermal detrapping of carriers (electrons/holes) in a preexcited solid specimen, under the action of a DC field applied across the specimen layer, may change its conductivity (called thermally stimulated conductivity or current, TSC) if the carriers cross over to the C-band or V-band; this may also give rise to TL emission if these carriers recombine radiatively at the recombination centers. In such circumstances of detrapping, TL and TSC will occur simultaneously in the specimen and there will be some correlation between them. If TL emission occurs via detrapping of carriers and their subsequent radiative recombination, through a tunneling process, without the carriers entering either the C-band or the V-band, then TSC will not occur along with TL. It is worth noting here that in case of the simultaneous occurrence of TL and TSC, their glow curves are usually found to be similar and the methods used for TL are also employed to analyze the TSC peaks (Chen and Kirsh, 1981). There may also be a TSC peak in an unexcited specimen corresponding to the dipole orientation, which cannot be correlated with the TL peak as it does not arise from carrier detrapping.

Simultaneous measurements of TL and TSC have been carried out by various researchers (Vij and Mathur, 1969; Broser and Broser-Warminsky, 1968; Fields and Moran, 1974; Böhm *et al.*, 1994; Castiglioni *et al.*, 1994). TSC measurements may also yield useful information on the nature of defects affecting TL and the traps and dynamics of traps and recombination centers.

5.2. Thermally Stimulated Exoelectron Emission

Thermally stimulated exoelectron emission (TSEE) is also considered to be an effective technique for interpreting the effects underlying TL and for understanding the nature and distribution of the trapping states that the electrons are released from when a preirradiated sample is heated. These released electrons (called exoelectrons) get raised to the C-band. After they reach high enough energies to overcome the potential barrier at the surface and the attraction of the image charges on the surface, they leave the solid sample, and can then be collected by a detector with a positive potential electrode in place above the sample. The curve of the yield of this emission obtained as a function of temperature is also called a glow curve. The same thermally detrapped electrons might recombine radiatively at a recombination center giving rise to TL emission. A correlation between a TL peak and TSEE peak at the same temperature may indicate the presence of electron traps contributing to the TL at that temperature (Becker, 1973). TL–TSEE correlation studies have been done by various researchers (Fukuda *et al.*, 1987; Tomita *et al.*, 1976; Holzapfel and Krystek, 1976; Spurny and Kaambre, 1985; Uchrin, 1993). Usually, the evaluated parameters of the kinetics, such as E, s, and peak position and shape, for TL and TSEE agree extremely well, which is a strong indication of an identical origin of emitted and recombined electrons. The Randall and Wilkins (1945) theory, which accounts successfully for TL, has also been used to explain TSEE, but many authors (Holzapfel, 1969; Kamada *et al.*, 1981) have attempted to formulate a more comprehensive theory of TSEE through other models. Any absence of correlation between TL and TSEE, could be due to the fact that TSEE is a result of surface traps (arising from the adsorbed atoms–molecules and/or surface lattice imperfections) as well as the volume traps (existing within the solid), whereas only volume traps contribute to TL emission.

5.3. Electron Spin Resonance

Electron spin resonance (ESR), provides information regarding charge trapping and its release mechanism in TL. The trapped electrons give rise to characteristic ESR lines. The height of the ESR signal, and hence the number of trapped electrons, is directly proportional to: (i) the number of traps, (ii) the duration of the irradiation, and (iii) the dose rate. In principle, ESR can give a more direct view of the crystal than TL, as ESR measures electrons in the traps. ESR studies of TL samples can provide information about the mechanism of luminescent centers, the structure of their surrounding nuclei, and characteristics of any paramagnetic centers created during the course of irradiation. It is difficult to make qualitative inferences, as the sensitivity factors of the two methods (TL and ESR) bear a ratio of nearly 10^6 (TL being more sensitive). There is a strong correlation between TL and ESR, which fact supports the idea of the existence of defects in solid phosphors such as NaCl, CaF_2, $CaSO_4$, and LiF (Takeuchi *et al.*, 1979; Stoebe *et al.*, 1980; Nambi *et al.*, 1974; Gupta *et al.*, 1975; Grün, 1991; Tatumi *et al.*, 1993).

5.4. Optical Absorption

Optical absorption and TL correlation studies have given rise to significant results identifying the types of centers formed and their relationship with glow peaks after they have been subjected to stepwise annealing (Sathyamoorthy and Luthra, 1978; Mayhugh *et al.*, 1968; Lakshmanan, 1994). The effect of deliberate incorporation of an impurity on the TL process can be easily observed if it is absorbed in the region of interest. Position, area, number, and relative intensity of various components of the optical absorption band can help determine the kind and concentration of defects that give rise to a particular band.

6. MATERIALS: PREPARATIONS AND CHARACTERISTICS

There has been a phenomenal growth in research concerning TL dosimetry using man-made phosphors because of their comparatively simple TL systems and the fact that they are easily reproduced in the laboratory. TL materials can be prepared either by mixing a controlled trace amount of impurity in the host material or by precipitating the reagent in a solution containing an appropriate concentration of the impurity. Heating of the mixture to 1000°C or above results in the diffusion of the impurity into the lattice of the host material. The host material must be very pure and is prepared mostly by such methods as vacuum distillation and zone refinement (Rao, 1986; Brightwell, 1993). The impurity is chosen either as the element itself or a convenient compound of the element, both in the pure form.

6.1. Materials Characteristics

TL characteristics of artificially made materials are being utilized for measuring different kinds of radiation doses, whereas those of minerals and rocks have been used for age estimates. Such TL materials are called *thermoluminescent dosimeters* (TLDs). TLDs are quite sensitive, highly economical, possess a reasonably accurate and linear response over a wide range of doses with large-scale, rapid, and on-the-spot dose evaluation. Before the detailed characteristics of the individual TL materials are discussed here, several of the properties such as the glow curve and TL spectral emission peak positions, stability, thermal treatment (annealing), dose response, energy dependence, and phototransferred-TL, that need to be examined for a phosphor when choosing it as a TL dosimeter will be mentioned briefly.

6.1.1. Peak Position of the Glow Curve and Thermoluminescence Light Emission Curves

Several factors have to be considered when selecting a TLD. Firstly, the material should be a highly sensitive TL-emitting phosphor for the radiation to be monitored, with a single or well-separated glow peak. The glow peak should exist at a temperature substantially higher than room temperature ($\sim$200–250°C) to avoid fading. Secondly, the wavelength of the TL emission spectrum peak should match the quantum efficiency peak of the photomultiplier tube, which mostly lies in the blue-green region of the visible spectrum. This, usually, eliminates the interference of any incandescence (infrared) emission from the heated phosphor and its immediate surroundings (sample holder, heater, etc.). High-temperature glow peaks (above 250°C) should be avoided, as in such cases the infrared emission interferes while measuring low doses.

6.1.2. Fading and Stability

Fading is defined as the ratio I_{TL_t}/I_{TL_o}, where I_{TL_o} is the TL yield of the phosphor measured immediately, and I_{TL_t} after a period t following the removal of the irradiation. Fading in a TLD is caused by a decrease in the number of trapped carriers during the storage period t (i.e., the time between the stopping of irradiation and the readout of TL), and is also influenced by the thermal and optical stimulation of the trapped carriers. Fading is one of the important characteristics to be noted for a TLD as it is one of the undesirable factors in radiation dose measurements and is difficult to control, especially in environmental radiation monitoring. The fading is less for high-temperature glow peaks than those at lower temperatures, since at storage ambient temperatures, especially when the TLD operates in a high-temperature environment, there may be a detrapping of carriers, decreasing the number of trapped carriers with an increase in time as well as storage temperature.

Optical fading of the irradiated TLD is also an important factor that influences the dose-monitoring accuracy. This is due to the fact that the optical illumination of a TLD phosphor, after it has been preexposed to the radiation dose (with some source of light whose wavelength corresponds to the optical absorption peak of the TLD phosphor), decreases the number of color centers, and hence the TL yield is also decreased (Nakajima, 1985). It is, therefore, advisable that the irradiated TLD be protected from daylight and other light sources, and be stored in the dark to avoid emptying of traps by optical stimulation prior to the TL readout.

6.1.3. Dose Response and Annealing

Dose response is the relationship between TL output (in terms of the area under the glow curve or peak height) and the amount of radiation (e.g., the dose) to which a TL phosphor has been exposed. A good TLD should exhibit a linear relationship between the dose received and the subsequent TL output. The dosimeter should also be most sensitive in the dose range of interest and not show any signs of saturation within this range. Usually, TL materials exhibiting many peaks in their curves show supralinear behavior with the dose rather than sublinear, whereas in the case of single-peak glow curves the dose response is nearly linear until the middle-dose range but becomes sublinear in the large-dose range. Both supralinearity and sublinear growth approaching saturation can lead to problems of under- or overestimation of the dose, respectively. The mechanism of sensitization (increase in TL sensitivity) following the absorption of radiation is shown by many TL phosphors and is closely related to the supralinearity. To overcome the problems of sensitization and supralinearity, annealing (heat treatment at constant high temperature) of the phosphor is carried out after each irradiation. The defect equilibrium is regained after the annealing, thus rendering the material ready for reuse. A schematic curve showing the variation in TL behavior with absorbed dose is shown in Fig. 7. It has also been established that the cooling rate after annealing is an important variable. The process of sensitization, combined with an ultraviolet/thermal anneal, has been utilized by many workers to achieve maximum sensitivity and minimum possible detectable dose (Mayhugh and Fullerton, 1975; Jain, 1980; Kitis and Furetta, 1994). Supralinearity and sensitization also depend on the linear energy transfer (LET) of radiation. The LET of a material for charged particles is the quotient of dE/dl, where dE is the energy loss by a charged particle in traversing a distance dl owing to collisions with electrons. In fact, the ionization capacity of radiation in a medium is usually signified by the linear energy transferred to the medium by the incident

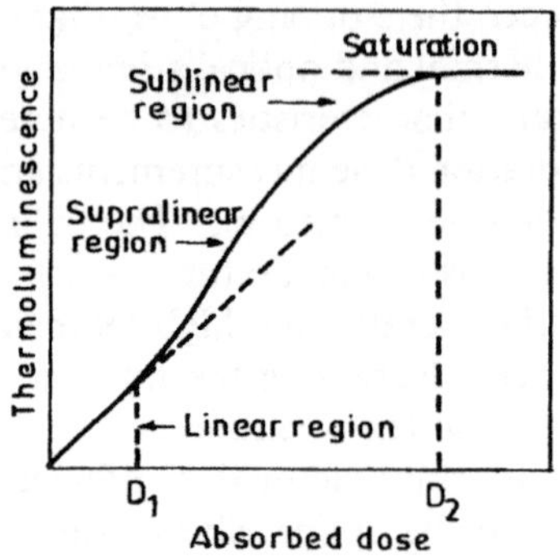

Figure 7. A schematic representation of a TL growth curve with the absorbed dose (McKeever, 1985, with kind permission from Cambridge University Press).

energy (LET expressed in keV/μm). Generally, the relative TL response, i.e.,

$$\left(\frac{\text{TL intensity}}{\text{Absorbed dose}}\right)_{\text{LET}} \Bigg/ \left(\frac{\text{TL intensity}}{\text{Absorbed dose}}\right)_{^{60}\text{Co}}$$

and supralinearity decrease with an increase in LET, but the variation is more complex (Bartlett and Edwards, 1979). The range of the linear response to dose in high-LET radiation is said to be wider than for low-LET radiation.

6.1.4. Energy Response

It is also desirable for a TLD to have a constant response over a wide range of energies. In fact, when a TLD phosphor is irradiated with photons, especially x-rays, the energy dependence of TL sensitivity (TL intensity per unit exposure for x- and γ-rays and per unit absorbed dose for high-energy electrons and charged particles) is drastically affected by the Z_{eff} (effective atomic number) of the phosphor. If the dose is to be delivered to the body tissue, *tissue-equivalent* TLDs are needed to evaluate the exposure. It may be noted that the energy responses for neutrons and β-particles are different, and can be modified, if necessary, by the use of filters whose optimum characteristics can be calculated in advance (Pradhan and Bhatt, 1979; Trousil and Kokta, 1982). For thermal neutron dosimetry, a pair of ^{6}LiF and ^{7}LiF TLDs can be used. These TLDs have the same response to photons, but exhibit different sensitivity to thermal neutrons because of different cross sections. ^{6}LiF is TL-sensitive to both γ-rays and thermal neutrons, whereas ^{7}LiF is TL-sensitive to γ-rays but not to thermal neutrons. Hence, the thermal neutron fluence can be predicted from the difference in their TL-intensity observations.

6.1.5. Phototransferred Thermoluminescence

Almost all TL phosphors exhibit residual TL after the usual dosimetry glow peaks have been read out. The intensity of this residual TL (RTL) appearing above 300°C, being very low, often goes undetected. High-temperature RTL peaks can be recorded using specially selected heat filters to cut out the background thermal signal of the heater. The trapped charges of the high-temperature residual peaks can also be transferred to those of the lower-temperature TL peaks. The transfer from residual deep traps to empty shallow traps can be easily achieved by light exposure of appropriate photon energy sufficient to photoionize the residual trap (Sunta, 1979). The TL thus produced is termed phototransferred thermoluminescence (PTTL).

In the phototransfer process in a previously irradiated sample that has been subsequently annealed at a temperature higher than 400°C, some of the high-temperature traps (residual traps) still remain filled. The electrons move from deep traps into shallow traps by illumination of the sample with ultraviolet light. The filled shallow traps then give TL glow peaks, which had been thermally bleached (annealed) previous to any ultraviolet irradiation. The ultraviolet-induced glow-peak intensity is seen to be directly proportional to detrapping of the deep traps, which makes PTTL useful in the reestimation of the dose received earlier by the TLD. The PTTL technique is of particular value in those TLD materials in which the low-temperature dosimetric peaks exhibit excessive fading and so cannot be used for TL readout. Stable deeper traps are, therefore, utilized for the purpose.

6.2. Thermoluminescent Dosimeter Phosphors

6.2.1. Lithium Fluoride (LiF)

LiF has been found to be the most popular and useful among the TL materials for use in clinical and radiobiological dosimetry. Advantages such as flat response over a wide range of γ-energies, tissue equivalence, inertness to environments, and long-term dose retention (less fading) outweigh the disadvantages of TL behavior complexities.

Daniels *et al.* (1953) were the first to investigate the LiF reagent, but it was first used as a commercial TLD by Cameron *et al.* (1961) and named TLD-100 (with isotopic components of ^{6}LiF and ^{7}Li as 7.4% and 92.6%, respectively). Later, two more LiF phosphors (TLD-600 and TLD-700) were also produced commercially with different Li isotope concentrations. TLD-600 (with ^{6}Li as 4.4% and ^{7}Li as 95.6%), having a large thermal cross section, can detect both neutron and γ-radiations whereas TLD-700 (with ^{6}Li as 0.007% and ^{7}Li as 99.993%) can monitor the γ-component only.

A good LiF-based phorphor has Mg (100–300 ppm) as the main dopant and Ti (10–20 ppm) as the coactivator. A LiF:Mg,Ti phosphor is produced by mixing optical grade LiF with a combination of activators (i.e., MgF_2, Li Cryolite, and LiTi Fluoride). Other activators, such as Al, Eu, Cu, Ag, Na, and the rare-earth elements, have also been tried by various workers and have been found to be of better quality than LiF:Mg,Ti. For more details the reader is referred to the reviews by Jain (1982) and Niewiadomski (1993).

More than ten glow peaks have been noted between room temperature and 400°C (see Fig. 8) (Jain, 1982). Peak V at around 200°C, being stable, is normally used for dosimetric purposes. The positions and relative intensities of peaks I to V at about 60, 100, 130, 170, and 200°C, however, change with the concentration of

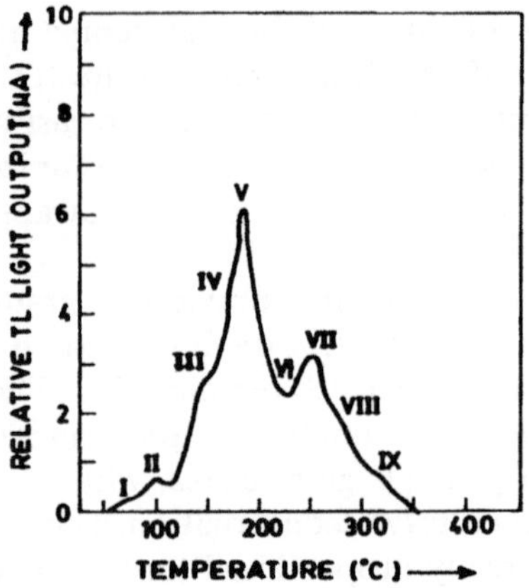

Figure 8. TL glow curve in LiF after ^{60}Co γ-irradiation (Jain, 1982).

dopants Mg and Ti. Peaks I to V in TLD-100 are believed to be associated with the Mg impurity (Christy *et al.*, 1967), peak II and III with the dipoles (called dimers) consisting of Mg^{2+} and cation vacancy (Jackson and Harris, 1970) in the LiF lattice, and peaks IV and V with aggregate (trimers) of Mg^{2+}–vacancy dipoles (Jain, 1982). Ti, however, is assumed to be involved in the emission process as shown in Fig. 9 (Nink and Kos, 1980). There are distinct absorption bands with their peaks at 250 nm (*F*-band, which are due to electrons trapped at negative ion vacancies and are unaffected by annealing at 400°C), 310 nm (in unannealed crystals doped with impurities), and 380 nm (after annealing the impurity-doped crystals at 400°C for 1 h) in LiF, which exist irrespective of the impurity available.

The standard annealing process for LiF phosphors is heating for 1 h at 400°C (during which time there is bleaching of dose information) followed by rapid cooling and then heating for 22 h at 80°C (which helps to retain the original sensitivity of the dosimeter) (Hsu *et al.*, 1978). Postirradiation annealing for 15 min at

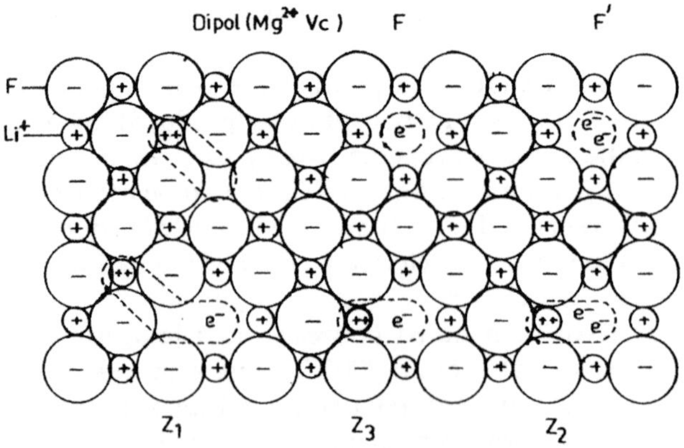

Figure 9. Lattice defects in Li:Mg,Ti relevant for trapping and thermal liberation of electrons (Nink and Kos, 1980, with kind permission from Elsevier Science).

100°C is also done to eliminate the effect of low-temperature peaks on dosimetric peak V.

Thermal fading of the light sum of dosimetric peak V is less than 5% in the case of a virgin and preannealed (at 400°C for 1 h and 80°C for 22 h) non-post-irradiated LiF dosimeter when it is stored at room temperature. Further annealing after irradiation (for 15 min at 100°C) shows almost no fading at a storage temperature of 50°C.

Dose response for the 200°C peak of LiF (TLD-100) has been found to be linear in the range 0.3–300 R (Marron and Attix, 1964). The crystal structure of LiF starts getting damaged at about 300 R resulting in a gradual increase of TL efficiency, which gives rise to supralinearity of the response curve with a slope of 1.2 on a log-to-log plot up to 5×10^5 R, after which saturation begins. Damage occurring within this range of exposure can be reversed by annealing the phosphor at 400°C for 15 min. Above 5×10^4 R, there is permanent damage after a few exposures that cannot be annealed out at 400°C (Jain *et al.*, 1974).

Under high γ-dose irradiations, the TL response becomes supralinear and TL sensitivity, also called radiation-induced sensitization, increases in this region of supralinearity, although it remains constant in the linear portion of the response curve. A number of models have been proposed to explain the mechanism of supralinearity and sensitization in LiF phosphors (Attix, 1975; Jain 1984; Piters and Bos, 1993; Mische and McKeever, 1989; Sunta *et al.*, 1994).

The response of peak V for γ-doses (low LET) is initially linear followed by supralinearity, whereas, for α- and thermal neutrons (high LET), the response is linear throughput. No dose-rate dependence of response in LiF TLDs was detected over the range of exposures (from 10^2 to 2×10^7 R) and exposure rates up to about 10^9 rads/s.

Owing to the high tissue-equivalence of LiF ($Z_{eff} = 8.2$), the influence of photon energy on the dose density (TL/unit dose) is less than with other solid state detectors. The response per röentgen of LiF (TLD-100) is nearly flat over a wide range of γ-energies (see Fig. 10). There is an error of 25% at 25 keV relative to ^{60}Co for powder samples, whereas for teflon rods it is 20% (Hsu *et al.*, 1978), which can be practically eliminated by suitable shielding.

6.2.2. Calcium Fluoride (CaF_2)

CaF_2 is available in the mineral form as a fluorite with a cubic crystalline structure and exhibits strong radiation-induced TL, which can be attributed to transition metal dopant ions such as Mn and rare-earth ions such as Sm, Ce, Tb, Tm and Dy. TL in pure CaF_2 crystals, first developed by Schayes *et al.* (1967) as a TL phosphor for dosimetry, is believed to be due to the trace impurities invariably found in commercially available crystals.

Single crystals of CaF_2 doped with rare-earth impurities have been grown in the past by horizontal growth, crystal-pulling, and Stockbarger methods (Guggenheim, 1963). Bangert *et al.* (1982) have ion-implanted Ce, Dy, and Mn in the CaF_2. Mathur (1993) prepared CaF_2:Mn by reacting $CaCO_3$ (25 g) with HF (25 ml of 48% HF in 250 ml of water). The settled CaF_2 is washed by decantation and dried at a low

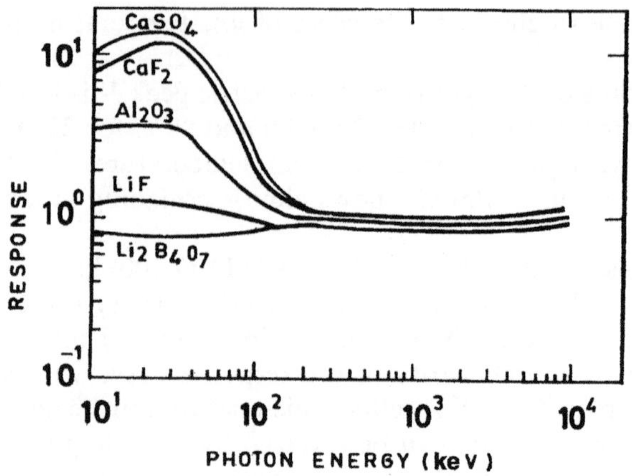

Figure 10. Energy response of various TLD phosphors.

temperature in a vacuum to avoid any hydrolysis. Mn was added in the form of NH_4MnF_3, obtained by precipitation of MnF_2 by NH_4F. These fluorides are thoroughly dry-mixed and heated at about 1200°C in a platinum crucible and inert atmosphere of purified He or Ni.

Synthetic CaF_2:Mn exhibits a single TL (dosimetric) peak around 260°C, and a TL emission spectrum extending from 450 to 600 nm with a maximum at 500 nm (Cameron *et al.*, 1968; Dekker, 1976). The position of the TL peak, however, depends upon both its Mn content and the presence of other foreign ions. Natural CaF_2 shows two more glow peaks at about 70–100°C and about 150–190°C (Schulman *et al.*, 1960). CaF_2:Dy (TLD-200) exhibits six glow peaks at about 120, 140, 200, 240, 340, and 400°C, which all show sharp emission bands at 460, 484, and 577 nm (Binder and Cameron, 1969). Dubourg *et al.* (1989) have attributed the dosimetric peak to a rare-earth Ce^{3+} ion coordinated with an oxygen ion whereas TL peak positions in the case of CaF_2 doped with rare-earth impurities have been found to be independent of the rare-earth ions, while the TL spectral emission is characteristic of the particular rare-earth impurity (Upadeo *et al.*, 1994; Holgate *et al.*, 1994).

CaF_2 phosphors, either read up to 450°C or annealed at 450°C for about 1 h, can be considered to be the virgin phosphors, and their sensitivity is not affected (Chandra *et al.*, 1981). The effects of temperature treatment on sensitivity and TL emission of CaF_2:Dy (TLD-200) to ultraviolet and ionizing radiations have been observed by Bassi *et al.* (1977) and Pradhan and Bhatt (1981); γ-ray irradiation (~1000 Gy) followed by a 30-min anneal at 450°C results in an increase in TL sensitivity. CaF_2:Dy also becomes ultraviolet-sensitive after a high-temperature (~900°C) anneal in humid air (Bassi *et al.*, 1977). The dosimetric peak at about 260°C exhibits negligible fading (6–10% in first 16 h) when the low-temperature peaks have been annealed at 100°C for 15 min. For natural CaF_2, the fading of the main peak has been reported to be less than 3% in 9 months.

Dose response for the 260°C peak of CaF_2:Mn (for ^{60}Co γ-ray) is linear over the range 5 mR to 10^5 R (Marron and Attix, 1964; Schulman *et al.*, 1960), after which the response is nonlinear. There is some degree of supralinearity in CaF_2 phosphors, the magnitude of which depends upon the phosphor configuration and the method of readout. For example, the thinner the phosphor layer during readout, the more pronounced is the apparent supralinearity in the neighborhood of 10^5 R (Gorbics *et al.*, 1973). No damage effects have been found in CaF_2:DY after high preexposure, and the response is not changed up to a preexposure of 2×20^6 R (Burgkhardt *et al.*, 1977). Hence this highly sensitive detector can be the one of choice at both lower and higher dose ranges.

CaF_2:Mn is not a tissue-equivalent material (Z_{eff} for CaF_2 is 16.57; Z_{eff} for tissue is 7.4) and exhibits an energy-dependence response. The response per röentgen is about 13-fold greater at 40 keV than for ^{60}Co γ-rays, but a tin filter can produce an essentially flat response over this range of energies, which does not differ from ^{60}Co response by more than 30% (Schulman *et al.*, 1960). The low neutron sensitivity of CaF_2:Mn makes it possible to use it as a γ-dosimeter in mixed $(n + \gamma)$ fields. No dose-rate dependence has been observed, at least over the range from 10 mR/min to 7000 R/min (Schulman *et al.*, 1960).

Natural CaF_2 also exhibits PTTL, which can be utilized as a potential technique for ultraviolet dose measurement (McCullough *et al.*, 1972; Jain, 1984; Sunta, 1979, 1984). The response is linear to low-ultraviolet exposures, and depends on the temperature of the sample during exposure.

The absorbed dose can also be measured by another technique known as optically stimulated luminescence (OSL), the so-called *cold process*, in which the preexposed phosphor is irradiated by ultraviolet radiation of suitable wavelength and not heated as in the case of TL. This irradiation detraps the carriers without degradation of the matrix (Henninger *et al.*, 1982). In a modified OSL technique, called cooled OSL (COSL), TL in the CaF_2:Mn phosphor can be read out by irradiating the phosphor at room temperature with γ-rays and cooling it to liquid nitrogen temperature. It is further exposed to ultraviolet radiation and then warmed up to room temperature; the light emitted during warming is proportional to the γ-dose (Miller *et al.*, 1988).

6.2.3. Calcium Sulfate ($CaSO_4$)

$CaSO_4$:Mn has long been known as a TL detector of ionizing radiation and was used for measuring ultraviolet and x-rays in the upper atmosphere by rocket flights (Watanabe, 1951). This phorphor is about 100 times more sensitive than the synthetic CaF_2 phosphors, and has been used mostly for low-dose measurements. The principal impurities for $CaSO_4$ phosphors for dosimetric purposes are Mn, Tm, Dy, and Sm. $CaSO_4$, being very sensitive to x-rays and γ-radiation, possessing good postirradiation stability and being easy to prepare, is widely used for dosimetric applications such as environmental monitoring where low-dose measurements are required (Nagpal *et al.*, 1979).

Preparation of $CaSO_4$:Mn can be carried out by mixing appropriate amounts of pure $CaSO_4$ and $MnSO_4$ in dilute H_2SO_4, letting the mixture stand for a few hours (to grow $CaSO_4$ crystals), evaporating it to dryness (to remove H_2SO_4), and heating the residue (for complete evaporation of H_2SO_4). The crystals are then ground into crystalline powder of 100–200 mesh. $CaSO_4$:Dy and $CaSO_4$:Tm were also prepared by this process. In order to introduce Dy or Tm as impurities, their oxides (Dy_2O_3 or Tm_2O_3) are added to the $CaSO_4$ host material (Yamashita *et al.*, 1971; Watanabe, 1951; Weng, 1993). Iga *et al.* (1977) developed TLDs based on $CaSO_4$:Tm for γ-, x-, and β-rays and thermal neutrons. $CaSO_4$:Dy, $CaSO_4$:Tm, $CaSO_4$:Dy,Tm, $CaSO_4$:Dy,Mn, $CaSO_4$:Na,Dy were all prepared using a similar procedure but in a sealed system (Azorin and Gatierrez, 1989; Lakshmanan *et al.*, 1989; Azorin *et al.*, 1993).

The glow curve peak of $CaSO_4$:Mn is found at 110°C with maximum emission of TL at 495 nm. Glow curves of both $CaSO_4$:Dy and $CaSO_4$:Tm are simple with their peaks at 220°C and shoulders at 80 and 120°C (Yamashita *et al.*, 1971). The emission spectrum of $CaSO_4$:Tm consists of a sharp main peak at 450 nm with minor peaks at 360, 470, and 520 nm, whereas $CaSO_4$:Dy has two main peaks at 480 and 570 nm without any minor ones. The activator concentration of Dy and Tm has been observed to affect the TL sensitivity, glow-curve structure, and fading characteristics of $CaSO_4$ TLDs (Hsu *et al.*, 1987).

Annealing the $CaSO_4$:Dy phosphor at 600°C for 20 min, prior to irradiation, enhances the dosimetric peak by 15% and suppresses its quite fast initial fading (about 20% when read after 20 h), which is considered to be due to lower-temperature peaks (occurring at 80 and 120°C). Postirradiation annealing at 100°C for 5 min can also erase the initial fast fading. Postirradiation annealing at 700°C is found to remove the sensitization effect in $CaSO_4$:Dy phosphor completely. Fading at room temperature for $CaSO_4$:Dy and $CaSO_4$:Tm is 3 to 6% and 4 to 6%, respectively, after 6 months. Fading is also enhanced considerably by exposure to light.

Dose response of $CaSO_4$:Mn TLD for γ-rays has been observed to be linear from about 0.1 mR to 10^4 R and gets saturated above about 3×10^4 R (Attix *et al.*, 1968). The minimum detectable dose for $CaSO_4$:Dy and $CaSO_4$:Tm is lower than 200 mR, with a standard deviation of $\pm 20\%$. Hsu *et al.* (1977), however, observed the γ-ray lower limit detection at about 6×10^{-3} mR. Dose responses of $CaSO_4$:Dy and $CaSO_4$:Tm powders are linear up to 3000 and 300 R, respectively, above which they behave supralinearly. The linear response range and starting point of supralinearity differ in the case of teflon-embedded or ribbon phosphors when they are preexposed to ^{60}Co γ-rays (Bacci *et al.*, 1980).

High-LET irradiations (α-rays) give rise to glow peaks in $CaSO_4$:Dy at 210 and 375°C possessing linearity in the range $1–10^4$ Gy, and the deep traps are not damaged by high doses of α-particles as is the case with γ-doses. Moreover, a transition from supralinearity for low-LET radiations (γ-rays) to linearity for high-LET radiations (α-particles, heavy ions) has been observed (Lakshmanan *et al.*, 1981).

Energy dependence in $CaSO_4$ phosphor powders is large for photon energies below 200 keV, which is attributed to the high Z-value of $CaSO_4$, and can be reduced by using metal filters such as Sn, Pb, Cu, and Al (Vohra *et al.*, 1980). The graphite-loaded $CaSO_4$:Dy teflon disk is one of the possible ways to measure β-radiation. Weng and Chen (1974) investigated TL response versus thermal and fast neutron fluence at high dose levels in a mixed neutron–γ-field using $CaSO_4$:Dy.

Although the γ-ray TL sensitivity of $CaSO_4$:Dy and $CaSO_4$:Tm phosphors is the same, TL sensitivity of $CaSO_4$:Tm for ultraviolet radiation is about 60 times higher than that of $CaSO_4$:Dy (Lakshmanan and Bhatt, 1979). The TL sensitivity also increases by a factor of 300 as the wavelength is decreased from 250 to 200 nm, and the response is negligibly small above 350 nm (Chandra *et al.*, 1976).

6.2.4. Lithium Borate ($Li_2B_4O_7$)

Schulman *et al.* (1967) were the first to propose the possibility of using $Li_2B_4O_7$:Mn for medical therapy dosimetry in place of LiF. As the former's Z_{eff} (7.3) is slightly smaller than that of tissue, one may design a completely tissue-equivalent $Li_2B_4O_7$ phosphor by adding a small amount of additive (Cameron *et al.*, 1968*a,b*). Moreover, it possesses a single glow peak with good trap stability and a low residual dose after evaluation. It does not need any preirradiation annealing and exhibits no change in response after high-dose preexposure (Burgkhardt *et al.*, 1978*a,b*). Because the TL emission of $Li_2B_4O_7$:Mn is in a region (605 nm), that does not match the spectral sensitivities of commercial photomultipliers, this phosphor does not have enough sensitivity to low doses, especially during personnel and environmental monitoring. $Li_2B_4O_7$:Cu phosphor prepared by Takenaga *et al.* (1980) has a high sensitivity to low doses as its TL emission at 360 nm matches well with the photomultiplier's spectral response. The main limitation of using $Li_2B_4O_7$:Mn as a TLD is its supralinear nature in the clinical range (over 100 rads), and here it cannot be a substitute for LiF even if it possesses ideal energy characteristics.

For preparing $Li_2B_4O_7$:Mn, reagent grade Li_2CO_3 and H_3BO_3 are mixed mechanically in the desired ratio, moistened with a solution containing $MnCl_2,2H_2O$ to yield a product with 0.1 mole % Mn, dried for 12 h at 100°C, fired at 950°C in a platinum crucible until it melts, mixed thoroughly, fired for another 20 min, and stirred again. $Li_2B_4O_7$:Cu phosphor prepared by a sintering method (Takenaga *et al.*, 1980) showed the best properties for use as a dosimeter. In this case, Li_2CO_3 and H_3BO_3 are mixed, melted at 950°C, and then rapidly cooled at room temperature. The resulting glass is reheated at 650°C to complete crystallization. To this raw $Li_2B_4O_7$ powder is added a solution of $CuCl_2$ in acetone and alcohol, and the mixture is then stirred and dried. The dried mixture is treated in air in a platinum boat for about 1 h at about 900°C (just below the melting point, (917°C) of $Li_2B_4O_7$).

There are two glow peaks for a $Li_2B_4O_7$:Mn phosphor, one at 55°C (fading fast) and the other at 200°C (stable and useful for dosimetry), whereas in case of $Li_2B_4O_7$:Cu these peaks lie at 120 and 205°C, respectively. The γ-ray sensitivity of $Li_2B_4O_7$:Cu has been found to be 20 times higher than that of $Li_2B_4O_7$:Mn. Although no annealing is needed in $Li_2B_4O_7$ phosphors, Furetta and Pellegrini (1981) have observed that in case of $Li_2B_4O_7$:Mn (TLD-800) in ribbon form, optimum annealing for producing high sensitivity is for 10 min at 275°C.

Fading in these phosphors at room temperature is about 10% in 1 month. Under fluorescent light, the fading is 85% in 3 days, which necessitates the storage of these phosphors in the dark during irradiation and readout (Tomaru *et al.*, 1983).

Dose response for γ-rays is linear from about 10 mR (lower limit of detection) to 300 R, above which it becomes supralinear with saturation occurring at about 10^6 R. At about 10^5 R, where the phosphor is sensitized, prior to irradiation annealing for 15 min at 300°C can be used to improve the linearity. TL output of $Li_2B_4O_7$:Cu is linear with exposure up to about 10^5 R, above which it is sublinear (Takenaga *et al.*, 1980).

The energy-dependent response curve is almost flat for $Li_2B_4O_7$:Mn, as the TL response is only 10% less for 30-keV x-rays than for 1 MeV. It shows very little dependence at low photon energies (10–100 keV) (Thompson and Ziemer, 1972), and depends critically on the activator concentration.

For thermal neutron–γ-ray mixed-field dosimetry, phosphors with mixed concentrations of $^7Li_2\,^{11}B_4O_7$:Cu and $^6Li_2\,^{10}B_4O_7$:Cu were prepared that are very sensitive to thermal neutron dose but not to fast neutrons. The albedo-type dosimeters were developed for fast neutrons (Yamashita, 1993).

6.2.5. Aluminum Oxide (Al_2O_3)

Al_2O_3 is abundantly available in natural form as alumina (ruby, sapphire). Because of their low cost, chemical inertness, and ease of preparation and handling; Al_2O_3 dosimeters can be made mechanically hard and nonfragile and can be easily washed and cleaned. These features combined with other properties (linearity or reproducibility), give them an advantage over other types of dosimeters. An Al_2O_3:Mg,Y dosimeter can be used in γ-ray and mixed neutron-γ-ray accident dosimetry, since its sensitivity to fast neutrons is low. The feasibility of Al_2O_3 is to be expected in industrial irradiation techniques, in accidental dosimetry, and in ultraviolet detection and measurement (Osvay and Biro, 1980; Salas *et al.*, 1993).

Mehta and Sengupta (1976) prepared pure Al_2O_3 from commercially available alumina powder purified by repeatedly boiling with concentrated HCl for 1 h, and further washing with distilled water. The purified powder was subjected to different heat treatments: (i) 1 h at 1100°C, (ii) 1 h at 1400°C, and (iii) 10 min at 1800°C. Flaming in an oxygen-burshane flame forms melted granules, which are crushed to powder of the requisite mesh size.

The prominent impurities responsible for TL in Al_2O_3 are: Na, Si, Ti, Ca, Fe, Cr, and Mg, which can be incorporated by thoroughly mixing the oxide or a carbonate of the impurity element in pure Al_2O_3 powder, and heat-treating in air in a furnace or in a reducing atmosphere such as H_2. The pellets can be formed by cold-pressing the mixture and sintering at 1500–2000°C in an oxidizing atmosphere (Ranogaject-Komor *et al.*, 1983). Al_2O_3:Li powder (a TSEE material) was prepared by slow heating, in air, of a mixture of Al_2O_3 and Li_2CO_3 and annealing for 4 h at 1200°C, whereas thin films of Al_2O_3:Li were prepared by flame-spraying the powder on graphite disks (Chryssou and Holzapfel, 1980).

Osvay and Biro (1980) observed two peaks of dosimetric importance in Al_2O_3:Mg,Y ceramic disks at 250 and 470°C when irradiated with ^{60}Co γ-rays. The 470°C peak can be observed only above an exposure of 100 Gy and it grows more rapidly than the 250°C peak. Mehta and Sengupta (1976) have observed glow peaks in Al_2O_3:Si,Ti upon irradiation with γ-rays, at 50, 125, 250, 325, 475, and 625°C with emission in the blue region. The 250 and 475°C peaks can be used for dosimetric purposes from a few mR to 10^5 R because these peaks show a linear response up to 40,000 R. Details concerning crystal structures and TL mechanism of Al_2O_3 have been reviewed by Osvay and Biro (1993) and Summers (1984).

Thermal annealing (preirradiation) from 100 to 2000°C of pure Al_2O_3 powder enhances its γ-induced TL response (Mehta and Sengupta, 1976). It has also been suggested (Cooke *et al.*, 1981) that annealing at 800°C for 1 h makes a Al_2O_3 phosphor reusable and restores the crystal to its virgin state. Fading is almost nil for 250, 475, and 625°C peaks in γ-irradiated Al_2O_3:Si,Ti at room temperature even after a 2-week storage period (Mehta and Sengupta, 1976). Comparative fading studies on Al_2O_3:Mg,Y, LiF, CaF_2, and MgB_4O_7:Dy TLDs have been carried out after ^{60}Co γ-irradiations at a dose of 0.2 Gy by Ranogajec-Komor and Osvay (1986).

Mehta and Sengupta (1976) have noted that the dosimetric peak (250°C) of Al_2O_3:Si,Ti grows linearly with exposure from a few mR to 400 R and at a supralinear rate of $(\text{exposure})^{1.4}$ up to 15,000 R, followed by saturation up to 20,000 R. They have also observed dose responses for 475 and 625°C glow peaks. The TL response in pellets for a sensitizing exposure of 10^5 R and a partial annealing at 525°C shows that supralinearity is almost removed in the growth of the 250°C peak.

The Al_2O_3 phosphor with $Z_{eff} = 10.2$ is a tissue-equivalent down to 140 keV and in the range 140–33 keV, and it gradually rises to three times the tissue-equivalent response, thus lying between the response of LiF and CaF_2 (see Fig. 10) (Mehta and Sengupta, 1975). Thermal neutron sensitivity has been found to be much lower than that reported for TLD-700, and is competitive with $CaSO_4$ and BeO. Al_2O_3 is almost insensitive to fast-neutron irradiation. A relatively low response to thermal and fast neutrons up to 7 MeV and the identical glow curves after exposure to neutrons make it possible to determine the γ-dose in mixed $(n + \gamma)$ fields (Osvay and Biro, 1993).

Although Al_2O_3:Si,Ti has been found to be insensitive to visible light, an ultraviolet-irradiated phosphor exhibits a stable peak at about 250°C with TL emission at 410 nm, thus, making these TL phosphors ideally suited for ultraviolet dosimetry (Mehta and Sengupta, 1978, Cooke, 1974). Cooke (1974) noted the enhancement in ultraviolet sensitivity, by a factor of six, in the samples that were first exposed to ^{60}Co γ-irradiation. This enhancement can be attributed to the fact that deep traps are filled during γ-irradiation and the subsequent ultraviolet irradiation photostimulates the electrons to shallow traps exhibiting PTTL.

6.2.6. *Beryllium Oxide (BeO)*

BeO displays strong TL together with tissue-equivalent properties, which underscore its applicability for TL dosimeters for x- and γ-rays.

Preparation of a BeO TL phosphor can be made from the undoped form of commercially available BeO, Thermalox 995, with a total content of impurities not exceeding 0.5% (Spurny and Hobzova, 1977). The major impurities, in ppm by weight, are 2150 of Si, 995 of Mg, 100 of Fe, 61 of Ca, and 54 of Al. Yamashita *et al.* (1974) prepared efficient and reliable BeO:Li and BeO:Na TLDs with more advanced dosimetric characteristics by mixing purified BeO powder with various dopants, including SiO_2, GeO_3, $Al_2(SO_4)_3$, and Na_2SO_4, sintering at 1500°C, and cooling slowly. The ceramics are ground into small grains (100–300 μm in diameter) and sealed into glass ampoules.

Glow peaks of BeO:Li and BeO:Na occur at 180°C (Yamashita *et al.*, 1974). The glow-curve shape does not change for exposure levels between 1 mR and several hundred R. At more than 1000 R, a new glow peak appears at about 210°C. The first dominant peak in the TL of 'Thermalox 995' has been found around 200°C, while a second peak has been observed at about 280°C (Hobzova, 1974).

The TL emission spectrum of BeO lies between 200 and 450 nm, the peak value being from 330 to 340 nm, depending on the nature of the glow curve. Therefore, to obtain maximum sensitivity, BeO must be read using ultraviolet-sensitive photomultipliers with quartz windows and ultraviolet-transmitting filters (Spurny and Hruska, 1976; Lembo and Scarpa, 1993).

Various methods of thermal annealing have been adopted to remove the residual dose. Crase and Gammage (1975) annealed 'Thermalox 995' at 650°C for 30 min followed by a slow cooling to room temperature. Scarpa *et al.* (1971) used annealing at 600°C for 5 min. Tochilin *et al.* (1969) suggested an annealing cycle of 30 min at 550°C. Busuoli *et al.* (1984) have obtained good results by annealing at 400°C for 15 min.

Fading in BeO phosphors depends on the type of material used, on the annealing treatments, and on the techniques followed to carry out the measurements (Böhm *et al.*, 1984; O'Hagan *et al.*, 1986; Busuoli, 1981). Fading is much faster under exposure to fluorescent light or daylight, i.e., 50% in 20–30 min (for Thermalox 995) with light intensities of 1500–3000 luxes (Benincasa *et al.*, 1974). Crase and Gammage (1975) and Gammage and Cheka (1977) also observed fading, and similar values were obtained for high-temperature TL peaks.

Dose response for γ-irradiation ranges from 10–10,000 rads, with supralinearity starting at levels around 100 rads, and tends to saturate above 10,000 R. Sublinearity seems to be present in the very low-exposure region (Scarpa, 1970; Tochilin *et al.*, 1969). Crase and Gammage (1975) have observed linearity with γ-ray exposure from 1 mR to 10 R above which the behavior is supralinear. Yamashita *et al.* (1974) have found that BeO:Li and BeO:Na irradiated with γ-rays are linear from 1 mR to 500 R and 100 R, respectively, and the linear region can be extended to higher dose levels by increasing the amount of impurity.

Energy response of ceramic samples of BeO:Na and BeO:Li have roughly a flat response (within an experimental error of 15%) when exposed to x-rays of 30–115 keV and γ-rays of ^{60}Co (Yamashita *et al.*, 1974). Any discrepancies between the theoretical (flat) and experimental energy response can be attributed to the LET dependence (Tochilin *et al.*, 1969). BeO seems to be the least thermal-neutron sensitive. This property makes it useful for $(n + \gamma)$ mixed-field dosimetry. Gibson (1986) has reviewed all the data on neutron sensitivity of TL materials extensively.

6.2.7. Magnesium Silicate (Mg_2SiO_4)

Mg_2SiO_4 occurs abundantly in nature as forsterit or olivin. Mg_2SiO_4:Tb is known to be a γ- and ultraviolet-sensitive phosphor (Lakshmanan, 1979), exhibits a similar glow curve when exposed to ultraviolet or γ-radiation, and has a large measurable dose range and high storage stability.

Jun and Becker (1975) prepared Mg_2SiO_4 by mixing MgO, SiO_2, and Tb_2O_3 in HCl and HNO_3. The resulting paste is heated at 400–600°C to dryness, powdered, and heated again at 1700°C for several hours in a platinum crucible. It is further powdered and sieved. Bhasin *et al.* (1976) prepared this phosphor from the mixture of MgO and SiO_2 with Tb_4O_7 as the activator by heating at a much higher temperature (about 2750°C).

The glow-peak maximum from an x-irradiated phosphor appears near 190°C (Jun and Becker, 1975), the peak position of the dosimetric peak shifts gradually to 240°C with the increase of pre-γ exposure (Lakshmanan and Vohra, 1979).

TL emission spectrum peaks of Mg_2SiO_4 phosphors occur mainly around 552 and 380 nm, and the weak peaks around 420, 480, 580, and 650 nm, which are typical of Tb^{3+} ions.

Annealing for 10–30 min at 400–500°C, prior to reuse of the irradiated phosphor has been suggested by Jun and Becker (1975). Lakshmanan and Vohra (1979) reported that no complicated annealing is involved, but found that Mg_2SiO_4:Tb phosphor shows enhanced sensitivity to ultraviolet, γ-, and x-radiation after high exposures (>100 R) and a postannealing treatment at 300°C for 1 h.

Fading is almost insignificant at room temperature ($\sim$25°C) up to an interval of 20 days (Lakshmanan, 1979), and also during storage of the γ-irradiated phosphor up to 3 months (Jun and Becker, 1975).

Dose response is nearly linear for both powdered and teflon-embedded phosphors up to about 400 rads of γ-radiation, supralinearly at higher doses, followed by saturation at 7×10^6 rads. According to Yamazaki *et al.* (1978) the response is linear up to 35 rads, supralinearity occurs between 50 and 15,000 rads, and the response curve in the region obeys a power law $I(D) = 0.024D^{1.23}$, where $I(D)$ is the TLD response for a dose D (in rads). Sensitization of the phosphor can be obtained by preexposure to high γ-radiation levels followed by partial annealing at 250–300°C (Nakajima, 1971).

The energy dependence is less pronounced than in CaF_2 and $CaSO_4$, but more than in LiF, BeO, and $Li_2B_4O_7$ phosphors. An overall sensitivity varies with the energy of incident γ- or x-rays in the energy range from 23 keV to 1.25 MeV. The response for 38-keV radiation is about four times that for ^{60}Co γ-rays (Nakajima, 1993). By embedding the phosphor in teflon, or using a correlation filter, the energy response is substantially reduced to nearly the same value as that of LiF.

7. THERMOLUMINESCENCE MEASUREMENT METHODS

TL measurements can be made with a very simple and relatively inexpensive experimental setup, which can be easily assembled in a research laboratory, even on a limited budget. The TL intensity of a phosphor (preexposed to a radiation dose) in which the dose has been accumulated can be recorded in two ways: (i) TL intensity versus wavelength, and (ii) TL versus temperature. In the first case, recording of spectral distribution of the TL emitted light corresponding to each glow curve can reveal the band structure. Different mechanisms leading to emission, such as band-to-band transitions, transitions involving trapping via tunneling, or simultaneous trapping and annhiliation of electrons and holes at the recombination centers, can also be interpreted. The TL emission spectrum of a glow peak can be obtained by maintaining the temperature of the TLD material a little below the reference peak temperature and scanning the emission by means of a monochromator (Wintle, 1985; Delucas *et al.*, 1988).

A second investigation of TL intensity versus temperature is done to measure the radiation dose. The experimental setup meant for this purpose, the so-called TLD reader, may be a manual, semiautomatic, or fully automated system. The TLD reader essentially consists of (i) a heating system for controlled variation of the temperature of a TLD sample, (ii) a light collection and detection system, and (iii) a system for recording the signal and a display.

7.1. Basic Thermoluminescence Apparatus

An experimental arrangement, in its simplified form, consists of a light-tight box in which a small sample of the TLD specimen (powder, disk, or ribbon) is heated. The TLD reader system is designed in such a way that the window of the light-detecting photomultiplier (PM) fronts on the specimen, whereas the other heated areas face it as minimally as possible to avoid background incandescent emission from the heating system. To achieve maximum TL intensity, the spectral response of the PM tube should also match the spectrum of the TL emission. A schematic block diagram of a TLD reader is shown in Fig. 11.

Many types of TLD readers that can be suitably utilized to measure radiation dose are commercially available. Details of the equipment for measuring the TL has been described in the literature (Bjarland and Toivonen, 1980; Vij and Mahesh, 1985; Sankran *et al.*, 1979; Szabo *et al.*, 1984). TLD readers can be equipped with an output recorder, such as a simple printer, teletype, or similar device. Modern TLD

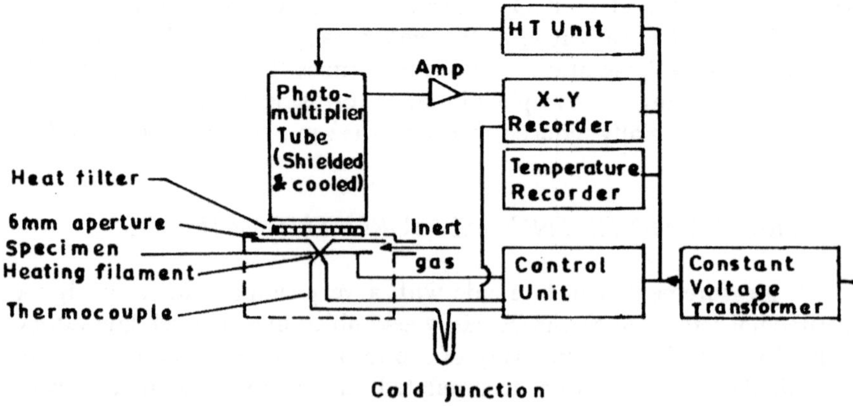

Figure 11. Schematic diagram of a TL apparatus (Mills *et al.*, 1977, with kind permission from IOP Publishing Ltd., UK).

readers come with additional computer-controlled systems, which are capable of such operations as glow-curve plotting, output writing and response handling, and automation. (Vana *et al.*, 1988).

7.2. Heating Systems

As the rate of heating plays an essential role in shaping the glow curve, the rate at which the TL specimen is heated should preferably be linear and reproducible to obtain reproducible and consistent results (Piters and Bos, 1994).

Heating of a TLD sample can be accomplished in various possible ways depending upon the form in which the specimen is available (i.e., a microcrystalline powder, single crystal, sintered powder pellets, extruded chips, or ribbons). The main types of heating systems are: (i) electrical resistance (ohmic) heating, (ii) gas heating, and (iii) optical heating.

Electrical resistance heating is the easiest and most widely used system: a metal planchet is either heated indirectly through a heater or a heavy current is passed through platinum/nichrome/graphite strip. A linear heating rate of a preset value can be obtained by adjusting the rate of change of the heater current, either mechanically by attaching a synchronous motor to the variac supplying the voltage to the heater or electronically by a temperature controller.

Gas heating can be carried out either indirectly by hot nitrogen gas or directly by injecting the hot air or nitrogen gas through a nozzle. This is a convenient and a quick heating system (up to 350°C in 4–8 s) without the necessity of contact between the heating source and the specimen. The signal-to-noise ratio is also increased owing to quick heating and the noise (blackbody radiation) owing to heat radiation from the heated planchet is reduced. (Soares and Ehrlich, 1981).

Optical heating, like gas heating, is a noncontract heating process, which is usually done through infrared, microwaves, and lasers (Bräunlich *et al.*, 1981). Microwaves are more penetrating than infrared rays and heat the interior part of the TL specimen uniformly, but laser heating provides a drastic increase in the heating rate (up to 2×10^4 deg K/s). Moreover, this enables one to heat a very small quantity of TL phosphor (20 μg). Infrared lasers are preferred over ultraviolet lasers.

7.3. Light Detection, Recording, and Display

PMs are widely used to detect TL light emitted by a phosphor. For obtaining maximum response from a PM tube, its spectral response should nearly match the TL emission spectrum of the specimen under investigation, which means different PM tubes have to be used for different TLDs. Moreover, the operating parameters of the PM tube, such as spectral response, sensitivity, linearity of response, and dark current, need to be optimized. For good linearity between the TL intensity as input and the current of the PM tube as output, the voltages at dynodes must be high enough to operate the PM tube under saturated electron-current conditions to eliminate voltage dependence of the response of the PM tube. Signal-to-noise ratio can be greatly increased and the dark current reduced by cooling the PM tube. Heat-suppressing filters (thick quartz plate) can be placed between the sample and the PM tube to eliminate the background blackbody radiation from the heater while detecting TL emitted from the deep trapping state at high temperatures, especially during archaeological and geological dating experiments. Light collection efficiency can be improved by using light guides such as light pipes or fiber optic cables. Close proximity of sample and detector can also increase the efficiency.

Silicon diodes or even vidicons can also be employed to detect the TL light emission (Castagne *et al.*, 1978) because of their broad, flat response between 0.35 and 1.2 μm.

The simplest and often adequate method for recording of TL intensity is with an X-Y chart recorder (DC method), which displays TL intensity as a function of time and may plot a glow curve directly (i.e., $I(T)$ vs. T curve). CRT displays are also employed in cases where there is some problem with the recorder response time. For short intervals the TL spectrum can also be obtained by counting the number of emitted photons (photon-counting method) displaying them in one of the channels. By using a multichannel analyzer (MCA) system, pulses can be counted and analyzed automatically. The complete characterization data analysis can be carried out using computers.

8. APPLICATIONS

Considerable progress has been made during the last two decades in determining the trapping parameters and describing the defect structures in solids. In addition, the TL technique has been widely developed and applied for detecting ultraviolet and ionizing radiation in the areas of health physics, radiation protection and control (such as environmental and personnel monitoring around nuclear power plants), and biomedical sciences. With this technique one can establish ages of archaeological materials and geological formations. TL has also been used in forensic science for the characterization of suspect materials and identification of forged objects, and for authentications artifacts and ancient artworks.

8.1. Radiation Dosimetry

In radiation monitoring of personnel working with reactors, particle accelerators, x-ray machines, and radioactive sources, or in surroundings where such facilities are in use, the need for radiation protection and control arises. This is even a statutory requirement for protection against radiation hazards. TL can be used as an effective experimental tool to measure radiation dose (Cameron *et al.*, 1961, 1968; McKinlay, 1981; Horowitz, 1984; Mahesh and Vij, 1985; Oberhofer and Scharmann, 1981). Quite sensitive and highly economical TLDs having reasonably accurate and linear responses over a wide dose range have been prepared in different laboratories all over the world. They can be made in small sizes and different shapes.

The typical dose ranges in which TLDs can be utilized are 10^{-4} to 10^{2} Gy. However, there are complications in heat treatment and anneal with the TLDs (prior to reuse) that have to be tolerated. TL dosimetry can be used for monitoring the trunk of the body as well as the extremities (e.g., skin, gonads, lungs, thyroid, breast, fingers, and nails) as these parts of the human body are affected by exposure to radiation. Skin dosimeters can assess the dose delivered to a depth of 5–10 mg/cm^2, whereas body dosimeters measure to a depth of 300–1000 mg/cm^2.

TLDs are mostly used to measure doses of γ-rays (DeWerd, 1979), β-rays (Lakshmanan *et al.*, 1980), β-rays in mixed (β–γ) fields (Piesch, 1980), neutrons (Pradhan *et al.*, 1978), mixed (n–γ) (Piesch and Burgkhardt, 1980), and heavy charged particles such as alphas, protons, and pions (Kalef-Ezra and Horowitz, 1982).

In radiation therapy and diagnosis, the estimation and planning of the dose delivered to a cancer patient are of major importance. During external beam radiotherapy, one needs to know the relative values of the doses absorbed by the cancerous tissue and the surrounding healthy tissue so that the dosage delivered during treatment does not destroy the healthy tissue. (Weng *et al.*, 1981). This can easily be done with the help of TLDs rather than the complex, but more sophisticated, computerized tomography technique, since the delivered dose can either be measured indirectly with a phantom or directly with *in vivo* techniques (McKlveen, 1980).

As TLDs are available in small sizes, they can be sent by mail to such places as reactor sites and clinics, exposed to the personnel working in or near the nuclear plants or other radioactive sources or to the patients or doctors in the clinics, and then mailed back to the monitoring facility for readings. This type of mail monitoring is being practiced on a large scale. Moreover, the small size of the TLDs makes it possible to insert them directly into body cavities or to use them to monitor small regions such as the cornea of the eye (Wochos *et al.*, 1978).

In a case of accidental exposure to radiation, where no preplanned use of a dosimeter is possible, the extent of the exposure can be estimated, e.g., by the accumulated dose in the subject's nails, teeth, or bones (Willhoit and Poland, 1967), or from the condition of the jewels of a watch worn at the time (Watanabe and Nakajima, 1973).

8.2. Environmental Monitoring

Monitoring of the level of environmental radiation has become essential owing to increased environmental radioactivity arising from cosmic rays, radioactive materials, atomic explosions, leakages from nuclear plants, and nuclear testing. Persons traveling in spacecraft or aircraft may also be exposed to cosmic rays, solar ultraviolet radiation and/or radioactive cargo (Waschmann and Regulla, 1978). Radionuclides in building materials can also increase the radiation level in homes. TLDs for environmental radiation dose measurements must be highly sensitive with high thermal and optical stability, since the measured dose levels are much smaller and the exposure periods much longer. They must also be durable and moisture- and chemical-resistant, and they should be protected from light with the necessary filters.

It may also be noted that TLDs have also been used to estimate the γ-radiation from the atomic bomb explosions at Hiroshima and Nagasaki through roof tiles that absorbed γ-rays during the explosions but were not burned (Ichikawa *et al.*, 1966).

8.3. Thermoluminescence Dating

Dating of noncarbonaceous archaeological samples is not possible by the C^{14} method. Similarly, the age of Quaternary period geological samples is difficult to assess by most of the classical radiometric methods. However, dating of both archaeological and geological materials can be carried out successfully by a TL technique. TL dating of such archaeological (mainly ancient pottery) and/or geological materials (ocean sediments), volcanic events, calcareous deposits (loess and till) is generally carried out on minerals separated from these materials. Natural TL exhibited by minerals and rocks is related directly to the nuclear radiation absorbed from radioactive U, K or Th present within the material and the cosmic background at the burial site, which the material has been subjected to from the time the pottery was buried or the ocean sediments/and rocks were formed.

For determining the age, the accumulated TL (natural TL) in the specimen owing to radiation from the radio-active elements present in the clay and the surroundings is measured and expressed in terms of absorbed dose by proper calibration techniques. The dose rates for different types of radiation are measured separately by placing TLDs at the sample sites for a specified time. The archaeological age can be obtained by simply dividing the absorbed dose by the dose rate. The details of the TL technique for dating an archaeological specimen are reviewed by Aitken (1984), Fleming (1979), Aitken and Fleming (1973), and Mejdahl and Aitken (1980).

Among geological materials, sedimentary rocks, being of recent origin, are amenable to TL dating. In the case of the other two forms (i.e., igneous and metamorphic) correct age determination is not possible because of saturation effects in these very old rocks. TL rock minerals such as calcite, dolomite, quartz, aragonite, and magnesite form part of the sedimentary rocks. The presence of radioactive U, Th, and K in these rock-forming minerals slowly builds up the natural TL with time. Their age is obtained if one can determine the annual dose rate received by the samples as in case of archaeological samples (Singhvi and Mejdahl, 1985; Wintle and Huntley, 1982). Geological age of a lava flow and/or the thermal flow profile around the flow can also be determined by TL. This is because of the fact that at the time of a volcanic eruption TL is released from nearby rock owing to the very high temperature of the lava that touches it. After the lava has cooled, the natural TL is accumulated with time because of the natural radioactivity present in the surroundings as well as the radioactive elements in the lava. With the measurements of natural TL intensity and the local radiation rate, one can estimate the age of the lava flow.

TL has also been used to determine the ages of meteorites (Sears, 1988), lunar materials (McDougall, 1968), sand dunes (Aitken, 1984), and biological materials (Jasinska and Niewiadomski, 1970; Christodoulides, 1993), including bone and marine shells (Driver, 1979), enamel and dentine (Willhoit and Poland, 1967), and fossil shells (Johnson and Blachard, 1967).

REFERENCES

Aitken, M. J. (1984), *Thermoluminescence Dating* (Academic, London).

Aitken, M. J., and Fleming, S. J. (1973), in: *Topics in Radiation Dosimetry* (F. H. Attix, Ed.) (Academic, New York).

Aramu, F., and Maxia, V. (1975), *J. Lumines.* **10**, 77.
Aramu, F., Maxia, V., and Rucci, A. (1975), *J. Lumines.* **10**, 277.
Attix, A. H. (1975), *J. Appl. Phys.* **46**, 81.
Attix, F. H., West, E. J., Nash, A. E., Gorbics, S. G., and Johnson, T. L. (1968), N.R.L. Report No. 20390.
Azorin, J., and Gatierrez, A. (1989), *Health Phys.* **56**, 551.
Azorin, J., Furetta, C., and Scacco, A. (1993), *Phys. Stat. Sol.* (a)**138**, 9.
Bacci, C., Calicchia, A., Furetta, C., Prigliani, L., and Salvadon, P. (1980), *Health Phys.* **38**, 21.
Balarin, M. (1979), *J. Thermal Analysis*, **17**, 319.
Bangert, U., Thiel, K., Ahmed, K., and Townsend, P. D. (1982), *Radiat. Effects*, **64**, 143; **64**, 153.
Bartlett, D. T., and Edwards, A. A. (1979), *Phys. Med. Biol.* **24**, 1276.
Bassi, P., Busuoli, G. and Rimondi, O. (1977), *Nucl. Inst. Meth.* **143**, 195.
Becker, K. (1973), *Solid State Dosimetry* (C.R.C. Press, Boca Raton, Fla.).
Benincasa, G., Ceravolo, L., and Scarpa, G. (1974), CNEN (Italy), RT/PROT(74), 1.
Bhasin, B. D., Sasidharan, R., and Sunto, C. M. (1976), *Health Phys.* **30**, 139.
Binder, W., and Cameron, J. R. (1969), *Health Phys.* **17**, 613.
Bindi, R., Lapraz, D., Iacconi, P., and Boutayeb, S. (1994), *J. Phys. D., Appl. Phys.* **27**, 2395.
Bjarland, B., and Toivonen, M. (1980), *IEEE Trans. Nucl. Sci.* **NS-27**, 969.
Böhm, M., Pitt, E., and Scharmann, A. (1984), *Radioat. Prot. Dosim.* **8**, 139.
Böhm, M., Iacconi, P., Kromm, K. D., and Scharmann, A. (1994), *Phys. Stat. Sol. (a)* **146**, 757.
Booth, A. H. (1954), *Canad. J. Chem.* **32**, 214.
Braunlich, P. (1967), *J. Appl. Phys.* **38**, 1221, 2516.
Braunlich, P. (1968), in: *Thermoluminescence of Geological Materials* (D. J. McDoughall, ed.) (Academic, New York), p. 61.
Braunlich, P. (Ed.) (1979), *Thermally Stimulated Relaxation in Solids* (Springer-Verlag, Berlin).
Braunlich, P., and Scharmann, A. (1964), *Z. Phys.* **177**, 320.
Braunlich, P., Gasiot, J., Fillard, J. P., and Castagne, M. (1981), *Appl. Phys. Lett.* **39**, 769.
Brightwell, J. W. (1993), in: *Thermoluminescent Materials* (D. R. Vij, Ed.) (Prentice Hall, Englewood Cliffs, NJ), p. 30.
Broser, I., and Broser-Warminsky, R. (1968), in: *Thermoluminescence of Geological Materials* (D. J. McDoughall, Ed.) (Academic, New York), p. 87.
Burgkhardt, B., Singh, D., and Piesch, E. (1977), *Nucl. Inst. Meth.* **141**, 363.
Burgkhardt, B., Herrera, H., and Piesch, E. (1978*a*), *Nucl. Inst. Meth.* **155**, 293, 299.
Burgkhardt, B., Piesch, E., and Singh, D. (1978*b*), *Nucl. Inst. Meth.* **148**, 613.
Busuoli, G. (1981), in: *Thermoluminescence Dosimetry* (Adam Hilger, Bristol), p. 83.
Busouli, G., Lembo, L., Nanni, R., and Sermenghi, I. (1984), *Radiat. Prot. Dosim.* **6**, 317.
Cameron, J. R., Daniels, F., Johnson, H., and Kenney, G. N. (1961), *Science* **134**, 333.
Cameron, J. R., Sunthralingham, N., and Kenney, G. N. (1968), in: *Thermoluminescent Dosimetry* (University of Wisconsin Press, Madison), p. 49.
Castagne, M., Gasiot, G., and Fillard, J. P. (1978), *J. Phys.* **E11**, 345.
Castiglioni, M., Martini, M., Spinolo, G., and Vedda, A. (1994), *Radiat. Meas.* **23**, 361.
Chandra, B., Ayyangar, K., and Lakshmanan, A. R. (1976), *Phy. Med. Biol.* **21**, 66.
Chandra, B., Bhatt, R. C., and Supe, S. J. (1981), *Nucl. Inst. Meth.* **184**, 549.
Chen, R. (1969), *J. Appl. Phys.* **40**, 570.
Chen, R. (1974), *J. Mat. Sci.* **9**, 345.
Chen, R. (1976), *J. Mat. Sci.* **11**, 1521.
Chen, R. and Kirsh, Y. (1981), *Analysis of Thermally Stimulated Processes* (Pergamon, Oxford).
Chen, R., and Winer, S. A. (1970), *J. Appl. Phys.* **41**, 5227.
Christodoulides, C. (1985), *J. Phys. D., Appl. Phys.* **18**, 1501, 1665.
Christodoulides, C. (1993), in: *Thermoluminescent Materials* (D. R. Vij, Ed.) (Prentice Hall, Englewood Cliffs, NJ), p. 392.
Christy, R. W., Johnson, N. M., and Wilbarg, R. R. (1967), *J. Appl. Phys.* **38**, 2099.
Chryssou, E., and Holzapfel, G. (1980), *Phy. Stat. Sol. (a)* **59**, K49.
Claffy, E. W., Klick, C. C., and Attix, F. H. (1968), Proc. 2nd Int. Conf. on Luminescent Dosimetry, Gatligberg, p. 302.
Cooke, D. W. (1974), *Health Phys.* **27**, 130.

Cooke, D. W., and Close, D. M. (1982), *J. Chem. Phys.* **76**, 2174.
Cooke, D. W., Payne, I. W., and Santi, R. S. (1981), *J. Appl. Phys.* **52**, 3606.
Crase, K. W., and Gammage, R. B. (1975), *Health Phys.* **29**, 739.
Curie, D. (1960), *Luminescence in Crystals* (Methuen, London).
Daniels, F., Boyd, C. A., and Saunders, D. F. (1953), *Science* **117**, 343.
Dekker, H. (1976), *Health Phys.* **30**, 399.
Delucas, A., Maxia, V., and Spano, G. (1988), *J. Lumines.* **36**, 373.
Dexter, D. L., Klick, C. C., and Russell, G. A. (1955), *Phys. Rev.* **100**, 603.
DeWerd, L. A. (1979), in: *Thermally Stimulated Relaxation in Solids* (P. Bräunlich, Ed.) (Springer-Verlag, Heidelberg), p. 275.
Driver, H. S. T. (1979), *PACT* **3**, 290.
Dubourg, R., Schvoerer, M., Berger, R., and Dumereg, B. (1989), *Phys. Stat. Sol. (a)* **115**, 335.
Fain, J., Sanzelle, S., Miallier, D., Montret, M., and Pilleyre, Th. (1994), *Radiat. Meas.* **23**, 287.
Fields, D. E., and Moran, P. R. (1974), *Phys. Rev. B* **9**, 1836.
Fleming, S. J. (1979), *Thermoluminescence Techniques in Archaelogy* (Clarendon, Oxford).
Fukuda, Y., Tomita, A., and Takeushi, N. (1987), *Phys. Stat. Sol. (a)* **99**, K135.
Furetta, C., and Pellegrini, R. (1981), *Radiat. Effects Lett.* **58**, 17.
Gammage, R. B., and Cheka, J. S. (1977), *Health Phys.* **32**, 189.
Garlick, G. F. J., and Gibson, A. F. (1948), *Proc. Phys. Soc.* **60**, 574.
Gartia, R. K., and Ratnam, V. V. (1975), *Ind. J. Pure Appl. Phys.* **13**, 82.
Gasiot, J., and Fillard, J. P. (1977), *J. Appl. Phys.* **48**, 3171.
Gibson, J. A. B. (1986), *Radiat. Prot. Dosim.* **15**, 253.
Gorbics, S. G., Attix, F. H., and Kerris, K. (1973), *Health Phys.* **25**, 499.
Grant, R. M., and Cameron, J. R. (1966), *J. Appl. Phys.* **37**, 3791.
Grossweiner, L. I. (1953), *J. Appl. Phys.* **24**, 1306.
Grün, R. (1991), *Nucl. Tracks Radiat. Meas.* **18**, 143.
Guggenheim, H. (1963), *J. Appl. Phys.* **34**, 2482.
Gupta, N. M., Luthra, J. M., and Nambi, K. S. V. (1975), *J. Lumines.* **10**, 305.
Hagebeuk, H. J. L., and Kivits, P. (1976), *Physica*, **83B**, 289.
Hagekyriakou, K., and Fleming, R. J. (1982), *J. Phys. D: Appl. Phys.* **15**, 1795.
Halperin, A., and Braner, A. A. (1960), *Phys. Rev.* **117**, 408.
Henninger, J., Horlbeck, B., Hubner, K., and Prokert, K. (1982), *Nucl. Instr. Meth.*, **204**, 209.
Hobzova, L. (1974), Proc. 4th Int. Conf. Luminescent Dosimetry, Krakow, Poland, p. 1081.
Holgate, S. A., Sloane, T. H., Townsend, P. D., White, D. R., and Chadwick, A. V. (1994), *J. Phys. Cond. Mat.* **6**, 9255.
Holzapfel, G. (1969), *Phys. Stat. Sol.* **33**, 235.
Holzapfel, G., and Krystek, M. (1976), *J. Phys. (Paris)* **37**, C-7-238.
Hoogenstraaten, W. (1958), *Philips Res. Report* **13**, 515.
Horowitz, Y. S. (Ed.) (1984), *Thermoluminescence and Thermoluminescent Dosimetry* (C.R.C. Press, Boca Raton, Fla.).
Hsu, P. C., Lan, C. Y., and Weng, P. C. (1977), *Nucl. Sci. J.* **14**, 13.
Hsu, P. C., Tseng, C. L., and Weng, P. S. (1978), *Nucl. Inst. Meth.* **157**, 147.
Hsu, P. C., Weng, P. S., Lin, Y. M., and Huang, J. Y. (1987), *Radiat. Prot. Dosim.* **8**, 237.
Ichikawa, Y., Higashimura, T., and Sidei, T. (1966), *Health Phys.* **12**, 395.
Iga, K., Yamashita, T., Takenaga, M., Yasuno, Y., Onishi, H., and Ikedo, M. (1977), Health Phys. **33**, 605.
Jablonski, A. (1935), *Z. Phys.* **94**, 38.
Jackson, J. H., and Harris, A. M. (1970), *J. Phys. C: Sol. St. Phys.* **3**, 1967.
Jain, V. K. (1980), *Phys. Stat. Sol. (a)* **60**, 351.
Jain, V. K. (1982), *Radiat. Prot. Dosim.* **2**, 141.
Jain, V. K. (1984), in: *Thermoluminescence and Thermoluminescent Dosimetry, Vol. 2* (Y. S. Horowitz, Ed.) (C.R.C. Press, Boca Raton, Fla.).
Jain, V. K., Kathuria, S. P., and Ganguly, A. K. (1974), *J. Phys. C., Sol. St. Phys.* **7**, 3810.
Jasinska, M., and Niewiadomski, T. (1970), *Nature* **227**, 1159.
Johnson, N. M., and Blachard, R. L. (1967), *Amer. Mineral.* **52**, 1297.
Jun, S. J., and Becker, K. (1975), *Health Phys.* **28**, 459.

Kalef-Ezra, J., and Horowitz, Y. S. (1982), *Int. J. Appl. Radiat. Isot.* **33**, 1085.

Kamada, M., Furukawa, K., and Tsutsumi, K. (1981), *Jap. J. Appl. Phys.* **20**, 71.

Kirsh, Y., Kristianpoller, N., and Chen, R. (1979), *Phil. Mag.* **35**, 653.

Kitis, G., and Furetta, C. (1994), *Nucl. Inst. Meth.* **B94**, 441.

Kitis, G., Papadopoulos, J. G., Charalambous, S., and Tuyn, J. W. N. (1994), *Radiat. Prot. Dosim.* **55**, 183.

Klasens, H. A. (1953), *J. Electrochem. Soc.* **100**, 72.

Krivits, P., and Hagebeuk, H. J. L. (1977), *J. Lumines.* **15**, 1.

Lakshmanan, A. R. (1979), *Bull. Radiat. Prot.* **2**, 67.

Lakshmanan, A. R. (1994), *Phys. Stat. Sol. (a)* **42**, 225.

Lakshmanan, A. R., and Bhatt, R. C. (1979), *Int. J. Appl. Radiat. Isot.* **30**, 375.

Lakshmanan, A. R., and Vohra, K. G. (1979), *Nucl. Inst. Meth.* **159**, 585.

Lakshmanan, A. R., Chandra, B., Pradhan, A. S., Kher, R. K., and Bhatt, R. C. (1980), *Int. J. Appl. Radiat. Isot.* **31**, 107.

Lakshmanan, A. R., Bhatt, R. C., and Supe, S. J. (1981). *J. Phys. D: Appl. Phys.* **14**, 1683.

Lakshmanan, A. R., Chandra, B., and Bhatt, R. C. (1989), *Radiat. Prot. Dosim.* **27**, 15.

Lambe, J., and Klick, C. C. (1955), *Phys. Rev.* **98**, 909.

Lembo, L., and Scarpa, G. (1993), in: *Thermoluminescent Materials* (D. R. Vij, Ed.) (Prentice Hall, Englewood Cliffs, NJ.), p. 264.

Lewandowski, A. C., Markey, B. G., and McKeever, S. W. S. (1994), *Phys. Rev. B: Cond. Mat.* **49**, 8029.

Lushchik, C. B. (1956), *Sov. Phys. JETP* **3**, 390.

Luthra, J. M. (1989), *Defects and Diffusion Forum* **62/63**, 183.

Mahesh, K., and Vij, D. R. (Eds.) (1985), *Techniques of Radiation Dosimetry* (Wiley Eastern, New Delhi).

Manam, J. (1994), *Ind. J. Phys. A.* **68**, 105.

Mandowski, A. and Swiatek, J. (1994), *Diff. Defects Data B: Sol. St. Phenom.* **37/38**, 249.

Marron, M. J., and Attix, F. H. (1964), *Health Phys.* **10**, 431.

Mathur, V. K. (1993), in: *Thermoluminescent Materials* (D. R. Vij, Ed.) (Prentice Hall, Englewood Cliffs, NJ.), p. 181.

Mayhugh, M. R. (1970), *J. Appl. Phys.* **41**, 4776.

Mayhugh, M. R., and Fullerton, G. D. (1975), *Health Phys.* **28**, 297.

Mayhugh, M. R., Christy, R. W., and Johnson, N. M. (1968), *Proc. 2nd Int. Conf. Luminescent Dosimetry*, Gatlinberg, p. 294.

McCullough, E. C., Fullerton, G. D., and Cameron, J. R. (1972), *J. Appl. Phys.* **43**, 77.

McDoughall, D. J. (Ed.) (1968), *Thermoluminescence of Geological Materials* (Academic, London).

McKeever, S. W. S. (1985), *Thermoluminescence of Solids* (Cambridge University Press, Cambridge), p. 8–10.

McKinlay, A. F. (1981), *Thermoluminescent Dosimetry* (Adam Hilger, Bristol).

McKlveen, J. W. (1980), *Health Phys.* **39**, 211.

Mehta, S. K., and Sengupta, S. (1975), *Proc. Natl. Symp. on TL and Its Applications*, Kallapakam, India, p. 580.

Mehta, S. K. and Sengupta, S. (1976), *Phys. Med. Biol.* **21**, 955.

Mehta, S. K., and Sengupta, S. (1978), *Phys. Med. Biol.* **23**, 471.

Mejdahl, V., and Aitken, M. J. (Eds.) (1980), *Proc. 2nd Specialist Seminar on TL Dating, Vol. 6* (PACT, Oxford).

Miller, S. D., Endres, G. W. R., McDonald, J. C., and Swinth, K. L. (1988), *Radiate. Prot. Dosim.* **25**, 201.

Mills, A. A., Sears, D. W., and Hearsey, R. (1977), *J. Phys. E.: Sci. Inst.* **10**, 51.

Mische, E. F., and McKeever, S. W. S. (1989), *Radiat. Prot. Dosim.* **29**, 159.

Moharil, S. V. (1981), *J. Phys. D: Appl. Phys.* **14**, 1677.

Nagpal, J. S., Kher, R. K., and Gangadharan, P. (1979), *Radiat. Effects Lett.* **43**, 181.

Nakajima, T. (1971), Proc. 3rd Int. Conf. Luminescent Dosimetry, Riso, Denmark, Report No. 249, p. 466.

Nakajima, T. (1985), in: *Techniques of Radiation Dosimetry* (K. Mahesh and D. R. Vij, Eds.) (John Wiley, Chichester), p. 156.

Nakajima, T. (1993), in: *Thermoluminescent Materials* (D. R. Vij, Ed.) (Prentice Hall, Englewood Cliffs, NJ), p. 322.

Nambi, K. S. V., Bapat, V. N., and Ganguly, A. K. (1974), *J. Phys. C: Sol. St. Phys.* **7**, 4403.

Nicholas, K. H., and Woods, J. (1964), *Brit. J. Appl. Phys.* **15**, 783.
Niewiadomski, T. (1993), in: *Thermoluminescent Materials* (D. R. Vij, Ed.) (Prentice Hall, Englewood Cliffs, NJ), p. 142.
Nink, R., and Kos, H. J. (1980), *J. Nucl. Inst. Meth.* **175**, 15, 24.
Oberhofer, M., and Scharmann, A. (Eds.) (1981). *Applied Thermoluminescence Dosimetry* (Adam Hilger, Bristol).
O'Hagan, J. B., Pearson, A. J., and Driscoll, C. M. J. (1986), *Radiat. Prot. Dosim.* **17**, 347.
Osvay, M., and Biro, T. (1980), *Nucl. Instr. Meth.* **175**, 60.
Osvay, M., and Biro, T. (1993), in: *Thermoluminescent Materials* (D. R. Vij, Ed.) (Prentice Hall, Englewood Cliffs, NJ), p. 243.
Partidge, J. A., and May, C. E. (1965), *J. Chem. Phys.* **42**, 1965.
Piesch, E. (1980), in: 3rd Information Seminar on the European Radiation Protection Dosimeter Intercomparisons Programme, Grenoble, France, 6-8 Oct.
Piesch, E., and Burgkhardt, B. (1980), *Nucl. Inst. Meth.* **175**, 180.
Piters, T. M., and Bos, A. J. J. (1993), *J. Phys. D: Appl. Phys.* **26**, 2225.
Piters, T. M., and Bos, A. J. J. (1994), *J. Phys. D: Appl. Phys.* **27**, 1747.
Pradhan, A. S., and Bhatt, R. C. (1979), *Nucl. Inst. Meth.* **166**, 497.
Pradhan, A. S., and Bhatt, R. C. (1981), *Int. J. Appl. Rad. Isot.* **32**, 179.
Pradhan, A. S., Bhatt, R. C., Lakshmanan, A. R., Chandra, B., and Shinde, S. S. (1978), *Phys. Med. Biol.* **23**, 723.
Pradhan, A. S., Bhatt, R. C., and Supe, S. J. (1984), *J. Phys. D: Appl. Phys.* **17**, 1699.
Prener, J. S., and Williams, F. E. (1956), *J. Electrochem. Soc.* **103**, 342.
Randall, J. T., and Wilkins, M. H. F. (1945), *Proc. Roy. Soc. Lond.* **184**, 366, 390.
Ranogajec-Komor, M., and Osvay, M. (1986), *Radiat. Prot. Dosim.* **17**, 379.
Ranogajev-Komor, Osvay, M., Dvornik, I., and Biro, T. (1983), *Nucl. Inst. Meth.* **212**, 383.
Rao, R. P. (1986), *J. Mat. Sci.* **21**, 3357.
Salas, R. P., Aceves, R., Mclendrez, R., and Barboza-Florea, M. (1993), *Appl. Phys. Lett.* **63**, 894.
Sankran, A., Kannan, S., and Gangadharan, P. (1979). *Med. Phys.* **7**, 73.
Sathyamoorthy, A., and Luthra, J. M. (1978), *J. Mat. Sci.* **13**, 2637.
Sathyamoorthy, A., Bhalla, K. C., and Luthra, J. M. (1976), *J. Lumines.* **11**, 35.
Scarpa, G. (1970), *Phys. Med. Biol.* **15**, 667.
Scarpa, G., Benincasa, G., and Ceravolo, L. (1971), *Proc. 3rd Int. Conf. Luminescent Dosimetry*, Riso, Denmark, p. 427.
Schayes, R., Brook, C., Kozlowitz, I., and Lheureus, M. (1967), *Proc. Int. Conf. Luminescent Dosimetry*, USAEC CONF-650637, p. 138.
Schön, M. (1942), *Z. Physik* **119**, 463.
Schulman, J. H., Attix, F. H., West, E. J., and Ginther, R. J. (1960), *Rev. Sci. Inst.* **31**, 1263.
Schulman, J. H., Kirk, R. D., and West, E. J. (1967), *Proc. Ist Int. Conf. Luminescent Dosimetry*, USAEC-CONF-650637, p. 113.
Sears, D. W. G. (1988), *Nucl. Tracks, Radiat. Meas.* **14**, 5.
Seitz, F. (1939), *Trans. Faraday Soc.* **35**, 680.
Singhvi, A. K., and Mejdahl, V. (1985), *Nucl. Tracks Radi. Meas.* **10**, 137.
Soares, C. G., and Ehrlich, M. (1981), *IEEE Trans. Nucl. Sci.* **NS-28**, 1614.
Spurny, Z., and Hobzova, L. (1977), *Radio. Chem. Radio. Ann. Lett.* **29**, 287.
Spurny, Z., and Hruska, J. (1976), *Phys. Med. Biol.* **21**, 439.
Spurny, Z., and Kaambre, H. (1985), in: *Techniques of Radiation Dosimetry* (K. Mahesh and D. R. Vij, Eds.) (John Wiley, Chichester), p. 281.
Stoebe, T. G., Wolfenstine, J. B., and Las, W. C. (1980), *J. Phys. (Paris), Suppl.* **7**, C6-265.
Summers, G. P. (1984), *Radiat. Prot. Dosim.* **8**, 69.
Sunta, C. M. (1979), *Phys. Stat. Sol. (a)* **53**, 127.
Sunta, C. M. (1984), *Radiat. Prot. Dosim.* **8**, 25.
Sunta, C. M., Okuno, E., Lima, J. F., and Yoshimura, E. M. (1994), *J. Phys. D: Appl. Phys.* **27**, 2636.
Szabo, P. O., Feher, I., Demer, S., Szabo, B., Vagvolgyi, J., and German, E. (1984), *Radiat. Prot. Dosim.* **6**, 100.
Takenaga, M., Yamamoto, O., and Yamashita, T. (1980), *Nucl. Inst. Meth.* **175**, 77.

Takeuchi, N., Adachi, M., and Inabe, K. (1979), *J. Lumines.* **18/19**, 897.

Tatumi, S. H., Nagamoto, T., Matsuoka, M., and Watanabe, S. (1993), *J. Phys. D: Appl. Phys.* **26**, 1482.

Thompson, J. J., and Ziemer, P. L. (1972), *Health Phys.* **22**, 399.

Tochilin, E., Goldstein, N., and Miller, W. G. (1969), *Health Phys.* **16**, 1.

Tomaru, T., Onai, Y., Irifune, T., Uchida, I., and Yamashita, T. (1983), *Jpn. Radiol. Phys.* **3**, 83.

Tomita, A., Hirai, N., and Tsutsumi, K. (1976), *Jpn. J. Appl. Phys.* **15**, 1899.

Townsend, P. D. (1994), *Radiat. Meas.* **23**, 341.

Trousil, J., and Kokta, L. (1982), *Radiat. Prot. Dosim.* **2**, 169.

Uchrin, G. (1993), *Radiat. Prot. Dosim.* **47**, 595.

Upadeo, S. V., Gundurao, T. K., and Moharil, S. V. (1994), *J. Phys. Cond. Mat.* **6**, 9459.

Urbach, F. (1930), *Wiener Berichte* **139**, 363.

Vana, N., Erlach, R., Fugger, M., Gratzl, W., and Reichhalter, I. (1988), *Nucl. Track Radiat. Meas.* **14**, 181.

Vij, D. R. (Ed.) (1993), *Thermoluminescent Materials* (Prentice Hall, Englewood Cliffs, NJ).

Vij, D. R., and Mahesh, K. (1985), in: *Techniques of Radiation Dosimetry* (K. Mahesh and D. R. Vij, Eds.) (John Wiley, Chichester), p. 182.

Vij, D. R., and Mathur, V. K. (1969), *Brit. J. Appl. Phys.* **2**, 624.

Vohra, K. G., Bhatt, R. C., Chandra, B., Pradhan, A. S., and Lakshmanan, A. R. (1980), *Health Phys.* **38**, 193.

von Hippel, A. (1936), *Z. Phys.* **101**, 101.

Waschmann, F., and Regulla, D. F. (1978), *Kerntechnik* **7**, 318.

Watanabe, K. (1951), *Phys. Rev.* **83**, 785.

Watanabe, S., and Nakajima, T. (1973), *J. Nucl. Sci. Tech.* (Japan) **10**, 202.

Weng, P. S. (1993), in: *Thermoluminescent Materials* (D. R. Vij, Ed.) (Prentice Hall, Englewood Cliffs, NJ), p. 213.

Weng, P. S., and Chen, K. M. (1974), *Nucl. Inst. Meth.* **117**, 89.

Weng, P. S., Hsu, P. C., Tsai, F. L., and Huang, C. C. (1981), *Proc. 15th Int. Cong. of Radiology*, Brussels, Belgium.

Willhoit, D. G., and Poland, A. D. (1967), *Health Phys.* **15**, 91.

Williams, F. E. (1955), *Brit. J. Appl. Phys. Suppl.* **4**, 97.

Wintle, A. G. (1985), *J. Lumines.* **33**, 333.

Wintle, A. G., and Huntley, D. J. (1982), *Quat. Sci. Rev.* **1**, 31.

Wochos, J. W., Fullerton, G. D., and DeWerd, L. A. (1978), *Amer. J. Roentgenol.* **131**, 617.

Yamashita, T. (1993), in: *Thermoluminescent Materials* (D. R. Vij, Ed.) (Prentice Hall, Englewood Cliffs, NJ), p. 298.

Yamashita, T., Nada, N., Onishi, H., and Kitamura, S. (1971), *Health Phys.* **21**, 295.

Yamashita, T., Yasuno, Y., and Ikedo, M. (1974), *Health Phys.* **27**, 201.

Yamazaki, T., Tomimasu, T., Mikado, T., and Chiwaki, M. (1978), *J. Appl. Phys.* **49**, 4929.

8

Lyoluminescence*

George T. Reynolds

1. INTRODUCTION

Lyoluminescence (LL) is the term applied to the emission of light when a material goes into solution. The phenomenon was first reported by Wiedemann and Schmidt (1895), who noted that in some cases light was emitted by materials that had been irradiated (with cathode rays) prior to dissolution, whereas these same materials did not emit light on dissolution if they had not been irradiated.

Some 65 years later, Ahnström and Ehrenstein (1959) in a brief note described experiments that were remarkable in their breadth of coverage of the phenomenon, calling attention to many of its aspects that have since been expanded upon in subsequent research. Using a photomultiplier to detect the luminescence, they found that:

1. Several (organic) substances luminesced for about 1 hr after irradiation with x-rays, γ-rays, or fast neutrons.
2. Among these, glucose, after the initial radiation completely disappeared, gave a "new luminescence" when dissolved in water at room temperature. After 1 min (of presumably relatively rapid decay) the light emission reached a comparatively constant level. One day later, the emission was about one-half that of the 1-min level.
3. Glucose that had not been irradiated did not show luminescence when dissolved.
4. The luminescence from glucose was enhanced by a factor of 20 when the pH of the solvent was increased from about 3 to about 12 with the addition of 0.1 *N* NaOH.

*This chapter has been reproduced from the *Journal of Luminescence* **54** (1992), with kind permission from Elsevier Science Publishers, B.V.

George T. Reynolds • Department of Physics, Princeton University, Princeton, New Jersey 08544, United States.

Luminescence of Solids, edited by D. R. Vij. Plenum Press, New York, 1998.

5. With ^{60}Co γ-rays at dose levels of 0.6 to 1.70 MGy, the luminescence was directly proportional to both the amount of glucose dissolved and to the radiation dose.

By considering various possible decomposition products owing to radiation, they were led to point out that "by means of electron spin resonance (ESR), it has been shown that long-lived free radicals are induced by x-rays in different substances and in living material." As will be discussed later, it was not until some years later that ESR was taken seriously as a potential dosimetric technique.

As can be appreciated from the nature and source of LL, much of the research has been reported in symposia proceedings, not easily found in the literature, on radiation effects, protection, detection, and health concerns (e.g., food preservation). Also as with sonoluminescence (SL) (Walton and Reynolds, 1984), chemiluminescence (CL) is involved in an essential manner. This brings in studies in chemistry as well as in physics and detection instrumentation technology, so that relevant publications are found in a wide range of journals.

Very shortly after the report of Ahnström and Ehrenstein (1959), Westermark and Grapengiesser (1960) described experiments in which not only organic phosphors, but also the inorganics NaCl, LiF, and KI were investigated. Further, the solvents included water, aqueous solutions, organic solvents, and organic solutions. Two aspects of this work, to be discussed in more detail later, are of particular interest:

1. The LL of substances identical to or related to the materials used in liquid scintillators were studied (Reynolds, 1960), and in general gave weak responses.
2. Irradiated NaCl dissolved in a solution of $TlCl_3$ in water showed a light output "a thousand times greater than that for water"; the effect has a "pronounced maximum at about 0.01 g/liter $TlCl_3$," i.e., about 4×10^{-5} *M*.

As in most of these earlier experiments, doses were high ($\sim$0.5 MGy) and, in this case, were achieved with 2-MeV electrons.

After these reports in 1960, relatively few papers appeared on the subject for a decade. Notable among the ones that did appear were two by Ahnström (1961, 1965). The first discussed the use of luminescence in the study of free radical reactions (mainly in organic substances, with electrolytically produced free radicals), and the second dealt with the role of *F*-centers in the luminescence of irradiated NaCl dissolved in a fluorescein solution. In the early 1970s interest in the phenomenon grew rapidly, directed mainly to its possible uses as a dosimetry technique, and generally (but not exclusively) concerned with high radiation doses, in some cases kilograys. Good reviews of work through 1980 can be found in two papers (Ettinger and Puite, 1982; Puite and Ettinger 1982). The first published reports of the use of LL as a dosimeter is in a paper by Atari *et al.* (1973).

As noted in studies of SL (Walton and Reynolds, 1984), CL involving free radicals is an important aspect, as it is in LL, and useful conclusions may result from studying the relationships between SL and LL, especially by means of spectroscopy.

As the organic materials in liquid scintillators respond to ionizing radiation (Reynolds, 1960), it might be that involving them in LL studies could be instructive; and similarly, information gained in LL studies might lead to better scintillators. The relation between LL and thermoluminescence (TL) is particularly important, especially for inorganic phosphors. Although this relationship is mentioned in the literature, it appears not to have been fully exploited.

In the early work of Ahnström and Ehrenstein (1959), attention was called to the use of ESR in identifying long-lived free radicals induced by x-rays in different substances, including living material. The possible use of ESR as a dosimetry technique was first suggested by Bradshaw *et al.* (1962) and was discussed in several articles some years later (Bartlett *et al.*, 1982; Ettinger 1989; Nakajima and Otsuki, 1990). As will be emphasized below, LL in organic substances is generally due to CL of free radicals. In this view, LL dosimetry is a form of free radical dosimetry with optical readout, making it potentially more sensitive and less complicated than ESR dosimetry.

Almost all of the more recent studies discussed in this chapter are concerned with observations of irradiated materials. It is natural to think of the phenomenon as a potential technique for dosimetry.

2. MATERIALS AND MECHANISMS

The now extensive literature discussing LL describes some studies of the phenomenon for its own sake and, more numerous, studies related to its usefulness as a dosimetric technique. The basic techniques in both cases are similar and are indicated schematically in Fig. 1. Many materials, organic and inorganic, have been investigated during the past two decades, mainly for applications to dosimetry. To be useful for this purpose the LL phosphor must have long-lived induced radicals in the case of organics and long-lived trapping centers in the case of inorganics. In addition, it must be readily soluble in a convenient solvent, and the energy transition operating must result in light detectable with reasonable efficiency by means of a conventional photomultiplier. It turns out that many organic and inorganic substances are available. Although early (pre-1980) work usually involved water or an aqueous solution as the solvent, such organic solvents as alcohols have come into use, as well as a variety of "enhancers," as will be described below.

Two curves are of importance in the discussion of LL: (1) The first is a curve showing the light emission as a function of time after the onset of dissolution. This is frequently called a "glow curve" by analogy with TL (Ettinger and Puite, 1982) (and, in fact, the energy transitions in LL observed in inorganic phosphors have a direct relation to those responsible for light emission in TL). Figure 2 shows a representative glow curve. (2) The second is a curve showing the light yield as a function of dose, called a "dose response curve". For low doses, this can be presented on a linear scale; for very high doses ($>10^2$ Gy) a log–log scale is generally used. To be useful the system should yield a nearly linear or straight-line relation. In

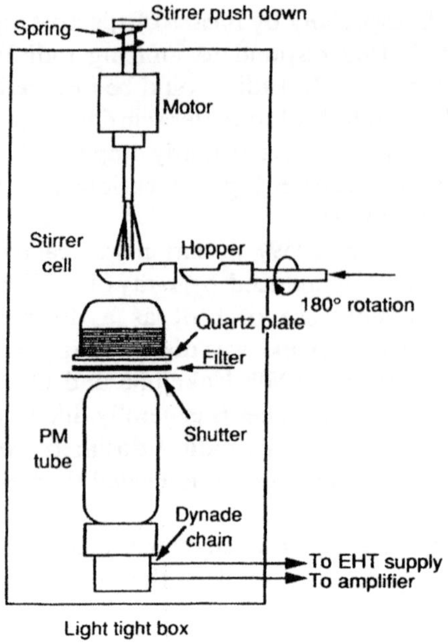

Figure 1. Representative LL "reader." (Puite and Ettinger, 1982; reprinted with kind permission from Elsevier Science, Ltd.).

organic phosphors saturation is finally observed in the light response when the dose is increased (and in some cases even a decrease), usually above some kilograys. A possible explanation of this saturation may be found in the role of oxygen in forming the peroxyl radicals essential for light emission. At low doses, the oxygen diffuses into an active zone near the solid–liquid interface, in excess of primary radicals. At higher doses the diffusing oxygen is insufficient to oxidize all of the available primary radicals. Figure 3 shows representative dose response curves.

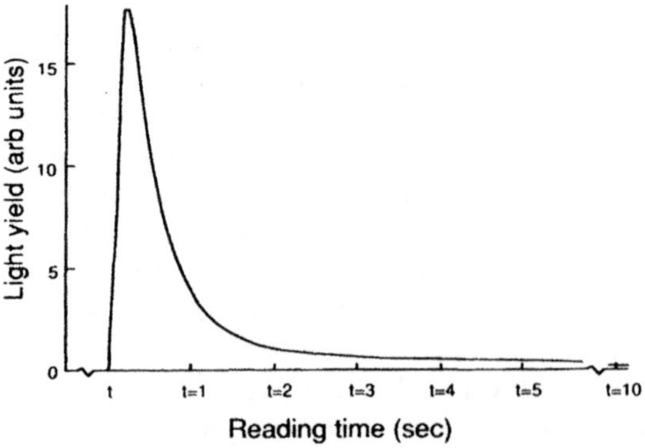

Figure 2. Glow curve of LL from a saccharide (mannose) (Puite and Ettinger, 1982; reprinted with kind permission from Elsevier Science, Ltd.).

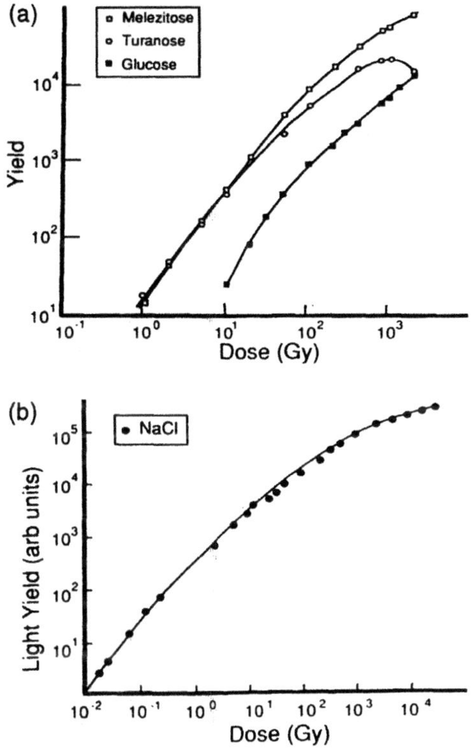

Figure 3. Representative dose response curves: (a) some saccharides in water, ^{60}Co irradiation; (b) NaCl in water, ^{60}Co irradiation (Ettinger and Puite, 1982, reprinted with kind permission from Elsevier Science Ltd.).

2.1. Organic Lyoluminescence Phosphors

The idea that reactions involving induced radicals are operative in LL observed with organic solutes is well established (Ahnström and Ehrenstein, 1959; Ahnström, 1961; Ettinger and Puite, 1982; Ettinger *et al.*, 1980). Experiments with a wide range of organic substances indicate that LL is ubiquitous: it is found in every commercially available sugar in all water-soluble amino acids, as well as in simple proteins, glycoproteins, enzymes, RNA, DNA, in some alcohols and in salts of organic acids; it is also found in such compounds as polymethyl methacrylate, polystyrene, and cellulose acetate. The solvents have included water, aqueous solutions, glacial acetic acid, acetone, methanol, and ethanol. For dosimetric techniques, mannose (a sugar) and glutamine (an amino acid: α-aminoglutaramic acid) appear to have received most attention. The composition of these compounds makes them nearly tissue equivalent for x-rays and γ-rays even though their structure is not similar.

2.1.1. Mechanisms

No single mechanism for radical formation in organics seems able to explain all of the observed effects. Many possibilities are discussed by Ettinger and Puite (1982). A representative process involves the abstraction of a hydrogen atom from a molecule, forming a radical trapped in the solid for a long period of time, depending on the temperature of storage. This can be represented by

$$RH^+ + RH^\bullet \rightarrow R^0 + RH_2^+$$

It remains to describe why light is emitted upon dissolution. Spectral analysis is particularly useful for identifying details, as will be discussed below. One early paper reports results for glutamine and saccharides (Hunter *et al.*, 1982). For this work, materials were given very high doses ("almost to saturation," i.e., kilograys). Since spectra may vary with dose, it would clearly be useful to obtain results from (the much weaker) emission associated with low doses, utilizing an image-intensified spectrometer (Reynolds and Gruner, 1975). It is found that the spectra of saccharides vary with pH as well as with dose. (The amount of light per dose is also found to vary with pH in the case of certain inorganics, as will be discussed later). In general, the spectrum of glutamine exhibits a broad peak centered at about 450 nm, with very little emission beyond 550 nm. In the case of saccharides, the emission extends to much longer wavelengths, especially for high pH (e.g., upon addition of 0.1 *M* NaOH to the water solvent).

An explanation proposed for the luminescence process in some organics is the Russell–Vassil'ev scheme, as described by Ettinger and Puite (1982). This initiates with

$$R^\bullet + O_2 \rightarrow RO_2^\bullet$$

leading finally to a triplet ground state O_2 and an excited carbonyl group, which has been observed to give off CL. Since oxygen is also known to quench LL in certain cases, the concentration of radicals relative to available oxygen has to be discussed, and the nature of the process leading to the emission of light is still open to discussion.

2.1.2. Enhancement

There is a variety of techniques for enhancing LL in organic phosphors.

Luminol (aminophthalichydrazid) has been known and studied exhaustively as a CL agent for decades. Its relation to bioluminescence (BL) (Harvey, 1952) and SL (Walton and Reynolds, 1984) has been discussed. [The details of the CL reaction(s) need not be discussed here, but for future remarks, it is useful to note that in certain circumstances Cu^{2+} and Co^{3+} have acted as catalysts.] It has been noted that LL from irradiated saccharides, amino acids, and some alcohols when the solvent is an alkaline aqueous luminol solution is much brighter than that with pure water as the solvent. For example, when luminol solution was used instead of water, the enhancement of light emission was 10^6 in the case of glucose, and 20 times in the case of the saccharide trehalose dihydrate (Atari and Ettinger, 1974*b*). Incidentally, the same work showed trehalose to yield about 10^3 times as much light as glucose and mannose.

Details of the luminol solution were not given in this paper, but other work (showing a gain of only about 10^4 for luminol solution over pure water) indicates that the enhancement is a sensitive function of luminol concentration, enhancement being a factor of 10 at 10^{-5} mg/g luminol to water to a factor of 10^4 at 10^{-2} mg/g ($\sim 6 \times 10^{-5}$ *M*), where it levels off (the radiation dose is 1 kGy). These same studies (Ettinger and Puite, 1982) describe the temperature dependence of LL for trehalose dehydrate, in all cases for highly dosed (kGy) phosphors. Generally speaking, yield increases as a function of temperature (10 to 60°C) and pH (7 to 11).

A comparison of LL from tris-(hydroxymethyl)-aminomethane in various solvents was reported by Mehta *et al.* (1983). The relative response for solutions (all at pH 6.5) of distilled water, fluorescein sodium, copper sulfate, thallous chloride and luminol are, respectively, 1, 3, 4, 40, and 2200. Later studies, such as that by Atari *et al.* (1985), which compare the LL response of five different saccharides, use luminol solutions as a standard procedure. The use of luminol to enhance LL has also been discussed by Chazhoor and Mishra (1979, 1982) and by Busoli and Montiventi (1984).

After luminol, the next most successful enhancer is lucigenin (dimethyl-biacridinum) in alkaline solution. The light enhancement peaks at a concentration of about 10^{-8} mol/ml. Dose response curves saturate at higher doses for lucigenin solutions than for pure water solution.

A distinct disadvantage in the use of luminol and lucigenin is that they exhibit self-glow in solution, which cannot be circumvented by reducing concentration, since the enhancement decreases proportionately. This calls to mind the advantage of image intensifiers over photomultipliers in the matter of cathode dark current, and indicates a means of implementing a suggestion of Ettinger and Puite (1982) to form luminolized phosphors, i.e., materials prepared by incorporating luminol in the solid matrix of the phosphor. Even with the dark current problem in photomultipliers, a significant advantage was found in the case of luminolized mannose. The idea is that during dissolution, the luminol (or lucigenin) is made available at high local concentration, without an excessive distributed self-glow. The effect is, naturally, dependent on grain size (more enhancement at greater grain size). A possibility for reducing the limitation of self-glow is that the emission spectrum of self-glow in some cases may differ sufficiently from that of LL (especially as a function of pH) that the use of filters might improve the signal-to-background level.

It is attractive to think of enhancing LL by means of an energy transfer mechanism, as has been done in liquid scintillator technology. An advantage could be achieved for two possible reasons: (1) transfer to an emitter whose wavelength (longer) is better suited to the spectral response of the detector than the emission wavelength (shorter) of the primary excited species, or (2) a rapid transfer to an efficient emitter, not subject to quenching via vibrational levels.

Enhancement of CL of organic systems by means of fluorescent dyes has been noted. As will be described later, it has been used successfully for LL by Kalkar *et al.* (1986) and by Atari and Ettinger (1974*a, b*) in the case of NaCl using a solution of luminol and fluorescein, improving the response by a factor of four over luminol alone. Atari and Ettinger (1974*a*) also found an improvement for NaCl but report that with saccharides these fluorescent systems have no effect.

An interesting means of enhancing LL is the addition of various positive ions to the solvent. The technique has been discussed or hinted at in many papers and apparently applies to organic as well as inorganic phosphors. A more detailed discussion will be given below for the case of inorganics, but a few comments here will indicate the effects found in organics.

Ettinger and Puite (1982) note, in the course of discussing luminol as a sensitizer, that "in an aqueous alkaline medium in the presence of oxygen, small amounts of H_2O_2 and a catalyst (e.g., Cu^{2+}, Co^{3+}), luminol CL takes place with an intensity depending on the concentrations of H_2O_2 and the catalyst." Ettinger and Anunuso (1981) report results using Eu^{3+} and Tb^{3+} ions in studies of the amino acids glutamine, threonine, and serine. Emission from Eu^{3+} is at 590–613 nm, thus requiring significant correction for photocathode response. Tb^{3+} emits at about 545 nm. The greatest enhancement was observed for Tb^{3+} from dissolved $Tb(NO_3)_3 \cdot 6H_2O$, the enhancement factors being 95 (glutamine), 490 (threonine), and 70 (serine), uncorrected for the photocathode response for the LL of the same materials unenhanced. This correction would increase the advantage by a factor of 3.5.

2.1.3. *Spectra*

Just as ESR is useful for identifying the radicals present in CL and LL processes (Bradshaw *et al.*, 1962; Bartlett *et al.*, 1982; Ettinger, 1989; Nakajima and Otsuki, 1990), it is also useful for studying the emission spectra of LL species. In these studies, techniques have ranged from coarse analyses using selected filters and photomultipliers to conventional spectroscopes. Corrections must be made for photocathode response and possible decay of the signal during the course of measurement. In general, either interest or necessity has resulted in studies being made on materials that have received massive doses. For the study of (weaker emission from) low-dose conditions, it would clearly be advantageous to use an image-intensified spectroscope (Reynolds and Gruner, 1975) or an on-line spectrofluorimeter system as developed by Wampler and DeSa (1971). Atari *et al.* (1985) report spectra obtained using the Princeton Applied Research intensified silicon-target optical multichannel analyzer (PARC 1205 A OMA) but the spectrum they report is notably lacking detailed characteristics (see Fig. 4). Atari (1980*a*) reports

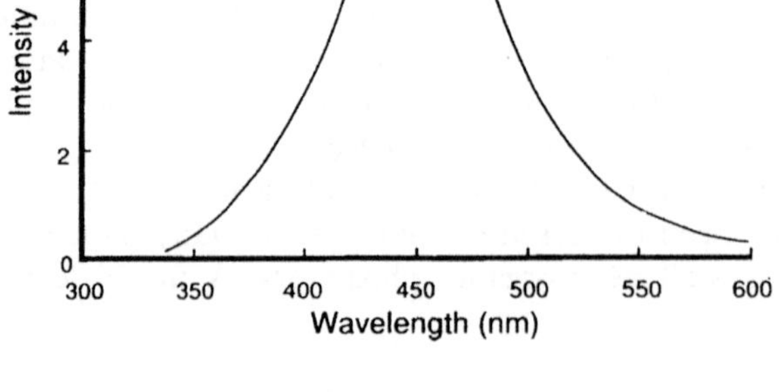

Figure 4. LL spectrum of irradiated saccharide dissolved in luminol solution. (Atari *et al.*, 1985; reprinted with kind permission from Elsevier Science, Ltd.).

spectra for a number of alkali halides in more detail. In any case, a number of spectral observations have been reported for organic phosphors.

Emission spectra of organics show differences, depending on the LL phosphor; in general, the spectra are found to be broad and in the blue-green region. This corresponds to the spectral sensitivity of the detectors used. It is interesting to note that significant energy may be emitted in the near ultraviolet (as in SL) or infrared (as in some BL). Selected examples of spectra for various organics are shown in Figs. 4, 5, and 6. A detailed study of the LL spectra of typical dosimeter materials was reported by Hunter *et al.* (1982). Points to be noted include the necessity of correcting for spectral response of the detector and possible time dependence of emission during the course of observation when certain (slow) detection techniques were used.

2.1.4. pH Dependence

Mention has been made above of the dependence of LL output on the pH of the solvent. Figure 7 gives a dramatic demonstration of the effect of pH on the spectral distribution for the case of the saccharide glucose monohydrate (glutamine does not show a significant spectral shift with pH). This spectral shift could easily result in an increase in light output if certain detectors were used, without concern for their spectral sensitivities.

As noted above one of the earliest observations of the effect of pH on LL was made by Ahnström and Ehrenstein (1959). Results support the hypothesis by Matyushkov *et al.* (1977) that there are two CL reactions. For irradiated glucose dissolved in an acid medium or distilled water, the light emitted is proportional to the dose, whereas for an alkaline solvent, this is no longer true. Stated another way, for strongly alkaline solutions, the quantum efficiency is an almost linear function of dose up to about 400 Gy. These features become important in the calibration and application to LL dosimetry.

2.1.5. Temperature Dependence

In discussing the use of luminol and lucigenin as enhancers of LL, attention was called to the limitation caused by the self-glow of these agents. Since this glow is a sensitive function of temperature (increasing roughly by a factor of about 3 per 10°C rise) it becomes an important consideration in applications (self-glow is also a function of pH).

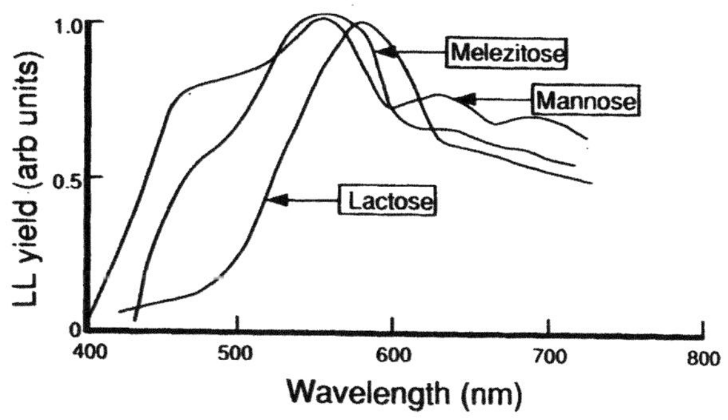

Figure 5. LL spectra of some saccharides dissolved in water at pH = 6 (Ettinger and Puite, 1982; reprinted with kind permission from Elsevier Science, Ltd.).

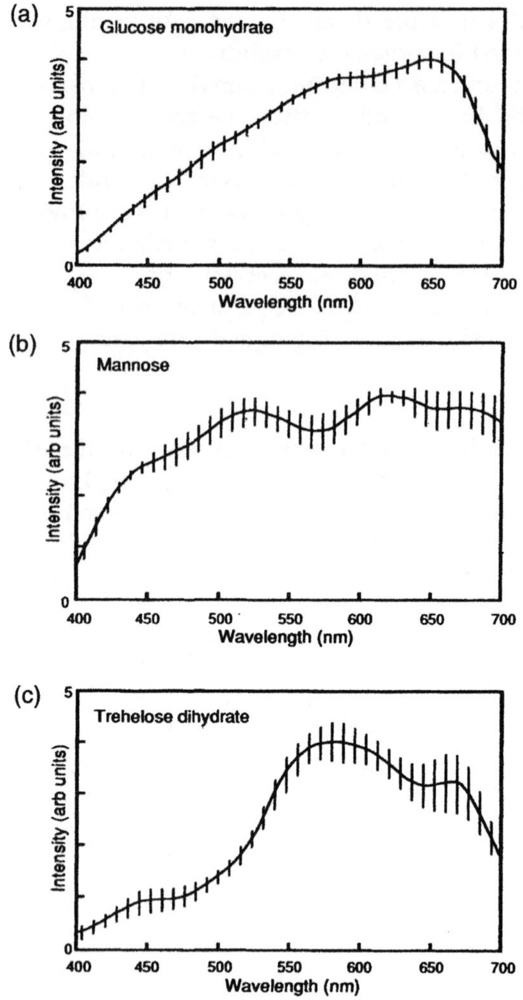

Figure 6. LL spectra of various sugars dissolved in water at pH = 5.5: (a) glucose monohydrate; (b) mannose; (c) trehelos dihydrate (Hunter *et al.*, 1982; reprinted with kind permission from Elsevier Science, Ltd.).

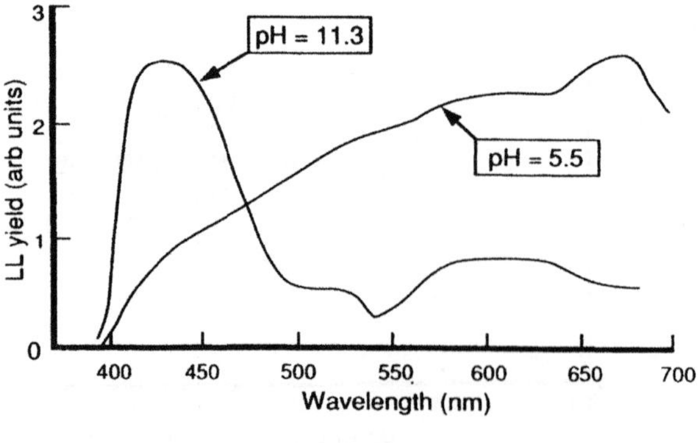

Figure 7. LL spectrum of irradiated glucose monohydrate dissolved in water at two pH values (Ettinger and Puite, 1982; reprinted with kind permission from Elsevier Science, Ltd.).

Temperton *et al.* (1984*a*) have studied the dependence of LL yield on several parameters, including temperature at radiation and solvent temperature. The decrease in LL yield with the temperature of the solvent (pure water) was reported by Puite and Ettinger (1982) for several saccharides. Glutamine has been selected for special study for several reasons related to its advantage in high-dose dosimetry and tissue equivalence. The work of Temperton *et al.* (1984*a*) showed that LL output decreased as (water) solvent temperature increased. On the other hand, when the solvent was a 0.1 *M* $TbNO_3$ water solution, a significant increase was observed as temperature increased. A study was also made of the effect of the temperature at which irradiation occurred. Results are shown in Fig. 8. Puite (1980) had already shown that a postirradiation heat treatment of 114°C for 5 h increased the LL yield from glutamine and the samples charted in Fig. 8 were subjected to this postirradiation heat treatment.

2.1.6. Miscellaneous

More recent studies appear to deal more with refinements of technique, both of detection and preparation of materials, aimed at achieving more reliable dosimetry,

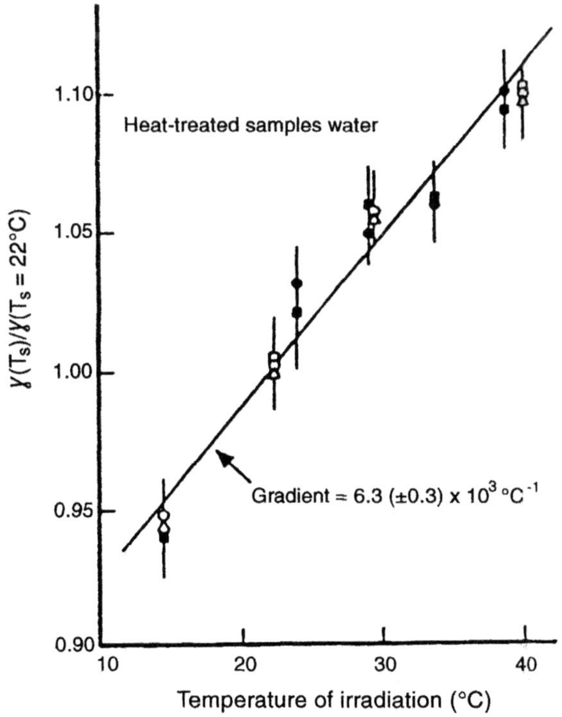

Figure 8. Dependence of LL of glutamine on the temperature of irradiation normalized at $T = 22$°C: (⊙) 100 Gy, temperature from 14.6 to 38.4°C; (□) 100 Gy, temperatures from 14.6 to 39.8°C; (△) 250 Gy, temperatures from 14.6 to 38.4°C; (□) 250 Gy, temperatures from 14.6 to 39.8°C. (Temperton *et al.*, 1984*a*; reprinted with kind permission from Elsevier Science, Ltd.).

and generally at high radiation doses (of interest in food preservation). However, a number of interesting special investigations are to be found in the literature.

Hanig (1984) reported results of enhancement of LL by radiation sensitization. The procedure was to prepare a 1 *M* solution of trehalose and irradiate it with ^{60}Co γ-rays (the sensitizing dose) at doses of about 0.3, 1.0, and 3.0 kGy. Then the irradiated solutions were crystallized (at 6°C) and the phosphors prepared for irradiation at 10^{-2}, 5×10^{-2}, and 10^{-1} Gy. The phosphors were then dissolved in an aqueous luminol solution (9×10^{-5} *M*) with 0.1 *N* Na_2CO_3 at pH 11. The results are shown in Fig. 9.

The effect of grain size was noted early (Ettinger and Puite, 1982). Some mention was made above about the role of grain size in regard to the accessibility of oxygen in free radical processes. A further factor is due to packing, i.e., the fraction of radiation energy deposited in the air in the sample compared with the energy deposited in the phosphor. Results reported by Balogun and Adensami (1989) showed that dissolution rate, light yield, dose response curve, and stability of the LL response are all grain-size-dependent in mannose. An interesting point was also made that light yield should be integrated for the time in which the response decays to a fixed fraction (say, one-tenth) rather than for a fixed time in evaluating LL sensitivity. Results showed lower yield for finer grain size, as did earlier results reported by Ettinger and Puite (1982). Another factor explaining grain-size dependence might be insensitive surface layers. Balogun and Adesanmi (1989) offered the interesting speculation that the increased packing ratio explains reduced sensitivity owing to the increase in nonradiative radical–radical reactions (i.e., reactions prior to dissolution) and also explains the observed increase in dissolution time (attributed to the increased viscosity).

It was noted earlier that, in many cases, LL increases with storage time for high doses (>100 Gy) (Bartlett, 1979). Storage times investigated were 1 to 9 months. This is, of course, a concern for dosimetry applications. Temperton *et al.* (1984*a*) found that the effect of storage time in the case of glutamine could be minimized by

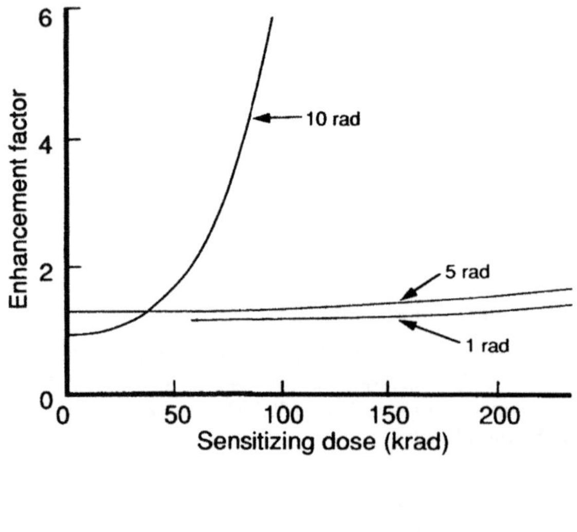

Figure 9. Enhancement factor for trehalose as a function of sensitizing dose. (Hanig, 1984; reprinted with kind permission from Elsevier Science, Ltd.).

heat treatment at 114°C for 5 h, 2 or 3 days after irradiation. Puite (1980) also studied storage time effects. Again, results are for high doses.

The effect of heat treatment, or annealing, has been investigated. Puite's (1980) work showed that LL was increased for glutamine with postirradiation heat treatment at 114°C for 5 h. The work of Bartlett *et al.* (1982) indicated a decrease in LL upon annealing for sugars. The work of Temperton *et al.* (1984*a*) confirmed the increase in LL for glutamine. In some early work to probe the (radical) chemical origin of LL, Ettinger *et al.* (1982) used spin-trapping techniques. In addition to the identification of radicals derived from γ-irradiation of certain amino acids and saccharides, the effect of thermal annealing after irradiation was studied. They found that although the average concentration of radicals (for glutamine, determined by spin-trapping techniques) decreased by 32%, the LL light yield increased by 50%. The observations were "rationalized" by assuming that "selective destruction of radical sites in the solid by heating increases the probability of reaction of the remaining radicals with oxygen or a trap on dissolution. Increase in the concentration of peroxyl radicals leads to an increase in light yields." Similar lines of reasoning surfaced in other references to explain annealing and storage effects, assuming two classes of radicals, $R_1^{\cdot}$ responsible for LL and more stable than $R_2^{\cdot}$, which quenches LL, but which decays more rapidly during storage (or annealing). To study the survival of the free radicals responsible for LL at high temperatures, Kundu and Mitra (1988) metlted γ-irradiated saccharides and kept them liquid for several minutes and then measured the LL. The results showed that about 20% of the LL signal survives the high temperature for 4 to 6 min. Radiation doses were high ($\sim$kilograys).

Most of the radiation leading to LL has been the result of ^{60}Co γ-rays, but other ionization sources have been used, including π^--beams (Bartlett *et al.*, 1981), electrons (Atari *et al.*, 1985), neutrons (Atari *et al.*, 1985; Ettinger and Miola, 1985) and x-rays (Ahnström and Ehrenstein, 1959; Atari *et al.*, 1985). An interesting comparison of effects was reported by Ettinger *et al.* (1984).

2.1.7. Application to Dosimetry

Much of the reported work concerning LL has been driven by concerns about applications to dosimetry, with the goal of providing a detector with good accuracy, reproducibility, sensitivity, and linear response that is useful for relatively high doses (a few grays to kilograys). A good review of the state of the art up to early 1982 was given by Puite and Ettinger (1982). A report on the performance of commercial luminometers as LL dosimetry readers was published by Temperton *et al.* (1984*b*).

2.1.8. Low-Dose Detection

Although most application is to high-dose dosimetry, some work has been on low-dose detection (Busuoli and Montiventi, 1984). Again emphasizing the advantage of certain organics as tissue equivalents in dosimetry, Busuoli and Montiventi (1984) studied mannose, glucose, and *N*-acetyl-D-glucosamine at low doses using luminol as a sensitizer. In these studies grain size was kept below 150 μm, and the quantity of irradiated material was 0.1 ml (~50 mg). The luminol solution consisted of 125 mg/liter in bidistilled water, giving an LL advantage of 10^3 over pure water.

A compromise between the influence of temperature on LL yield (increasing with temperature) and autoluminescence of luminol (also increasing with temperature) led to a choice of 30°C for the solvent. As LL also depends on oxygen concentration and pH, 1.25 mg of Na_2CO_3 and 2.5 mg of haemin (per liter) were added to maintain the pH between 10.5 and 11.5. It is interesting to note that "the experimental data are not reproducible during the first two weeks after the solution is prepared; an acceptable stability is only reached after an elapsed period of more than 20 days." Conclusions were that the detection limit for *N*-acetyl-D-glucosamine and glucose is of the order of 2 mGy (0.2 rad) and that reproducibility is ±10% for doses of about 50 mGy for glucose.

A semitheoretical analysis of limiting sensitivity was given by Matyushkov *et al.* (1977). If the LL output is considered to be S_L photons for a dose of D_S [eV] and a linear relation vs. D_S and S_L is assumed, then

$$D_S = KS_L$$

The sensitivity of the detector is given by

$$q = S_L/D_S = 1/K$$

with a CL indicator of quantum yield ϕ (photons/eV). The limiting sensitivity results when every radiation defect (free radical in the case of organics and *F*- or *V*-center in the case of inorganics) interacts with the indicator. If N_{rd} is the number of stable radiation defects, and G is the radiochemical yield (i.e., number of stable radiation defects per 100 eV of absorbed energy), then

$$S_L = N_{rd} = D_s(G/100)$$

and the limiting sensitivity is

$$q = (G/100)\phi$$

for organics where $1 < G < 4$. A representative sample size in LL observations is 0.05 to 0.10 g. If a sample size of 0.1 g is assumed, a dose $D_S = 100$ eV in this sample is 1000 eV/g; 1 rad is an absorbed dose of 100 ergs (10^{-5} J)/g; and 1000 eV/g is 1.6×10^{-16} J/g or 1.6×10^{-11} rad. Thus, taking $G = 4$ (the most optimistic case), 1 rad will produce

$$N = \frac{4}{1.6 \times 10^{-11}} = 2.5 \times 10^{11} \text{ radicals/rad}$$

A typical ϕ value is 10^{-2}, so

$$S_L = 2.5 \times 10^9 \text{ photons/rad } (= 2.5 \times 10^{11} \text{ photons/Gy})$$

With modern electronics, a limiting value of q of this order corresponds to the reliable detection of absorbed doses with $D_{min} < 10^{-3}$ rad (Matyushkov *et al.*, 1977). However, such limitations as background light (self-glow of optimizing sensitizers), nonlinearities of S_L vs. D_S, and effects of pH have as yet not resulted in realizing this limit by a factor of at least 10^2. (The figure $D_{min} \approx 10^{-3}$ rad in the ideal case is not inconsistent with a photomultiplier cathode efficiency of 20% collecting about one-third of the emitted light in an integration time of 10 s, resulting in somewhat fewer than 10^3 cathode electrons cm^2 s.) These numbers should be compared with results to be discussed below when LL of NaCl in luminol is described.

2.2. Inorganic Lyoluminescence Phosphors

Although investigations of the LL of organic phosphors have been extensive because of their simulation of tissue composition and their relation to familiar and interesting CL processes, the study of LL of inorganic phosphors has also yielded interesting and useful results. Some of the very early research involved alkali halides, and, in particular, NaCl. In fact, the discovery of LL came from the work of Wiedemann and Schmidt (1895) using chlorides of Li, Na, and K. Westermark and Grapengiesser (1960) appear to have been the next to investigate the LL of NaCl, and in fact noted that the light output for a solution of TlCl in water (showing a pronounced maximum at about 0.01 g/liter, i.e., 4×10^{-5} *M*) was 1000 times greater than for pure water. Figure 10 is reproduced from their paper. Enhancement methods for LL of inorganics will be discussed in detail below. Other early work on LL from NaCl was reported by Lelievre and Adloff (1964) and Ahnström (1965).

2.2.1. Mechanisms

In discussing the mechanisms for LL in inorganic phosphors, the relationship to those of TL (Templer, 1985, 1986; Clark and Templer, 1988; McKeever, 1984; Böhm et al., 1986) should be emphasized. Very likely, much of value concerning LL could be learned from TL physics, because the creation of lattice defects by ionizing radiation plays a primary role in both phenomena. The importance of this possible relationship does not appear to have been fully exploited by experiment, although Ettinger and Puite (1982) made reference to it.

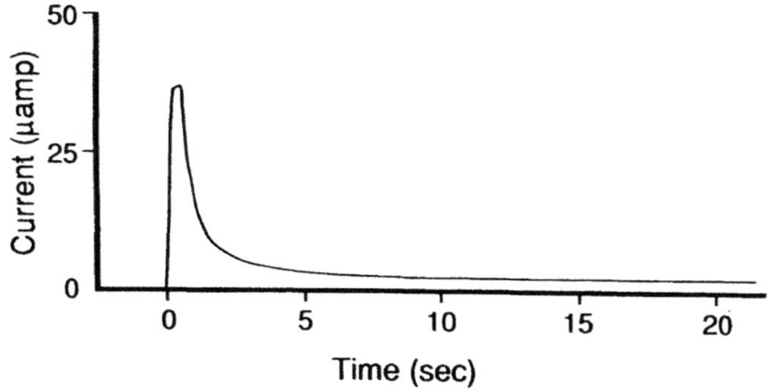

Figure 10. Glow curve of LL from NaCl after a dose of 50 Mrads with 2-MeV electrons, after 12 months of storage. Solvent was 3.2×10^{-5} *M* TlCl in water (Westermark and Grapengiesser, 1960; reprinted with permission from Macmillan Magazines, Ltd.).

When ionizing radiation is absorbed by matter, both excitation and ionization can occur. Crystals of alkali halides develop lattice defects. Excited electrons usually return to the ground state in a short time, either directly (energy emitted as light-fluorescence or phosphorescence) or via vibrational level transitions (energy converted to heat). An electron that escapes from the original atom leaves behind a hole, trapped in an isolated positive ion vacancy (*V*-center), and the electron may be trapped at some negative ion vacancy (*F*-center). The concentration of *F*-centers is a monotonic function of dose, but generally not linear. Irradiation produces both types of centers simultaneously.

Etzel and Allard (1959) found significant differences between naturally occurring NaCl (halite) and reagent grade crystals formed in the laboratory. Ahnström (1965) also noted that, "It is well known that the origin of the material has a great influence on the number of *F*-centers produced per unit dose as well as the maximum number of *F*-centers that can be obtained." These authors also noted that the saturation dose for *F*-centers exceeded 250 kGy. By supplying energy to the crystal (by heat as in TL or by light as in photoluminescence (PL) (Rao *et al.*, 1984; Mathur *et al.*, 1986), the trapped electron can be lifted into the conduction band and then may recombine with a hole by the emission of light or via vibrational levels (heat). For the phenomenon of LL, the electron may be released from the trap when the irradiated material is dissolved in a suitable solvent. In this case the electron may never be free, resulting in a hydrated electron e_{aq}, which then interacts to produce an excited state of an atom or molecule in the solvent. Light is emitted when this excited entity returns to its ground state or some lower energy level. In this case it can be expected that the energies released by these recombination processes will be different from those in the solid state, and the emission spectra will be correspondingly different.

Attention focuses on the role of the hydrated electron as a cause of LL in inorganic phosphors. The role of hydrated electrons in CL processes is well known (Hart and Anbar, 1970), resulting in excited-state products (Martin *et al.*, 1972). It is now useful to distinguish two different situations: (1) where the solvent is pure water, and (2) where some agent has been introduced into the water as an enhancer.

1. In the first case, Ahnström (1965) and Arnikar *et al.* (1970, 1971) suggested that the source of LL results from the reaction

$$e^- \text{ trapped in lattices} + H_2O \rightarrow e_{aq}^-$$

$$e_{aq}^- + Cl^{\bullet} \rightarrow Cl^{-*} \rightarrow Cl^- + h\nu$$

   where e_{aq}^- is the hydrated electron.
2. The energy of Cl^{-*} may alternatively be converted to heat unless, the energy can be transferred to a fluorescent molecule (or complex) and finally is emitted as light.

Reaction (1) would occur with pure water as the solvent; reaction (2) would be possible with the use of enhancers.

Even in case (1) the situation can be complicated by possible surface effects (which will also lead to a grain size effect). When dissolution first starts, the e^- and

$Cl^{\bullet}$ will exist as partially hydrated forms in the crystal–liquid interphase. Again, it is useful to distinguish between reactions where both e^- and $Cl^{\bullet}$ are in the solution, where deactivation of Cl^{-*} competes with the emission of light, and reactions where the partially hydrated electronically excited $Cl^{\bullet}$ or e^- is at the liquid surface. The interphase reactions will take place only in a limited space (cage), which includes the surface and region of the fully hydrated species close enough to react with their surface counterpart. Since an excited state is more protected from a deactivation encounter in the interphase than in the solution, there may be a better chance for energy transfer leading to reaction (2).

Diffusion and stirring would destroy the cage.

Ahnström (1965) proceeded to analyze the consequences of reactions taking place in the crystal surface, the solvent part of the cage, and the solution outside the cage. With the use of "reasonable" values for dissolution rates, recombination rate constants, and quantum efficiencies of energy transfer to a fluorescent molecule in the three different regions, fairly good agreement was found between observed light output and concentration of e_{aq}^- and $Cl^{\bullet}$.

The efforts to fit experimental curves to the sort of theoretical calculations described above bring to light several experimental problems. When an enhancer such as fluorescein is used, the optimum concentration appears to be dose-dependent. Mechanical disturbances or stirring irregularities appear to change local concentrations significantly. F-center concentration is not uniformly distributed through the particles. It was found that a higher yield of F-centers was obtained by irradiation of the fine-grain NaCl, which might be the result of structural defects in the surface of pulverized crystals. This hypothesis is consistent with the fact that when fluorescein solution was slowly added to irradiated crystals, a considerable part of the material remained undissolved, in spite of which the total light output was nearly as great as in cases where complete dissolution occurred.

V-bands in alkali halides have long been the subject of study (Casler *et al.*, 1950). Atari (1980*a*) investigated the recombination of released F-centers with trapped holes (V_2-centers) at the water–solid interface, in LL of NaCl, NaBr, KBr, CsCl, and CsBr in pure water. He found that exceptionally pure additively colored crystals having only F-centers produced no light when dissolved in pure water, whereas impure crystals emit light on dissolution, presumably owing to recombination of F-centers with traces of impurity ion centers (V-centers). Irradiation produces both trapped electron color centers (F) and trapped hole color centers (V). This then is taken as the LL mechanism (Atari, 1980*a*):

1. The presence of both F- and V_2-centers together is essential for exhibiting the LL phenomenon. Neither center is capable of producing LL on its own.

2. Afterglow (AG) and LL phenomena take place primarily as a result of recombination of electrons and holes in the crystal (AG) or at the water–solid interface (LL). AG is essentially due to thermal release of trapped electrons (F-centers) to the conduction band followed by their recombination with holes in V_2-centers or via a tunneling process. On the other hand, LL is mainly due to the liberation of trapped electrons (F-centers) by the hydration process to form hydrated electrons (e_{aq}^-) followed by recombination with their counter-parts (V_2-centers) at the water–solid interface.

3. Because the emission spectra of both AG and LL for studies of alkali-halide crystals were almost identical, the recombination process between a hydrated electron and a V_2-center must take place mainly at the water–solid interface and partially in solution. The latter possibility requires that both the F- and V-centers be hydrated before recombination. If the V_2-center is hydrated it will behave chemically as a halogen atom ($X^{\cdot}$). The recombination of e_{aq}^- and $X^{\cdot}$ atoms will give rise to $(X^-)^*$ followed by the emission of a photon of light. There is evidence that such a process is possible but its probability as compared with recombination at the water–solid interface is small. If such a probability were high, the emission spectra for both AG and LL would be slightly different owing to the modification of the energy levels of excited halide ions by the presence of a dielectric (water).

4. The large quenching effect of hydrated electron scavengers suggests that the released F-center must be hydrated before recombination with the V_2-center. Such a hydration process takes place in a very short time ($\sim 10^{-9}$ s) after liberation of the trapped electron by dissolution of the crystal. The rate at which light is produced depends on the diffusion constant of the hydrated electron (9×10^{-5} cm^2/s) and the availability of its counterpart (the V_2-center) at the water–solid interface.

5. The above LL processes can be represented as follows:

$$F\text{-center} \xrightarrow{\text{Hydration}} e_{aq}^-,$$

$$e_{aq}^- + V_2\text{-center} \xrightarrow{\text{Recombination}} h\nu,$$

(water–solid interface).

6. The observed light emission on dissolution of additively colored crystals in pure water is due just to the presence of trace impurities in these crystals. The released F-centers after undergoing hydration recombine with these acceptors (A) at the water–solid interface:

$$F\text{-center} \xrightarrow{\text{Hydration}} e_{aq}^-,$$

$$e_{aq}^- + A \rightarrow (A^-)^* \rightarrow A^- + h\nu$$

The literature appears to attribute to Ahnström work compatible with V-centers, whereas Ahnström (1965) did not consider this process. Further, the literature reference to Ahnström is in error, and turns out to be to a paper by Erikson (1962) on the quenching effect of oxygen on LL from NaCl. LL observed in crystals without V-centers (i.e., contrary to the hypothesis) was "merely due to the presence of trace impurities." Papers published after this one still seem to promote the role of $Cl^{\cdot}$ (rather than V-centers) in the case of dissolution in pure water, although, of course, the two mechanisms may be very closely associated.

Atari (1980*b*) reported further studies of enhancement of LL in alkali halides in fluorescent and CL solutions.

The use of LL as a means of defect concentration measurement was described by Avotinsh *et al.* (1985). In this technique, the alkali halide single crystals are in the form of disks of 5-mm radius and 2-mm thick. They are mounted on a disk that is rotated at constant angular rates so that the Reynolds number is in the range 100–5000. The disk surface then dissolves (uniformly) and the thickness of the dissolved part is calculated according to a formula that depends on several crystal–solvent parameters, speed of rotation, and time. In this way it is claimed that the depth in the original crystal at which dissolution is occurring at any given time is known, and that the LL at that time provides a measure of the concentration of F-centers at that depth. Correlation between LL and spectrophotometric measurements appear to confirm this assertion. The essential advantage of the LL method is claimed to be its spatial resolving power (1 μm).

2.2.2. Enhancement and Spectra

In the case of the inorganics, it is desirable to discuss enhancement techniques and spectra together, since spectral analysis elucidates the mechanism of enhancement. The spectral results of Atari (1980*a*) for LL from alkali halides dissolved in pure (triple-distilled, pH 6.9) water provide a useful basis for comparison. NaCl, NaBr, KCl, KBr, CsCl, and CsBr were studied (ultrapure). Atomic absorption measurements indicated that there were fewer than 0.1 ppm impurities Cu^{2+}, Zn^{2+}, Mn^{2+}, and Cd^{2+}. Exposure doses were about 6 kGy. AG experiments were conducted 15 min after irradiation. A comparison of the time decay of (normalized) AG and LL is given in Fig. 11. Figure 12 shows a comparison of the spectral distribution for AG and LL (normalized). Again in terms of the mechanisms discussed above, the similarity of these spectral distributions is taken as evidence that light is a result of recombination of F- and V-centers in LL.

Enhancement of LL in alkali halides has been achieved by the use of (1) solutions containing CL and fluorescent agents (Atari, 1980*b*); and (2) solutions containing certain positive ions.

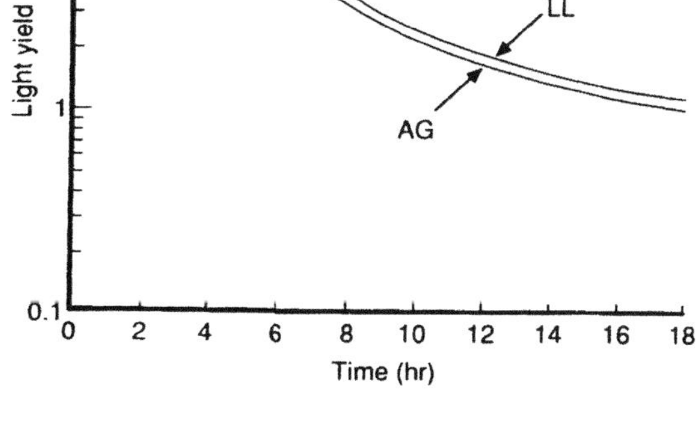

Figure 11. LL and AG decay curves of NaCl exposed to 100 krad dose of γ-radiation (Atari, 1980*a*; with kind permission from Elsevier Science, Ltd.).

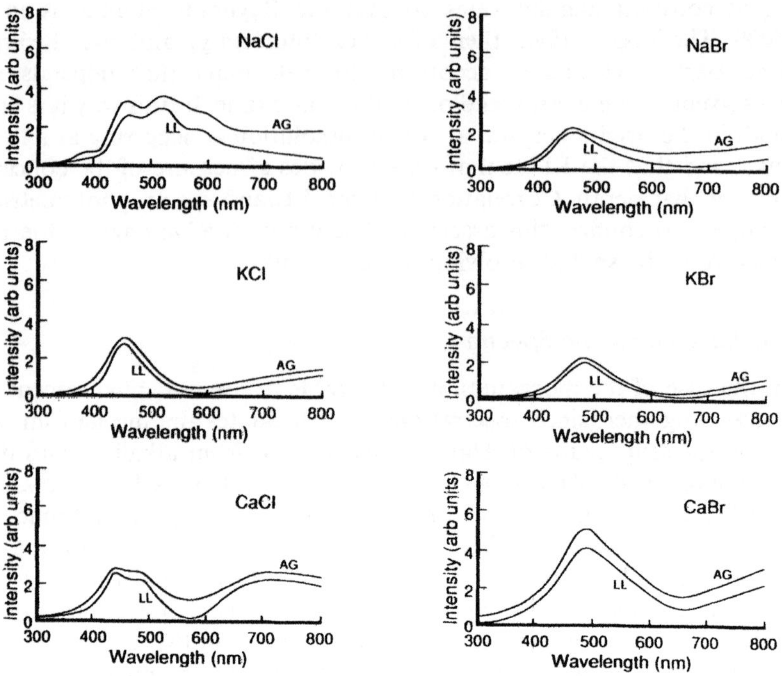

Figure 12. LL and AG spectra of γ-radiated (600 krad) alkali halides (Atari, 1980*a*; reprinted with kind permission from Elsevier Science, B.V.).

(a) Solutions Containing Chemiluminescent and Fluorescent Agents. Westermark and Grapengiesser (1960) appear to have been the first to report enhancement of LL from NaCl using solutions of organic substances instead of pure water. Dissolution in a liquid scintillator gave less light than in water itself. Enhancement of nearly a thousand times was obtained using a solution of components known to be fluorescent to ultraviolet light or x-rays (diphenyloxazole and naphthalene sulfonic acid condensed with formaldehyde).

Ahnström (1965) reported that light emission is considerably increased by the addition of certain fluorescent substances to the solvent, but no quantitative enhancement factor was given, and only fluorescein (10^{-4} *M* at pH 7) and Rhodamine B were discussed. Atari and Ettinger (1974*a*) reported that when a fluorescein solution was used for NaCl, enhancement was observed but only at "higher doses."

Arnikar and Kalkar (1977) reported results on the LL spectra of irradiated NaCl dissolved in fluorescein- and eosin-doped water. A conventional spectrograph was used to obtain spectra from heavily dosed samples (2 h at 80 Gy/min). Asymmetric broadband spectra were observed (compare with Arnikar *et al.* (1971) for pure water). The results for fluorescein (10^{-4} *M*) showed a maximum at about 530 nm and a half-width at about 50 nm; for eosin (5×10^{-6} *M*) the maximum was at about 560 nm and a half-width at about 70 nm. No enhancement factor was reported.

Atari (1980*b*) reported LL results for irradiated and additively colored NaCl, NaBr, KCl, and KBr dissolved in fluorescent and CL aqueous solutions. LL spectra (obtained using the PARC 1205 A OMA) were similar to the fluorescence spectra of the solutions. Samples were subjected to a dose of 6 kGy. "Additive coloration" was achieved by exposure of the crystals to the respective alkali vapors for 48 h (these were not irradiated). The results are shown in Table 1 and Fig. 13.

The optimum luminol solution was 125 mg luminol, 1.25 Na_2CO_3, 2.5 (g?) chlorohemine and 15 mg fluorescein and eosin per liter of triple-distilled water; and 5×10^{-3} *M* Cu^{2+}. No details were given for TlCl. It was found that oxygen, in general, had a marked quenching effect. The role of *V*-centers was investigated [see Casler *et al.* (1950) for terminology]. For fluorescein and eosin the enhancement was greater for $F + V_2$ centers than for *F*-centers only or $F + V_3$ centers. The presence of V_2-centers in the Cu^{2+} solution had very little effect. The greatest enhancement was found for luminol solutions, and the presence of V_2-centers had no additional effect. However, whereas oxygen is a quencher in the other solutions, its presence is essential for stimulating luminescence in the luminol solution. The proposed LL mechanisms are as follows (Atari, 1980*b*) (D ≡ dye; A ≡ Acceptor):

1. LL IN PURE WATER

$$F\text{-center} \xrightarrow{\text{Hydration}} e^-_{aq},$$

$$e^-_{aq} + V_2\text{-center} \xrightarrow[\text{Water-solid interface}]{\text{Recombination}} h\nu$$

The wavelength of the photon emitted is dependent on the nature of the gamma colored crystal. No LL was detected for pure additively colored crystals.

Table 1. Relative LL Yields per $10^{15}/cm^3$ F-Centers of Alkali Halides in Different Solutions (Pure Water pH = 6.9)[a]

Alkali halide coloration of abs. bands	NaCl Additive *F*	NaCl Gamma *F*, V_2	KCl Additive *F*	KCl Gamma *F*, V_2	KCl Gamma *F*, V_3	KBr Additive *F*	KBr Gamma *F*, V_2, V_3
Pure water	0	0.9	0	1.2	0.5	0	0.8
Fluorescein	3.6×10^1	5×10^3	2.7×10^1	7.5×10^2	9×10^1	5.1×10^1	1.3×10^2
TlCl	4.6×10^1	3.8×10^2	1.5×10^2	3.4×10^2	1.6×10^2	8	2.3×10^1
Cu^{2+}	1.9×10^2	2×10^2	3.2×10^1	5.8×10^1	1.3×10^2	5.3×10^1	1.2×10^1
Luminol	4.6×10^4	4.5×10^4	4.8×10^4	4.5×10^4	4.7×10^4	4.9×10^4	4.6×10^4

[a]Atari, 1980*b*; reproduced with kind permission from Elsevier Science Publishers, B.V.

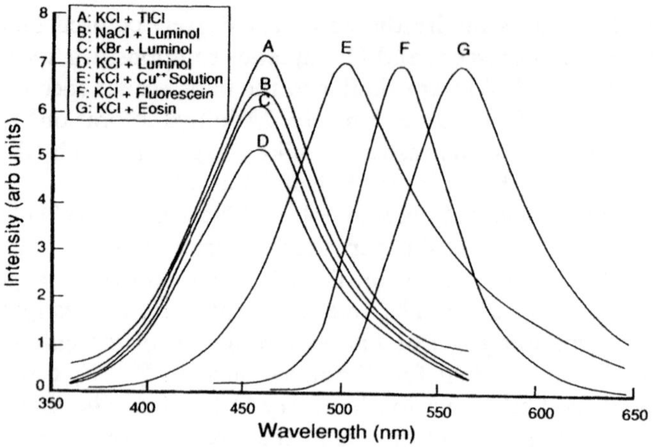

Figure 13. LL spectra of several alkali halides in different solutions (Atari, 1980*b*; with kind permission from Elsevier Science, Ltd.).

2. LL IN FLUORESCEIN AND EOSIN SOLUTIONS:

$$F\text{-center} \xrightarrow{\text{Hydration}} e^-_{aq}$$

$$e^-_{aq} + \mathrm{D} \cdots \rightarrow |\mathrm{D}|^* \rightarrow \mathrm{D} + h\nu$$

$$e^-_{aq} + \underset{\substack{\text{solvated}\\ \text{hole}}}{|\mathrm{X}^{\bullet}|} \xrightarrow{\text{Recombination}} |\mathrm{X}^-|^* \times \xrightarrow[\text{Energy transfer}]{\text{Dye}} |\mathrm{D}^*| \rightarrow \mathrm{D} + h\nu$$

where ... indicates that the occurrence of a series of reactions leading to the excited product is possible.

3. LL IN TlCl SOLUTION:

$$F\text{-center} \xrightarrow{\text{Hydration}} e^-_{aq}$$

$$e^-_{aq} + |\mathrm{TlCl}_n^{1-n}| \cdots \rightarrow |\mathrm{A}|^* \rightarrow \mathrm{A} + h\nu,$$

$$e^-_{aq} + \underset{\substack{\text{Solvated}\\ \text{hole}}}{|\mathrm{Cl}^{\bullet}|} \xrightarrow{\text{Recombination}} |\mathrm{Cl}^-|^* \xrightarrow[\text{Energy transfer}]{|\mathrm{TlCl}_n^{1-n}|} |\mathrm{A}| \rightarrow \mathrm{A} + h\nu$$

The wavelength of the emitted photon had a maximum at 455 ∓ 4 nm which corresponds to the fluorescence emission of $|\mathrm{TlCl}_4^{-3}|$. It is very likely that this chlorocomplex is the species responsible for light emission.

4. LL IN Cu^{2+} ION SOLUTION:

$$F\text{-center} \xrightarrow{\text{Hydration}} e^-_{aq},$$

$$e^-_{aq} + \mathrm{Cu}^{2+} \xrightarrow{\text{Recombination}} |\mathrm{Cu}^+|^* \rightarrow \mathrm{Cu}^* + h\nu$$

5. LL IN LUMINOL SOLUTION:

$$F\text{-center} \xrightarrow{\text{Hydration}} e_{aq}^{-}$$

$$e_{aq}^{-} + O_2 \rightarrow O_2^{-}$$

$$2O_2^{-} + 2H_2O \rightarrow O_2 + 2OH^{-} + H_2O_2 \qquad \text{[oxidant]}$$

$$OH^{\bullet} + OH^{\bullet} \rightarrow H_2O_2 \qquad \text{[oxidant]}$$

These are the possible reactions by which the H_2O_2 (oxidant) can be produced when colored alkali halides are dissolved in aqueous solutions. The radical ($OH^{\bullet}$) is supposedly present in the system. Consequently, the CL of luminol may proceed as follows:

$$\text{Luminol} + O_2 + \text{Oxidant } (H_2O_2) \cdots \rightarrow |L|^* \rightarrow L + h\nu$$

where the symbol |L| represents the dinegative aminophthalate ion, which is responsible for light emission preceded by a series of reactions involving oxygen that leads to its formation.

The role of Cu^{2+} and $TlCl_n^{1-n}$ will be discussed in more detail below.

Kalkar (1983) investigated the enhancement and spectra of LL [which he calls aqualuminescence (AL)] from irradiated NaCl as a function of the concentration of the xanthene dyes fluorescein, eosin Y, and pyronin G. Doses were the order of 1 kGy. A detail, lacking in other papers, notes that even with this high dose, the use of the spectrograph with ASA 400 film required a continuous dissolution process and an exposure of 4 h. The intensity of irradiated NaCl dissolved in pure water was too feeble to generate spectral data. Table 2 gives the optimum concentrations and enhancement factors (over pure water) as determined by photomultiplier–scaler detection. All of the spectra were broadband (half-width ~30–60 nm) with peaks between 530 and 555 nm. Comparison of dye emission and LL spectra leads to the following proposed mechanism:

$$D + e_{aq}^{-} \rightarrow (D^{-})^* \rightarrow D^{-} + h\nu$$

$$D^{-} + Cl_{aq}^{\bullet} \rightarrow D + Cl_{aq}^{-}$$

Kalkar *et al.* (1985) described observations of LL from irradiated NaCl

Table 2. Concentration Dependence of AL (i.e., LL) Intensity in the Presence of Some Xanthene Dyes[a]

Dye	Optimum concentration (mg/liter)	Relative AL intensity
Fluorescein	6	580
Eosin Y	20	760
	80	540
Pyronin G	20	250

[a]After Kalkar (1983).

dissolved in alkaline luminol. Previously Atari (1980*b*) had claimed that both LL and CL result from the same luminol process, namely,

$$F\text{-center} \rightarrow e^-_{aq}$$

$$O_2 + e^-_{aq} \rightarrow O_2^-$$

$$2H_2O + 2O_2^- \rightarrow O_2 + 2OH^- + H_2O_2$$

$$\text{Luminol(L)} + O_2 + H_2O_2 \rightarrow A^* \rightarrow A + h\nu$$

where A represents the dinegative (dianion) aminophthalate ion.

From a study of the spectra of CL and LL, and the work of others on hole scavengers, Kalkar *et al.* (1985) claimed that the mechanism of LL is "entirely different from that of CL" and proposed the following (L^{2-} is the dianion of luminol):

$$L^{2-}_{aq} + (Cl_2^{\bullet-})_{aq} \rightarrow L^-_{aq} + 2Cl^-_{aq}$$

$$L^-_{aq} + e^-_{aq} \rightarrow (L^{2-}_{aq})^* \rightarrow L^2_{aq} + h\nu$$

Deciding between the two mechanisms [since A* and $(L^{2-}_{aq})^*$ are the same] would require careful studies of rates, quenching agents, and hole scavengers.

In a further paper Kalkar *et al.* (1986) repeated the above proposed mechanism based on observations of LL from irradiated NaCl dissolved in a solution of fixed concentration of luminol ($3.5 \times 10^{-4}\,M$) at various concentrations of fluorescein. Figure 14 shows approximately a fivefold increase at a concentration of about $2 \times 10^{-6}\,M$ of fluorescein.

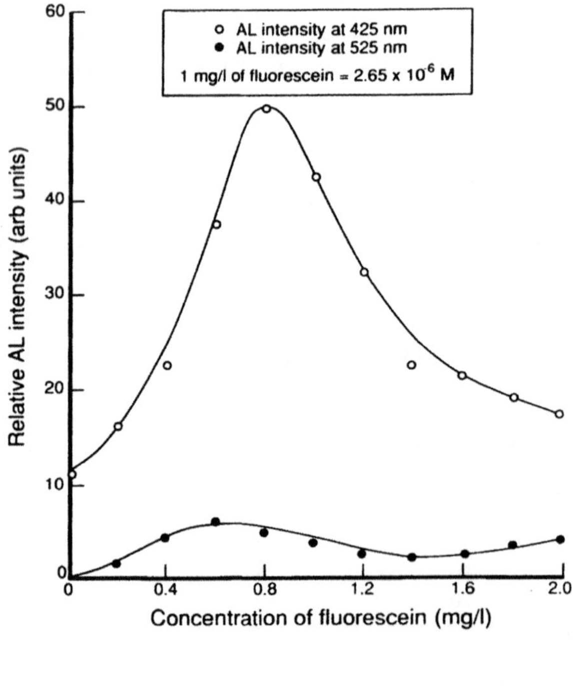

Figure 14. Relative AL (i.e., LL) intensity of luminol and fluorescein with concentration of fluorescein, (Kalkar *et al.*, 1986; reprinted with kind permission from Elsevier Science, Ltd.).

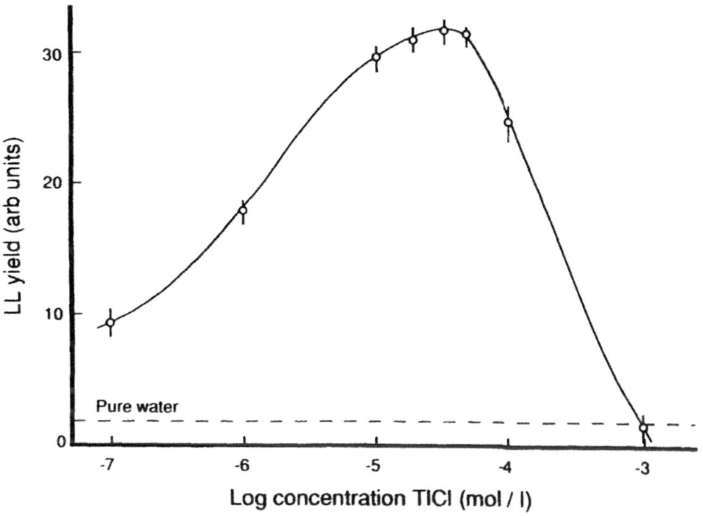

Figure 15. Dependence of LL from NaCl on concentration of TlCl in aquous solution. (Lelievre and Adloff, 1964).

(b) Solutions Containing Certain Positive Ions. Enhancement effects achieved in inorganics by the addition of certain positive ions are of particular interest. This was not treated at much length by Ettinger and Puite (1982), perhaps because interest in applied dosimetry focused on procedures involving phosphors simulating body tissue. Primary attention has been given to enhancement of LL from irradiated NaCl by the addition to the solvent (usually water) of small concentrations of salts of Tl, Cu, and Co. Westermark and Grapengiesser (1960) reported an increase by a factor of "about a thousand times" in the LL of irradiated NaCl when a solution of 0.01 g/liter (4×10^{-5} M) of TlCl was used instead of pure water.

Lelievre and Adloff (1964) used an aqueous solution of TlCl and found two types of light pulse: the first of 30-ms duration and the second lasting several seconds. They concluded that the LL results from "la formation et l'excitation du complexe $TlCl_2^-$," and the measurement of the emission provides a means of following the formation of F-centers in the irradiated NaCl. From a photomultiplier pulse recorded on an oscilloscope, they found that the peak of the 30-ms pulse was about three times that of the slower pulse, but that the major part of the light output was in the slow pulse, which was then studied in detail as a function of various parameters (grain size, temperature, pH). Doses were about 3 kGy. Light output as a function of TlCl concentration is shown in Fig. 15. The enhancement was found to be a maximum at a concentration of about 4.5×10^{-5} M. The spectral distribution was found to be broad (FWHM $\sim$130 nm) with a peak at about 435 nm. The LL was attributed to emission from the complex $TlCl_2^-$, formed during dissolution by

$$Tl^* + 2Cl^- \rightarrow TlCl_2^-$$

after which energy exchange with electrons from F-centers results in $(TlCl_2^-)^*$. Studies of absorption and emission spectra of $TlCl_2^-$ led to the conclusion that the emission is due to the transition

$$^3P_1 \rightarrow {}^1S_0[4350\,\text{Å}]$$

The diminution of light for Tl^+ concentrations above $4.1 \times 10^{-5}\,M$ is attributed to

$$(TlCl_2^-)^* + TlCl_2^- \rightarrow 2TlCl_2^- + \text{Heat}$$

consistent with observation of PL of TlCl. If it is assumed that the LL is proportional to the number of F-centers, one can determine the latter as a function of dose.

Atari and Ettinger (1974*a*) made a brief qualitative reference to the enhancement of LL in NaCl using a TlCl solution, saying that the effect is seen only at higher (>1 Gy) doses. Ettinger and Puite (1982) stated that Arnikar and Kalkar (1977) studied the enhancement owing to thallous ions, and that analysis of the emitted spectrum points to LL being due mainly to excitation of $(Tl^+_{aq})^*$. Subsequently, Kalkar and Ramani (1981) described enhanced LL from irradiated NaCl in solutions containing Cu^{2+} and Co^{2+} ions. They referred to previous work as having shown that thallous ions increased the LL by a factor of 60, whereas Cd^{2+}, Zn^{2+}, and NO_3^- quenched it nearly completely.

Reagent grade NaCl was irradiated with a 370-Gy dose, to study the effects of Cu^{2+} and Co^{2+} ions in concentration ranges of 10^{-3} to $10^{-4}\,M$. Total intensity and spectra were observed. The spectrum was crudely measured by means of a set of narrow transmission filters and a photomultiplier. The variation of LL with the concentration of Cu^{2+} ions is shown in Fig. 16. A factor of about 4.3 was observed at an optimum concentration of $3 \times 10^{-4}\,M\ Cu^{2+}$, and $CuCl_2$ and $CuSO_4$ showed the same enhancement factor. The Na_2SO_4 results showed that SO_4^{2-} ions quenched the emission slightly, so the indication is that the enhancement is due to the Cu^{2+} ions. Co^{2+} introduced through $CoSO_4$ showed an enhancement of a factor of about

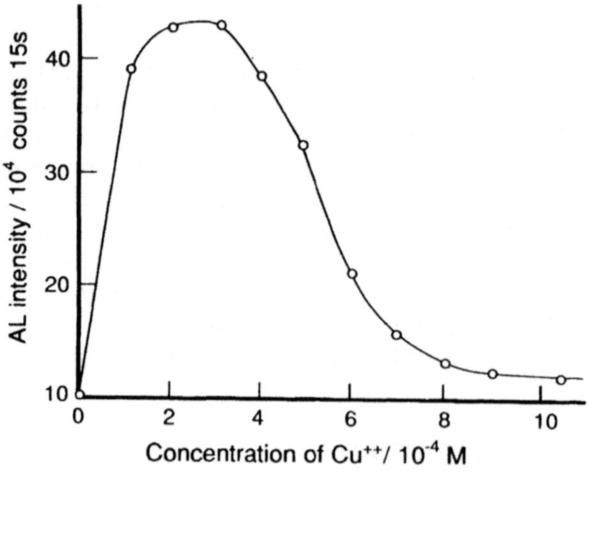

Figure 16. Dependence of AL (i.e., LL) intensity from NaCl with Cu^{2+} concentration in aqueous solution (Kalkar and Ramani, 1981).

2.5. The spectra observed were characteristic of the activators (Cu^{2+} and Co^{2+}). The authors attributed the enhanced LL to the interaction between a hydrated electron and the metal ion. Cu ions in solution form a complex $Cu(H_2O)_6^{2+}$ and the mechanism of emission is taken to be

$$Cu(H_2O)_6^{2+} + e_{aq}^- \rightarrow Cu(H_2O)_6^{+*} \rightarrow Cu(H_2O)_6^+ + h\nu$$

The formation of such an excited complex by reaction with a hydrated electron was proposed by Martin *et al.* (1972) in the CL of tris-(bipyridyl)-ruthenium(III). The reaction in pure water,

$$Cl_{aq}^{\bullet} + e_{aq}^- \rightarrow Cl_{aq}^{-*} \rightarrow Cl_{aq}^- + h\nu$$

is weaker owing to the competing reactions

$$e_{aq}^- + H_2O \rightarrow H + OH_{aq}^-$$

$$e_{aq}^- + e_{aq}^- \xrightarrow{H_2O} H_2 + 2OH_{aq}^-,$$

$$Cl^{\bullet} + Cl^{\bullet} \rightarrow Cl_2$$

As in other work, NH_3 is observed to quench LL, presumably because of the reaction

$$Cu(H_2O)_6^{2+} + 4NH_3 \rightarrow Cu(NH_3)_4^{2+} + 6H_2O$$

It should be noted that the net effect of e_{aq}^- is to reduce Cu^{2+} and Co^{2+} to Cu^+ and Co^+. The role of Tl^+ in TlCl cannot be explained in the same way.

The decrease in LL with Cu^{2+} ion concentration above 3×10^{-4} *M* is taken as being due to a decrease in the reaction rate of e_{aq}^- with an increase in electrolyte concentration (Anbar and Hart, 1965), which is a consequence of the fact that the arrangement of the solvent molecules around the Cu^{2+} ions is partially inhibited in concentrated electrolyte solutions.

Westermark and Grapengiesser (1960) stated that no enhancement in LL of NaCl was observed from the (known to be fluorescent) lanthanide ions (Gd, Nd, Sm, Tb, and Eu). Although the work involved organic rather than inorganic phosphors it is interesting to recall experiments by Ettinger and Anunuso (1981) showing enhancement of LL by energy transfer to rare-earth ions. The enhancement was achieved using Tb^{3+} ions at a pH of 2.3–3.9, mainly in the form of nitrates. The enhancement factors were as discussed above. Some enhancement was noted for Eu^{3+} and Dy^{3+}, but there was no significant enhancement from Tb and Sm ions. The results were characterized as preliminary.

2.2.3. Effect of pH

Much of the work discussed does not make specific reference to the pH of the solution: pH ~6.9 or 7 can be inferred. Some reports note that optimum results were obtained at pH = 6.9. However, specific pH has been reported in several instances. Lelievre and Adloff (1964) reported a drastic falling off of LL for NaCl from pH 4 to 1. A systematic study of pH dependence for NaCl, NaBr, KCl, and KBr was made by Arnikar *et al.* (1971). Their results show that the optimum for NaCl is at pH $\sim$4. They call attention to the fact that this optimum pH range (4–10) is also optimum for the stability of the hydrated electron.

2.2.4. Effect of Grain Size

As is to be expected from the discussion of the work of Ahnström (1965) concerning the interphase model, surface effects and thus grain size will be important parameters in LL from inorganic phosphors, as it was in organics. ("...surface localized F-centers will evidently give a higher local effective concentration than that calculated from the mean F-center concentrations"). Lelievre and Adloff (1964) show an optimum grain size of about 80 μm for NaCl and also discuss the rate of dissolution as a function of grain size for grain diameters from 50 to 400 μm.

2.2.5. Quenching

Quenching of LL from alkali halides has been noted in many reports. For example, Ahnström (1965) noted that O_2, NO_3^-, and SO_3^{2-} as well as $NADP^+$ quenched NaCl LL in a fluorescein solution at concentrations where no effect could be seen on the light-induced fluorescein fluorescence. Erikson (1962) studied the quenching effect of oxygen on LL from NaCl. He found that water saturated with nitrogen gave light pulses with areas 10–15 times larger than those for oxygen-saturated water. Saturation with air reduced the light emission to about one-quarter that of nitrogen-saturated water. Among the possible explanations suggested are:

(i) $H_2O + e^- \rightarrow H_2O^- \rightarrow H + OH^{-\cdot}$
(ii) A self-trapped electron (polaron)
(iii) $O_2 + e^- \rightarrow O_2^-$

Lelievre and Adloff (1964) noted quenching owing to NO_2^- in Tl-enhanced LL. Arnikar *et al.* (1971) found that the addition of Cd^{2+} ions (as $CdCl_2$) quenched LL, and attributed it to the reduction by the hydrated electrons (needed for LL) by the reaction

$$Cd^{2+}_{aq} + e^-_{aq} \rightarrow Cd^+_{aq}$$

Progressive quenching was found as the $CdCl_2$ concentration went from 0.002 M to 0.012 M. Other hydrated electron scavengers have been found to be CrO_4^- and MnO_4^-. A claim by Mittal (1971) that LL is due to a triplet–triplet energy transfer in alkali halides was based in part on supposed quenching by Co^{2+}, which has not been verified (Kalkar *et al.*, 1986) and Mittal's proposal has not been referred to in subsequent literature.

2.2.6. Miscellaneous

In contrast to experiments on organic LL materials, the temperature effect on inorganic phosphors is not much discussed in the literature. Ettinger and Puite (1982) said that there is a profound effect on the yield of NaCl, LL increasing by a factor of three as the temperature rises from 20 to 60°C. Atari *et al.* (1973) discussed the relation from NaCl in more detail. They stated that the temperature of the solvent was an important factor, but did not specify a quantitative dependence or optimum temperature. All the work was done at 20°C. They presented TL glow curves (for different temperatures 100, 200, 300, 400, 500, 600, 700, and 760°C), and stored all samples for 24 h at room temperature before dissolution. They found that heat treatment before irradiation had a considerable effect. "The higher the temperature of annealing the more sensitive but less stable became the material."

As in TL, storage time of irradiated NaCl becomes a factor affecting LL. The effect has been studied for times from minutes (Lelievre and Adloff, 1964) to hours (Atari, 1980*a*) to months (Atari and Ettinger, 1975). The AG and LL decay together (Atari, 1980*a*) to 1% after about 20 h. The decay of NaCl over long periods is a function of dose and storage temperature (Atari and Ettinger, 1975). At 20°C and a dose of 50 Gy, the decay is up to 90% in 0.5 month, 70% in 3 months, and 60% in 7 months.

There have been isolated studies of LiF. The disk work of Avotinsh *et al.* (1985) was designed to stabilize the rate of dissolution and provide efficient stirring. Ettinger and Puite (1982) gave Avotinsh's results for LL from LiF dissolved in a 2×10^{-5} *M* of fluorescein, showing the response up to 10^6 Gy before saturating. Avotinsh (1982) described application of the disk technique to LL observations of dissolution of LiF in H_2SO_4, showing linear response in the range 10^4–10^7 Gy, and "the upper limit appears to be some two orders of magnitude higher yet." Other advantages claimed for the single-crystal disk method of LiF in H_2SO_4 are:

1. If the dose turns out to be less than the lower detection limit ($\sim 10^4$ Gy) then, because the dissolution is very slow, the same crystal can be used for TL dose readout.
2. Fading of LL appears only in the low-dose region. At doses higher than 5×10^4 Gy, the AG is less than 1% of LL, and there is no decrease in LL within a year of storage.
3. There seems to be little effect from impurities in the H_2SO_4.

Interestingly, water is found to be an effective quencher of LL in fluorides.

3. APPLICATIONS AND FUTURE TRENDS

There are several directions in which research and applications of LL can be extended, both in the interest of learning more about the physics and chemistry of the process and in providing more effective dosimetry techniques.

3.1. Organic Materials

Some of the earliest work of LL reported by Westermark and Grapengiesser (1960) described results using a liquid scintillator (dioxane, diphenyloxazole, naphthalene, and water) as a solvent for NaCl that gave less output than water alone.

They then dissolved diphenyloxazole in water and observed a strong light emission. They also found some LL when polystyrene was dissolved in the liquid scintillator NE213 (xylene + POPOP).

In the years since these experiments were done, there have been many investigations centered on materials useful for liquid and solid organic scintillators, i.e., materials that when exposed to ionizing radiation experience energy-level changes that result in light. A systematic study of these materials might result in identification of compounds that yield significant LL. One particularly interesting possibility is suggested by the fact that in scintillators utilizing organic components, the light output per ionization loss is a function of the rate of ionization. For example, a heavily ionizing 5-MeV α-particle stopping in a typical organic scintillator will give only about one-tenth as much light as a minimum ionizing particle (electron, meson) depositing 5 MeV of energy as it passes through the scintillator. In contrast, the energy response of inorganic crystal scintillators, such as NaI(Tl) is the same for either type of particle. It is possible that this difference, the so-called α/β ratio, in organic scintillators is due to the dissociation of the scintillator molecules by the high density of ionization (i.e., rapid energy loss rate) of the α-particle. Thus, in view of the important role played by radicals in the LL or organic materials, it may be that some of them will be found to be particularly useful for dosimetry involving α-particles, as in radon detection. The α/β difference has been noted in TL (Zimmerman, 1972), and TL instrumentation for radon detection has been discussed (Niewiadomski and Ryba, 1984).

3.2. Wavlength Shifter

A common technique in liquid scintillator detection is to provide a wavelength shifter, some high-quantum-yield agent that when present in very small concentrations (e.g., POPOP in a solution of terphenyl in xylene) shifts the short-wavelength primary emission to wavelengths better matched to the photomultiplier cathodes generally used to detect the light. Some of the spectral distributions reported in the literature indicate that wavelength shifters would be useful in LL detection.

3.3. Enhancement

As noted above, enhancement of LL by the introduction of certain positive ions has been reported frequently. The probability is that these ions form complexes that are then reduced by hydrated electrons e^-_{aq} in the LL process. Since experiments demonstrate that irradiated NaCl, e.g., provides hydrated electrons upon dissolution, the use of irradiated alkali halides might be a valuable source of this reducing agent for studies of CL (Hammermaier *et al.*, 1986). Kalkar and Ramani (1981), in the course of reporting enhanced LL from solutions containing Cu^{2+} and Co^{2+}, noted that Cd^{2+} and Zn^{2+} quench LL. Nevertheless, other 2^+ (or greater) ions appear to be worth investigating, to follow up on the hypothesis that the e^-_{aq} reduce the aquacomplexes formed by such ions (Martin *et al.*, 1972). The object would be to find suitable salts, soluble to the order of 10^{-3} *M*, with anions that do not quench

LL. Potential candidates would be:

$FeCl_2$, $FeCl_3 \cdot 6H_2O$, Fe_2Cl_6, $FeBr_2$, $FeBr_3 \cdot 6H_2O$

$MgCl_2$, MgI_2

$CoBr_2$

$NiBr_2$, $NiCl_3$, NiI_2

$SnCl_2$, $SnCl_4$, SnI_2

$TiCl_3$, $TiCl_4$

$MnBr_2$, $MnCl_2$, $MnCl_3$ (in alcohol), MnF_2, MnI_2

3.4. Image Intensification

The use of image intensifiers has proved useful in the study of many low-light-level emission processes, for spatial, temporal, and spectral data. These include bioluminescence (Reynolds, 1978), fluorescence (Reynolds, 1980), sonoluminescence (Crum and Reynolds, 1985), thermoluminescence (Walton and Debenham, 1980), and crystalloluminescence (Gibbon *et al.*, 1988). Similar advantages for LL could be realized by the use of image intensifiers in several areas. As low-dose LL is generally weak, spectra could be obtained by the use of an intensified spectroscope (Reynolds and Gruner, 1975). Ahnström (1965) considered in detail the role of the interphase and cage at and near the grain surface in the LL process. The spatial information provided by the image intensifier could be useful in such detailed studies, as it has been in thermoluminescence and crystalloluminescence.

The time characteristic of LL emission has been frequently discussed in the literature. Reasonably enough, it appears to vary with, e.g., the amount of phosphor dissolved, and the mixing procedures. There is generally a distinction between a "fast" component (~ 1 s) and "slow" component (many seconds). In a few instances, a component significantly shorter than 1 s is commented upon. In particular Hanig (1984) noted that the problem of reproducibility at low doses may be a result of inconsistency in the flash of light produced at variable solid–liquid interfaces. These flashes last milliseconds as compared to the entire LL emission cycle, which may be from 2 to 30 s. Many LL observations (and in particular those on spectra) are based primarily on the longer components of emission. Image intensification provides spatial and temporal information on LL, as it does in other areas.

3.5. Further Instrumentation

In descriptions of the application of LL to commercial dosimetry devices, it is frequently pointed out that the time required to make an observation is undesirably long. Replacing the conventional photomultipliers by the photon detectors now available could improve this aspect of applied LL:

1. An image intensifier with a fiber optics input window provides an area detector with a number of discrete outputs. An array of irradiated phosphor

samples distributed on the cathode fiber optics would give separated outputs at the anode that could be treated individually (but in parallel).

2. Simpler multielement detectors for simultaneous measurement of a number of LL samples are available in several forms, all involving fiber optics input windows to photosensitive cathodes. The photodetector could be a proximity focused image intensifier with an individual multianode readout. Another possible detector is the mesh dynode intensifier. Relevant references are given (Boutot *et al.*, 1987; Smith *et al.*, 1988; Suzuki *et al.*, 1988; Vallerga *et al.*, 1988).

ACKNOWLEDGMENTS

The author acknowledges with thanks helpful suggestions by Prof. K. Ettinger. This work was supported by DOE Contract DEFG02-87ER 60522-A000.

REFERENCES

Ahnström, G. (1961), *Acta Chem. Scand.* **15**, 463.
Ahnström, G. (1965), *Acta Chem. Scand.* **19**, 300.
Ahnström, F., and Ehrenstein G. (1959), *Acta Chem. Scand.* **13**, 855.
Anbar, M., and Hart, E. J. (1965), *J. Phys. Chem.* **69**, 1244.
Arnikar, H. J. and Kalkar, C. D. (1977), *J. Lumines.* **15**, 227.
Arnikar, H. J., Damte, P. S., and Chaure, B. D. (1971), *J. Chem. Phys.* **55**, 3668.
Arnikar, H. J., Damle, P. S., Chaure, B. D., and Madhav Rao, B. S. (1970), *Nature* **228**, 357.
Atari, N. A. (1980*a*), *J. Lumines.* **21**, 305.
Atari, N. A. (1980*b*), *J. Lumines.* **21**, 387.
Atari, N. A., and Ettinger, K. V. (1974*a*), *Nature* **247**, 193.
Atari, N. A., and Ettinger, K. V. (1974*b*), *Nature* **249**, 341.
Atari, N. A., and Ettinger, K. V. (1975), *Radiat. Effects* **26**, 39.
Atari, N. A., Ettinger, K. V., and Fremlin, J. H. (1973), *Radiat. Effects* **17**, 45.
Atari, N. A., Malik, S. R., and Ramani, R. (1985), *J. Appl. Radiat. Isot.* **36**, 517.
Avotinsh, J. (1982), *J. Appl. Radiat. Isot.* **33**, 781.
Avotinsh, Yu. E., Dzelme, Yu. R., Tiliks, Yu., Bugaenko, L. T., Gotlib, V. I., and Kreishmanc, T. Ye (1985), *J. Appl. Radiat. Isot.* **36**, 789.
Balogun, F. A., and Adesanmi, C. A. (1989), *J. Appl. Radiat. Isot.* **40**, 243.
Bartlett, D. T. (1979), *Phys. Med. Biol.* **24**, 63.
Bartlett, D. T., Ettinger, K. V., and Reading, D. H. (1981), *J. Appl. Radiat. Isot.* **32**, 603.
Batlett, D. T., Brown, J. K., and Durrani, S. A. (1982), *Radiat. Effects* **66**, 233.
Böhm, J., Piesch, E., and Regulla, D. R. (1986), *Radiat. Prot. Dosim.* **17**, 25.
Boutot, J. P., Laboute, P., and Eshard, G. (1987), *IEEE Trans. Nucl. Sci.* **NS-34**, 449.
Bradshaw, W. W., Cadena, D. J., Crawford, G. W., and Spetzler, H. A. W. (1962), *Rad. Res.* **17**, 11.
Busuoli, G., and Montiventi, F. (1984), *Radiat. Prot. Dosim.* **6**, 172.
Casler, R., Pringsheim, P., and Yuster, P. (1950), *J. Chem. Phys.* **18**, 1564.
Chazoor, J. S., and Mishra, U. C. (1979), *J. Lumines.* **20**, 397.
Chazoor, J. S., and Mishra, U. C. (1982), *J. Lumines.* **26**, 419.
Clark, P. A., and Templer, R. H. (1988), *Archaemometry* **30**, 19.
Crum, L. A., and Reynolds, G. T. (1985), *J. Acoust. Soc. Amer.* **78**, 137.
Erikson, L. E. G. (1962), *Acta Chem. Scand.* **16**, 2113.
Ettinger, K. V. (1989), *J. Appl. Radiat. Isot.* **40**, 865.

Ettinger, K. V., and Anunuso, C. I. (1981), *J. Appl. Radiat. Isot.* **32**, 673.
Ettinger, K. V., and Miola, U. J. (1985), *Radiat. Prot. Dosim.* **10**, 319.
Ettinger, K. V., and Puite, K. J. (1982), *J. Appl. Radiat. Isot.* **33**, 1115.
Ettinger, K. V., Fairchild, R. G., Arunuso, C. I., Srirath, S., and Vigna-Filho, E. (1980), *Nucl. Inst. Meth.* **175**, 119.
Ettinger, K. V., Forrester, A. R., and Hunter, C. H. (1982), *Cnd. J. Chem.* **60**, 1549.
Ettinger, K. V., Miola, U. J., Hunter, C. H. Temperton, D. H., Dixon, S. M., and Forrester, A. R. (1984), *Radiat. Prot. Dosim.* **6**, 113.
Etzel, H. W., and Allard, J. G. (1959), *Phys. Rev. Lett.* **2**, 452.
Gibbon, M. A., Sopp, H., Swanson, J., and Walton, A. J. (1988), *J. Phys. C. Sol. St. Phys.* **21**, 1921.
Hammermaier, A., Peich, E., and Bögl, W. (1986), *Radiat. Prot. Dosim.* **17**, 171.
Hanig, R. E. (1984), *J. Appl. Radiat. Isot.* **35**, 987.
Hart, E. J., and Anbar, M. (1970), *The Hydrated Electron* (Wiley, New York).
Harvey, E. N. (1952), *Bioluminescence* (Academic, New York).
Hunter, C., Temperton, D. H., Ettinger, K. V., and Forrester, A. R. (1982), *J. Appl. Rad. Isot.* **33**, 1291.
Kalkar, C. D. (1983), *Radiochem. Radioanal. Lett.* **58**, 313.
Kalkar, C. D., and Ramani, R. (1981), *Radiochem. Radioanal. Lett.* **47**, 203.
Kalkar, C. D., Arnikar, H. J., Doshi, S. V., and Varkhede, R. S. (1985), *J. Appl. Radiat. Isot.* **36**, 51.
Kalkar, C. D., Doshi, S. V., and Pendse, M. (1986), *Appl. Radiat. Isot.* **37**, 41.
Kundu, H., and Mitra, B. (1988), *J. Appl. Rad. Isot.* **39**, 179.
Lelievre, B., and Adloff, J. P. (1964), *J. de Phys.* **25**, 789.
Martin, J. E., Wart, E. J., Adamson, A. W., Gafney, H., and Halpern, J. (1972), *J. Amer. Chem. Soc.* **94**, 9238.
Mathur, V. K., Gasiot, J., Abbundi, R. J., and Brown, M. D. (1986), *Radiat. Prot. Dosim.* **17**, 333.
Matyushkov, V. V., Mikalchenko, G. A., Sigaev, V. Ya., Yudin, I. V., and Lymarev, A. V. (1977), *Izv Akad. Nauk SSSR, Ser. Fiz.* **41**, 1363.
McKeever, S. W. S. (1984), *Radiat. Prot. Dosim.* **8**, 3.
Mehta, S. K., Oommen, I. K., Sengupta, S., and Soman, S. D. (1983), *Nucl. Inst. Meth.* **213**, 393.
Mittal, J. P. (1971), *Nature*, **230**, 160.
Nakajima, T., and Otsuki, T. (1990), *J. Appl. Radiat. Isot.* **41**, 359.
Niewiadomski, T., and Ryba, E. (1984), *Radiat. Prot. Dosim.* **6**, 67.
Puite, K. J. (1980), *Nucl. Inst. Methods* **175**, 122.
Puite, K. J. and Ettinger, K. V. (1982), *J. Appl. Radiat. Isot.* **33**, 1139.
Rao, R. P., DeMurcia, M., and Gasiot, J. (1984), *Radiat. Prot. Dosim.* **6**, 64.
Reynolds, G. T. (1950), *Nucleonics* **6**, 68.
Reynolds, G. T. (1978), *Photochem. Photobio.* **27**, 405.
Reynolds, G. T. (1980), *Microsc. Acta* **83**, 55.
Reynolds, G. T., and Gruner, S. M. (1975), *IEEE Trans. Nucl. Sci.* **NS-22**, 404.
Smith, G. R., Mayer, J. K., and Durocher, J. J. G. (1988), *IEEE Trans. Nucl. Sci.* **NS-35**, 365.
Suzuki, S., Matshita, T., Kimura, S., and Kumi, H. (1988), *IEEE Trans. Nucl. Sci.* **NS-35**, 382.
Temperton, D. H., Dixon, S. M., and Ettinger, K. V. (1984*a*), *J. Appl. Radiat. Isot.* **36**, 655.
Temperton, D. H., Dixon, S. M., Shentall, G. S., and Ettinger, K. V. (1984*b*), *J. Appl. Radiat. Isot.* **35**, 139.
Templer, R. H. (1985), *Nucl. Tracks* **10**, 531.
Templer, R. H. (1986), *Rad. Prot. Dosim.* **17**, 493.
Vallerga, J., Hull, J. and Lamprton, M. (1988), *IEEE Trans. Nucl. Sci.* **NS-35**, 539.
Walton, A. J., and Debenham, N. C. (1980), *Nature* **284**, 42.
Walton, A. J., and Reynolds, G. T. (1984), *Adv. Phys.* **33**, 595.
Wampler, J. E., and DeSa, R. J. (1971), *Appl. Spectrosc.* **25**, 623.
Westermark, T., and Grapengiesser, B. (1960), *Nature* **188**, 395.
Wiedemann, E., and Schmidt, G. C. (1895), *Ann. Phys.* **54**, 604.
Zimmerman, D. W. (1972), *Radiat. Effects* **14**, 81.

9

Sonoluminescence

Sean M. Cordry and Lawrence A. Crum

1. INTRODUCTION

Sonoluminescence (SL) literally means "light from sound." In practice, it means that a sound field of sufficient intensity applied to a liquid can cause the liquid to emit light—both visible and invisible wavelengths. This conversion of mechanical energy to electromagnetic energy results in an energy density amplification of 10^{12}. The facilitator of this effect is the ubiquitous bubble.

The results of SL were first observed and reported by Marinesco and Trillat (1933), who had been studying the effects of ultrasound on the development of photographic emulsions. They attributed latent spots of light on the photographic plates to intense chemical activity caused by the ultrasound. Frenzel and Schultes (1934) performed similar experiments and concluded that the images were due to a light source.

There are two different classes of SL phenomena. The first is called multibubble sonoluminescence (MBSL) and the second is known as single-bubble sonoluminescence (SBSL). As the name implies, MBSL has to do with cavitation fields filled with a large number of bubbles. Cavitation is a necessary (but not sufficient) condition for SL. SBSL refers to the case of one bubble emitting its light far from the influences of any boundaries or objects. The observed properties of the two classes are similar in some ways, yet quite different in others. The possible reasons for the similarities and differences will be addressed after each class is discussed separately.

This chapter will give a brief tutorial on bubble dynamics, discuss historical and contemporary SL research, present theories of the SL mechanism, and describe some applications of SL.

Sean M. Cordry • Department of Physics and Astronomy, University of Mississippi, University, Mississippi 38677, United States *Lawrence A. Crum* • Applied Physics Laboratory, University of Washington, Seattle, Washington 98105.

Luminescence of Solids, edited by D. R. Vij. Plenum Press, New York, 1998.

2. BUBBLE DYNAMICS

Bubbles are complicated mechanical systems, which can experience complex and nonlinear behavior when oscillating at even modest displacement amplitudes. At the root of their unique behavior is the huge difference in compressibility between water and air. Because of this difference, bubbles react very easily to changes in the pressure around them. If, say, the pressure is reduced, the bubble will expand and increase its radius. Or, alternatively, it will shrink if the pressure is increased. Other factors affecting a gas bubble's motion include surface tension, viscosity, sound radiation by the bubble, and gas and heat flow in and out of the bubble.

2.1. Radial Motion

The most commonly used equation of motion to describe the bubble's behavior is the Rayleigh–Plesset equation (Noltingk and Neppiras, 1950; Neppiras and Noltingk, 1951):

$$R\frac{d^2R}{dt^2} + \frac{2}{3}\left(\frac{dR}{dt}\right)^2 = \frac{1}{\rho}\left[\left(P_0 + \frac{2\sigma}{R_0}\right)\left(\frac{R_0}{R}\right)^{3\gamma} - \frac{2\sigma}{R} - \frac{4\mu(dR/dt)}{R} + P_\infty\right] \quad (1)$$

where R is the radius of the bubble, ρ the density of the liquid, P_0 the ambient pressure, σ the surface tension, R_0 the equilibrium (resting) radius of the bubble, γ the ratio of specific heats of the gas, μ the viscosity, and P_∞ the pressure in the liqud. The radius of a bubble can change nonlinearly in response to the pressure fluctuations of an acoustic field. Flynn (1964) solved the differential equation of motion for a bubble in a sound field numerically and calculated this radius–time curve with the parameters: $P_A = 0.33$ bar, $f = 83.4$ kHz, and $t_0 = R_0/c$. Figure 1 illustrates how the radius changes as a function of time. Equation (1) neglects thermal and radiation losses as well as gas diffusion and vapor pressure. To include these additional effects, more sophisticated equations are needed. These will not be presented here since the goal of this section is only to familiarize the reader with elementary aspects of bubble behavior. For details concerning bubble dynamics, see Young (1989) and Neppiras (1980).

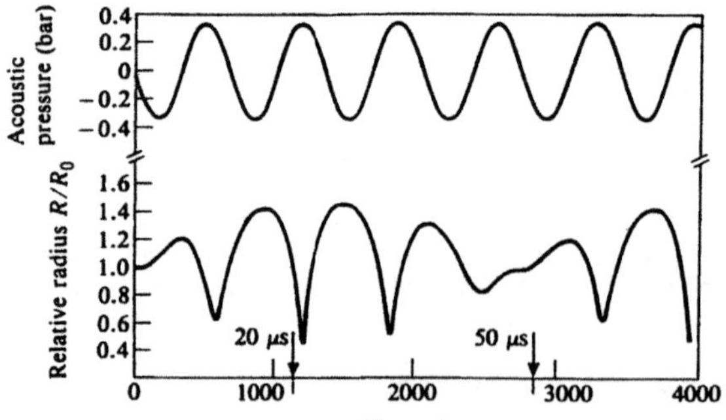

Figure 1. Variation of R/R_0 and acoustic pressure as a function of time (Flynn, 1964; with kind permission from Academic Press).

Oscillating bubbles have resonance frequencies and characteristic response curves. Unlike simple Hookes' law springs, bubbles are nonlinear oscillators, which can lead to quite complex behavior. However, it is possible to obtain an equation for the linear resonance frequency or "natural" frequency at which freely oscillating bubbles will vibrate. In general this equation is

$$\rho\omega_r^2 R_0^2 = 3\gamma\left(P_0 + \frac{2\sigma}{R_0}\right) - \frac{2\sigma}{R_0} \tag{2}$$

where the resonance frequency is $f_r = \omega_r/2\pi$. For air bubbles in water that are not too small ($R_0 > 1\ \mu$m), Eq. (2) can be approximated by the simple relation

$$f_r R_0 \approx 3.28 \tag{3}$$

where the values for ρ, γ, and P_0 are $10^3\ \text{kg m}^{-3}$, 1.40, and $10^5\ \text{N m}^{-2}$, respectively. For Eq. (3), f_r is in hertz and R_0 is in meters.

For the case of a bubble driven by an oscillating pressure field, P_∞ from Eq. (1) is given by

$$P_\infty = P_0 + P_A \sin(\omega t) \tag{4}$$

Because of nonlinearity, there are many resonances for the bubble. It is useful to define a ratio α such that $\alpha = f/f_r$, where f_r is the resonance frequency for a bubble of a given size and f is the frequency of the driving pressure field. The main resonance of the driven bubble has $\alpha = 1$. If $\alpha = n$, where n is an integer, the bubble is operating in a subharmonic resonance. If $\alpha = 1/m$, where m is also an integer, then

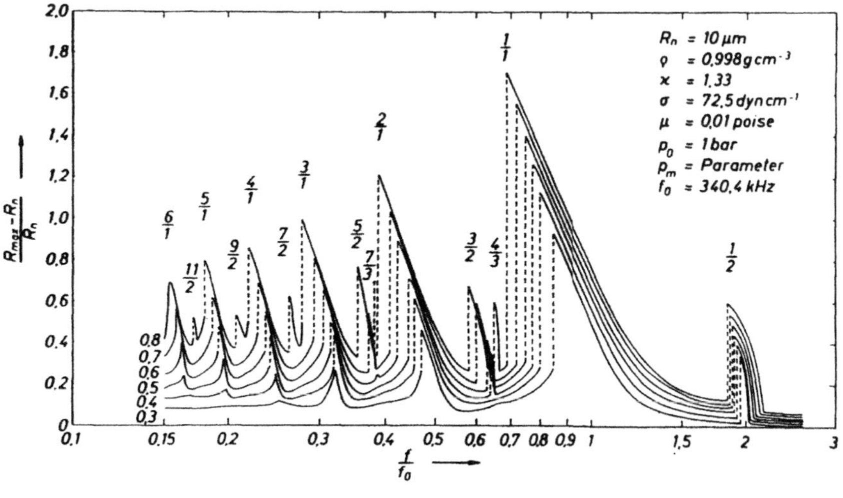

Figure 2. The computed maximum radius of a 10-μm bubble as it is driven at different frequencies and varying amplitudes of acoustic pressures. The maximum radius and the driving frequency have been normalized in terms of the bubble's equilibrium radius and linear reasonance frequency, respectively (Cramer, 1980; with kind permission from Springer-Verlag).

the bubble is operating in an harmonic resonance. It also happens that the bubble could have a resonance such that $\alpha = n/m$. This last case is called an ultraharmonic resonance.

In order to examine a bubble's behavior over a range of driving frequencies, the differential equation of motion must be solved. Figure 2 shows the normalized maximum radial response for a bubble with a radius R_0 of 10 μm ($f_r = 340.4$ kHz), driven at various pressure amplitudes with the driving frequency ramped from $\alpha = 0.15$ to 2.5. The labels on the peaks correspond to the fraction m'/n': the number of maxima m' in n' acoustic periods. It should be noted that this fraction is nearly equal to the value of $1/\alpha$ but deviates from this value as the pressure is increased. The bubble response is a function of both α and the acoustic pressure. It must be pointed out that Fig. 2 was not calculated using Eq. (1), but rather a more complicated one based on the enthalpy of the bubble, which takes into account the compressibility of the fluid [see Cramer (1980) for details].

2.2. Rectified Diffusion

A simple expression of Henry's law states that the partial pressure of a gas must be the same on either side of an interface for a state of equilibrium to exist (Carroll, 1991). Gas diffusion, in many cases, can significantly influence the dynamics of the bubble's behavior. Thus, if a liquid is undersaturated, gas diffuses into the liquid across the liquid–gas interface in order to restore equilibrium. Alternatively, if a liquid is supersaturated, gas again diffuses across the interface, but out of the liquid. A bubble, of course, has a liquid–gas interface, so it is expected that gas diffusion across this interface will occur for off-equilibrium conditions.

If one considers a bubble in a liquid that is free from surfaces and does not rise through buoyancy and, further, allows this bubble to have zero surface tension, then if the liquid is supersaturated with gas, gas must diffuse across the bubble wall into the bubble, resulting in bubble growth. Alternatively, if the liquid is undersaturated, gas inside the bubble diffuses across the bubble wall, resulting in bubble dissolution. If the liquid were exactly at saturation, an equilibrium condition would exist between the gas within the bubble and the gas dissolved in the liquid. If one considers next that the bubble has nonzero surface tension, the total pressure inside the bubble is increased by the Laplace pressure, given by $2\sigma/R$. If the liquid is at saturation, the partial pressure of the gas within the bubble is higher than that in the liquid and the bubble will dissolve. Equilibrium will only exist if the liquid is supersaturated with gas. However, this is a point of unstable equilibrium, since the Laplace pressure is an inverse function of the bubble's radius. If the bubble is slightly smaller or larger than the required equilibrium radius, then the bubble dissolves or grows, respectively.

For the case of a bubble being driven by a sound field, there are two additional effects to be considered. First, it should be noted that since the total pressure inside the bubble is varying with time, there will be gas flow out of the bubble during its compression phase and gas flow in during its expansion phase. The first additional effect is called the "shell effect," which is associated with the expansion and contraction of a layer of fluid immediately surrounding the bubble. When the bubble is compressed, the shell is stretched radially, thus reducing the partial pressure of gas immediately outside the bubble wall. When the bubble is expanded, the shell is compressed, thus increasing the partial pressure of gas in the liquid immediately outside the bubble wall. The net effect is to compound the effects of the varying internal pressure. The second additional effect is the "area effect." The total amount of gas to diffuse across an interface is proportional to the area of the interface. Thus, more molecules of gas diffuse across the interface when the bubble is large than when the bubble is small, given the same difference in partial pressures.

How then can a stable bubble be made to exist? From the previous paragraph it should be noted that if a bubble's radial response is too small, diffusion out of the bubble will be greater than diffusion into the bubble. Also, if a bubble is driven and its response is too large, diffusion into the bubble will be greater than diffusion out. In the former case there is net dissolution, and in the latter case, net growth. This latter case is commonly referred to as rectified diffusion. A driven bubble is at the threshold for rectified diffusion if gas diffusion into the bubble exactly matches gas diffusion out. However, driving pressures and the frequencies at which the threshold for rectified diffusion occurs, are complicated functions of the liquid's dissolved gas concentration and the bubble's nonlinear response.

Thresholds for rectified diffusion can be both easily measured and calculated; an example of a calculation is shown in Fig. 3. For this curve, threshold pressure is plotted as a function of bubble radius for a fixed driving frequency. For a bubble of any given size, if the acoustic pressure amplitude driving it is below the threshold curve, the bubble will dissolve. If, however, the bubble is driven by a pressure amplitude above the line, it will grow. It is then seen that all points on the rectified diffusion threshold curve that have a positive slope will be stable, which means that a slight variation in the bubble radius will return it to its original position. The concept of rectified diffusion is important for understanding the conditions that lead to the stability of SBSL. Additional information about rectified diffusion can be found in Eller and Flynn (1965), Crum (1984), and Church (1988).

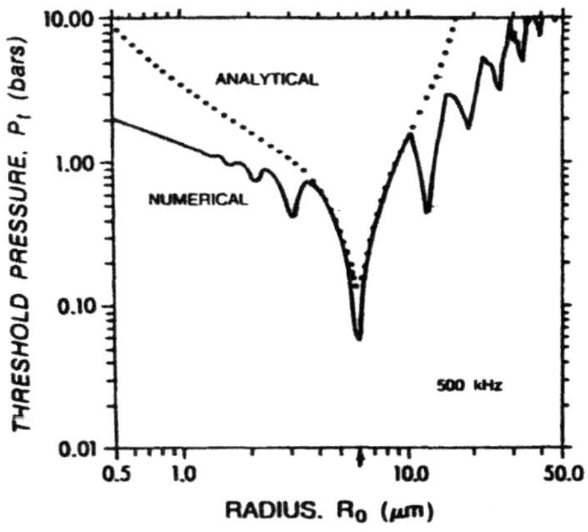

Figure 3. Plot of threshold pressure vs. bubble radius. The dotted line represents an analytical solution for the onset of rectified diffusion, while the solid line represents the numerically calculated solution (Church, 1988; with kind permission from the American Institute of Physics).

2.3. Surface Instabilities

The easiest problem to treat mathematically in this context is the one of spherically symmetric bubble motion, or purely radial motion. Naturally, nature is not always so accommodating. If a bubble has a boundary nearby, such as a solid obstacle or another bubble, or if the acoustic pressure incites large radial excursions, it will not oscillate in a purely radial mode. These motions are very complicated and will not be discussed here in depth.

For an oscillating bubble, it is possible to express the surface of the bubble as

$$r_s = R + aY_n \tag{5}$$

where Y_n is a spherical harmonic of order n, and a is an amplitude term (Plesset, 1954). One can then substitute Eq. (5) into the equations of motion, separate out R behavior, and study how a behaves. What is found is that a is negligible until a threshold driving pressure is reached. Then, there are well-defined surface waves on the bubble. If the driving pressure is increased further, a can get larger until the surface is so severely perturbed that surface waves on the bubble become large enough to encapsulate microbubbles and microdrops (see also Plesset and Mitchell, 1956, and Neppiras and Fill, 1969).

3. MULTIBUBBLE SONOLUMINESCENCE

MBSL is usually so weak that it must be observed with photomultiplier tubes (PMTs) or other similar low-light-level detectors. By studying the light emissions from cavitation regions with PMTs it was learned that SL light emission was composed of discrete flashes synchronized with the sound field. Meyer and Kuttruff (1959) were able to take a series of photographs of cavitation on the end of a vibrating nickel rod and demonstrate that the SL flashes occur near the end of the bubbles' collapse.

Many investigators have made significant contributions to the various parameters that affect the "performance" of MBSL. Jarman (1959) discovered that the amount of light emitted is related to σ^2/P_V, where P_V is the vapor pressure. His results showed that for the case of mercury, this ratio is 75,000 times that of water. Indeed, Kuttruff (1962) examined the SL light emissions from several liquid metals and found that mercury had the highest emission rates. Jarman (1959) also found that SL intensities decrease linearly with increasing temperature. This result is consistent with the previous observation since P_V tends to decrease with temperature. Young (1965) found that increasing liquid viscosity yielded a higher intensity by studying MBSL in water–glycerin mixtures.

The effects of the type of gas used in MBSL has also been examined. Among the different investigators, there was consistent agreement for noble gases (Young, 1976; Prudhomme and Guilmart, 1957). The trend is that decreasing the thermal conductivity increases light intensity (Hickling, 1963). This seems reasonable since less energy would be lost to the liquid through heat conduction and thus more energy would be available for the production of light. For nonnoble gases, in particular, air, nitrogen, and oxygen, there was disagreement among the various investigators concerning levels of light intensity. The resolution of the discrepancies may lie in the sensitivities of the light detectors used or experimental conditions such as frequency or sound pressure amplitude, the latter rarely being stipulated.

It is important to note that several factors can affect both the measured SL intensity and its spectrum. The number and sizes of light-emitting bubbles present would influence the actual number of photons entering the detector. Furthermore, both the frequency response of the light detector and the absorption spectrum of the host liquid limit the number of photons that can be counted. Of course these effects can largely be corrected through imaginative experimental techniques; however, they present difficulties in understanding the underlying physics of SL.

Along with the intensity of the light emitted by MBSL, the spectral components of the emitted light have been examined. The typical method for obtaining a spectrum is to immerse an ultrasonic horn into the test liquid and use a monochrometer–PMT combination to measure intensity as a function of wavelength.

Figure 4 shows several spectra obtained from MBSL in air- and argon-saturated water (Taylor and Jarman, 1970). The most noticeable feature is the large peak around 310 nm. This is an emission line associated with the relaxation of OH*. (Here, the asterisk refers to the molecule being in an excited state.) Other components thought to contribute to the water emission include H_2O^* and $H + OH$ recombination. The former process is associated with light production below 310 nm, and the latter is a three-body type and is recognized as the source of the continuum above the 310 nm OH* band.

Flint and Suslick (1989) have examined SL from nonaqueous liquids. An example of a spectrum obtained from argon-saturated dodecane shows spikes, which are attributed to the $d^3\Pi_g - a^3\Pi_u$ Swan transition of C_2. The broad continuum is attributed to emissions from C_2H^*. Flint and Suslick (1989) also indicate that the features seen in this dodecane SL spectra are also seen in alkanes, arenes, and Dow Corning 200 silicone oil.

4. SINGLE-BUBBLE SONOLUMINESCENCE

One of the many difficulties involved in the study of SL produced by a field of bubbles is the number of individual bubbles involved in the process. When some aspect of SL is investigated, only averaged or group behavior can be measured. In analogy, this is like trying to learn English by hanging a microphone over New York. It is possible to learn about when the New Year begins and various other festivals and major events, but individual behavior remains obscured. In a given cavitation field, there are bubbles of different sizes. All of these different-sized bubbles respond differently to the driving sound field, as has been shown earlier. This "discord" then affects all aspects of the SL. Thus, trying to understand SL on a more fundamental level is difficult at best in a MBSL context. Obtaining SL from a single bubble would greatly increase the control aspect of experimental SL research. Figures 5 and 6 show visually the different environments for MBSL and SBSL.

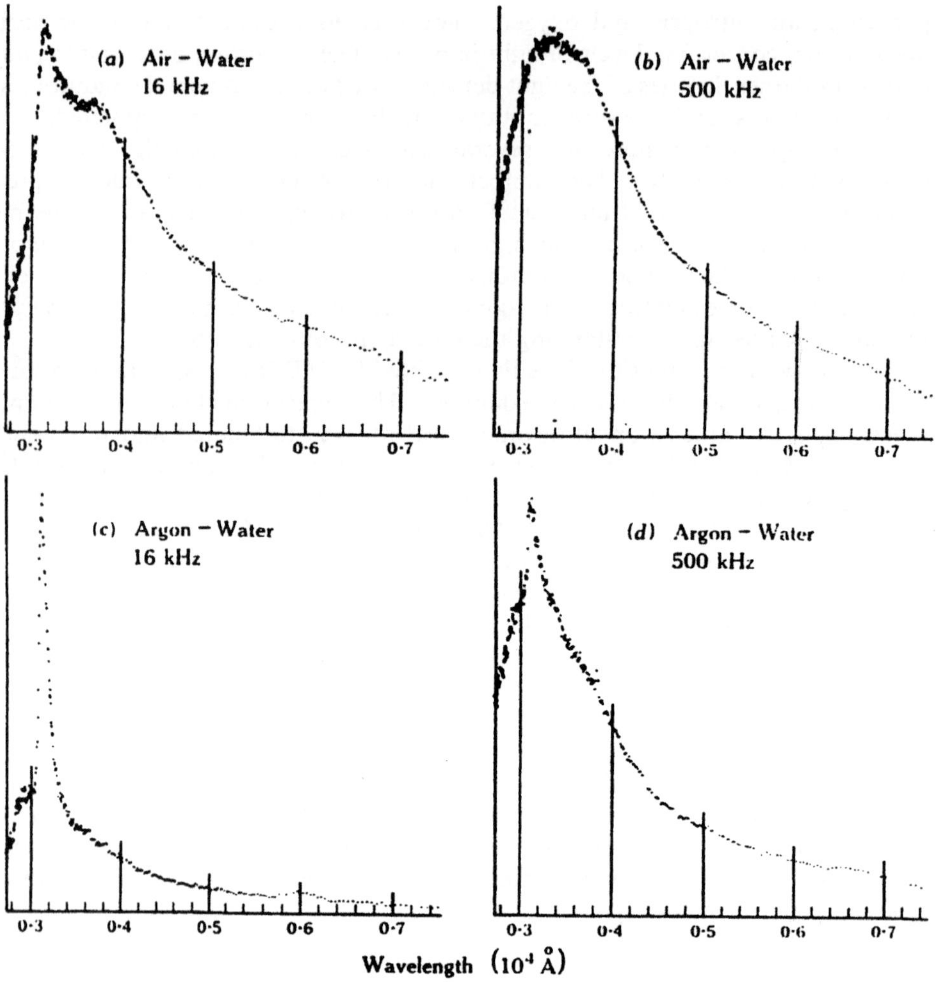

Figure 4. Light-emission spectra from a multibubble SL field in triple-distilled air- and argon-saturated water at 16.0 and 500 kHz. The vertical axis for these figures is arbitrary units of intensity (Taylor and Jarman, 1970; with kind permission from CSIRO, Australia).

4.1. Radius–Time Curves

In 1990, Gaitan *et al.* (1992) discovered the "hydrogen atom" of SL. He was able to obtain SL from a single bubble by acoustically levitating it in a standing wave. Then, by scattering laser light off the bubble, he was able to obtain a radius–time curve. His results are presented in Fig. 7 and show that a single flash of light is emitted each acoustic period. The motion of the bubble wall was also obtained via Mie scattering of polarized laser light.

Obtaining radius–time (R–t) curves is important because it serves to verify the derived equations of motion and tests the theoretical understanding of bubble

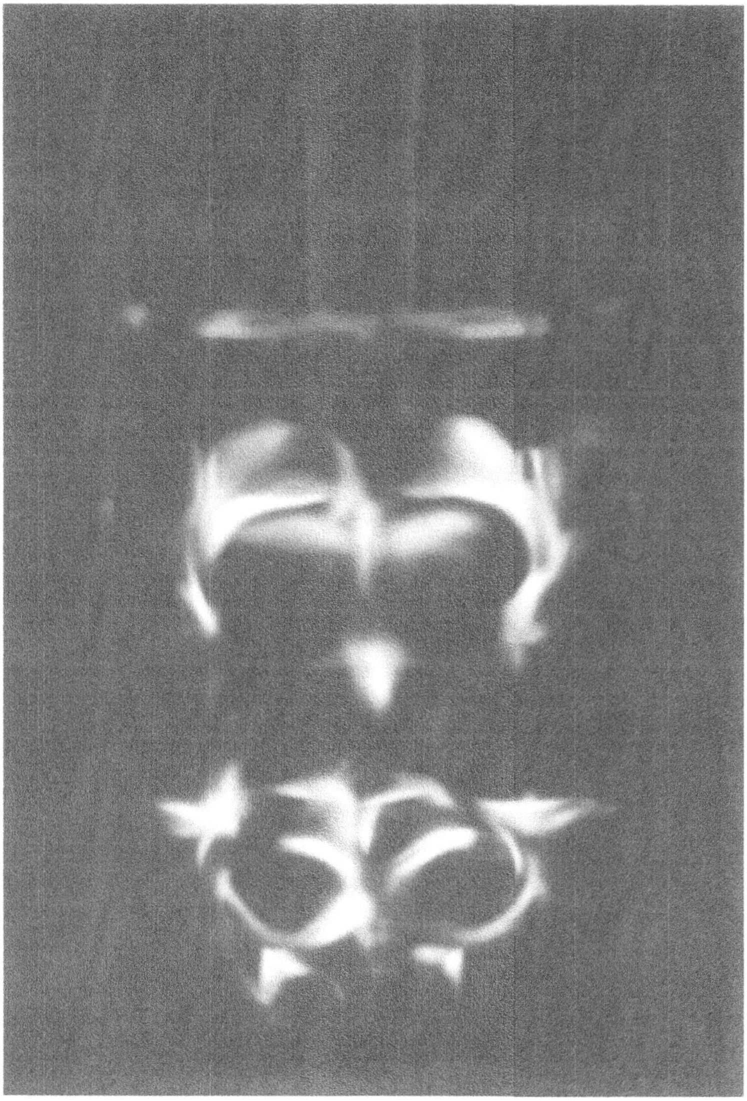

Figure 5. Photograph of the MBSL produced by the vibrating tip of an ultrasonic homogenizer. Luminol has been added to the pure water in order to enhance the light production (Photo by L. A. Crum, appeared in Crum, 1994*b*).

behavior. Gaitan *et al.* (1992) used aqueous glycerin solutions in his experiments. Barber and Putterman (1992) studied techniques used by Gaitan and were able to duplicate and expand upon these early results. They further refined the laser scattering results by using a time-averaging technique to obtain smoother R–t curves. Working in water, they showed a marked difference in the behavior of a

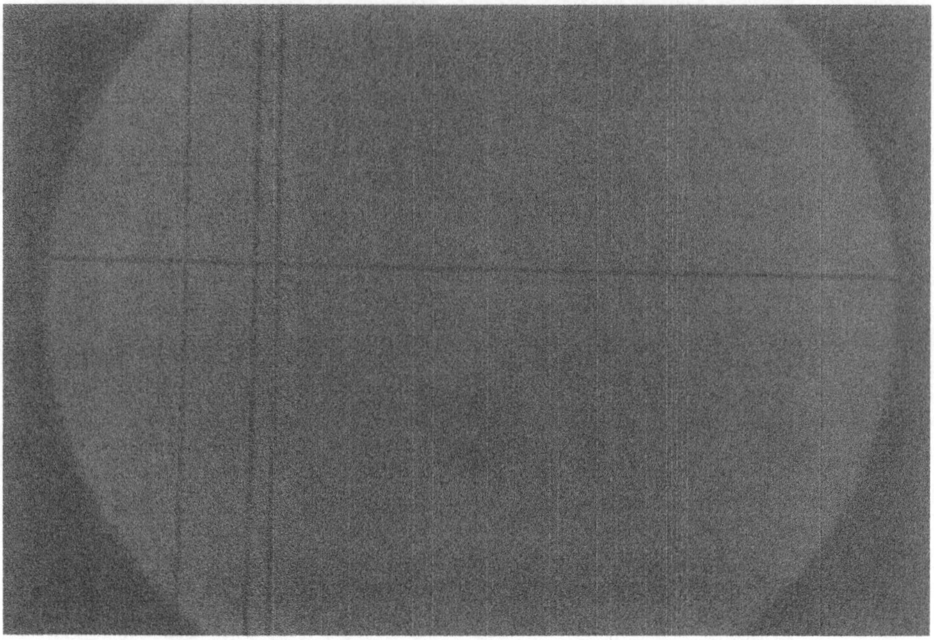

Figure 6. A bubble in a state of SBSL as viewed through a microscope. The bubble is back-illuminated and shows the maximum radius (here about 100 μm) and the light emitted (Photo by S. M. Cordry, appeared in Crum, 1994*b*).

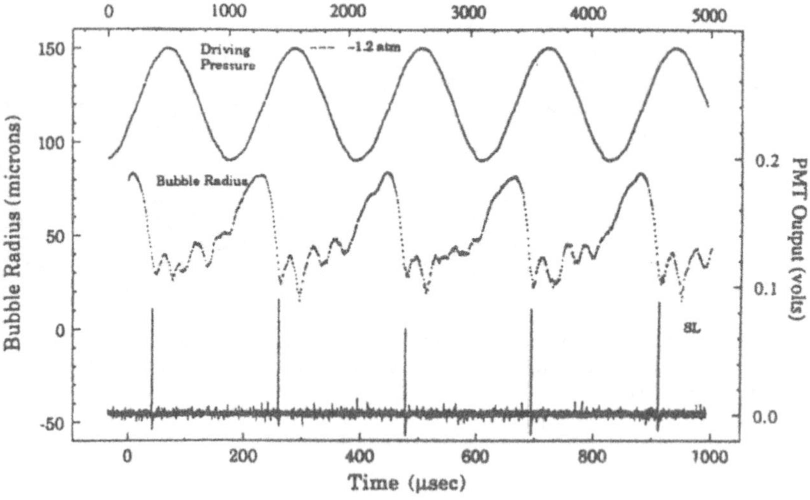

Figure 7. The driving acoustic pressure, the radial motion of the bubble and the PMT response shown simultaneously for SBSL. The light flashes are synchronous with the bubble collapse and occur once every period (Gaitan *et al.*, 1992; with kind permission from the Acoustical Society of America).

bubble undergoing SL and one that was not. This effect is shown in Figure 8. The SL bubble with an equilibrium radius of about 4 μm has a much larger normalized maximum radius than the non-SL bubble with an equilibrium radius of about 10 μm.

4.2. Light Emission

Barber *et al.* (1992) studied the pulselength and synchronization of SBSL flashes. The results of their study are quite remarkable. They show that the pulselength of the SBSL flashes is of the order of 50 ps. By using the fastest available PMT, the SBSL flash could not be differentiated from a 35-ps-pulselength laser pulse. Furthermore, the jitter in the time interval between pulses was also 50 ps, yielding a remarkable level of synchronicity. The synchronicity of the SBSL flashes is about six orders of magnitude higher than the acoustic period. Furthermore, the electronic oscillator used for creating the sound field has more uncertainty to it by several orders of magnitude. The reason for this synchronicity is still not understood.

Measurements have been made (Crum and Cordry, 1994; Barber *et al.*, 1994) of the light emission from the initial stages of SBSL. In order to make these measurements, a few bubbles are released in the levitation chamber with the sound field deactivated. The sound field is activated and the various bubbles coalesce into a single bubble; soon thereafter the bubble begins to emit light, and the intensity of the emissions undergoes a progression (Crum and Cordry, 1994). It is seen that the maximum intensity is not the most stable one. The time scale of events noted is relatively long. At 20 kHz, the 3-s timescale corresponds to 60,000 acoustic cycles. It

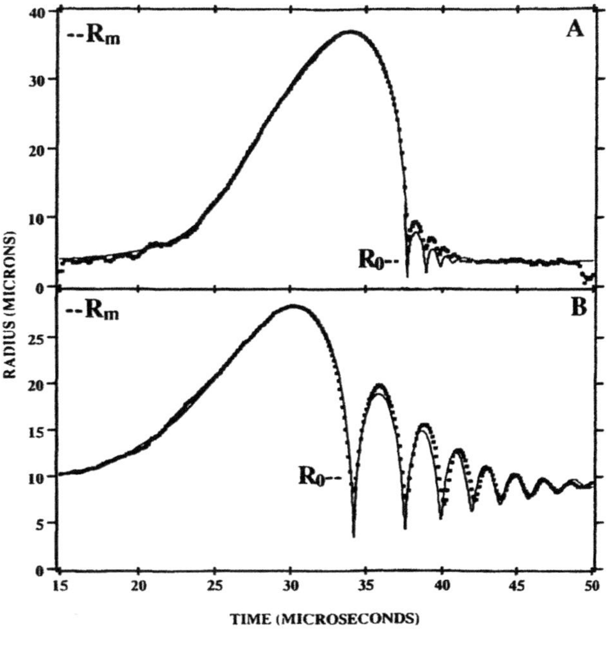

Figure 8. R–t curves for a single bubble undergoing luminescence (upper graph) and one that is not (lower graph) (Barber and Putterman, 1992; reprinted with permission from the American Physical Society).

is suggestive from these measurements that SBSL must exist within some "stability well."

An example of the spectra of light emission from SBSL obtained by Hiller *et al.* (1992) in pure water shows that this spectrum is very different from those observed in cases of MBSL. The most striking feature is the lack of features. The spectra of SBSL in water shows no pronounced peaks—a feature associated with blackbody radiation. Comparisons between SBSL emission spectra and blackbody emission spectra yield fairly good matches for temperatures as high as 30,000 K. At these extreme temperatures, only the "tail" of the blackbody curve could be seen because of the strong ultraviolet absorption by water. This lack of a complete spectrum has challenged the notion that the radiation is blackbody in nature. The uncertainties associated with the spectra will be discussed in more detail below.

4.3. Factors Affecting Single-Bubble Sonoluminescence

When Gaitan *et al.* (1992) made the initial discovery of SBSL, it was found that it could only be obtained by significantly "degassing" the host liquid. Reducing the amount of dissolved gas in the liquid can be accomplished in a number of different ways. The role of dissolved gas can be understood by examining the rectified

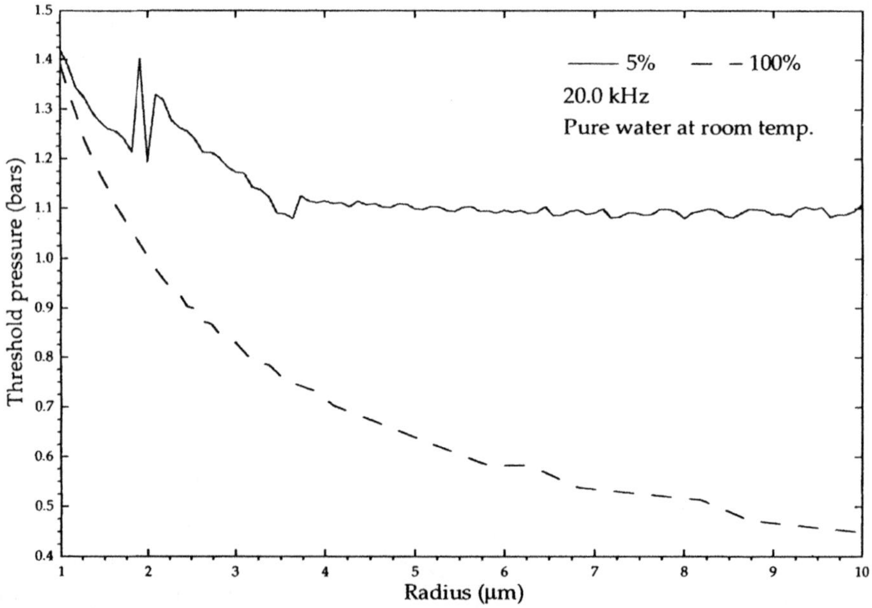

Figure 9. Rectified diffusion thresholds for air bubbles in water with dissolved gas concentrations of 5% and 100% for a frequency of 20.0 kHz. The "notches" appear as the dissolved gas concentration is reduced and the pressure required for stability increases. SBSL, can occur in water with the 5% concentration, but not at the higher level (Crum and Cordry, 1994).

diffusion thresholds (RDTs). Figure 9 shows two RDTs, one for SBSL conditions and the other not. The one for SBSL has a dissolved gas concentration of 5% saturation, the other 100%. It can be seen that for 100% gas concentration there are no stable points (i.e., positive slopes) in this range of radii. Barber and Putterman (1992) report SBSL bubble radii on the order of 3.5 μm.

Arakeri (1993*a*) has reported that the intensity of SBSL increases with decreasing gas concentration. Also the type of gas in the liquid affects the SBSL intensity. Arakeri (1993*b*) reports that nitrogen-filled bubbles yield less light than those that are air-filled, which yield less light still than those that are oxygen-filled. Hiller (1993) reports that the nitrogen-filled bubbles are approximately one-tenth as bright as those that are air-filled.

Variations in the SBSL intensity with temperature have also been reported by Arakeri (1993*b*) and Hiller *et al.* (1992). Both reports find that SBSL intensity increases with decreasing temperature. The question to ask then is how does the temperature affect the water? Decreasing temperature increases the affinity of water for dissolved gas. Thus, if a liquid is degassed to some small percent of gas concentration, say 5%, at 20°C and then cooled to near zero, the dissolved gas concentration would be lowered. Another effect associated with a reduction in temperature is corresponding reduction of the vapor pressure of the host liquid. Whether or not the "σ^2/P_V rule" still holds for SBSL has not yet been determined.

5. MECHANISMS OF SONOLUMINESCENCE

The mechanisms responsible for the conversion of mechanical energy into electromagnetic energy have been speculated about since the initial discovery of SL. In this final section, differences between the spectra (and indirectly the temperature) of MBSL and SBSL will be discussed as the current theories of light production are analyzed.

5.1. Thermal Origins

By far the most successful theory, historically speaking, has been the "Hot-Spot" theory (Noltingk and Neppiras, 1950; Neppiras and Noltinkg, 1951), which basically states that the internal contents of the bubble are adiabatically heated during the final stages of bubble collapse. This heating results in light production. However, from this point, the exact method by which the hot contents are able to emit light is a point of controversy.

The discrete spectra seen in cases of MBSL indicate electronic and/or vibrational/rotational excitation. These discrete lines suggest that part of the light emission comes from atomic or molecular deexcitation and chemical recombination processes. The mystery, though, is why these discrete lines are not observed to be more pronounced in the case of SBSL.

The most obvious explanation for the smoothness of the SBSL spectra would be that the emissions are due to hot gas within the bubble, i.e., blackbody radiation.

Indeed Wu and Roberts (1993) and Greenspan and Nadim (1993) have proposed that imploding shockwaves could heat the contents to the point of ionization of the gas, creating a light-emitting plasma. Preliminary calculations have also shown that such a shock wave could also account for the extremely brief light-emission times that have been observed. An interesting extrapolation of the shock-wave theory is that with the proper deuterium–tritium mixture of gas in the SL bubble, Barber *et al.* (1994) have shown that fusion-induced neutron production within the bubble may be possible.

Another idea that has been proposed for the smoothness of the water spectra is that the light emission is due solely to chemiluminescence (CL). In this theory, the primary light emitters are emissions H_2O^*, OH^*, $H + OH$, and $H + HO_2$. Furthermore, reaction rates can be calculated; by following the approach of Saksena and Nyborg (1970), the decay constant for $H + OH$ for a SBSL temperature of 10,000 K is about 0.5 ps, which implies very short emission times. However, it should be pointed out that at temperatures of 10,000 K, all the molecules should be completely ionized, thus favoring the plasma hypothesis.

Suslick and Hammerton (1986) have demonstrated the possibility of two sites for sonochemical reactions: a hot gas phase inside the bubbles and a hot liquid layer around the bubble. Using comparative-rate thermometry to estimate the effective temperatures, they determined interior temperatures to be as high as 5000 K and exterior temperatures to be 1900 K. It should be pointed out, however, that these experiments were performed in MBSL conditions.

Crum (1994a) has pointed out that there are large differences between bubble behavior in MBSL and SBSL conditions that could explain the differences seen in spectra from the two. Bubbles in MBSL situations are not likely to undergo spherically symmetric collapses (Tomita and Shima, 1990). In fact, it is more likely that the bubbles will collapse asymmetrically and produce an impacting liquid jet. Furthermore, Crum (1994*a*) has obtained high-speed photographs that clearly show that drops of liquid can be introduced into the center of a cavitating bubble.

Suslick and Hammerton (1986) have evidence that there are two reaction sites. It was assumed that the liquid phase for the reaction was a layer immediately outside the bubble. With Crum's (1994*a*) hypothesis and evidence, it would appear that the two sites could be coexistence, i.e., liquid introduced into the center of the bubble. This stark difference between MBSL and SBSL and the possibility that different processes may be occurring in each case has led Crum (1994*a*) to introduce the nomenclature of sonochemistry and sonophysics for MBSL and SBSL events, respectively.

5.2. Electrical Discharge

There have been several theories proposing electrical discharges as the mechanism for light production and chemical behavior inside a SL bubble. Margulis (1990) has presented the most recent electrical discharge theory, in which ζ potentials and microbubble ejection play key roles.

The ζ potential arises from charge accumulation on the surface of the bubble. Charges arise from the dipole orientation of water molecules on the surface of the bubble. The negative side of the dipole points toward the bubble. Negative ions in the liquid are then attracted by the outward-facing charges of the water dipole. Thus, there is a double layer of charge surrounding the bubble.

As the bubble collapses, instabilities are formed on the surface and microbubbles are ejected, causing a flow of charge toward the fragmenting bubble via fluid flow and a simultaneous discharge via conduction. The net result is a large, localized charge buildup of anions in the vicinity of the pinch-off point of the bubble fragment and cations opposite the bubble fragment. Electric fields produced by this local charge buildup can overcome the dielectric strength of the interior of the bubble and a discharge results. This discharge then is responsible for the light emission and the chemistry that results.

It should be noted that the electrical discharge theory has not been put to the test with regard to sonophysics (SBSL) situations. Whether or not single bubbles in a state of SL undergo asymmetric collapses has not been experimentally verified. This theory does have some success in predicting certain chemical reactions. In particular, many sonochemical processes closely resemble those of choronachemistry (LePoint and Mullie (1994).

6. APPLICATIONS

The mechanisms of MBSL are also the mechanisms of sonochemistry. The high temperatures, pressures, and rapid heating and cooling taking place in SL conditions provide a unique environment in which chemical processes occur. New applications of sonochemistry are emerging quickly. Because ultrasound is easily applied to large volumes, it is well-suited to industrial applications (Suslick, 1990).

The actions of MBSL can act as a catalyst in many situations. In some cases production times of chemical products can be reduced from hours to seconds (Zanetti, 1992). The potential savings in cost are obvious. Interactions between solids and liquids are enhanced by cavitation bubbles. Additionally the rapid quenching times of MBSL could lead to the synthesis of new chemical products. The spectrum of applications for sonochemistry (MBSL) is worthy of an entire chapter in itself. (The interested reader is encouraged to consult the text by Suslick, 1990.)

7. CONCLUSIONS

SL is an exciting, important area of research. The resolution of the differences between SBSL and MBSL, as well as an understanding of the exact mechanisms responsible for the production of light lie in the future. Definitive experiments and better theories need to be developed. There is still uncertainty as to the true origins of SL. What is more certain is the new science and technology that will result from the exploration of this phenomenon.

ACKNOWLEDGMENTS

The authors would like to acknowledge the valuable input from Ron Roy, John Allen, and Tom Matula. Also the authors would like to thank the Office of Naval Research, Physics Programs for support of this research. SMC wishes especially to acknowledge support for ONR's AASERT program.

REFERENCES

Arakeri, V. H. (1993*a*), *Pramana–J. Phys.* **40**, L145.
Arakeri, V. H. (1993*b*), *Pramana–J. Phys.* **41**, L291.
Barber, B. P., and Putterman, S. J. (1992), *Phys. Rev. Lett.* **69**, 3839.
Barber, B. P., Hiller, R., Arisaka, K., Fetterman, H., and Putterman, S. J. (1992), *J. Acoust. Soc. Amer.* **91**, 3061.
Barber, B. P., Wu, C. C., Lofstedt, R., Roberts, P. H., and Putterman, S. J. (1994), *Phys. Rev. Lett.* **72**, 1380.
Carroll, J. J. (1991), Chem. Eng. Prog., September, p. 48.
Church, C. C. (1988), *J. Acoust. Soc. Amer.* **83**, 2210.
Cramer, E. (1980), in: *Cavitation and Inhomogeneities in Underwater Acoustics* (W. Lauterborn, Ed.) (Springer-Verlag, Berlin).
Crum, L. A. (1984), *Ultrasonics*, September, p. 215.
Crum, L. A. (1994*a*), *J. Acoust. Soc. Amer.* **95**, 559.
Crum, L. A. (1994*b*), *Physics Today* **47**, 22.
Crum, L. A., and Cordry, S. M. (1994), in: *Bubble Dynamics and Interface Phenomena* (J. R. Blake, J. M. Boulton-Stone, and N. H. Thomas, Eds.) (Kluwer, Dordrecht), pp. 287–297.
Elker, A., and Flynn, H. G. (1965), *J. Acoust. Soc. Amer.* **37**, 493.
Flint, E. B., and Suslick, K. S. (1989), *J. Am. Chem. Soc.* **111**, 6987.
Flynn, H. G. (1964), in: *Physical Acoustics, Vol. 1B* (W. P. Mason, Ed.) (Academic, New York), Ch. 9.
Frenzel, J., and Schultes, H. (1934), *Z. Phys. Chem.* **B27**, 421.
Gaitan, D. F., Crum, L. A., Church, C. C., and Roy, R. A. (1992), *J. Acoust. Soc. Amer.* **91**, 3166.
Greenspan, H. P., and Nadim, A. (1993), *Phys. Fluids A.* **5**, 1065.
Hickling, R. J. (1963), *J. Acoust. Soc. Amer.* **35**, 967.
Hiller, R. (1993), Private Communication.
Hiller, R., Putterman, S. J., and Barber, B. P. (1992), *Phys. Rev. Lett.* **69**, 1182.
Jarman, P. (1959), *Proc. Phys. Soc.* **73**, 628.
Kuttruff, H. (1962), *Acoustica* **12**, 230.
LaPoint, T., and Mullie, F. (1994), *Ultrason. Chem.* **1**, S13.
Margulis, M. A. (1990), *Advances in Sonochemistry* **1**, 39.
Marinesco, M., and Trillat, J. J. (1933), *Comp. Rend. Acad. Sci.* **196**, 858.
Meyer, E., and Kuttruff, H. (1959), *Z. ang. Phys.* **11**, 325.
Neppiras, E. A. (1980), *Phys. Rep.* **61**, 159.
Neppiras, E. A., and Fill, E. E. (1969), *J. Acoust. Soc. Amer.* **44**, 1264.
Neppiras, E. A., and Noltingk, B. E. (1951), *Proc. Phys. Sec. B (London)* **64B**, 1032.
Noltingk, B. E., and Neppiras, E. A. (1950), *Proc. Phys. Sec. B (London)*, **63B**, 674.
Plesset, M. S. (1954), *J. Appl. Phys.* **25**, 96.
Plesset, M. S., and Mitchell, T. P. (1956), *Q. Appl. Math.* **13**, 419.
Prudhomme, R. D., and Guilmart, T. (1957), *J. Chem. Phys.* **54**, 336.
Saksena, T. K., and Nyborg, W. L. (1970), *J. Chem. Phys.* **53**, 1722.
Suslick, K. S. (1990), *Science* **247**, 1373.
Suslick, K. S., and Hammerton, D. A. (1986), *IEEE Trans. Ultrason., Ferroelect., Freq. Control.* **33**, 143.
Taylor, K. J., and Jarman, P. D. (1970), *Aust. J. Phys.* **23**, 319.
Tomita, Y., and Shima, A. (1990), *Acoustica* **71**, 161.

Wu, C. C., and Roberts, P. H. (1993), *Phys. Rev. Lett.* **70**, 3424.
Young, F. R. (1965), *Nature* **206**, 706.
Young, F. R. (1976), *J. Acoust. Soc. Amer.* **60**, 100.
Young, F. R. (1989), *Cavitation* (McGraw-Hill, London).
Zanetti, R. J. (1992), *Chem. Eng.*, October, p. 37.

10

Mechanoluminescence

B. P. Chandra

1. INTRODUCTION

Mechanoluminescence (ML) is a type of luminescence induced during any mechanical action on solids. It can be excited by grinding, rubbing, cutting, cleaving, shaking, scratching, compressing, or crushing of solids. ML can also be excited by thermal shocks caused by drastic cooling or heating of materials or by the shock waves produced during exposure of samples to powerful laser pulses. ML also appears during the deformation caused by the phase transition or growth of certain crystals as well as during separation of two dissimilar materials in contact. The phenomenon of ML has also been called by many other names, such as trennugslicht, triboluminescence, piezoluminescence, deformation luminescence, and stress-activated luminescence. Since the prefix "mechano" is correlated in a general way with different mechanical concepts, such as deformation, piezo, tribo, stress, fracto, plastico, elastico, cutting, cleaving, grinding, rubbing, compressing, and crushing, "mechanoluminescence" has been the preferred nomenclature (Chandra and Shrivastava, 1978; Sodomka, 1978; Krauya *et al.*, 1978, 1981; Sodomka *et al.*, 1980; Molotskii, 1983; Mukhopadhyay, 1984; Tokhmetov and Vettegren, 1990; Batylin *et al.*, 1992).

The first recorded discovery of ML was made by Francis Bacon (1605), who reported in his writings, *Advancement of Learning*, that "hard sugar being nimbly scraped with a knife would afford a sparkling light." Robert Boyle (1664) reported that a particular diamond when "pressed on with a steel bodkin produced a very vivid but exceedingly short-lived splendor." Waller (1684) reported that when substances such as white sugar, loaf sugar, and salt gemme (rocksalt) were crushed in a mortar, they "gave such intense light that the sides of the mortar and shape of the pestle could distinctly be displayed." By the end of the 1700s, a number of minerals and pottery materials were known to exhibit ML. Wedgwood (1792) reported the ML of many substances including quartz, diamond, and ruby. The

---

B. P. Chandra • Department of Postgraduate Studies and Research in Physics, Rani Durgavati University, Jabalpur 482001, India.

Luminescence of Solids, edited by D. R. Vij. Plenum Press, New York, 1998.

observations of light emission associated with rock failure in mines and earthquakes have been reported for several centuries (Brady and Rowell, 1986). In the more recent past, the phenomenon of ML has been reviewed by Meyer *et al.* (1970*a*,*b*), Walton (1977); Zink (1978), Chandra (1985, 1996), and Molotskii (1989).

The physical processes involved in inducing ML in solids indicate that basically there should be two types: (1) deformation-ML (DML) and (2) tribo-ML (TML). The former is produced owing to the physical processes induced during deformation of solids, whereas, the latter is due to contact phenomena such as triboelectricity, tribochemical reactions, and tribothermal generation induced during the contact or separation of two dissimilar materials.

DML depends only on the material under deformation and is independent of both the material used to produce the deformation and the contact phenomenon. TML, on the other hand, depends on the nature of the material under deformation as well as on the material used to produce it, and arises solely owing to contact phenomena. For example, the luminescence produced during the rubbing of sugar or ZnS crystals with steel or any other rod is DML, as it is independent of the material of which the rod is made; however, the luminescence produced during low-energy impact of a steel or sapphire needle on alkali halide crystals (AHC), is TML, as its intensity depends on the material of the rod used for impact and on the contact potential between the two materials. Similarly, the ML produced during separation of a tape from a substrate depends on the nature of both the tape and the substrate and occurs owing to triboelectrification. Hence, it is TML.

DML may further be subdivided into three types, namely, fracto-ML (FML), plastico-ML (PML), and elastico-ML (EML). In FML, the luminescence is produced owing to the creation of new surfaces during the fracture of solids; in PML, it is produced during plastic deformation of solids, where fracture is not required; and in EML, it is produced during elastic deformation of solids, where neither plastic deformation nor fracture is required. A large number of organic and inorganic solids exhibit FML (Meyer *et al.*, 1970*a*,*b*; Walton, 1977; Chandra, 1985). In several solids, charged surfaces are produced during fracture owing to processes such as piezoelectrification, defective-phase piezoelectrification, charged-dislocation movements, and charged-defect barodiffusion. In this case, the surface charge is neutralized by the charge carriers and ions produced owing to the dielectric breakdown of the intervening gases and the solids. Thus, depending on the prevailing conditions, light emission that resembles a gas discharge or luminescence of a solid or combined characteristics of both can be obtained. In this connection, it is worthwhile to note that the crystals of, e.g., sucrose, tartaric acid, and rochelle salt exhibit a FML that resembles a gas discharge; those of, e.g., coumarin, resorcinol, and phenanthrene exhibit a FML that resembles their photoluminescence (PL); and those of, e.g., uranyl nitrate hexahydrate, saccharin (impure), and chlorotriphenylmethane exhibit a FML that contains both gas discharge and PL emissions (Chandra, 1981). FML can also be excited by the chemical reaction, heat production, stress perturbation, mechanical excitation, deep-trap production, or thermal excitation that occurs during the creation of new surfaces (Chandra *et al.*, 1991).

PML has been observed in the crystals of colored alkali halides, certain noncolored alkali halides, II–VI compounds, and alkaline-earth oxides, and in metals (Butler, 1966; Alzetta *et al.*, 1970; Williams and Turner, 1978; Bredikhin and Shmurak, 1979; Chandra, 1984; Poletaev and Shmurak, 1984; Zakrevskii and Shuldiner, 1984; Al-Hashimi *et al.*, 1984; Molotskii, 1989; Hayashiuchi *et al.*, 1990). Certain varieties of rubber also exhibit PML (Grabec, 1974, 1975; Parfeev *et al.*, 1988). PML can be excited by, e.g., mechanical or electrostatic interaction of dislocations with defect centers, electrification of crystal surfaces by the movement of charged dislocations, or thermal excitation in the stressed regions of solids.

EML is exhibited by the single crystals of x- or γ-irradiated alkali halides (Butler, 1966; Alzetta *et al.*, 1970; Chandra, 1984) and may be produced by, e.g., the mechanical or electrostatic interaction of dislocations with defect centers, or thermal excitation in the stressed regions of solids.

The well-known examples of TML are the light produced during, e.g., separation of a tape from a substrate, movement of mercury on luminescent layers, rotation of a steel rod in blood, and low-energy impact of a sapphire needle on AHC (Meyer *et al.*, 1970*a,b*; Keszthelyi and Bard, 1973; Zhenyi *et al.*, 1988; Orel *et al.*, 1989).

2. MECHANOLUMINESCENT MATERIALS

Nearly one-half of all inorganic solids and from one-fourth to one-third of all organic solids both crystalline and noncrystalline exhibit the phenomenon of ML. It has been observed in insulators and semiconductors as well as in conductors. ML solids may be divided into the following six groups: crystalline, amorphous, polymeric, ceramic, composite, and tribo–mechanoluminophors.

Generally, all noncentrosymmetric crystals, i.e., piezoelectric crystals (PC), exhibit ML and those that do not are nonpiezoelectric. Certain centrosymmetric crystals also exhibit ML, but it has a nonpiezoelectric origin. Wolff *et al.* (1952) compiled a list of about 450 ML crystals including inorganic compounds, minerals, and aliphatic and aromatic organic compounds. Chandra and his co-workers reported the spectroscopy, intensity, and crystal structure–ML correlation of some 400 organic and inorganic crystals (Chandra and Elyas, 1979; Chandra and Shrivastava, 1979; Chandra *et al.*, 1980; Chandra and Verma, 1980; Chandra, 1981, 1985; Chandra and Tutakne, 1981; Chandra and Zink, 1980*a,b,c*, 1981*a,b*, 1982; Chandra *et al.*, 1982*a*).

Amorphous materials such as silicates, quartz, borate, and other glasses exhibit ML (Sodomka, 1971; Avilov *et al.*, 1982; Zink *et al.*, 1982; Chapmn and Walton, 1983*a*; Tokhmetov and Vettergren, 1990). Krauya *et al.* (1978) observed ML during fracture of human body tissues. Chandra *et al.* (1982*b*) reported that bones emit light when they fracture.

ML appears during the fracture of the polymers such as polyethylene, polyvinylidene fluoride, polyvinylchloride, polypropylene, and polystyrene. ML of *N*-isopropylcarbazole and its derivatives was reported by Nowack *et al.* (1982), Kitamura *et al.* (1986), Grabec (1974, 1975) and Parfeev *et al.* (1988) reported ML during the plastic deformation of certain varieties of rubber. Tokhmetov and Vettergren (1990) reported ML during bond-breaking in polymethylmethacrylate. Nakayama *et al.* (1992) observed ML during stretching of nylon-6 and polytetrafluoroethylene.

Ceramics of lead zirconate titanate (PZT), SiC, ZrO_2, Al_2O_3, and Si_3N_4 (Dickinson and Jensen, 1986; Nakayama *et al.*, 1992) and composite materials such as glass-filled polybutadine and polycarbonate, Kevlar-filled epoxy, fiber-reinforced epoxy, glass epoxy composite, boron epoxy composite, and angle-ply glass-fiber composite (Dickinson *et al.*, 1982; Crasto *et al.*, 1987; Tamuzh *et al.*, 1985; Dickinson *et al.*, 1993) also exhibit ML.

The well-known examples of tribo–mechanoluminophors include, among others, air and mercury, luminescent layer and mercury, human body and silky cloth, tape and substrate, steel (or sapphire) rod and AHC, blood and steel rod, and diamond-impregnated saw blade and glass (Meyer *et al.*, 1970*a,b*; Keszthelyi and Bard, 1973; Chapman and Walton, 1983*a*; Zhenyi *et al.*, 1988; Orel *et al.*, 1989).

ML of PC, noncolored and colored AHC, and II–VI compounds has been studied in detail (Meyer *et al.*, 1970*a,b*; Walton, 1977; Chandra, 1985, 1996; Molotskii, 1989). Ditriphenyl phosphine oxide–$MnBr_2$. Eu-tetrakis (dibenzoylmethide) triethyl-NH_4^+, impure saccharin. Mn-doped ZnS, acenaphthene. *N*-acetyl anthranilic acid, coumarin, phenanthrene, cholesteryl salicylate, uranyl nitrate hexahydrate, *N*-isopropyl carbazole, and sugar are examples of intense ML materials. The ML of the first three is so intense that it can be seen in daylight.

3. MECHANOLUMINESCENCE MEASUREMENT DEVICES

Two types of devices are generally needed for ML measurements; one for deforming the samples and the other for the spectral measurements. Techniques used for deforming the crystals in ML measurements are: (1) compression (Chandra and Zink, 1980*a,b*); (2) bending (Alzetta *et al.*, 1970); (3) stretching (Crasto *et al.*, 1987; Grabec, 1974; Parfeev *et al.*, 1988); (4) loading (Chandra and Elyas, 1979; Atari, 1982; Atari and Ramani, 1986; Eid *et al.*, 1986); (5) piston impact or impulsive (Chandra, 1983); (6) needle impact (Meyer and Polly, 1965); (7) cleaving and cutting (Longchambon, 1925; Batylin *et al.*, 1992; Dickinson *et al.*, 1990; Meyer *et al.*, 1970*b*; Chapman and Walton, 1983*a,b*); (8) laser (Hardy *et al.*, 1979); (9) shaking (Copty-Wergles *et al.*, 1990); (10) air-blast (Longchambon, 1925; Sodomka, 1968; Meyer and Obrikat, 1969); (11) scratching (Nakayama *et al.*, 1992); (12) grinding and milling (Inoue *et al.*, 1939); and (13) tribo or rubbing (Zhenyi *et al.*, 1995; Meyer *et al.*, 1970; Keszthelyi and Bard, 1973; Orel *et al.*, 1989).

The compression technique and the piston impact or impulsive techniques are the most popular for ML measurements. Chandra and Zink (1980*a,b*) as well as several others have made the simultaneous measurement of ML–strain and stress–strain curves by using the table model Instron testing machine shown in Fig. 1a. In this technique, ML intensity is measured using a photomultiplier tube, stress is measured with a load cell, and strain with a linear variable differential transducer. Figure 1b shows the schematic diagram of the piston impact technique used by Chandra (1983), in which ML is excited by the impact of a moving piston on the

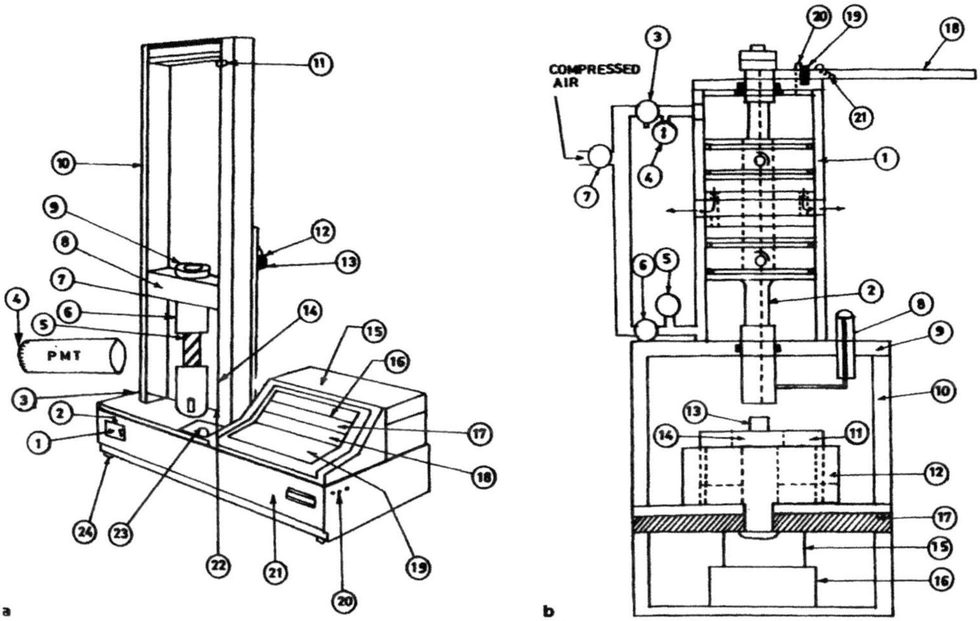

Figure 1. (a) A table model Instron testing machine for the simultaneous measurement of the stress–strain and ML–strain curves of a solid: (1) emergency stop push button, (2) on/off switch, (3) base grip adaptor, (4) photomultiplier tube, (5) sample, (6) upper clip adapter, (7) preload levers, (8) moving cross-head, (9) load cell, (10) load frame, (11) upper limit stop, (12) load cell socket, (13) extensometer connection, (14) lower limit stop, (15) control console, (16) limits panel, (17) display panel, (18) recorder panel, (19) basic control console, (20) console lock, (21) base unit, (22) cross-head limit switches, (23) leveling page, (24) cross-head control panel. (b) Schematic diagram of the piston impact technique for measuring ML: (1) cylinder, (2) piston rod, (3) three-way valve, (4 & 5) gauges, (6 & 7) regulators, (8) velocity transducer, (9) piston guiding plate, (10) spacers, (11) sample holder, (12) sample holder housing, (13) crystal, (14) lucite plate, (15) photomultiplier tube with housing, (16) photomultiplier housing platform, (17) rubber ring, (18) trigger handle, (19) pin stop, (20) center pin axis, (21) spring (Chandra, 1983).

crystal, with the velocity of the moving piston being measured by a velocity transducer.

Belyaev and Martyshev (1969), Hardy and Zink (1976), and Chandra (1983) among others used the monochromator-based spectrometer for measuring ML spectra, and Walton and Botos (1978) and Atari and Ramani (1986) used an image-intensifier spectrometer.

4. MECHANOLUMINESCENCE CHARACTERISTICS

4.1. Deformation Characteristic

The deformation characteristic is the curve of ML intensity vs. strain (or stress) of the sample at different strain (or stress) rates. The ML-vs.-strain curve indicates whether the ML occurs in an elastic or a plastic region or during fracture. Figure 2

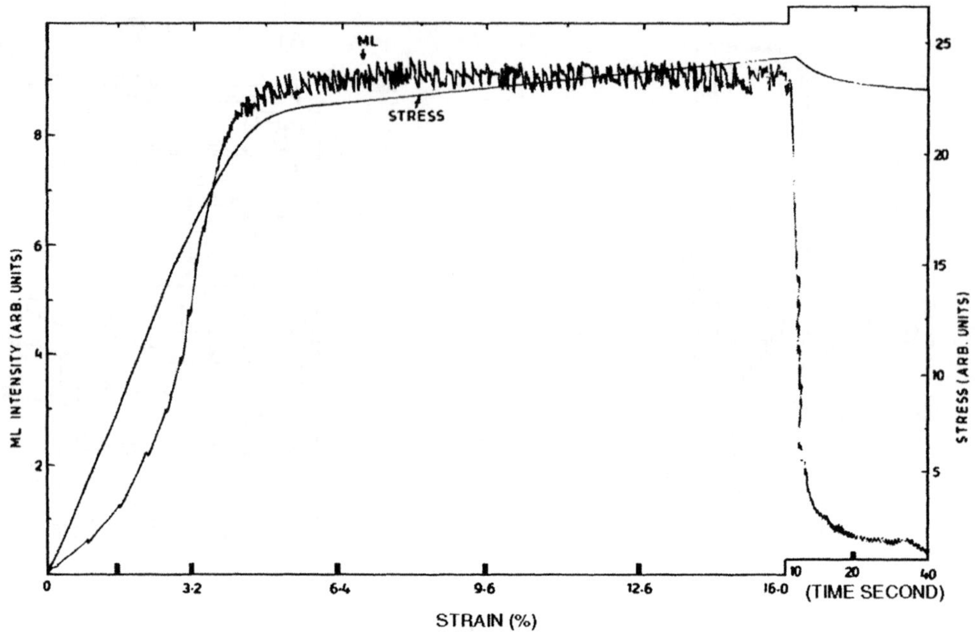

Figure 2. ML-vs.-strain and stress-vs.-strain curves of a γ-irradiated KCl crystal ($5 \times 5 \times 5\,\text{mm}^3$, $n_F \approx 10^{17}\,\text{cm}^{-3}$, $d\varepsilon/dt = 10^{-4}\,\text{s}^{-1}$).

shows the ML-deformation characteristic of a γ-irradiated KCl crystal, in which it is seen that the ML appears during its elastic and plastic deformations. It has been found that ML in these crystals also appears during fracture. ML in rubber appears during elongation as well as during shrinkage affecting the length. Figure 3 shows that the ML in a coumarin crystal appears during steps in its stress–strain curve. Since a step in the curve corresponds to the movement of a crack, it is evident that the ML occurs only during fracture. It has been found that the number of ML pulses produced per unit time increases with the strain rate. The ML intensity has been found to increase directly with the area of newly created crystal surfaces.

4.2. *Temporal Characteristic*

The temporal characteristic of ML is determined by cleaving a crystal, applying a fixed load, or impacting a moving piston on to the crystal. Figure 4a shows the time dependence of ML produced during cleavage of LiF and NaCl crystals. It is seen that initially the ML intensity increases with time, attains a peak value, and then decreases with time. The intensity I_m and time t_m, corresponding to the peak of

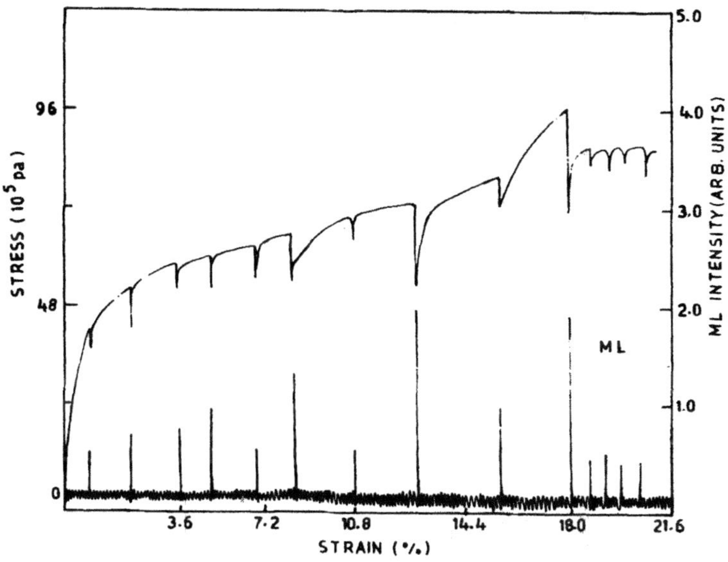

Figure 3. ML-vs.-strain and force-vs.-strain curves of a coumarin crystal ($5 \times 4 \times 2\ mm^3$).

an ML intensity-vs.-time curve, increase with increasing size of the crystals. After t_m, the intensity decreases exponentially with time.

Figure 4b shows the time dependence of both the ML intensity obtained during impact of a moving piston on fluorescent coumarin crystals and the velocity of the piston used to deform the crystal. It is evident that ML appears only during the deformation of the crystal. It is seen that I_m increases and t_m shifts toward shorter time values with increasing impact velocity of the piston. Generally, I_m increases linearly with the impact velocity v_0 of the piston and the total ML intensity I_T, defined as the area below the curve for ML intensity vs. deformation time, initially increases and then tends to attain a saturation value for higher impact velocities. The value of I_T increases linearly with the volume or mass of the crystals, and the value of I_m increases linearly with the area of their cross section. The value of time t_m corresponding to the peak of ML intensity-vs.-time curve increases with crystal thickness.

Figure 5a shows that ML in phosphorescent crystals appears during as well as after deformation. Figure 5b shows the plot of $\log I$ vs. $\log t$ for Kaolinite clay materials. It is seen that prolonged ML in these materials appears for several days after deformation. In this case the ML was excited either by grinding the sample with mortar and pestle or by fracturing it with a hammer.

ML kinetics of γ-irradiated LiF crystals were obtained during the application and release of a fixed load (Eid *et al.*, 1986). In this case, initially the ML intensity increases with time, attains a peak value, and then decreases with time. ML also

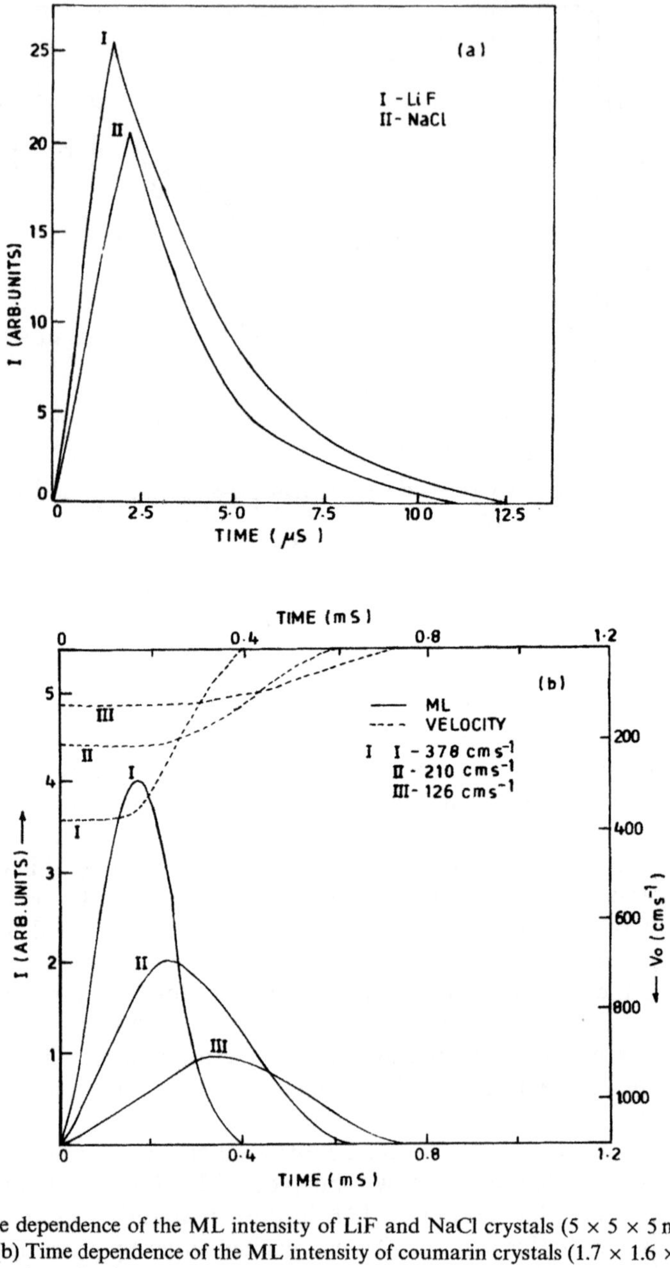

Figure 4. (a) Time dependence of the ML intensity of LiF and NaCl crystals ($5 \times 5 \times 5\,mm^3$) produced during cleavage. (b) Time dependence of the ML intensity of coumarin crystals ($1.7 \times 1.6 \times 1.2\,mm^3$) and the velocity of the piston (top).

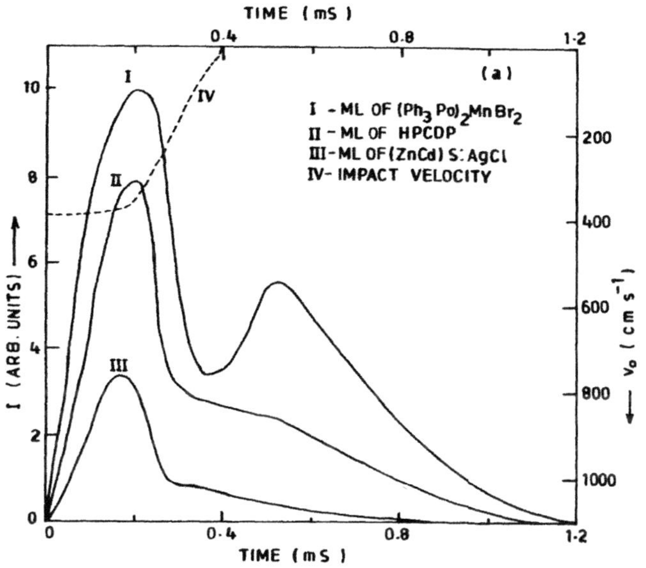

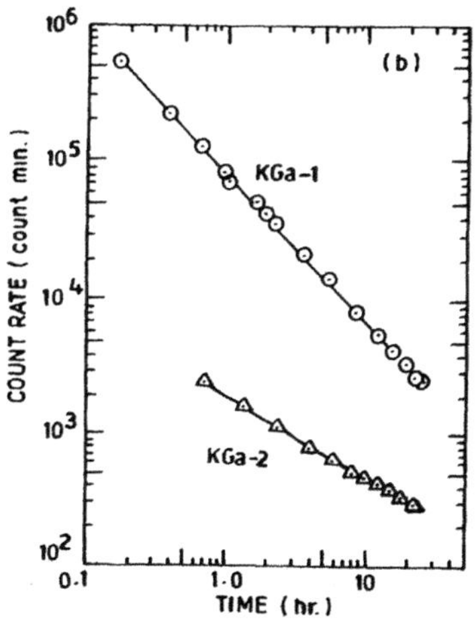

Figure 5. (a) Time dependence of the ML intensity (bottom) of $(Ph_3PO)_2MnBr_2$ and hexaphenylcarbodiphosphorane (HPCDP) crystals ($2.5 \times 2.5 \times 2.5\ mm^3$), and (ZnCd)S:Ag,Cl phosphors (5 mg) and the impact velocity of piston (top). (b) ML intensity as a function of time elapsed from the end of grinding, well-crystallized Kaolinite KGa-1 and poorly crystallized Kaolinite KGa-2 (Lahav *et al.*, 1982).

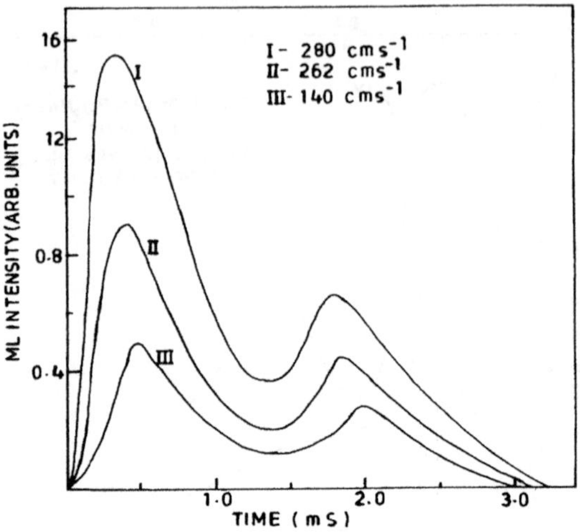

Figure 6. Time dependence of the ML intensity of γ-irradiated KCl crystals for different impact velocities.

occurs during the release of the applied load, where its intensity is several times less as compared to that obtained during the applications of a load. The value of I_m increases with the applied load, crystal size, and radiation dosage. It is seen that during both application and release of the load, ML intensity decays rapidly at first and then more slowly.

Figure 6 shows the time dependence of the ML intensity of γ-irradiated KCl crystals obtained during the impact of a moving piston on the crystals. It is seen that there are two peaks in the ML intensity-vs.-time curve: the first lies in the deformation region and the second in the postdeformation region. The intensities I_{m_1} and I_{m_2}, corresponding, respectively, to the first and second peaks, increase with the impact velocity of the piston, while the respective times, t_{m_1} and t_{m_2}, decrease.

4.3. Thermal Characteristic

Generally, the ML intensity of nonirradiated crystals decreases with temperature and disappears beyond a particular point (Chandra *et al.*, 1987). Figure 7a shows the effect of temperature on the ML intensity of coumarin, triphenylamine, and phenanthrene crystals, whose melting points are 70, 126, and 99.5°C, respectively. ML of ferroelectric crystals has been found to change drastically near the ferroelectric transition temperature.

The ML intensity of thermoluminescent (TL) crystals, such as γ-irradiated AHC, initially increases with temperature, attains a maximum value, and then decreases with temperature. Figure 7b shows that the temperature dependence of the ML of colored AHC follows an Arrhenius plot, and decreases at higher temperatures because of thermal bleaching.

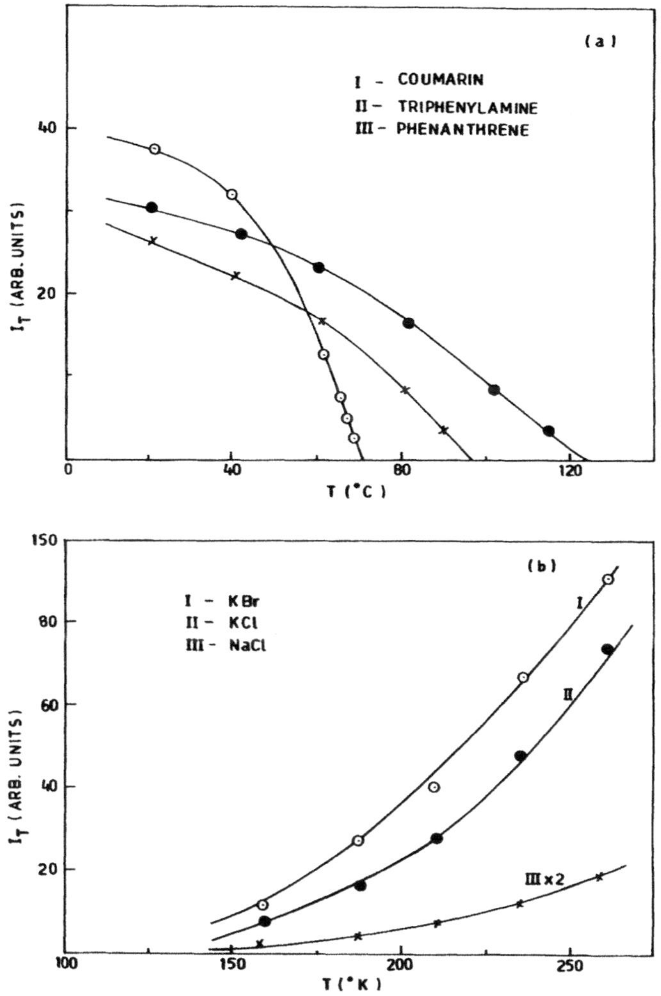

Figure 7. (a) Temperature dependence of the total ML intensity of coumarin, triphenylamine, and phenanthrene crystals. (b) Temperature dependence of the ML intensity of γ-irradiated KCl, KBr, and NaCl crystals.

4.4. Spectral Characteristic

Depending on the prevailing conditions, ML spectra that resemble the light emission that comes from a gas discharge or another type of luminescence of a solid or that combine characteristics of both may be obtained (Chandra, 1981; Hardy *et al.*, 1977, 1979, 1981). Figure 8a shows that the ML spectra of sugar crystals are similar to the discharge spectra of the surrounding gases. Figure 8b shows that the ML spectra of coumarin and ditriphenylphosphineoxide $MnBr_2$ crystals are similar

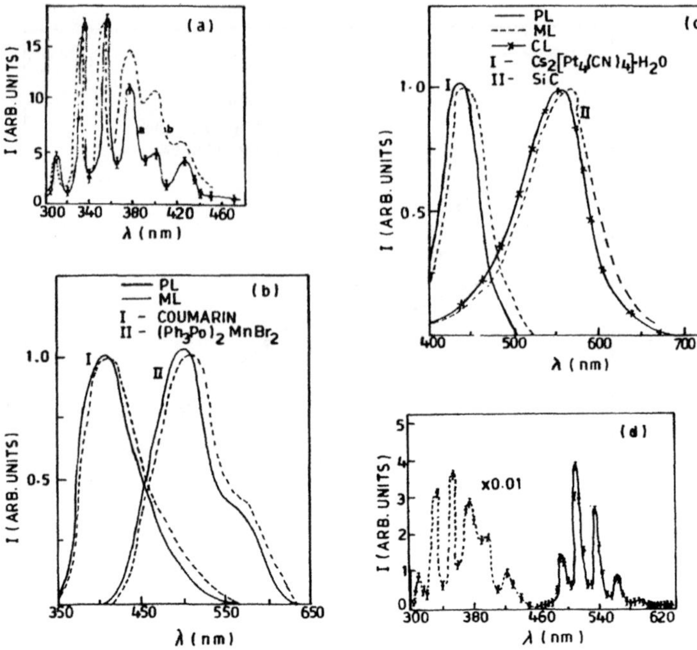

Figure 8. Spectra of the ML of sucrose crystals and the gas discharge at atmospheric pressure (a); ML and PL spectra of coumarin and $(Ph_3PO)_2MnBr_2$ crystals (b); ML and PL spectra of $Cs_2[Pt(CN)_4].H_2O$, and ML and CL spectra of SiC crystals (c); ML and PL spectra of uranyl nitrate hexahydrate crystals (d).

to their PL spectra. Figure 8c shows that the ML spectra of cesium platinocyanide monohydrate crystals resemble their PL spectra and that the ML spectra of SiC crystals are similar to their cathodoluminescence (CL) spectra. Figure 8d shows that the ML spectra of uranyl nitrate hexahydrate crystals consist of both the gas discharge and PL spectra.

Figure 9a shows that for impurity-doped ZnS crystals, the ML spectra shift slightly as compared to their PL spectra. Figure 9b shows that the ML spectra of γ-irradiated KCl and KBr crystals are similar to their TL spectra, where there is a slight shift in the ML spectra toward the shorter-wavelength side. The shift in ML spectra of II–VI compounds and colored AHC, as compared to other types of spectra, is primarily due to the stress resulting from the deformation of the crystals.

5. MECHANOLUMINESCENCE MECHANISMS

The excitation mechanisms of FML in PC and metals, PML and FML in noncolored AHC, and PML in colored AHC and II–VI compounds are now well understood, and are described below.

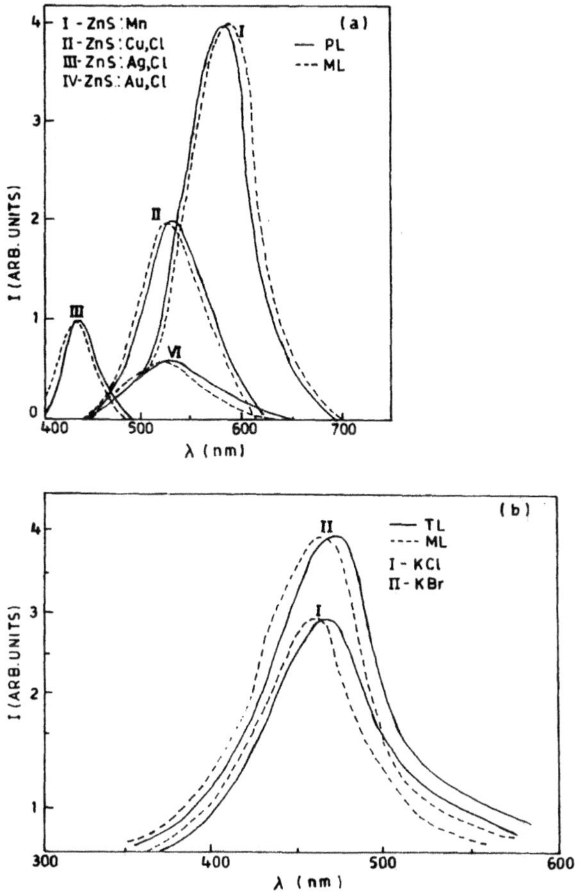

Figure 9. ML and PL spectra of ZnS phosphors (Chandra *et al.* 1991) (a), and ML and TL spectra of γ-irradiated KCl and KBr crystals (Atari, 1982) (b).

5.1. Fracto-Mechanoluminescence in Piezoelectric Crystals

Figure 10a shows the Langevin model for the creation of charged surfaces during the movement of a crack in a PC (Langevin, 1921). The piezoelectric constants and the stress needed to separate the surfaces of crystals are generally of the order of 10^{-12}–10^{-11} coulombs CN^{-1} and 10^8 N m^{-2}, respectively. Thus, the charge density γ of the newly created surfaces is of the order of 10^{-4}–10^{-3} C m^{-2}. The electric field between the oppositely charged surfaces will be γ/ε_0, where ε_0 is the permittivity of free space, equal to 8.85×10^{-12} $C^2 N^{-1} m^{-2}$. Thus, an electric field of the order of 10^7–10^8 V m^{-1} may be generated between the newly created oppositely charged surfaces. This field may cause the dielectric breakdown of the intervening gases and in turn may give rise to gaseous discharge ML. The field may

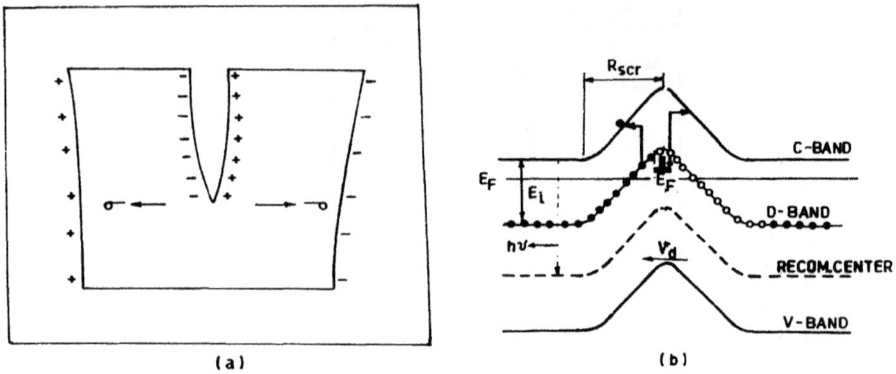

Figure 10. Electric field produced during movement of a crack in a piezoelectric crystal (a); Bending of bands near a dislocation and schematic diagram of the electronic transitions (b).

also cause the dielectric breakdown of the crystals, and the recombination of free carriers may give rise to recombination luminescence. Furthermore, the accelerated electrons moving from negatively charged surfaces toward the positively charged surface may excite CL.

5.2. Noncolored Alkali Halide Crystals

5.2.1. Mechanism of FML

In noncolored AHC, charged surfaces are produced during fracture because of the asymmetric movement of charged dislocations (Molotskii, 1989) as well as the barodiffusion of charged defects (Alekseev, 1992). The electric field generated by the newly created charged surfaces causes dielectric breakdown of the gases, so that gas-discharge ML is observed when certain noncolored AHC are broken in air.

In vacuum the relaxation of surface charges produced during fracture of noncolored AHC takes place in the following way (Molotskii, 1989). Electrons that have been accelerated in the field of fracture penetrate into the positively charged wall and excite a large number of excitons. Most of the excitons decay within the crystal but some of them emerge onto the surface. As defects can be generated in the cationic sublattices of the crystal during the nonradiative decay of excitons, free cations are formed during the decay of the surface excitons. These cations, which are accelerated in the field of the fracture, bombard the negatively charged wall and give rise to secondary-ion and electron emission.

The chain processes, including the electronic, excitonic, and ionic stages, are then repeated and luminescence may be produced owing to radiative electron–hole recombination or radiative decay of excitons. If there are alkali atoms present between the oppositely charged walls of the crack, their excitation may give rise to the emission of resonance lines. Meyer *et al.* (1970*a*) reported that ML in noncolored AHC obtained during fracture in vacuum may be due to blackbody radiation of hot spots produced during fracture.

5.2.2. Mechanism of Pulsed PML

Poletaev and Shmurak (1984) detected flashes of nonstationary luminescence during the plastic deformation of LiF crystals. These flashes arise when a stress of the order of magnitude of the yield point is imposed on the crystal and increase monotonically until they reach a stationary level, the magnitude of which depends on the rate of plastic deformation. The dependence of the number of flashes on the rate of deformation is of threshold nature. When $d\varepsilon/dt < 8 \times 10^{-6}\,\mathrm{s}^{-1}$, there are no flashes but at high deformation rates the number of flashes is linearly dependent on $d\varepsilon/dt$.

The positions of the main maxima in the flash luminescence of LiF are identical to the surface electroluminescence (EL) spectra of the crystals. This matching is evidence of the fact that pulsed ML is attributable to the luminescence of air molecules that have been adsorbed on the surface and excited during surface breakdown of the sample. The mechanism of surface EL of LiF is associated with the transfer of charged dislocations onto the surface. In order for pulsed ML to occur, the dislocation current, which creates the potential on the edges of the crystal, must exceed the leakage current through the sample.

5.2.3. Mechanism of Continuous PML

Zakrevskii and Shuldiner (1984) detected continuous ML in noncolored LiF and NaF crystals during the intersection of a forest of immobile and mobile dislocations. They proposed a thermal excitation mechanism to explain PML in noncolored AHC. According to this mechanism, there is a possibility that thermally activated electronic transitions can occur in regions of high stress when, upon fluctuational displacements along the configuration coordinate, the potential curves of the ground and excited states cross each other. This effect is the inverse of thermal radiationless quenching of excited states.

Low-energy excitons like electronic excitations that are formed can dissociate at dislocations, giving rise to electrons and holes in the dislocation bands. In order to explain the emission phenomena, one must assume that the centers that are formed are mobile and can move away from where they are formed along the dislocation bands. These dislocation electrons and holes will then be captured by anion and cation vacancies, respectively, to form F^- and V_k-centers. The recombination of these centers leads to the excitation of luminescence. The formation of F-centers has been recorded by Zakrevskii and Shuldiner (1985) in experiments on thermally and photostimulated emission of plastically deformed noncolored LiF and NaCl crystals.

5.3. Colored Alkali Halide Crystals

5.3.1. Mechanism of PML

During the plastic deformation of x or γ irradiated AHC, the moving dislocations interact mechanically with the F-centers and subsequently capture electrons from those nearby (Chandra, 1996). The captured electrons move with the dislocations, and they also move freely along the dislocation line at the same time. The dislocation electrons have a finite lifetime τ_d. At $t > \tau_d$, they either recombine with the holes or are captured by other traps in the crystal. If the moving dislocation containing electrons encounters the defect centers containing holes, the dislocation electrons may be captured by these centers and luminescence may arise owing to radiative electron–hole recombinations. From a comparison of ML spectra with those of other types of luminescence in x- or γ-irradiated pure AHC, it has been proved that ML arises owing to the recombination of electrons from F-centers with the holes in V_2-centers. Schematically, the ML process can be described by the following equations:

$$F + D \rightarrow e_d + [-]$$

$$V_2[=X^- + X^- + X^0] + e_d \rightarrow V_2^-(=3X^-) + D + h\nu$$

where F and D represent an F-center and a dislocation, respectively, e_d is the dislocation electron, i.e., the electron captured by a dislocation, $[-]$ is the negative ion vacancy, X^- is the halogen ion, X^0 is a self-trapped hole and V_2^- is the V_2-center with a captured electron.

In x- or γ-irradiated impurity-doped AHC, some of the holes may be in the impurity and their recombination with the dislocation electrons may induce impurity emissions (Ossipyan and Shmurak, 1981).

5.3.2. Mechanism of EML

During elastic deformation, the dislocations bend between pinning points and interact with the nearby F-centers. Subsequently, they may capture F-center electrons and transfer them to holes, whereby luminescence may occur.

5.4. Plastico-Mechanoluminescence in II–VI Compounds

5.4.1. Mechanism of Pulsed PML

Loading of high-ohmic ZnS crystals ($\rho > 10^{12}\,\Omega$-cm) results in short glow pulses (40 ns) (Bredikhin and Shmurak, 1974). The pulse amplitude and duration are independent of the deformation rate and temperature. The number of luminescence pulses per unit time is proportional to the strain rate $d\varepsilon/dt$. The spectrum of pulsed ML in ZnS:Mn crystals consists of a broad peak centered at 585 nm and a set of lines between 318 and 400 nm. The investigations of spectra and temperature as well as of kinetic and several other characteristics of pulsed ML in II–VI compounds imply that it has dislocation origin (Bredikhin and Shmurak, 1974, 1975, 1977).

During plastic deformation, charged dislocations of opposite signs move in opposite directions and bring charges out to the surface of the crystal. The potential difference across the crystal increases as deformation proceeds until it reaches a breakdown value at which a pulse of surface EL is emitted, and this process is repeated continuously with increasing deformation. A similar kind of luminescence can be produced without any deformation by the direct application of an external electric field greater than about $1.2 \times 10^6\ \mathrm{V\,m^{-1}}$.

5.4.2. Mechanism of Continuous PML

Pulsed ML dominates the light output from deformed II–VI compounds, but if it is suppressed by immersing the crystals in a conducting fluid, using a crystal of insufficiently high resistivity, or short-circuiting the lateral faces of the crystal, continuous ML is observed in ZnS, ZnSe, and CdS crystals (Bredikhin and Shmurak, 1977, 1979, Bredikhin *et al.*, 1982). The ML intensity increases rapidly with increasing external stress at $\sigma = \sigma_{el}$; then, as deformation proceeds at a constant rate of strain it attains a constant value whose magnitude is a linear function of the strain rate (Bredikhin and Shmurak, 1979). Continuous PML in II–VI compounds arises owing to the ionization of electron traps caused by tunneling in the presence of the strong field of the linear charge of a moving dislocation (Fig. 10b). The radiative recombination of excited electrons with luminescence centers leads to the appearance of luminescence bands. Veliev (1986) and Vardanyan *et al.* (1989) have also studied plastico-induced continuous ML in II–VI compounds.

5.5. Fracto-Mechanoluminescence in Metals

According to Molotskii (1989), ML emission in metals may be attributed to the generation of holes during the decay of mobile dislocations on to the surface and their subsequent recombination with electrons. The nonadiabatic process that leads to the generation of holes in deep bands in Cu during the decay of mobile dislocations can be understood in the following way: There are dangling *d*-states close to the core of the edge dislocations in Cu, which are similar to the corresponding Shockley–Read bonds on dislocations in covalent semiconductors. At the instant a dislocation breaks up on the surface, a single atom step arises, and the resonant *d*-states on the dislocation are converted into *d*-states on the edge of the step, which lie above the top of the *d*-band.

During the breakup of a dislocation, there is rupture of the old bonds and formation of new ones. The local electronic structure is strongly distorted during the rupture process. Certain states lying above the Fermi level are depressed below the *d*-levels of the decaying dislocations and intersect with them. Thus, a nonadiabatic transition leading to the formation of *d*-holes on the dislocation-generated step becomes possible. During the formation of the step, the energy levels of a *d*-hole are continuously depressed below the dislocation states that lie above the bulk *d*-band. This hole can tunnel into the metal, where the tunneling probability is proportional to the density of the states. Since the density of states in Cu increases rapidly close to the top of the *d*-band, the formation of holes in the L_3 state has the highest probability.

6. MECHANOLUMINESCENCE THEORY

The theories of FML in PC and PML in x- or γ-irradiated AHC have been well-understood for some time. These theories are described in the following paragraphs.

6.1. Fracto-Mechanoluminescence in Piezoelectric Crystals

If a crystal having length L, width W, and thickness H is cleaved along a plane parallel to its width, then the rate of the creation of new surfaces is given by $2Wv$, where v is the velocity of the separation of the cleavage plane or, in other words, the velocity of crack propagation. If γ is the charge density of the newly created surface, then the rate of generation of surface charge is given by

$$g_s = 2\gamma W v \tag{1}$$

It should be noted that the charged surfaces may be created owing to such phenomena as piezoelectrification, movement of charged dislocations, and barodiffusion of charged defects. When the charged surfaces are created, the surface charge density may get relaxed, as a result of both the charge carriers produced by the breakdown of the crystal and the electrons and ions produced by the breakdown of intervening gases. If α_1 and α_2 are the rate constants for the relaxation of charges on the newly created surfaces by the crystal and gaseous breakdown processes, respectively, one may write the following rate equation:

$$dQ/dt = g_s - \alpha_1 Q - \alpha_2 Q$$

or

$$dQ/dt = g_s - \alpha_3 Q \tag{2}$$

where $\alpha_3 = (\alpha_1 + \alpha_2)$, and Q is the surface charge at any time t.

As $Q = 0$ at $t = 0$, the solution of Eq. (2) gives

$$Q = (g_s/\alpha_3)[1 - \exp(-\alpha_3 t)] \tag{3}$$

The ML intensity I will be directly proportional to the rate of discharge $(-dQ/dt)$. If η_1 is the luminescence efficiency related to the movement of carriers produced by the dielectric breakdown of the crystals and η_2 is the luminescence efficiency related to the movement of electrons and ions produced by the dielectric breakdown of intervening gases, then the ML intensity may be expressed as

$$I = \eta_1 \alpha_1 Q + \eta_2 \alpha_2 Q \tag{4}$$

From Eqs. (3) and (4), one gets

$$I = [(\eta_1 \alpha_1 + \eta_2 \alpha_2) g_s/\alpha_3][1 - \exp(-\alpha_3 t)] \tag{5}$$

6.1.1. Kinetics of ML

For $\alpha_3 t < 1$ Eq. (5) may be written as

$$I = (\eta_1 \alpha_1 + \eta_2 \alpha_2) g_s t$$

or

$$I = 2(\eta_1 \alpha_1 + \eta_2 \alpha_2)\gamma W v t \tag{6}$$

Equation (6) shows that when a crystal is cleaved, initially the ML intensity should rise linearly with time t. After the completion of cleavage at $t = t_m$, $v = 0$ and $g_s = 0$.

Thus, from Eq. (2),

$$dQ/dt = -\alpha_3 Q \tag{7}$$

Taking $Q = Q_0$ at $t = t_m$ and integrating Eq. (7), one gets

$$Q = Q_0[-\alpha_3(t - t_m)] \tag{8}$$

As t_m is very short, negligible discharge occurs within a period of time 0 to t_m, and therefore one may take $Q_0 = 2WH\gamma$, i.e., the total charge of the newly created surfaces.

From Eqs. (4) and (8), one gets

$$I = [\eta_1\alpha_1 + \eta_2\alpha_2]Q_0 \exp[-\alpha_3(t - t_m)] \tag{9}$$

Equation (9) shows the exponential decay of the ML intensity after cleavage of the crystals.

6.1.2. Estimation of t_m

For the thickness H of the crystal, the time t_m taken for cleaving is

$$t_m = H/v \tag{10}$$

Equation (10) shows that the crack propagation velocity can be given by

$$v = H/t_m \tag{11}$$

Thus, the ML may provide a new tool for determining the velocity of crack propagation in crystals.

6.1.3. Estimation of I_m

From Eqs. (6) and (10), the value of the peak ML intensity at $t = t_m$ is given by

$$I_m = 2(\eta_1\alpha_1 + \eta_2\alpha_2)\gamma WH \tag{12}$$

As $2WH$ is the area A of the newly created surfaces, one has

$$I_m = (\eta_1\alpha_1 + \eta_2\alpha_2)\gamma A \tag{13}$$

Equation (13) shows that I_m should increase linearly with A and γ.

6.1.4. Estimation of I_T

The total ML intensity I_T, i.e., the integrated area below the ML intensity-vs.-time curve, may be given by

$$I_T = \int I\,dt \tag{14}$$

It is found experimentally that the integrated light emitted during the rising portion of the ML intensity-vs.-time curve is negligible compared to that emitted during the decaying portion. Thus, from Eqs. (9) and (14), one may express the total ML intensity as

$$I_T = [\eta_1\alpha_1 + \eta_2\alpha_2]Q_0 \int_{t_c}^{\infty} \exp[-\alpha_3(t - t_c)]dt$$

or

$$I_T = (\eta_1\alpha_1 + \eta_2\alpha_2)\gamma A/\alpha_3 \tag{15}$$

Equation (15) shows that the total ML intensity should be directly proportional to A. It is also evident that I_T should be higher for crystals having higher values of γ.

6.1.5. ML Spectroscopy

For $\alpha_2 = 0$, there will be no gaseous discharge ML and emissions will be due primarily to bulk ML. In this case, $\alpha_3 = \alpha_1$ and Eqs. (13) and (15) may be written as

$$I_m = \eta_1\alpha_1\gamma A \tag{16}$$

and

$$I_T = \eta_1\gamma A \tag{17}$$

For $\alpha_1 = 0$, there will be no bulk ML and emissions will be due primarily to the gaseous discharge ML. In this case, $\alpha_3 = \alpha_2$ and Eqs. (13) and (15) may be written as

$$I_m = \eta_2\alpha_2\gamma A \tag{18}$$

and

$$I_T = \eta_2\gamma A \tag{19}$$

For crystals, where α_1 is comparable with α_2, emissions should consist of both the solid state and gaseous discharge ML, and in this case Eqs. (13) and (15) should be obeyed.

6.1.6. Temperature Dependence

Equations (13) and (15) show that the temperature dependence of both I_m and I_T should be due primarily to the effect of temperature on η_1, η_2, and γ. Since all three decrease with increasing crystal temperature, I_m and I_T should also decrease. Beyond a specific temperature, γ may decrease to a point where the breakdown of gases and solids becomes impossible, so that ML no longer appears.

The proposed theory may also be applied to the FML of noncolored AHC, where the surface charge density owing to piezoelectrification is replaced by one that is due to the asymmetric movement of charged dislocations or to the barodiffusion of charged point defects.

6.2. Plastico-Mechanoluminescence in x- or γ-Irradiated Alkali Halide Crystals

In a crystal having N_d dislocations of unit length per unit volume, if r_F is the radius of interaction of a dislocation with F-centers and v_d is the average velocity of the dislocations, then in unit time, N_d dislocations may interact with the F-centers lying in the volume $N_d v_d r_F$. If n_F is the density of F-centers, then for a crystal of unit volume the rate of interaction of F-centers with dislocations may be given by (Chandra, 1996)

$$g_i = N_d v_d r_F n_F = \frac{(d\varepsilon/dt) r_F n_F}{b} \tag{20}$$

where b is the Burgers vector and $d\varepsilon/dt = N_d b v_d$ is the strain rate.

In the expansion region of dislocations, the average energy E_i of the F-centers interacting with dislocations is higher as compared to the noninteracting F-centers (Chandra, 1996). Thus, E_i will lie in between the normal ground state of F-centers and the dislocation band. If β_1 is the rate constant for the jumping of interacting F-center electrons to the dislocation band, and β_2 is the rate constant for the dropping back to the normal F-level, then one may write the following rate equation:

$$dn_i/dt = g_i - (\beta_1 + \beta_2) n_i \tag{21}$$

where $1/(\beta_1 + \beta_2) = \tau_i$ is the lifetime of the interacting F-centers and n_i is the number of interacting F-center electrons at any time t.

Integrating Eq. (21) and taking $n_i = 0$, at $t = 0$, one gets

$$n_i = \frac{g_i}{\beta_1 + \beta_2} \{1 - \exp[-(\beta_1 + \beta_2)t]\} \tag{22}$$

From Eqs. (21) and (22), the rate of generation of electrons in the dislocation band may be given by

$$g = \frac{\beta_1 g_i}{\beta_1 + \beta_2} \{1 - \exp[-(\beta_1 + \beta_2)t]\} \tag{23}$$

For $(\beta_1 + \beta_2)t \gg t$, i.e., in equilibrium, one gets

$$g = \frac{\beta_1 g_i}{\beta_1 + \beta_2} = N_d v_d p_F r_F n_F = \frac{(d\varepsilon/dt) p_F r_F n_F}{b} \tag{24}$$

where $p_F = \beta_1/(\beta_1 + \beta_2)$ is the dislocation capture probability of F-center electrons.

6.2.1. ML Intensity

When the dislocations containing captured electrons move in a crystal, the dislocation electrons may recombine with the defect centers containing holes causing luminescence. If τ_d is the lifetime of dislocation electrons, then one may write the following rate equation:

$$dn_d/dt = g - n_d/\tau_d \tag{25}$$

where n_d is the number of dislocation electrons at any time t.

Since τ_d is short (Molotskii and Shmurak, 1990), the factors on the right-hand side of Eq. (25) will be large and dn_d/dt may be neglected. Thus, one gets

$$n_d = g\tau_d \tag{26}$$

From Eqs. (23) and (26),

$$\begin{aligned} n_d &= p_F g_i \tau_d\{1 - \exp[-(\beta_1 + \beta_2)t]\} \\ &= \frac{p_F r_F n_F (d\varepsilon/dt)\tau_d}{b}[1 - \exp(-t/\tau_i)] \end{aligned} \tag{27}$$

where

$$\tau_i = 1/(\beta_1 + \beta_2) \tag{28}$$

The rate of recombination of dislocation electrons may be given by

$$R_d = n_d/\tau_d = \frac{p_F r_F n_F d\varepsilon/dt}{b}[1 - \exp(-t/\tau_i)] \tag{29}$$

If η is the probability of radiative recombination, then the deformation-induced ML intensity for a crystal of volume V containing N_dV dislocations of unit length may be written as

$$I_{\text{DI}} = \frac{\eta p_F r_F n_F (d\varepsilon/dt)V}{b}[1 - \exp(-t/\tau_i)] \tag{30}$$

Equation (30) shows that for a given strain rate, the ML intensity should initially increase with time and then attain a saturation value I_{DIS}, where

$$I_{\text{DIS}} = \frac{\eta p_F r_F n_F (d\varepsilon/dt)V}{b} \tag{31}$$

Equation (31) shows that I_{DIS} should depend linearly on n_F, $(d\varepsilon/dt)$, and V of the crystals.

6.2.2. Temperature Dependence of ML Intensity

For $\beta_2 > \beta_1$, p_F may be written as

$$p_F = \beta_1/(\beta_1 + \beta_2) \approx \beta_1/\beta_2$$

Since the activation process is involved in transferring the electrons from interacting F-centers to the dislocation band, the temperature dependence of the rate constant β_1 may be given by

$$\beta_1 = \beta_1^0 \exp(-E_a/kT) \tag{32}$$

where E_a is the energy gap between the bottom of the dislocation band and the average ground state of the interacting F-centers, and β_1^0 is a constant. Thus, the temperature dependence of p_F can be expressed as

$$p_F \approx p_F^0 \exp(-E_a/kT) \tag{33}$$

where $p_F^0 = \beta_1^0/\beta_2$.

From Eqs. (31) and (33),

$$I_{\text{DIS}} = \frac{\eta p_F^0 r_F n_F (d\varepsilon/dt) V}{b} \exp(-E_a/kT) \tag{34}$$

It is evident from Eq. (34) that for a given n_F, the ML intensity should increase with temperature and follow an Arrhenius plot, where the activation energy is the energy difference between the bottom of the dislocation band and the average ground-state energy of the interacting F-centers.

When the ML in x- or γ-irradiated AHCs is observed above room temperature, it initially increases with temperature, attains a maximum value at a critical temperature, and then decreases, and eventually disappears. The ML intensity increases initially because p_F increases with temperature, and at higher temperatures it decreases and then disappears because of thermal bleaching of the F-centers.

6.2.3. Decay of ML Intensity

If a crystal is being deformed at constant $d\varepsilon/dt$, and the cross-head is stopped at a time $t = t_c$, at which the ML intensity had reached I_{DIS}, experiments show that the stress decreases slowly from its value at $t = t_c$. If one assumes that the dislocation velocity decreases exponentially from its value v_{d_0} at $t = t_c$ and follows the relation $v_d = v_{d_0} \exp[-\gamma_1(t - t_c)]$, where v_d is the dislocation velocity at any time $(t - t_c)$ and γ_1 is the rate constant, the following two cases can be considered.

Case I: when $(t - t_c)$ is small. If one substitutes the value of g from Eq. (24) and uses $v_d = v_{d_0} \exp[-\gamma_1(t - t_c)]$, Eq. (25) may be written as

$$\frac{dn_d}{dt} = p_F r_F n_F N_{d_0} v_{d_0} \exp[-\gamma_1(t - t_c)] - \frac{n_d}{\tau_d} \tag{35}$$

where N_{d_0} is the density of dislocations at $t = t_c$.

Equation (35) may be expressed as

$$\frac{dn_d}{dt} = \frac{p_F r_F n_F (d\varepsilon/dt)}{b} \exp[-\gamma_1(t - t_c)] - \gamma_0 n_d \tag{36}$$

where $d\varepsilon/dt = N_{d_0} v_{d_0} b$ is the strain rate at $t = t_c$ and $\gamma_0 = 1/\tau_d$.

The solution of Eq. (36) gives

$$n_d = \frac{p_F r_F n_F (d\varepsilon/dt)}{b(\gamma_0 - \gamma)} \{\exp[-\gamma_1(t - t_c)] - \exp[-\gamma_2(t - t_c)]\} \tag{37}$$

As $\gamma_0 \gg \gamma_1$ and $p_F r_F n_F (d\varepsilon/dt)/b\alpha_3 = n_{d_0}$, Eq. (37) gives

$$n_d = \frac{p_F r_F n_F (d\varepsilon/dt)}{b\gamma_0} \exp[-\gamma_1(t - t_c)] \tag{38}$$

Thus, the postdeformation recombination rate is given by

$$R_{\mathrm{PD}} = \gamma_0 n_d = \frac{p_F r_F n_F (d\varepsilon/dt)}{b} \exp[-\gamma_1(t - t_c)] \tag{39}$$

The decay of postdeformation ML intensity in a crystal of volume V may be given by

$$I_{\mathrm{PD}} = \frac{\eta p_F r_F n_F (d\varepsilon/dt) V}{b} \exp[-\gamma_1(t - t_c)] \tag{40}$$

Thus, the relaxation time $\tau_r = 1/\gamma_1$ of the mobile dislocations can be determined from the fast-decay portion of the postdeformation ML intensity.

CASE II: WHEN $(t - t_c)$ IS LARGE. The deformation of colored AHC causes redistribution of electon and hole centers. As a matter of fact, a large number of F-centers may come very close to the hole centers causing diffusion of holes toward nearby F-centers producing luminescence (Chandra, 1995). If γ_2 is the rate constant for the diffusion of interacting holes, then one may write

$$dp_i/dt = -\gamma_2 p_i \tag{41}$$

where p_i is the number of interacting holes at any time t.

The decay of ML intensity can be expressed as

$$I'_{\mathrm{PD}} = I'_0 \exp[-\gamma_2(t - t'_c)] \tag{42}$$

where t'_c is the time at which v_d becomes negligible. I'_{PD} is the postdeformation ML intensity at any time $t - t'_c$, and I'_0 is the ML intensity at $t = t'_c$. Thus, the diffusion time $\tau_D = 1/\gamma_2$ of the interacting holes can be determined from the slow-decay portion of the postdeformation ML intensity.

It follows from the above discussion that ML in γ-irradiated AHC should decay with two rate constants, with the first, γ_1, being related to stress relaxation and the second, γ_2, to the diffusion of interacting holes in the deformed crystals. The experimental observations of Kruglov *et al.* (1966) give $\gamma_1 = 0.15\,\mathrm{s}^{-1}$ and $\gamma_2 = 0.02\,\mathrm{s}^{-1}$.

6.2.4. ML Spectroscopy

It has been proposed that ML emission in x- or γ-irradiated AHC is due to the recombination of F-center electrons with the V_2-centers. Thus, the energy corresponding to the peak of the ML spectra should reflect the energy difference between the bottom of the conduction band, E_C, and the energy level of V_2-center, E_{V_2}. The wavelength corresponding to the peak of ML spectra is calculated from the relation $\lambda_m = [ch/(E_C - E_{V_2})]$, where c is the velocity of light and h is Planck's constant. It can be seen from Table 1 that the calculated value of the emission peak is nearly the same as that found from experimental observations.

Table 1. Value of λ_m for Irradiated Alkali Halide Crystals

Crystal	Bandgap[a] $(E_c - E_{v_2})$ (eV)	λ_m(nm) Calculated	Experimental[b]
NaCl	2.74	451	450
KCl	2.70	458	455
KBr	2.61	476	463

[a]Atari (1980).
[b]Atari (1982).

7. APPLICATIONS

ML has many potentially important applications. It provides a self-indicating method of monitoring the microscopic and macroscopic processes occurring during deformation and fracture of solids. With the fast film now available, it is possible to photograph the active areas of crystals by their own ML-created light. The light produced provides information related to the initiation, propagation, and interaction of dislocations and cracks in solids.

FML has been found useful in the detection of fractures and microscratching during deformation. It can also be used for measuring the crack velocity, fracture-initiation time, local stress and temperature near a moving crack-tip, and phase-transition temperature (Chandra *et al.*, 1977). Since the measurement of ML intensity can be easily done at time intervals as short as a few nanoseconds, it may provide a sensitive optical tool for studying time-resolved fracture dynamics. The FML of a sample may display rapid fluctuations well in excess of the detector and amplifier noise (Langford *et al.*, 1989). FML may be used for investigating the electronic structure of metal surfaces. ML can also be used to study the kinetics of the emergence of mobile dislocations onto the metal surfaces (Molotskii, 1989).

ML can be used for the study of recombination kinetics (Dickinson *et al.*, 1993). Brady and Rowell (1986) reported that FML can be used successfully for the investigation of the electrodynamics of rock fracture in mines and earthquakes. It has also been found useful for studies of development and interfacial fracture at embedded surfaces of composite materials (Zhenyi and Dickinson, 1991), where the observed signal provides time-resolved information on the fracture sequence. Such results have been used to model the process of fiber/matrix debonding and fiber pullout in brittle matrix composites.

It has been found that, generally, noncentrosymmetric crystals exhibit ML and crystals that do not show ML are centrosymmetric, i.e., nonpiezoelectric. This fact has helped in verifying the point group and space group of crystals. Glass *et al.* (1975) and Glass (1977) reported that the FML of II–VI compounds can be used in a wireless fuse-system for army warheads. Haneman and McAlpine (1991) and Li *et al.* (1994) found that the FML of semiconductors can be used to determine their energy bandgaps and to understand the surface defects and the electronic structure of the newly created surfaces. Tokhmetov and Vettergren (1990) used FML to determine the activation energy of fluctuation-induced breaking of chemical bonds in solids. It has also been found useful in improving the design of fluid energy mills (Bukhari, 1977).

PML of colored AHC can be used to determine the lifetime of interacting defect centers with dislocations. It can also be used to determine the relaxation time of mobile dislocations and the diffusion time of holes. With the use of ML the effective radii for the interaction of mobile dislocations with *F*-, *V*-, and impurity-centers have been estimated; the first of them has been found to be close to the Burgers vector *b*, and the two others are equal to 5*b* and 16*b*, respectively (Ossipyan and Shmurak, 1981). The radius of interaction of moving charged dislocations with the traps purposely introduced into ZnS:Cu,Cl and removed 0.7–0.8 eV from the conduction band bottom has been found to be 50 Å (Bredikhin and Shmurak, 1979). From ML measurements the energy gap between the bottom of the electron dislocation band and the ground state of *F*-centers has been found to be 0.08, 0.12, 0.05, and 0.10 eV for γ-irradiated KCl, KBr, KI, and NaCl crystals, respectively (Molotskii, 1989). ML has also proved useful in understanding the dislocation excitons in AHC.

Since the PML intensity of TL crystals depends linearly on the radiation dosage, similarly to TL, the ML may also be used for radiation dosimetry (Copty-Wergles *et al.*, 1990; Chandra, 1996). EML may also be useful in radiation dosimetry (Chandra and Bisen, 1992). When an image is projected onto a γ-irradiated CsBr crystal using *F*-light and the crystal is squeezed sometime later, the part that had been exposed to *F*-light shows quite bright ML and reproduces the image (Meyer and Winnacker, 1982). This shows that with proper development this technique may be useful in photography as well as in the writing of secret messages.

The linear charge density of dislocations can be determined from ML measurements, and it has been found to be 1×10^{-2} cgs esu/cm for ZnS crystals (Bredikhin and Shmurak, 1975, 1977) and 1.5×10^{-4} cgs esu/cm for LiF crystals (Poletaev and Shmurak, 1984). Parfeev *et al.* (1988) reported that ML provides a suitable tool for determining the damage in filled organosilicon rubber under pulsating loading, and it can be used to determine which materials are unsuitable for use in long-term cyclic handling, as well as to optimize the composition of materials in production.

Using a TML technique, one can write messages that can be seen in the dark (Dickinson and Donaldson, 1995). Polly *et al.* (1967) calculated surface charge densities using TML produced during low-energy impact of a needle onto the surface of AHC. The triboluminometer is finding most promising applications in radiology, toxicology, pediatrics, and cardiology (Orel *et al.*, 1989). It has been reported that like acoustic emission, ML measurements may also be useful for nondestructive testing of solids (Sodomka *et al.*, 1980; Chandra, 1984; Crasto *et al.*, 1987). Sodomka (1978) reported that ML can be used to interpret and understand erosion.

Finding suitable EML materials that will show repeated ML with undiminished intensity is the current challenge in this field.

REFERENCES

Alekseev, D. V. (1992), *Sov. Phys. Sol. St.* **34**, 1085.

Al-Hashimi, A., Eid, A. M., Ettinger, K. V., and Mallard, J. R. (1984), *Radiat. Prot. Dosim.* **6**, 203.

Alzetta, G., Chudacek, I., and Scarmozzino, R. (1970) *Phys. Stat. Sol.* (*a*) **1**, 775.

Atari, N. A. (1980), *J. Lumines.* **21**, 305.

Atari, N. A. (1982), *Phys. Lett.* **90A**, 93.

Atari, N. A., and Ramani, R. (1986), *Phys. Stat. Sol.* (*a*) **97**, 461.

Avilov, A. M., Volovik, V. D., Evtukh, A. A., and Ivanav, S. I. (1982), *Sov. Phys. Technol.* **27**, 1271.

Bacon, F. (1605), Cited in *Of the Advancement of Learning* (G. W. Kitchin, Ed.) (J. M. Dent and Sons, London, 1915).

Batylin, V. N., Molotskii, M. I., and Shmurak, S. Z. (1992), *Sov. Phys. Sol. St.* **34**, 437.

Belyaev, L. M., and Martyshev, Yu. N. (1969), *Phys Stat. Sol.* **34a**, 157.

Boyle, R. (1664), *Experiments and Consideration Touching Colours* (London), p. 413.

Brady, B. T., and Rowell, G. A. (1986), *Nature* **321**, 488.

Bredikhin, S. I., and Shmurak, S. Z. (1974), *JETP Lett.* **19**, 367.

Bredikhin, S. I., and Shmurak, S. Z. (1975), *JETP Lett.* **21**, 156.

Bredikhin, S. I., and Shmurak, S. Z. (1977), *Sov. Phys. JETP* **46**, 768.

Bredikhin, S. I., and Shmurak, S. Z. (1979), *Sov. Phys. JETP* **49**, 520.

Bredikhin, S. I., Glushchuk, O. A., and Shmurak, S. Z. (1982), *Sov. Phys. Sol. St.* **24**, 1278.

Bukhari, S. J. H. (1977), *The Nucleus* **14**, 3.

Butler, C. T. (1966), *Phys. Rev.* **141**, 750.

Chandra, B. P. (1981), *Phys. Stat. Sol.* (*a*) **64**, 395.

Chandra, B. P. (1983), *Pramana* **19**, 455.

Chandra, B. P. (1984), *J. Phys. D: Appl. Phys.* **17**, 117.

Chandra, B. P. (1985), *Nuclear Tracks* **10**, 225.

Chandra, B. P. (1996), *Radiat. Effects and Defects Sol.* **138**, 119.

Chandra, B. P., and Bisen, D. P. (1992), *Phys. Stat. Sol. (a)* **132**, K101.

Chandra, B. P., and Elyas, M. (1979), *J. Phys. C: Sol. St. Phys.* **12**, L695.

Chandra, B. P., and Shrivastava, K. K. (1978), *J. Phys. Chem. Sol.* **39**, 939.

Chandra, B. P., and Shrivastava, K. K. (1979), *Pramana* **13**, 611.

Chandra, B. P., and Tutakne, P. R. (1981), *Ind. J. Phys.* **55A**, 114.

Chandra, B. P., and Verma, R. D. (1980), *Krist. Tech.* **15**, 1197.

Chandra, B. P., and Zink, J. I. (1980*a*), *Phys. Rev.* B **21**, 816.

Chandra, B. P., and Zink, J. I. (1980*b*), *J. Chem. Phys.* **73**, 5933.

Chandra, B. P., and Zink, J. I. (1980c), *Inorg. Chem.* **19**, 3098.

Chandra, B. P., and Zink, J. I. (1981*a*), *J. Phys. Chem. Sol.* **42**, 529.

Chandra, B. P., and Zink, J. I. (1981*b*), *J. Lumines.* **23**, 363.

Chandra, B. P., and Zink (1982), *J. Phys. Chem.* **82**, 4138.

Chandra, B. P., Periasamy, N., and Das, J. N. (1977), *Pramana* **8**, 395.

Chandra, B. P., Elyas, M., Shrivastava, K. K., and Verma, R. D. (1980), *Sol. St. Comm* **36**, 931.

Chandra, B. P., Verma, R. D., Jaiswal, A. K., Majumdar, B., Chandrakar, T. R., and Deshpande, S. V. (1982*a*), *Phys. Stat. Sol. (a)* **72**, 361.

Chandra, B. P., Elyas, M., Jaiswal, A. K., and Majumdar, B. (1982*b*), *Curr. Sci.* (*India*) **52**, 673.

Chandra, B. P., Deshmukh, N. G., and Jaiswal, A. K. (1987), *Mol. Cryst. Liq. Cryst.* **142**, 157.

Chandra, B. P., Tiwari, S., Ramrakhiani, M., and Asari, M. H. (1991), *Cryst. Res. Tech.* **26**, 767.

Chapman, G. N., and Walton, A. J. (1983*a*), *J. Phys. C: Sol. St. Phys.* **16**, 5542.

Chapman, G. N., and Walton, A. J. (1983*b*), *J. Appl. Phys.* **54**, 5961.

Copty-Wergles, K., Nowotny, R., and Hille, P. (1990), *Radiat. Prot. Dosim.* **33**, 339.

Crasto, A. S., Corey, R., Dickinson, J. T., Subramaniam, R. V., and Eckstein, Y. (1987), *Compos. Sci. Technol.* **30**, 35.

Dickinson, J. T., and Donaldson, E. E. (1995), Autographs from peeling pressure sensitive adhesives, *J. Adhesion* (in press).

Dickinson, J. T., and Jensen, L. C. (1986), *J. Amer. Ceramics Soc.* **68**, 235.

Dickinson, J. T., Jensen, L. C., and Langford, S. C. (1990), *J. Mater. Res.* **5**, 109.
Dickinson, J. T., Langford, S. C., and Jensen, L. C. (1982), *J. Vac. Sci. Technol.* **20**, 436.
Dickinson, J. T., Langford, S. C., and Jensen, L. C. (1993), *J. Mater. Res.*, **8**, 2921.
Eid, A. M., Moussa, A., El-Adl, E. H., and Ettinger, K. V. (1986), *Egypt. J. Sol.* **8**, 148.
Glass, C. (1977), A Light Activated Fuse System, BRL MR 2726 (US Army Report).
Glass, C., Dante, J., Cjalella, C., and Golaski, S. (1975), An Electro-Optical Fuse System, BRL MR 2552 (US Army Report).
Grabec, I. (1974), *Polym. Phys. Lett.* **12**, 573.
Grabec, I. (1975), *Nondestr. Test.* **8**, 258.
Haneman, H., and McAlpine, N. (1991), *Phys. Rev. Lett.* **66**, 758.
Hardy, G. E., and Zink, J. I. (1976), *Inorg. Chem.* **15**, 3061.
Hardy, G. E., Baldmwin, J. C., Zink, J. I., Kaska, W. C., Liu, P. H., and Dubois, L. (1977), *J. Amer. Chem. Soc.* **99**, 3552.
Hardy, G. E., Chandra, B. P., Zink, J. I., Adamson, A. W., Fukuda, R. C., and Walters, R. L. (1979), *J. Amer. Chem. Soc.* **101**, 2787.
Hardy, G. E., Kaska, W. C., Chandra, B. P., and Zink, J. I. (1981), *J. Amer. Chem. Soc.* **103**, 1074.
Hayashiuchi, Y., Hagihara, T., and Okada, T. (1990), *Phys. Lett.* **147**, 245.
Inoue, T., Kunitomi, M., and Shibata, E. (1939), *Hiroshima J. Sci.* **9**, 129 [in Japanese].
Keszthelyi, C. P., and Bard, A. J. (1973) *J. Electrochem. Soc.* **120**, 1726.
Kitamura, N., Saravari, O., Kim, H. B., and Tazuke, S. (1986), *Chem. Phys. Lett.* **125**, 360.
Krauya, U. E., Knets, I. V., and Laizan, V. B. (1978), *Polym. Phys.* **201**, 157.
Krauya, U. E., Upitis, Z. T., Rikards, R. B., Teters, G. A., and Yansons Ya., L. (1981) *Compos. Mater.* **17**, 236.
Kruglov, A. S., El-Shanshonry, I. A., and Matta, M. K. (1966), *J. Phys. Soc. Jpn* **21**, 2147.
Lahav, N., Coyne, L. M., and Lawless, J. G. (1982), *Clays Clay Min.* **30**, 73.
Langevin, M. (1921), Rapports et discussions du contact de physique tenua Bruxellesen 1913, 'Sovs les auspices de I' Institut Solvay (Paris, Gassthier–Villars).
Langford, S. C., Zhenyi, M., and Dickinson, J. T. (1989), *J. Mater. Res.* **4**, 1272.
Li, D. G., McAlpine, N. S., and Haneman, D. (1994), *Surf. Sci.* **303**, 171.
Longchambon, H. (1925), *Bull. Fr. Soc. Mineralog.* **48**, 130.
Meyer, K., and Polly, F. (1965), *Phys. Stat. Sol.* **8**, 441.
Meyer, K., and Obrikat, D. (1969), *Z. Phys. Chem.* **240**, 309.
Meyer, K., and Winnacker, A. (1982), *Rad. Effects* **64**, 135.
Meyer, K., Obrikat, D., and Rossberg, M. (1970*a*), *Krist. Tech.* **5**, 5.
Meyer, K., Obrikat, D., and Rossberg, M. (1970*b*), *Krist. Tech.* **5**, 181.
Molotskii, M. I. (1983), *Fiz. Met. Metall.* **55**, 43.
Molotskii, M. I. (1989), *Sov. Sci. Rev. B: Chem.* **13**, 1.
Molotskii, M. I., and Shmurak, S. Z. (1990), *Phys. Stat. Sol.* (*a*) **120**, 83.
Mukhopadhyay, G. (1984), *Bull. Mater. Sci. India.* **6**, 755.
Nakayama, K., Suzuki, N., and Haschimato, H. (1992) *J. Phys. D: Appl. Phys.* **25**, 303.
Nowack, R., Krajewek, A., and Samoc, M. (1982), *Chem. Phys. Lett.* **94**, 270.
Orel, V. E., Opov, Ya. Z., Goraiskii, E. K., Leshchinskii, I. V., and Khazanovish, D. M. (1989), *Medits. Tech.* **4**, 34.
Ossipyan, Yu. A., and Shmurak, S. Z. (1981), *Proc. Intern. Conf. on Defects in Insulating Crystals* (V. M. Turchkevich and K. K. Schvarts, Eds.) (Springer-Verlag, Berlin), p. 135.
Parfeev, V. M., Tutan, M. Ya, Krauya, U. E., and Scrukovskis, A. A. (1988), *Mekhanika Kompozitnykh Materialov*, No. **1**, p. 125.
Poletaev, A. V., and Shmurak, S. Z. (1984), *Sov. Phys. Sol. St.* **26**, 2147.
Polly, F., Obrikat, D., and Meyer, K. (1967), *Krist. Tech.* **2**, 425.
Sodomka, L. (1968), *Czech. Casopis Fys. A* **18**, 497 [in Czech].
Sodomka, L. (1971), *Phys. Stat. Sol. (a)* **1**, K65.
Sodomka, L. (1978), *Proc. 3rd Intern Seminar on Energy Transfer in Condensed Matter* (Charles University Press, Prague), p. 286.
Sodomka, L., Leistner, M., Tessmar, H., and Karel, F. (1980), *Krist. Tech.* **15**, 1101.
Tamuzh, V. P., Krauya, U. E., and Kalninsh, P. P. (1985), *J. Physique* **46**, C5-661-C5.

Tokhmetov, A. T., and Vettergren, V. I. (1990), *Sov. Phys. Sol. St.* **32**, 17.
Vardanyan, R. A., Veselko, S. G., and Kirakosyan, G. G. (1989), *Sov. Phys. Sol. St.* **3**(1), 12.
Veliev, Z. A. (1986), *Sov. Phys. Sol. St.* **28**, 1632.
Waller, R. (1684), *Essays of Natural Experiments Made in the Academic des Cimento* (London).
Walton, A. J. (1977), *Adv. Phys.* **26**, 887.
Walton, A. J., and Botos, P. (1978), *J. Phys. E: Sci. Instr.* **11**, 513.
Wedgwood, T. (1792), Cited in *A History of Luminescence from the Earliest Times until 1900* by E. N. Harvey (The American Philosophical Society, Philadelphia, 1957), p. 378.
Williams, G. P., and Turner, T. J. (1978), *Sol. St. Comm.* **29**, 201.
Wolff, G., Gross, G., and Stranski, I. N. (1952), *Z. Electrochem.* **56**, 420.
Zakrevskii, V. A., and Shuldiner, A. V. (1984), *Sov. Tech. Phys. Lett.* **10**, 56.
Zakrevskii, V. A., and Shuldiner, A. V. (1985), *Sov. Phys. Sol. St.* **27**(10), 1826.
Zhenyi, M., and Dickinson, J. T. (1991), *J. Appl. Phys.* **70**, 4787.
Zhenyi, M, Jiawen, F., and Dickinson, J. T. (1988), *J. Adhesion* **25**, 63.
Zink, J. I. (1978), *Acc. Chem. Res.* **11**, 289.
Zink, J. I., Beese, W., Schindler, J. W., and Smiel, A. T. (1982), *Appl. Phys. Lett.* **40**, 110.

11

Bioluminescence and Chemiluminescence

N. N. Ugarova, L. Yu. Brovko, and E. I. Dementieva

1. INTRODUCTION

Emission of energy in the form of visible light during chemical or biochemical processes, so-called chemiluminescence (CL) or bioluminescence (BL), is not a rare phenomenon in nature. There are numerous organisms, both terrestrial and aquatic (freshwater and marine forms), that emit light (Hastings, 1983).

Light emission in chemical reactions requires the following conditions:

1. The energy released should be higher than that of the electronically excited product or intermediate molecule.
2. The product must be a fluorescent molecule so that the transformation of the excited molecule to the ground state is accompanied by visible light emission, or the reaction mixture has to include energy acceptor molecules with fluorescent properties.

As a rule, CL appears during oxidation reactions or free-radical recombination. In most cases the quantum yield of CL, Φ_{CL}, which is the ratio of the number of light quanta emitted to the number of product molecules formed, does not exceed 0.1–0.001%. For several other CL systems, e.g., the special substituted oxamide or dioxetane intermediates, it can reach as high as 30% (for a review, see Adam and Cilento, 1982).

CL of living organisms is a result of oxidation reactions that can be either nonspecific or specific in nature. Nonspecific CL results from inadvertent release of $O_2^{-\cdot}$, H_2O_2, or $^*O_2^{\cdot}$ precursors and their subsequent CL reactions with suitable molecules in their vicinity. A specific, although nonfunctional, and presumably

N. N. Ugarova, L. Yu. Brovko, and E. I. Dementieva • Department of Chemistry, Lomonosov Moscow State University, Moscow, Russia.

Luminescence of Solids, edited by D. R. Vij. Plenum Press, New York, 1998.

low-quantum yield, CL can be observed in peroxidative reactions, in oxidative decarboxylation, and in dioxygenase reactions that involve dioxetane intermediates, and in epoxidative reactions where rearrangement of perepoxides to dioxetanes is possible. This CL is specific since it results from oxygenation of a substrate molecule involved in a specific biochemical pathway.

There are several specific CL systems in living cells with high quantum yields (from 1 to 100%), where BL appears during oxidation of the highly specific substrate luciferin, catalyzed by the specific enzyme luciferase. Luciferin and luciferase are collective-functional rather than structure-functional, and denote substrates and enzymes whose interaction produces visible light. The characterizing feature of BL systems is the participation of specific proteins in the reactions. Proteins play the role of a matrix covalently or noncovalently bound to the emitting nonprotein chromophore. They can also act as biocatalysts without which the formation of an electronically excited reaction product is less probable (for a review see DeLuca and McElroy, 1981).

The present chapter concerns the basic science and applications of CL and BL. The systems that have been used for the study of a variety of biological processes and for the analytical assay will be discussed in detail.

2. MECHANISMS OF CHEMILUMINESCENCE AND BIOLUMINESCENCE

Molecules can be excited by a variety of energy sources, the most familiar involving electromagnetic fields and the energy of highly exothermic chemical reactions. When an excited molecule is created through a chemical reaction, the relationship between singlet and triplet states and electronic and vibrational levels will depend on the excitation energy source. However, in a condensed phase the initially created excited states relax very quickly (in picoseconds) without radiation to the lowest thermal level of the excited state with the same multiplicity. Thus, in solution the luminescent spectra of a molecule must be independent of the excitation source (photo- or chemiexcitation) (Seliger and McElroy, 1965; Seliger, 1975).

CL involves several consecutive processes, the first of which is the formation of a product whose subsequent transformations are sufficiently exothermic to emit visible light (400–700 nm). For this reason, the reaction enthalpy ΔH has to be more than 40 kcal/mol.

The second step in a CL process is a transition of some reaction intermediate to the electronically excited state, i.e., the real transformation of chemical potential energy into the energy of electron excitation. The final step is light emission from the excited state that has been formed. As the phosphorescence efficiency of organic substances in solution is very low, the concentration of singlet excited states must be high enough to realize efficient luminescence.

For organic peroxides, two different chemiexcitation mechanisms have been discussed. The first is monomolecular homolysis, which can be shown for alkylsubstituted dioxetane, in which simultaneous cleavage of the O—O and C—C bonds is accompanied by direct formation of carbonyl compounds. The second mechanism includes consecutive dissociation of the dioxetane cycle with intermediate biradical formation following the initial O—O bond cleavage and final formation of carbonylic compounds. While neither of these mechanisms has been proved directly, theoretical and experimental results weigh in favor of the latter. Theoretical analysis shows that the efficiency of chemiexcitation based on one-electron transfer processes may be up to 100% unlike 10^{-6}–10^{-3}% for thermoluminescence (TL).

Beginning with the important mechanism studies of McCapra, Koo, and Schuster it has been recognized that the electron exchange decomposition of dioxetanes to intramolecular biradicals is a major evolutionary pathway in BL reactions (Koo *et al.*, 1978; Schuster *et al.*, 1979). The radical recombination results in an electronically excited state of the monomer product whose formation is not governed by photoselection rules. This mechanism was called chemically initiated electron exchange luminescence (CIEEL) (Fig. 1). CIEEL can operate in systems where there are electron–acceptor (peroxide) and electron–donor (activator) groups, which can be parts of one or different molecules. When a peroxide–activator complex is formed, the stretching of the O—O bond results in electron transfer from the activator to the peroxide group, cleavage of the O—O bond and biradical formation: an anion radical on the peroxide and a cation radical on the activator group.

The first requirement for a CIEEL mechanism is related to the reduction potential of the peroxide, i.e., a peroxide should be readily reduced. This means structurally that the peroxide must be substituted with electron-accepting groups. The most available electron-accepting groups are carbonyls. For example, diacyl peroxides have two electron-accepting carbonyl groups, but secondary peroxyesters have one electron-accepting group. A second requirement for the operation of CIEEL is that an irreversible chemical reaction must occur upon the reduction of the peroxide. In the case of the dioxetanone this reaction is the cleavage of the O—O bond and decarboxylation. If the electron transfer were reversible, the electron would just go back to the activator. The third requirement for the CIEEL path is that there be a chemical route available for the release of energy. The chemical path in the case of the dioxetanone is the loss of CO_2 to generate a carbonyl compound. The fourth requirement is that there be sufficient energy released in the ion radical annihilation to form an electronically excited state (Koo and Schuster, 1977).

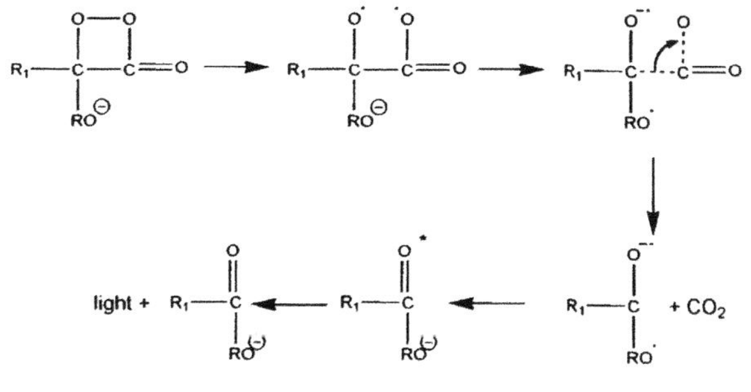

Figure 1. CIEEL mechanism of peroxide CL.

For tetramethyldioxetane, no direct evidence for electron-exchange CL was observed because it does not meet the first requirement. The reduction potential of the tetraalkyl dioxetanes is high and electron transfer does not compete with normal thermal decomposition. However, there was evidence of electron transfer for some dioxetanes in protonating media, such as acidic alcohol or silica gel. In these cases the reduction potential of the dioxetane is changed owing to the formation of intermolecular complexes.

3. CHEMILUMINESCENCE

Excessive energy of product molecules formed in chemical reactions usually dissipates upon collisions with surrounding molecules. However, in highly exothermic reactions, formation of product in an electronically excited state is highly probable, and this state may then emit a photon of light either as fluorescence or phosphorescence, depending on the spin orientation of the excited molecule, (direct CL) or enter into a chemical reaction. Alternatively, this electronically excited state may transfer energy to a suitable acceptor, and the excited acceptor thus formed can emit light (indirect CL).

The most effective CL reactions can be divided into three groups according to the mechanism of chemiexcitation.

3.1. Chemiluminescent Reactions of Electron Transfer

Recombination (annihilation) of free ion radicals, oxidation of anion radicals of aromatic hydrocarbons (carbo- or heterocyclic), or, rarely, reduction of cation radicals leads to the formation of neutral reaction products in an electronically excited state and the subsequent emission of light. For many substances the enhanced intensity of emitted light is observed if an appropriate energy acceptor is present in the reaction mixture.

3.2. Chemiluminescent Reactions of Singlet Oxygen

During oxidation of hydrogen peroxide by chlorine in alkali media the emission of red light is observed. The reason for this phenomenon is the formation of excited oxygen molecules. The lowest excited states of the oxygen molecule are singlets $^1\Delta_g$ and $^1\Sigma_g$ (although the ground state for oxygen is triplet). These excited states have energies of 92 kJ/mol (22 kcal/mol) and 160 kJ/mol (38 kcal/mol), respectively, which are sufficient for emission in the red region of visible light. More often the excited dimer of $^1\Delta_g$ emits light ($E = 185$ kJ/mol = 44 kcal/mol = 640 nm). A heterodimer of excited oxygen atoms ($^1\Delta_g + {}^1\Sigma_g$) emits light at 478 nm. The presence of a proper fluorophore in the reaction mixture, as well as in the case of electron transfer CL, provides energy transfer from excited oxygen to the acceptor followed by light emission. It is worth noting that singlet oxygen is a very active particle and it participates at a high rate in a large number of oxidative reactions, which is the main reason for the low quantum efficiency of

CL that involves singlet oxygen. However, upon the reaction of singlet oxygen with organic substances, endo- and cycloperoxides are formed. Thermal decomposition of the latter is a highly exothermic process and represents the third and the most effective type of CL reaction.

3.3. Chemiluminescent Reactions of Peroxide Decomposition

The most effective (in terms of the intensity of the light emission) CL reactions are the decompositions of organic peroxides. In most cases bright CL is observed when the intermediate cycloperoxides, 1,2-dioxetanes, decompose. As in other CL reactions, the enhanced luminescence appears in the presence of an energy acceptor. The scheme of the CL reaction of 1,2-dioxetane decomposition is shown in Fig. 2.

CL reactions of different types will be considered below in more detail. Mechanisms, time-courses, and structures of the emitters will be examined.

3.3.1. CL Reactions with Electron Transfer

For CL reactions of the first type (electron transfer) the participation of free ion radicals is a characteristic feature. There are different ways and sources for their formation during the reaction. Chemically it is a result of homolytic cleavage of compounds. There are also electrochemical paths to free radical generation upon oxidation or reduction of substances on the surfaces of electrodes. Free radicals can also be generated by physical means upon an electric discharge in the reaction mixture or irradiation with ultraviolet light. One of the examples of such a CL reaction is the oxidation of an anion radical of 9,10-diphenylanthracene by 9,10-dichlorodiphenylanthracene. Light emission observed during this reaction is spectrally identical to the fluorescence of neutral diphenylanthracene.

One of the well-known CL reactions of the first type is the oxidation of luminol (3-amino-phthalhydrazide) and its derivatives (Fig. 3), which proceeds in an alkali medium. In addition to hydrogen peroxide, sodium ferricyanide and hypochloride can serve as oxidizers in this reaction. Metal ions (such as Cu^{2+}, Co^{2+}, Fe^{3+}, and

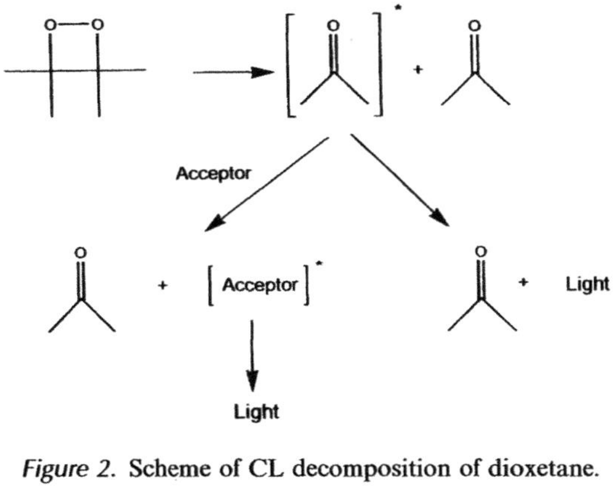

Figure 2. Scheme of CL decomposition of dioxetane.

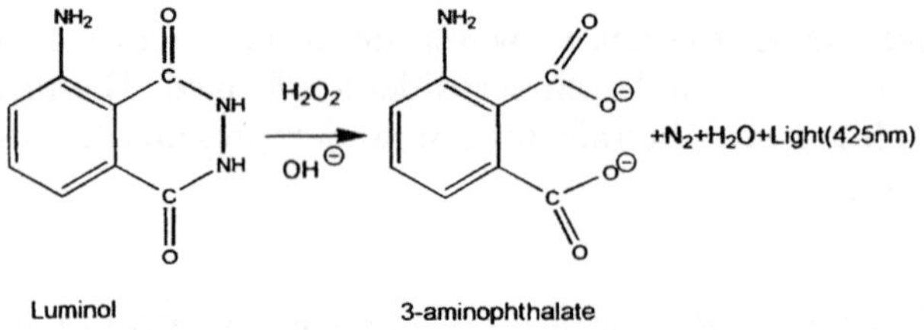

Figure 3. CL reaction of luminol oxidation.

Mn^{3+}), some organophosphorus compounds, heamatin, and heam-proteins (e.g., myoglobin, and peroxidase) are good catalysts for this reaction. During the last decades this reaction mechanism was investigated in detail, and it was shown that intermediate free radicals participate in the process. Recombination of the radicals leads to the formation of the excited product molecule followed by the emission of light. However, there are still many open questions concerning key stages of radical formation and the conversion of radicals during the course of the reaction. Many luminol derivatives also have CL properties. The quantum yield and time-course of the light-emitting reactions strongly depend on the substance structure and the components of the reaction mixture. For example, the replacement of the amino group with the nitro- or chloro-group leads to a decrease in the luminescence intensity, while replacement of hydrogen atoms in NH-groups with acyl residues does not have a substantial effect. Substitution of the same atoms with methyl-groups makes the derivative nonluminescent.

3.3.2. CL of Oxalates

CL of oxalates is one of the excellent examples of indirect CL. The principal scheme of the reaction is shown on Fig. 4. The emission spectra depend on the nature of the sensitizer used. Special porphyrines (uro-, copro- and protoporphyrines) are known as excellent sensitizers in oxalate CL systems. In this case intensive red luminescence is observed in the region of 610–650 nm. The compound 9,10-diphenylanthracene can be also used as an energy acceptor. In this case the emission maximum is at 400–440 nm. The efficiency (quantum yield) and time-course of the luminescence depend on the nature of the leaving group X.

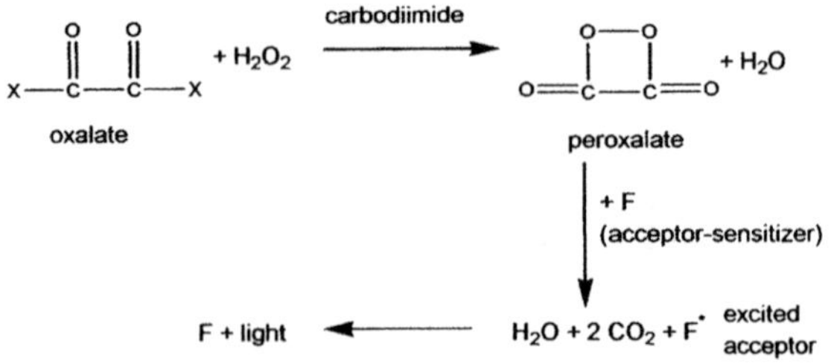

Figure 4. CL reaction of oxalate oxidation.

3.3.3. CL of Acridinium Compounds

The reaction scheme for CL of acridinium compounds is shown in Fig. 5. McCapra *et al.*, (1989) investigated both the influence of side groups on the efficiency and the time-course of the reaction in details. It was shown that to achieve a high quantum yield the leaving group should have a pK <11. Unfortunately in the case of a low pK of the leaving group the rate of the CL reaction is too high for experimental detection. To improve CL properties of the acridinium compounds it was proposed to use derivatives with bulky side groups. The best known CL substrate among the acridinium compounds is lucigenin (10,10′-dimethyl-9,9′-biacridinium dinitrate). Many new acridinium derivatives with improved properties have been synthesized recently. In all cases the emitting particle in these reactions is the same—*N*-methylacridone—and the emission spectra for the various acridinium compounds are all the same.

3.3.4. CL of Dioxetanes

The most advanced theories on the mechanisms of light-emitting reactions relate to CL thermal decomposition of 1,2-dioxetanes. In a large number of CL reactions the organic peroxides (including 1,2-dioxetanes) prove to be intermediates. They are usually extremely unstable and it was impossible to isolate and investigate them properly. In the 1970s it was shown that dioxetane stability depends strongly on the nature of the side groups.

The 1,2-dioxetanes decompose thermally into two carbonyl compounds, one of which is formed in an excited state. Unsymmetrically substituted dioxetanes with a single adamantyl group and an energy acceptor group show optimal stability and luminescence activity for both research purposes and commercial use. Bronstein *et*

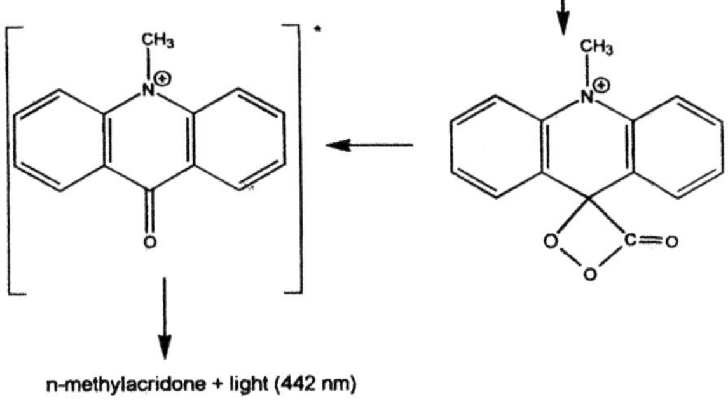

Figure 5. CL reaction of acridinium compounds.

al. (1989) proposed an intelligent approach to designing very effective CL substrates on the basis of dioxetanes (Fig. 6). The adamantyl group is used to give the designed compound stability. Cleavage of peroxide O—O bond releases approximately 100 kcal/mol, which results in the formation of the excited aryl carbonyl fragment, followed by the emission of light. This weak O—O bond thus acts as an "energy storehouse." Dioxetane decomposition is initiated by cleavage of a protecting group on an aromatic nucleus (more often a phenoxy group). It was shown that the substituted phenolates provide sufficient stability to the adamantyl 1,2-dioxetanes. On the other hand, phenol derivatives of adamantyl dioxetanes decompose rapidly at room temperature with bright light emission. This fact is consistent with the proposed CIEEL theory. The phenolate group proved to be good electron donor, which is necessary for the formation of the charge transfer intermediate. The phosphate group was shown to be good protector fragment, which can easily be removed by a special enzyme and thus initiate the CL reaction. As an example of such a design, the structure of the commercially available dioxetane CL substrate (AMPPD) is shown in Fig. 7. The quantum yield for the best CL reactions is as high as 20–30%, i.e., close to the efficiency of BL reactions, which are known to be the most effective.

3.3.5. Cellular CL

As early as in the beginning of the century Gurvich (1927) observed low-level light emission from plant roots, yeast cells, and blood cells in the range 200–1000 nm. It was later shown that the higher the metabolic activity of the cells, the stronger the light emission. The brightest, though very weak, CL was shown for dividing, proliferating cells and active phagocytes. This phenomenon, called non-specific low-level cellular CL (Allen, 1986), originates from the oxidative processes following the formation of reactive oxygen species (e.g., singlet oxygen, peroxide, and hydroxide radicals) and their interaction with cell components, mainly lipids. Ultraweak nonspecific cellular CL can be enhanced by adding effective CL substrates such as luminol and lucigenin. This technique is used for the investigation of the so-called "oxidative burst" in mammalian cells. Neutrophils are the most predominant white blood cells and are the mammalian body's first line of defence against invading microorganisms. After activation of neutrophils by external stimuli, the intracellular influx of reactive oxygen species proceeds. The release of these highly reactive metabolites causes damage to the surrounding tissues, but it results in a rapid

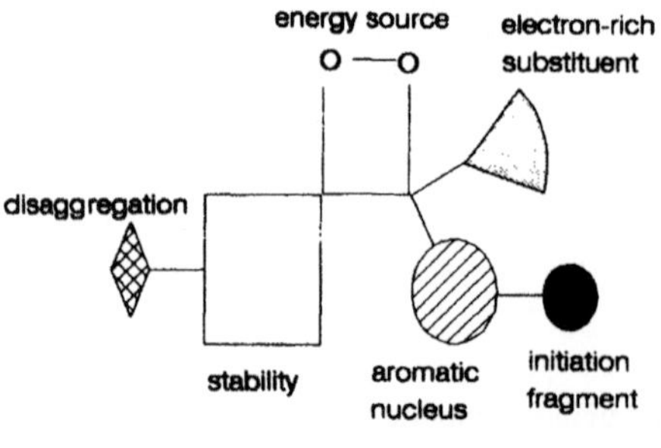

Figure 6. Schematic design of CL 1,2-dioxetane substrate (Bronstein *et al.*, 1994; with kind permission from Academic Press).

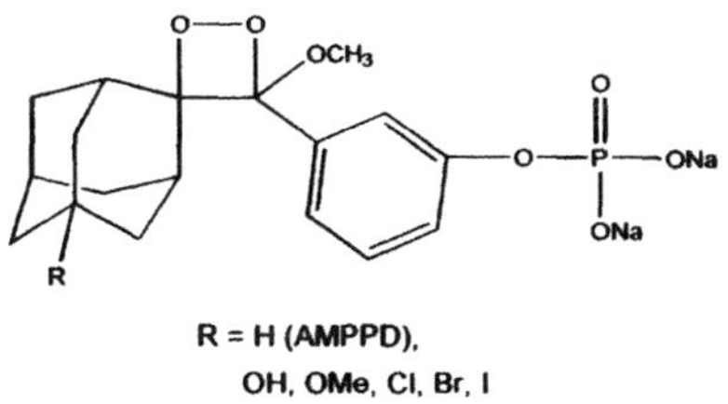

Figure 7. Structure of AMPPD [3-(2′-spiroadamantane)-4-methoxy-4-(3″-phosphoryloxy) phenyl-1,2-dioxetane disodium] salt.

increase of nonspecific cellular CL. Several CL mechanisms are included in this process. Among them are the recombination of oxygen radicals and the formation of peroxides and peroxide radicals followed by recombination or conversion to the CL products. Increased cellular CL was observed not only in cases of infection but in such diseases as rheumatoid arthritis, myocardial infarction, and kidney and lung disorders as well as under various conditions of stress. CL is the by-product of cell metabolic response to unfavorable factors, but it helps in investigations, diagnoses, and the treatment of patients.

4. BIOLUMINESCENCE

BL is the luminescence that occurs in or is derived from living organisms. The basis of BL is a CL reaction catalyzed by a specific enzyme. Thus, all the energy requirements and limitations for the CL reactions are valid for BL. In most cases organization of protein structures in BL systems is highly specific and imposes strict requirements on the structure of specific luciferase substrates. Obviously, this is a precondition for maximal concentration of energy on a chromophore sufficient for the formation of an electronically excited product that will release a quantum of light upon deactivation.

The participation of a protein in a chemical reaction leads to a high efficiency of luminescence, and the quantum yields of BL reactions usually lie between 0.1–0.9. The highest quantum yield (0.88) has been observed for the firefly BL reaction.

All protein catalysts of BL reactions isolated from many organisms are oxygenases. The energy necessary for light emission usually originates from the oxidation of a specific substrate by air, oxygen, or hydrogen peroxide. The substrates that undergo oxidation have different structures. For a number of BL systems such cosubstrates as $FMNH_2$, and ATP are necessary components of the reaction. Thus, the main features of BL reactions, which distinguish them from CL, are the participation of at least one protein component and a high quantum efficiency of the oxidative process. In some BL systems this protein acts as a catalyst (e.g., luciferin–luciferase systems, BL of fireflies, bacteria, and dinoflagellates). The other systems involve so-called photoproteins, which themselves undergo transformation in the course of a BL reaction (e.g., *Aequorea* and *Obelia* systems). Some 25% of the known BL systems include photoproteins, while the others are luciferin–luciferase systems.

4.1. Biological Functions and Evolution of Bioluminescence

Nowadays many chemically different luciferins and luciferases isolated from quite different luminous organisms have been investigated. BL is observed for many different phylogenetic groups of organisms. Luminous species were found in 13 phyla, including bacteria, unicellular protists, fungi, and animals ranging from jellyfish and brittle stars to scaleworms, fireflies, squid, and fish. In terms of the number of existing taxa, considering all organisms, BL species are quite rare (only 0.67% of genera are luminescent) but this property, when it is present, is functionally important for the organism. BL can play the following roles, among others: frightening of predators, providing communicative signals, including the mating signal, and attracting of prey.

In analyzing the wide and sporadic phylogenetic distribution of luminescent organisms, the very different structures of various luciferins and luciferases, including the genes that encode BL enzymes, it was postulated that the ability to emit light appeared independently 30 or more times during the process of evolution (Herring, 1987; Hastings, 1983).

4.2. The Primary Bioluminescent Systems: Structures of Luciferins

Table 1 summarizes the structures of the known luciferins and spectral characteristics of the seven of the most often studied BL systems. Apart from luciferin, most of the systems require other substrates and cofactors. Firefly BL proceeds in the presence of ATP, coelenterates use Ca^{2+} ions, and dinoflagellate BL is triggered by a pH change. Several systems need a second protein and a secondary emitter.

The luciferins can be divided into two structurally different classes: long-chain aldehydes and complex heterocyclic aromatic compounds. Aldehydes undergo oxidation to the corresponding acids. *Diplocardia* earthworm and bacterial BL systems are examples of this class. The other class of substrates involves the formation of ring peroxides. The cleavage, which is due to CIEEL, results in the formation of an excited molecule in the singlet state. Firefly and Cypridina BL reactions proceed according to this mechanism.

As was noted above, all BL systems require molecular oxygen. The coelenterazine system constitutes an apparent exception to this rule. So-called photoproteins can be isolated, e.g., from the jellyfish *Aequorea.* Aequorin can emit light simply on the addition of calcium, the oxygen being unnecessary here because it was part of the earlier reaction with the coelenterazine–protein complex that formed the stable peroxide. The addition of Ca^{2+} ions to this peroxide results in the light emission. Thus, the coelenterazine system, similar to other BL systems also needs oxygen to produce light.

The BL systems of bacteria and fireflies have been the most frequently studied. Although the enzymes were isolated in the 1950s, the mechanisms, kinetic characteristics of the reactions, and the role of protein in the production of light have not yet been elucidated completely. The methods of genetic engineering have been adapted during the last decade for the investigation of BL systems. The genes of firefly luciferases isolated from several species, as well as those of bacterial luciferase, aequorin, and *Renilla* and *Vargula* luciferases, have been cloned and sequenced (for a review see Campbell *et al.*, 1994; Devine *et al.*, 1993).

4.2.1. BL System of the Firefly

The firefly luciferin–luciferase system is an effective catalytic system that converts the energy of enzymatic reactions to light with a quantum efficiency near 0.9, which is the highest among all known CL and BL systems. Firefly luciferin is a benzothiazole derivative (structure 1 in Fig. 8), which reacts enzymatically with MgATP to form luciferyl adenylate (structure 2) and pyrophosphate. This step occurs prior to the oxidation with molecular oxygen. The luciferyl adenylate is then oxidized with the intermediate formation of the cyclic peroxide dioxetanone (structure 3) and a molecule of AMP. Oxidation of luciferin through dioxetanone formation is an excellent example of the CIEEL mechanism. The molecule of firefly luciferin contains the readily oxidizable heterocyclic fragment. The intramolecular electron transfer from this fragment to dioxetanone leads to the formation of the resonance charge transfer structure before decarboxylation. This process brings the system to the singlet excited state. Decarboxylation of the dioxetanone results in excited oxyluciferin in the enol or keto form (structure 5). The return to the ground state of the enol form is accompanied by emission of a quantum of visible light with $\lambda_{max} = 562–570$ nm (for a review see DeLuca, 1978; Ugarova, 1989).

The protein globule has a significant influence on the BL spectra. Luciferases isolated from different beetle species are distinguished by the wavelength maximum of the light emission: luciferases from *L. mingrelica*, and *P. pyralis* have l_{max} at 562–570 nm, whereas l_{max} of *L. lateralis* luciferase is at 552 nm. Four luciferases isolated from different organs of the click beetle *Pyrophorus plagiophthalamus* catalyze the same reaction as firefly luciferase but have different emission maxima: green (546 nm), yellow-green (560 nm), yellow (578 nm), and orange (594 nm). In all these cases the emitting particle has the same structure and is in the same electronic state, and changes in the spectrum are caused by differences in the properties of the microenvironment of the emitter in the complex with protein, such as the dielectric constant and polarization (Wood *et al.*, 1989).

The unique property of firefly BL is that ATP is an essential component of the reaction, together with luciferin and luciferase. Other than ATP and Mg^{2+} no cofactors or prosthetic groups are involved.

4.2.2. BL Systems of Bacteria

Luminous bacteria exist in the oceans as planktonic forms and in various association with higher organisms, e.g., in symbioses in special light organs and as intestinal flora. The bacterial luciferase catalyzes the monoxygenation reaction of long-chain aldehyde to the corresponding carboxylic acid; the cosubstrate is $FMNH_2$, which is oxidized to FMN:

$$FMNH_2 + O_2 + RCHO \rightarrow FMN + RCOOH + H_2O + h\nu(490\,nm)$$

Table 1. Characteristics of the Main Bioluminescent Systems

Organism (genera)	Structure of luciferin	Other factors	Emission maximum (nm)
Fireflies, click beetles (*Photinus*, *Photuris*, *Luciola*, etc.)	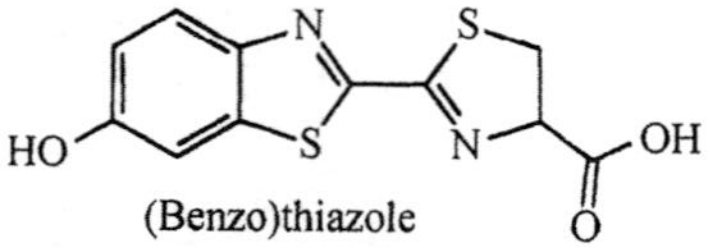 (Benzo)thiazole	ATP, Mg^{2+}	543–582
Bacteria (*Photobacterium*, *Vibrio*, *Beneckea*, etc)	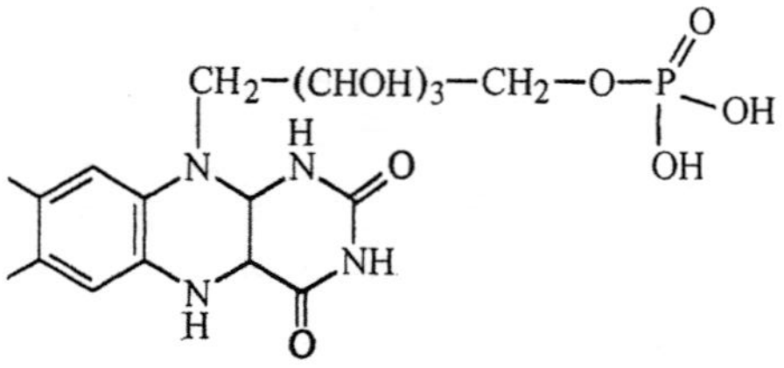 Reduced riboflavin phosphate ($FMNH_2$)	Long-chain aldehyde	478–505
Crustacea (*Vargula*, *Cypridina*)	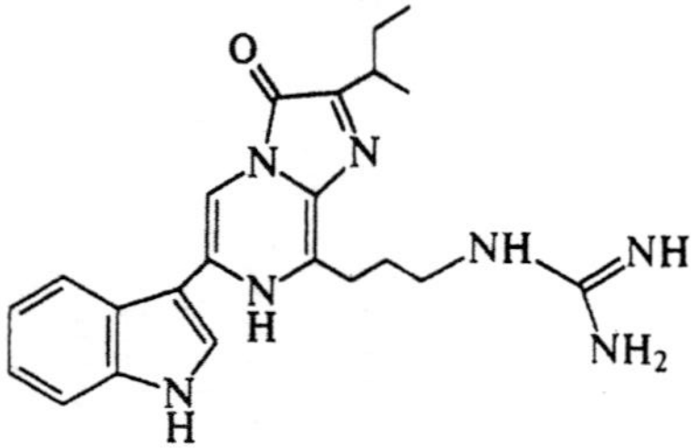 Imidazopyrazine derivative	—	465

Dinoflagellates (*Gonyaulax*, *Pyrocystis*, etc)	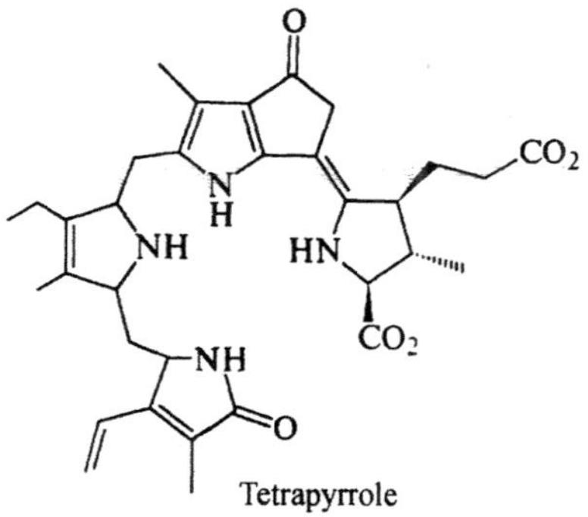 Tetrapyrrole	Luciferin-binding protein	4
Cnidarians (*Aequorea*, *Renilla*, *Obelia*, etc)	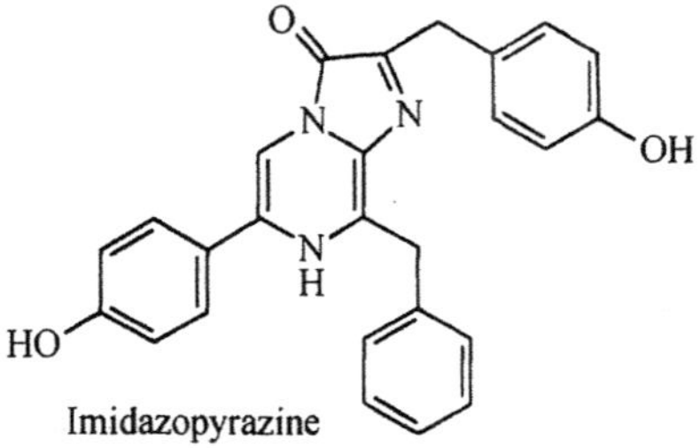 Imidazopyrazine	Ca^{2+}, substrate-binding protein, green fluorescent protein	460·
Annelids, earthworn *Diplocardia*	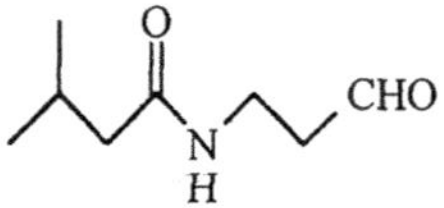 Aldehyde	H_2O_2	5(
Mollusks, freshwater snail *Latia*	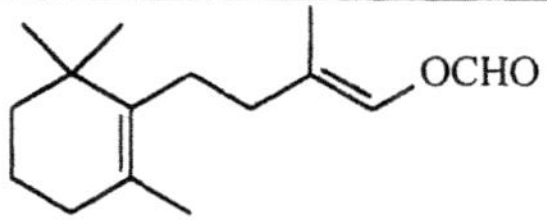 Aromatic or terpene aldehyde	—	5(

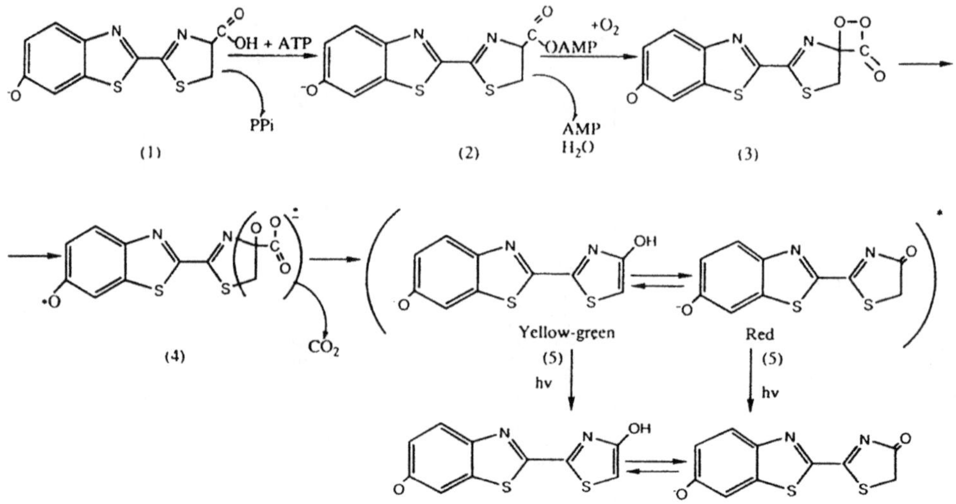

Figure 8. Scheme of the firefly luciferase catalyzed reaction.

Several mechanisms for bacterial BL have been proposed (Baldwin and Ziegler, 1992). The most probable one proved to be CIEEL adapted to the bacterial luciferase reaction. The linear peroxide intermediate (peroxyhemiacetal with long-chain aldehyde) is formed. It decays to the flavin-4*a*-hydroxide with the emission of blue light. Then dehydration occurs, resulting in the formation of oxidized FMN.

Bacterial luciferase *in vivo* (in the bacterial cell) operates together with FMN-reductase, which reduces the product of the BL reaction FMN to $FMNH_2$:

$$\mathrm{NAD(P)H + H^+ + FMN \xrightarrow{\text{FMN-reductase}} NAD(P)^+ + FMNH_2}$$

In addition to the bacterial luciferin, yellow fluorescent protein (YFP) was isolated from a strain of BL bacteria that emit yellow light. YFP, which contains noncovalently bound FMN, causes an acceleration in the rate of luciferase-catalyzed reaction with the strains of, e.g., *V. fischeri* and *P. phosphoreum*, and causes a shift in the emission spectrum from the blue-green to yellow with the maximum at 540 nm. The mechanism of this phenomenon is that YFP appears to be able to interact with a stable intermediate conformer of the luciferase, and the chromophore of the YFP becomes excited in such a way that the emission occurs at 540 nm.

4.2.3. Coelenterazine BL Systems

Imidopyrazine CL is ubiquitous in BL organisms inhabiting the seas (for a review see Szalay *et al.*, 1993). There are two types of coelenterazine systems: the luciferin–luciferase system from the sea pansy *Renilla* and the decapod *Systellaspis* and photoproteins such as aequorin (from the jellyfish *Aequorea victoria*), clytin, and obelin, where the luciferin is covalently bound to the protein. The reaction catalyzed by a Ca^{2+}-triggered photoprotein aequorin is shown in Fig. 9.

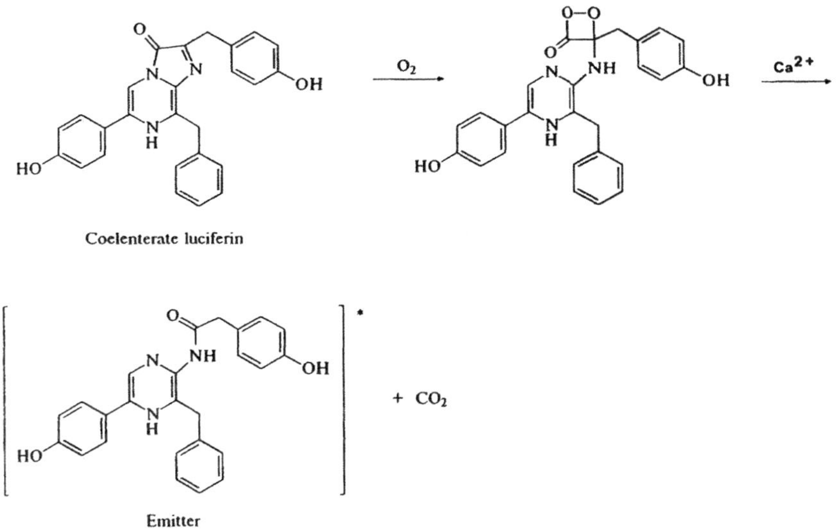

Figure 9. Scheme of the reaction catalyzed by Ca^{2+}-triggered photoprotein aequorin.

Prior to the emission of light, coelenterazine bound to the *Aequorea* luciferase (or apoaequorin) reacts with oxygen to form a stable peroxide with the oxygen attached to a 2-C carbon atom. The complex formed is isolated as aequorin (photoprotein) which emits light on the simple addition of Ca^{2+} ions:

$$\text{apoaequorin} + \text{coelenterazine} + O_2 \rightarrow \text{aequorin} \xrightarrow{Ca^{2+}} [\text{aequorin}]^*$$

$$\rightarrow \text{coelenteramide} + \text{apoaequorin} + h\nu$$

The chemistry of the BL reactions of the *Aequorea* and *Renilla* luciferin–luciferase systems are the same. *Renilla* produces blue light with $\lambda_{max} = 480$ nm. Examination of the BL *in vivo* reveals green emission with $l_{max} = 509$ nm. Green luminescence occurs owing to the presence of green fluorescent protein (GFP) with MW 54,000, which can form a complex with *Renilla* luciferase (MW 35,000). The spectral shift apparently occurs in a nonradiative energy transfer process from the excited oxyluciferin to GFP, which emits a quantum of visible light:

Similar to *Renilla*, *Aequorea* contains GFP (a monomer with MW 30,000), which acts as an energy transfer acceptor, but in this case the reaction appears to be primarily a radiative transfer.

4.2.4. *Cypridina* BL

Luciferin and luciferase in the ostracod crustaceans are found in different glands and react with each other when they are squirted into the seawater together (Hastings, 1983). The structure of *Cypridina* luciferin has much in common with those of *Renilla* and *Aequorea*. Although these structures are quite different from that of firefly luciferin (Table 1), the reaction mechanisms are rather similar. The BL reaction involves the formation of ring peroxide followed by oxidative decarboxylation, and then the formation of the excited carbonyl group. The quantum yield of the reaction is 0.31.

4.2.5. Dinoflagellate BL System

Dinoflagellates are unicellular phytoplanktonic organisms responsible for red tides and phosphorescence in the ocean. The light is emitted as a brief flash upon stimulation (Lee *et al.*, 1993). The luciferin molecule in this case is tetrapyrrole (Table 1). Dinoflagellate luminescence is triggered by a change in pH and involves in addition to luciferase the participation of a second, luciferin-binding, protein (MW 70,000). Dinoflagellate luciferase is a large protein (single polypeptide chain with MW $\sim$ 140,000). At alkaline pH the luciferase is completely inactive and the luciferin-binding protein binds tightly to the luciferin. Upon a shift in the pH, the substrate is released and rapidly oxidized by luciferase with light emission. There is not yet any available information about intermediates or products of the reaction, although this system is a very interesting example of the regulation of BL by means of a pH shift.

5. APPLICATIONS OF CHEMILUMINESCENCE AND BIOLUMINESCENCE

CL analysis belongs to the kinetic type of analysis, where analyte concentration is proportional to the reaction rate. In CL reactions the measuring parameter of the reaction rate is the rate of photon emission, i.e., luminescence intensity. According to the basic equation of chemical kinetics, the reaction rate is proportional to the concentrations of the participating substances. This provides simple methods for analyzing all the components of BL–CL reactions. Many thousands of papers have been published showing that one of the components of a CL or BL reaction can be coupled to a substance of biological interest, and thus provide an assay for it (for a review see Szalay *et al.*, 1993; Stanley and Kricka, 1990; Campbell *et al.*, 1994). Key features of BL–CL analysis are the following:

1. Sensitivity
2. Wide linear range
3. A signal can be detected and visualized from within the cells
4. Nonisotopic and nonhazardous
5. Rapid response

Theoretically good chemiluminometers can detect just a few hundred photons per second. Thus a CL reaction with a quantum yield of 0.1 and halflife of 0.2 s will allow as little as 10^{-21} mol ($\sim$6000 molecules) to be measured. This extremely high sensitivity can be achieved with the help of rather simple and inexpensive equipment. Commercially available CL substances can be detected with high sensitivity within seconds or so and they are stable for years.

There are five major areas in which these unique features of BL–CL analysis have been exploited:

1. Measurement of enzymes and metabolites in small portions of tissue or cell extract. This can be done either directly using suitable CL or BL reactions, or indirectly using coupled sequences of chemical and biochemical reactions with BL–CL detection in the final step.
2. Measurement of toxic reactive oxygen metabolites (O_2^-, H_2O_2, OCl^-, $^1O_2$, ROOH, etc.) produced by cells, tissues, or enzymes, monitoring ultraweak or indicator-dependent CL using luminol or lucigenin.
3. Measurement of the number of living cells in a sample and their viability and testing antibiotic and cytotoxic effects of chemical and physical factors. This has been called "rapid microbiology."
4. Nonisotopic labels in immunoassay and recombinant-DNA technology.
5. Using genes of BL proteins as reporter genes in genetic engineering studies for the detection of the expression level and regulation functions of promoters and repressors, and for the ultrasensitive detection of microbes with the help of engineered specific viruses or plasmids.

Below are the examples of applications of these methods.

5.1. Biochemiluminescent Analysis of Metabolites

Hydrogen peroxide can be assayed both in CL reactions of luminol and lucigenin oxidation. Sensitivity of the developed methods is 1 nmol/liter. The coupled three-enzyme system of mutarotase, glucose oxidase (GOD), and horseradish peroxidase (HRP) is used for glucose analysis in biological samples. Several protocols for analyses were proposed with a sensitivity of up to 0.01 nmol. The same principle was used for a cholesterol assay. As in the previous case hydrogen peroxide formed in the reaction of cholesterol with cholesterol oxidase was measured. Sensitivity of the method was about 50 pmol.

The reaction rate of luminol oxidation also depends on the concentration of the catalyst (metal ions and their complexes). Methods for CL determination of Co^{2+}, Fe^{3+}, Mn^{2+}, Cr^{3+}, ferricyanide, phthalocyanide, and other catalysts were developed. HRP was proved to be the best catalyst in the CL reaction of luminol oxidation. As low as 5 attomol/ml of peroxidase can be measured by the CL method.

A major advantage of enzymatic BL reagents is the high specificity of the catalyzed reaction. Any substrate required for luciferase reaction, e.g., ATP, NAD(P)H, O_2, or Ca^{2+}, can be detected directly. A linear response for ATP concentration in the range of 10^{-15}–10^{-6} M is observed for the analytical reagents based on firefly luciferase. The aequorin system permits the detection of Ca^{2+} over the range 10^{-7}–10^{-4} M.

Coupling other enzymatic reactions to a BL system greatly extends the range of analytical applications. Any reaction that produces or utilizes ATP can be coupled to the firefly system and a reaction producing or utilizing NAD(P)H to the bacterial BL system. Good examples are the measurement of creatine phosphokinase (e.g., for diagnosis of myocardial infarction), ATP-ase, glucose-6-phosphate, and pyruvate kinase, among others, using firefly luciferase.

5.2. Bioluminescent Express Methods for Control of Bacterial Contamination

The most intensively developing field of BL analysis, in particular, ATP-metry, is the so-called "rapid microbiology" for fast analyses of microbial contamination in food, technological materials, and environment.

All living cells contain ATP. After cell death, the ATP content decreases sharply. Thus, ATP content can serve as a criterion for biomass and cell viability. The sensitivity of the BL method is quite sufficient for many practical purposes, such as the analysis of food for bacterial contamination. A sensitivity of 20 cells at a concentration of 1 cell/ml can be achieved (for a review see Campbell *et al.*, 1994).

A different method for ultrasensitive determination of microorganisms was proposed by Ulitzur and Kuhn (1987), in which the main element is a DNA fragment encoding part of the *lux*-operon from BL marine bacteria. The application of the *lux* gene containing bacteriophage specific for *Salmonella* causes luciferase expression inside the cells and the appearance of luminescence that is proportional to the number of bacteria in the sample.

5.3. Bioluminescent Monitoring of Intracellular Processes

Cellular CL is a good indicator of metabolic processes. One of the most promising examples of such an approach was proposed and demonstrated experimentally by Allen (1986) for activation of polymorphonuclear leukocytes (PMNs) by infective agents. This metabolic activation takes place right after the interaction between the cell membrane and external stimuli. The net effect of oxidase activation is a huge production of highly reactive oxygen species: superoxide, singlet oxygen, hydroxyl radical, etc. Reaction between oxygen and the double bonds of cell membranes generates excited carbonyl compounds, which can relax to the ground state by photon emission. This low-intensity light can be increased by several orders of magnitude through the use of CL substrates such as luminol or lucigenin. As the intensity and time-course of cellular CL depend on the entire process of cell activation as well as on the metabolic status of the cells, it thus becomes a useful tool for medical diagnosis.

An important direction of BL analysis is the investigation of cellular metabolites and the changes induced in them by external stimuli. Tests for determination of the total toxicity of sewage and analyses of the effect of low doses of ionizing radiation on fibroblasts and bacteria have been developed. The second approach is the analysis of the response of a single cell to external influences. Changes in the concentration of Ca^{2+} in myocytes during contraction and oxidative phosphorylation in myofibrils have both been studied, and the main problem has been found to be introducing the components of a BL system into the cell.

5.4. Biochemiluminescent Methods for Immuno- and DNA Analysis

A wide range of CL substances, cofactors, and proteins involved in CL and BL reactions have been coupled to antigens or antibodies to provide nonisotopic immunoassays for haptens and polypeptides. Analytes include drugs, hormones, vitamins, microbial antigens (for a review see Collins, 1985; McCapra *et al.*, 1989). Three labels are presently used in clinical laboratories: (1) isoluminol derivatives, e.g., aminobutyl aminoethyl isoluminol (ABEI); (2) acridinium esters; (3) peroxidase detected by luminol CL enhanced by agents such as *p*-iodophenol. All these labels can be detected down to 10^{-15} mol. Signal-to-noise ratio is best with acridinium esters, where the sensitivity of analysis is about 10^{-18}–10^{-17} mol, which is close to the detection limit of ^{125}I labeled analytes. A particular feature of the peroxidase CL reaction is the continuous emission of bright light for an extended period of time. Peroxidase and acridinium esters have been used for DNA labeling both directly and via the biotin–avidin system. Detection sensitivity of these labels is rather high but still less than for ^{32}P.

Interest in the application of luciferases as enzymatic labels in nonradioactive immunoanalysis appeared in the 1970s. Bacterial luciferase was used in immunoanalysis of hormones. Conjugates of the luciferase with the derivatives of different haptens (estradiol and others) have been synthesized. Luciferase–hapten–antibody complexes have been detected by the BL method after regeneration of the active luciferase (Baldwin and Ziegler, 1992).

A new approach for BL–CL immuno- and DNA-assay was developed in the late 1980s by several groups. The main idea is to trigger BL–CL activity of substrates by enzymatic cleavage of the substituents. There are several enzyme-removable protecting groups that are suitable for this function, namely a phosphate and a galactopyranoside, which can be cleaved with alkaline phosphatase and β-glucosidase, respectively. Among dioxetane derivatives, 3-(2-spiroadamantane)-4-methoxy-4-(3″-phosphoryloxy)phenyl-1,2-dioxetane disodium salt (AMPPD) and 3-(2′-spiroadamantane)-4-methoxy-4-(3″-β-D-galactopyranosyloxy-phenyl)-1,2-dioxetane (AMPGD) were designed as "chemiluminogenic" substrates for alkaline phosphatase and galactosidase, respectively. Upon enzymatic cleavage of the phosphate or galactopyranoside groups, dioxetane derivatives become destabilized and decompose with light emission. Luminescence light intensity is dependent on enzyme concentration. The detection limit for alkaline phosphatase in such a reaction is 10^{-20} mol in solution or immobilized on various supports (Bronstein *et al.*, 1989).

New capabilities of firefly luciferase appeared several years ago when derivatives of luciferin with protected OH- and COOH-groups were synthesized as bioluminogenic substrates for enzyme markers:

$$\text{luciferyl phosphate} \xrightarrow{\text{alkaline phosphatase}} \text{luciferin} + \text{phosphate}$$

$$\text{luciferyl-}\beta\text{-galactoside} \xrightarrow{\beta\text{-galactosidase}} \text{luciferin} + \beta\text{-galactose}$$

With the use of luciferyl-β-galactoside, β-galactosidase has been detected down to 10^{-18} mol. By means of luciferin derivatives BL detection has been employed in immunoenzyme analysis of peptides and proteins and in DNA hybridization analysis (for a review see Stanley and Kricka, 1990).

5.5. Immobilization of Bioluminescent Enzymes

The increasing field of application of BL analysis requires the creation of stable and standard reagents. One of the most interesting approaches to this problem is immobilization of luciferases on the insoluble matrices (e.g., nylon tubes, BrCN-activated agarose, cellulose films). Immobilization of enzymes increases their stability and permits the repeated use of the reagents (for a review see Schölmerich *et al.*, 1987). In recent years methods for coimmobilization of polyenzymatic systems have been developed. As a rule, coimmobilization reduces the diffusion barriers for the reagents and the overall reaction rate increases as compared with the separately immobilized enzymes, which results in amplification of the sensitivity of the analysis. The coimmobilized specific dehydrogenase (or hexokinase), bacterial luciferase, and oxidoreductase were used for the analysis of bile acids, D-glucose. L-lactate, L-malate, L-glutamate, L-alanine in the range of 0.01–100 nmol. Coimmobilized system adenylate kinase–pyruvate kinase—firefly luciferase enabled more than a tenfold increase in the sensitivity of adenine nucleotide analysis, (Brovko *et al.*, 1994).

5.6. Chemiluminescent and Bioluminescent Reporter Gene Assays

Genes of some enzymes with unique specificity and high sensitivity of their determination are employed in molecular biology and genetic engineering as gene markers (for a review see Bronstein *et al.*, 1994). BL has been explored in reporter-gene studies in two ways: directly using genes that produce BL proteins, e.g., the gene for firefly luciferase or aequorin; and indirectly via BL–CL methods for detection of products of non-BL genes, e.g., alkaline phosphatase. Genes of bacterial and firefly luciferases are very promising in this respect. A few femtomoles of luciferases can be detected, i.e., a few orders of magnitude less than with the use of enzymes with chromogenic substrates. New imaging techniques using such devices as CCD cameras provides researchers with possibility of investigating both spatial and temporal distributions of gene expression. By combining a CCD camera and a microscope it is possible to image the light emission at the level of a single cell and to study gene expression in live intact mammalian cells.

REFERENCES

Adam, W. and Cilento, G. (Eds.) (1982), *Chemical and Biological Generation of Excited States* (Academic, New York).

Allen, R. C. (1986), in: *Methods in Enzymology Vol 33* (M. A. Deluca and W. D. McElroy, Eds.) (Academic, New York), p. 449.

Baldwin, T. O. and Ziegler, M. M. (1992), in: *Chemistry and Biochemistry of Flavoenzymes, Vol. III* (F. Müller, Ed.) (CRC Press, Boca Raton, FL.) p. 467.

Bronstein, I. Edwards, B., and Voyta, J. C. (1989), *J. Biolum. Chemilum.* **4**, 99.

Bronstein, I., Fortin, J., Stanley, P. E., Stewart, G. S. A. B., and Kricka, L. J. (1994), *Anal. Biochem.* **219**, 169.

Brovko, L. Yu., Romanova, N. A., and Ugarova, N. N. (1994), *Anal. Biochem.* **220**, 410.

Campbell, A., Kricka, L. J., and Stanley, P. E. (Eds.), (1994), *Bioluminescence and Chemiluminescence: Current Status* (Wiley, Chichester).

Collins, W. P. (Ed.) (1985), *Alternative Immunoassays* (Wiley, Chichester).

Deluca, M. A. (Ed.) (1978), in: *Methods in Enzymology, Vol. 57* (Academic, New York).

Deluca, M. A., and McElroy, W. D. (Eds.) (1981), *Bioluminescence and Chemiluminescence* (Academic, New York).

Devine, J. H., Kutuzova, G. D., Green V. A., Ugarova, N. N., and Baldwin, T. O. (1993), *Biochim. Biophys. Acta.* **1173**, 121.

Gurvich, A. G., and Frank, G. M. (1927), *C. R. Acad. Sci.* (Paris) **184**, 903.

Hastings, J. W. (1983), *J. Mol. Evol.* **19**, 309.

Herring, P. J. (1987), *J. Biolum. Chemilum.* **1**, 147.

Koo, J. Y., and Schuster, G. B. (1977), *J. Amer. Chem. Soc.* **99**, 6107.

Koo, J. Y., Schmidt, S. P., and Schuster, G. B. (1978), *Proc. Natl. Acad. Sci. (USA)* **75**, 30.

Lee, D. H., Mittag, M., Sczekan, S., Morse, D., and Hastings, J. W. (1993), *J. Biol. Chem.* **268**, 8842.

McCapra, F., Watmore, D., Sumun, F., Patel, A., Beheshti, I., and Branson, J. (1989), *J. Biolum. Chemilum.* **4**, 51.

Schölmerich, J., Andreesen, R., Kapp, A., Ernst, M., and Woods, W. G. (Eds.) (1987), *Bioluminescence and Chemiluminescence: New Perspectives* (Wiley, Chichester).

Schuster, G. B., Dixon, B., Koo, J. Y., Schmidt, S. P., and Smith, J. P. (1979), *Photochem. Photobiol.* **30**, 17.

Seliger, H. H. (1975), *Photochem. Photobiol.* **21**, 355.

Seliger, H. H., and McElroy, W. D. (1965), *Light: Physical and Biological Action* (Academic, New York).

Stanley, P. E. and Kricka, L. J. (Eds.) (1990), *Bioluminescence and Chemiluminescence: Current Status* (Wiley, Chichester).

Szalay, A. A., Kricka, L. J. and Stanley, P. E. (Eds.) (1993), *Bioluminescence and Chemiluminescence: Status Report* (Wiley, Chichester).

Ugarova, N. N. (1989), *J. Biolum. Chemilum.* **4**, 406.

Ulitzur, S., and Kuhn, J. (1987), in: *Bioluminescence and Chemiluminescence: New Perspectives* (Schölmerich, J. *et al.*, Eds.) (Wiley, Chichester) p. 463.

Wood, K. V., Lam, J. A., and Seliger, H. H. (1989), *Science* **244**, 700.

Index

GPSR Compliance

The European Union's (EU) General Product Safety Regulation (GPSR) is a set of rules that requires consumer products to be safe and our obligations to ensure this.

If you have any concerns about our products, you can contact us on ProductSafety@springernature.com

In case Publisher is established outside the EU, the EU authorized representative is:

Springer Nature Customer Service Center GmbH
Europaplatz 3
69115 Heidelberg, Germany

Batch number: 10755284

Printed by Printforce, the Netherlands